ENVIRONMENTAL MANAGEMENT

ENVIRONMENTAL MANAGEMENT

Science and Engineering for Industry

IYYANKI V MURALIKRISHNA AND

VALLI MANICKAM

BSP BS Publications
A Unit of BSP Books Pvt. Ltd., India.

ELSEVIER

Butterworth-Heinemann
An imprint of Elsevier
elsevier.com

Butterworth-Heinemann is an imprint of Elsevier
The Boulevard, Langford Lane, Kidlington, Oxford OX5 1GB, United Kingdom
50 Hampshire Street, 5th Floor, Cambridge, MA 02139, United States

Library of Congress Cataloging-in-Publication Data
A catalog record for this book is available from the Library of Congress

British Library Cataloguing-in-Publication Data
A catalogue record for this book is available from the British Library

ISBN: 978-0-12-811989-1

For information on all Butterworth-Heinemann publications visit our website at https://www.elsevier.com/

Publisher: Joe Hayton
Acquisition Editor: Ken McCombs
Editorial Project Manager: Peter Jardim
Production Project Manager: Kiruthika Govindaraju
Cover Designer: Victoria Pearson

Typeset by TNQ Books and Journals

CONTENTS

ABOUT THE AUTHORS

Dr. Iyyanki V. Muralikrishna, Presently at Defense Research and Development Organization, Government of India, as Dr. Raja Ramanna Distingushed Fellow. He has his PhD from Indian Institute of Science, Bangalore in Engineering Faculty and M Tech from IIT Madras. He was formerly, adjunct professor, Asian Institute of Technology, Bangkok. Previously, he was a faculty member and professor in spatial information technology during 1990–2008 at JNT University at Hyderabad and former director of Research and Development Centre at JNT University and is basically an Environment and Water Resources Engineer with expertise in geospatial technology, terrain mapping, environmental management, climate change, numerical modeling, and environmental impact assessment. Earlier, he worked as faculty in Ocean Engineering center at IIT Madras during 1978–79 and Scientist National Remote Sensing Centre/ISRO, Department of Space during the 1980s for a period of 8 years. He has executed several projects for government of India. He is a member of MOEF, NNRMS committee, Government of India and consultant to Administrative Staff College of India, Hyderabad from 2008 for four projects on Bhu Bharati land survey project evaluation, Manuals for EIA, Reorganization of Hazardous Waste Management Division of MOEF, GOI, and assessment of impact of climate change on agriculture. He has coauthored books including *Climate Change and Weather Modification Technologies* and chapters on *Global Data Base Design*, ISPRS publication edited by Ryutaro Tateishi, Chiba University, Japan, and remote sensing applications in Marine Science and Technology edited by AP Cracknell, University of Dundee, Scotland, published by European Space Agency. He has more than 100 papers in peer-reviewed journals to his credit and has guided 40 PhD studies to date.

Dr. Valli Manickam, Professor and Chairperson, Environment Area of Administrative Staff College of India (ASCI), has her PhD in Environmental Science and Technology with masters in both Computer Applications and Sciences, with over 10 years of teaching experience in the field of environmental science (to post graduate students). Before joining ASCI, she was at Jawaharlal Nehru Technological University, Hyderabad. She has worked for a number of environmental projects and activities, which include environmental impact assessment study for taking up industrial or tourism activity, study of lakes in and around Hyderabad, study of pollution in surface and ground waters, review of environmental statements, risk assessment, common effluent treatment plants, air pollution assessment studies, and weather modification operations. Dr. Valli has conducted training programs in ASCI for IFS officers in the area of Natural Resource Management, Environmental Indicators, and Environmental Economics besides organizing three international conferences and three workshops. She has coauthored five books in the field of environment, which include *Environmental Impact Assessment*, *Analytical Chemistry*, *C for Environmental Engineers*, *Climate Change*, and *Weather Modification Technologies*. She has published about 30 papers in national and international referred journals. She is presently executing projects sponsored by MoEF, DST, and other state governments. She has carried out consultancy assignments for various national and state organizations in areas of environmental importance. She is serving as Associate Editor for the journal of Air and Water Management Association of T&F. She has been awarded Fellow by the AP Academy of Sciences in 2015 for her contribution in the areas of environmental science and technology.

FOREWORD

In 1986, I was visiting Calgary Zoo, and in the zoo, there was a cage with a label saying "Calgary Zoo, world's most dangerous animal." Unlike other cages, it was open and visitors could go inside it and be photographed, and I took the opportunity; see photograph.

Though, obviously, this was a photo opportunity for visitors, the management of the zoo may have had a more serious intention as well. They may have intended people to think about the relentless extinction of botanical and zoological species that human beings have been causing for hundreds of years. They may have been thinking about the many violent deaths of human beings caused by murders, by wars, and by traffic and industrial accidents. But apart from these direct causes of death, there is yet another way in which human beings are dangerous to one another: that is by the damage that we cause to the general environment in which we live. Why is our environment important? The simplest explanation is that, as humans, the environment—the Earth—is our homeland. If we damage the environment, we threaten harm to ourselves.

This is a book on environmental management. There is need for a commitment to raising environmental standards, keeping our place of living safe from public health threats, and maintaining a clean, green, and livable environment. Today, industrialization and urbanization have led to increasing affluence and a growing population. This places enormous pressure on the environment, and we need to transform and improve ourselves to tackle these challenges. It is against this background that this book focuses on the various causes of pollution and damage to the environment. It addresses the national and international regulations intended to limit pollution of the environment, and it also considers technical solutions to reduce pollution and conserve resources.

We do not live in isolation; our entire life support system is dependent on the well-being of many other species living on Earth. The term biosphere was created to describe the totality of living things on Earth by Vladimir Vernadsky, a Russian scientist, in the 1920s. The biosphere refers to the one global ecological system in which all living things are interdependent to a greater or lesser extent. Within the overall biosphere, or ecosystem, there are smaller ecosystems like the rainforests, marine ecosystems, the desert, and the tundra. When any of these systems is threatened the entire planet may be affected. The various anthropogenic activities that take place may have widespread and unpredicted consequences for the health of our planet.

This book addresses issues on life cycle assessment and environmental impact assessment are crucial management tools for conserving the environment through the developmental process. This is followed by a discussion of environmental audit along with a detailed study of environmental accounting. The book covers environmental risk assessment, energy management and audit, and the ISO series, which are important for internal standardization of procedures and practices as with the safety norms in the OHSAS (Occupational Health and Safety Standards, 1801). Some other topics covered in the book, which add to its value, are related to water treatment technologies for polluted wastewater from domestic and industrial sources, technologies available for the treatment of air pollution and its sources, and solutions to noise pollution and solid waste management, hazardous waste management, and landfill technologies available for the disposal of hazardous wastes, etc.

Climate change in the form of global warming is a serious problem leading to increasing ocean temperatures and changes in ocean currents, which affect marine life, and to rising sea levels that threaten coastal communities worldwide. Global warming also melts the glaciers and the Greenland and Antarctic ice caps, leading to the input of fresh water into the oceans, thereby locally affecting the salinity; it also leads to the shrinking of the Arctic sea ice. The ice reflects light, and so as the ice disappears, the Earth will get darker and absorb more heat, thereby leading to further global warming.

Many of the natural resources that we consume are nonrenewable, so when they run out, we will be forced to find new alternatives. It is against this background that I welcome the appearance of this book on environmental management as a primer. There are, of course, several publications on the environment and its management from global publishing agencies written by international experts. This book is not like them. These books have been written for experts and often contain a tremendous amount of information. This can be overwhelming for anyone who simply wants to get up to speed on the basics. By contrast, this primer was written specifically for environmental management professionals as well as researchers and hobbyists. There is a saying that we must think globally and act locally when we are handling issues related to the environment and natural resources. I welcome this book as it relates these problems to south Asian issues and people while maintaining the flavor of global concerns.

Transformation of the environment is not possible without developing our people. We need to develop our people to their fullest potential, equipping them with the necessary skills and capabilities to handle both their current jobs and prepare them for the new challenges ahead. It is against this background that I personally feel a comprehensive book on environmental management at an affordable price helps students to promote research. I am confident that this book meets this requirement. There is certainly scope for the book in providing a major source of knowledge for students and researchers as well as practicing experts. The concepts and techniques presented here apply not only to a course for training students but also to other designers of public health and environmental systems and tasks as well. The goal here is to explain the environmental issues and blueprinting process in a format that is both easy to understand and apply. Rather than assuming any particular level of proficiency, concepts are explained from the point of view of the novice. With an easy-to-follow format, one can find the book as a single source of reference on a variety of environmental concerns and topics along the way.

Arthur P. Cracknell
Emeritus Professor, University of Dundee, Scotland

PREFACE

Is environment under threat? It is a very common perception. Yes, for many and No, for few. Why is our environment important? It is worth rephrasing its importance. We hear and read so much about protecting and saving the planet earth. The simplest explanation about why environment matters is that, as humans, the environment—the earth—is our homeland. All of us have major apprehension for climate change and are cognizant that global warming is a major progeny of concern. It is well known that our entire life support system is dependent on the well-being of all the species living on earth. Human beings are continuously making attempts to improve their quality of life. One of the ways to achieve this is by developing it in its social, economic, and cultural context, which involves manipulation of the complex, interrelated natural and man-made components, and processes of human environments. The concept of goods, services, and hazards opens and extends the concept of natural resources and links the concepts of environmental quality and life quality.

The purpose of this book *Environmental Management* is to provide practitioners with a scientific and practical understanding of ecological and social systems that can be applied in a policy or management context. In this book, very significant aspects of environmental protection are congregated and deliberated in 18 chapters starting from the notions of sustainable development and biodiversity, international laws on environmental protection and the environmental management practices followed in reducing air, water, and soil pollution. This book makes an endeavor to tackle the processes and governors, which are directed at environmental safeguarding. It also explains the rational in utilization of natural resources and how it is conceivable to augment the interrelations between society and the environment and the upgrading of quality of life of all incarnate beings. As this is a primer, the authors took all care to familiarize the themes in unassuming jargon with well-defined portrayal.

Dr. Arthur P. Cracknell, editor-in-chief, International Journal of Remote Sensing, a Taylor and Francis Publication, and Emeritus Professor, University of Dundee, Scotland, United Kingdom, provided the Foreword. The authors express their deep sense of gratitude to Dr. Cracknell for scribing a meticulous Foreword, in which, he not only expressed serious concern about environmental issues but also presented about the relevance of the topics for understanding the environmental issues either by a novice or a professional. Many thanks to him.

The authors envisage the book *Environmental Management—Science and Engineering for Industry* would oblige the needs of both individuals as well as institutions convoluted in the major mission of the espousing the complex concerns of environment and gazing for path-breaking approaches to model and alleviate the harms to all incarnate creatures on the earth.

Iyyanki V. Muralikrishna
Valli Manickam

ABOUT THE BOOK

Environmental pollution is a problem both in developed and developing countries. Factors such as population growth, technological advancement, and urbanization invariably place greater demands on the planet and stretch the use of natural resources to the maximum. This book on environmental management is an attempt to showcase the various issues in pollution, the international and national regulations in place to prevent and minimize pollution, and the various latest technological solutions to pollution reduction. The book consists of 18 chapters. Chapter 1 deals with a general introduction to environmental management, followed by a brief on sustainable development and natural resource management including biodiversity conservation in Chapters 2 and 3, respectively. Environmental Policy and legislation are dealt with in Chapter 4. Life cycle assessment and environmental impact assessment are crucial management tools for conserving the environment through the developmental process and are discussed in Chapters 5 and 6 of this book. Chapter 7 deals with environmental accounting, in which methods for accounting of natural resources are addressed. Environmental risk assessment is important to predict potential damage to the environment by a particular developmental activity, the details of which are discussed in Chapter 8. Energy management and audit and ISO series are important for internal standardization of procedures and practices as with the safety norms in OHSAS 18001. These issues are addressed in Chapters 9 and 10 of environmental management.

Chapter 11 deals with principals and design of water treatment technologies, while Chapters 12 and 13 deal with technological solutions to pollution of wastewater from domestic and industrial sources. Chapter 14 deals with the various technologies available for abatement of air pollution, and Chapter 15 discusses the various sources and solutions to noise pollution. Solid and hazardous waste management is a major problem across the world; Chapter 16 and Chapter 17 deal with the treatment methods for waste management. Chapter 18 deals with the assessment and analysis of the water, wastewater, soil, and air, which are essential and the starting point for all pollution abatement studies. The book on environmental management discusses all issues related to legislation, pollution assessment, prevention, and abatement.

ABBREVIATIONS

AAQ	Ambient air quality
AFF	Anaerobic fixed film
ALARP	As low as reasonably practical
ASEAN	Association of Southeast Asian Nations
BAT	Best available technique
BOD	Biological oxygen demand
BPEO	Best practical environmental option
BPM	Bureau of Indian Standards (BIS)
CBD	Convention on Biological Diversity
CFCs	Chlorofluorocarbons
CITES	Convention on International Trade in Endangered Species
CMB	Completely mixed batch
CMF	Completely mixed flow
COD	Chemical oxygen demand
CPCB	Central Pollution Control Board
CPU	Central Processing Unit
CRZ	Coastal Regulation Zone
CS	Cost savings
CSR	Corporate social responsibility
DO	Dissolved Oxygen
EA	Environmental accounting
EA	Environmental audit
EB	Expanded bed
ED	Electrodialysis
EFA	Environmental financial accounting
EIA	Environmental impact assessment
EMA	Environmental management accounting
EMP	Environmental Management Plan
EMS	Energy management systems
EMS	Environmental management systems
ENA	Environmental national accounting
ENNP	Environmentally corrected net national product
EOP	End of pipe
EPA	Environment Protection Act
EPD	Environmental product declarations
ERA	Environmental risk assessment
ERM	Environmental risk management
ESI	Environmental sustainability index
ESP	Electrostatic precipitator
FMEA	Failure mode and effect analysis
FML	Flexible membrane liners
GAC	Granular activated carbon
GATT	General Agreement on Tariffs and Trade

GC	Gas chromatography
GDP	Gross domestic product
GHGs	Green House Gases
GNP	Gross national product
GOI	Government of India
HAZOP	Hazard and operability
HCFCs	HydroChloroFluoroCarbons
HDPR	High-density polyethylene
HF	Hydrogen fluoride
HF	Hyperfiltration
HPLC	High-performance liquid chromatography
HSE	Health, safety, and environment
IAEA	International Atomic Energy Agency
ICJ	Court of Justice
ILM	Median inhibitory limit
IMD	Indian Meteorological Department
ISRO	Indian Space Research Organization
ITTA	International Tropical Timber Agreement
ITTO	International Tropical Timber Organization
LAC	Life cycle impact assessment
LC	Liquid chromatography
LC_{50}	Lethal concentration 50
LCA	Life cycle assessment
LCEA	Life cycle energy analysis
LCI	Life cycle inventory
LD_{50}	Lethal dose 50
LDO	Light diesel oil
LNWT	Low- and non-waste technologies
LPG	Liquefied petroleum gas
LSHS	Low-sulfur heavy stock
MFA	Material flow accounts
MIC	Methyl isocyanate
MLSS	Mixed liquor suspended solids
MLVSS	Mixed liquor volatile suspended solids
MoEF	Ministry of Environment and Forests
MPC	Maximum permissible concentration
MS	Mass spectrometry
MSF	Multistage flash distillation
MSW	Municipal solid waste
NAAQS	National ambient air quality standards
NEP	National environmental policy
NFPA	National Fire Protection Association
NGOs	Non-governmental organizations
NRA	Natural resource accounts
NRC	Noise reduction coefficient
ODS	Ozone-depleting substances
OECD	Organisation for Economic Co-operation and Development
OHSAS	Occupational Health and Safety Assessment Series

P	Pressure
PAC	Powdered activated carbon
PAHs	Polynuclear aromatic hydrocarbons
PCBs	Polychlorinated biphenyl
PEP	Plant energy performance
PET	PolyEthylene Terephthalate
PIC	Pesticides in international trade
PILs	Public interest litigations
PIOT	Physical input–output tables
PLIA	Public Liability Insurance Act
POPS	Persistent organic pollutants
PS	Polystyrene
QU	Quantity units
R	Range
RAS	Returned activated sludge
RBC	Rotating biological contactor
RCRA	Resource Conservation and Recovery Act
RO	Reverse osmosis
SC	Specific condition
SCR	Selective Catalytic Reduction
SDI	Sludge density index
SEA	Strategic Environmental Assessment
SEIAA	State Environmental Impact Assessment Authority
SNA	System of national accounts
SPADNS	Sodium-2-(parasulfophenylazo)-dihydroxy-3,6-napthal ene disulfonate
SPCBs	State pollution control boards
SVI	Sludge volume index
T	Temperature
TBL	Triple bottom line
TC	To Contain
TD	To Deliver
TFAP	Tropical Forestry Action Plan
TLM	Median threshold limit
TOR	Terms of reference
TPN's	Thematic program networks
TSDF	Treatment, storage, and disposal facilities
UASB	Upflow anaerobic sludge blanket reactor
UF	Ultrafiltration
UNFCCC	United Nations Framework Convention on Climate Change
UNCED	UN Conference on Environment and Development
UNCLOS	UN Convention on the Law of the Sea
V	Volume
VOCs	Volatile organic compounds
WAS	Waste-activated sludge
WHO	World Health Organization
WPA	Wildlife Protection Act
WQI	Water quality index
ZSI	Zoological Survey of India

CHAPTER ONE

Introduction

Environmental pollution is one of the most serious problems facing humanity and other life forms on our planet today. Environmental pollution is defined as "the contamination of the physical and biological components of the earth/atmosphere system to such an extent that normal environmental processes are adversely affected." Pollutants can be naturally occurring substances or energies, but they are considered contaminants when in excess of natural levels. Any use of natural resources at a rate higher than nature's capacity to restore itself can result in pollution of air, water, and land.

Generally, environmental pollution takes place when the environment cannot process and neutralize harmful by-products of human activities in due course without any structural or functional damage to its system. Although pollution has been known to exist since life began, it has seen a growth of truly global proportions since the onset of the industrial revolution during the 19th century. Environmental pollution is a problem not only in the developed countries but also in developing countries. Factors such as population growth, technological advancement, and urbanization invariably place greater demands on the planet and stretch the use of natural resources to the maximum.

Pollution is something that we face on an everyday basis; probably, this is something we may even be immune to in our fast-paced lives. There are three major types of environmental pollution: air, water, and land pollution. Air and noise pollution are increasing at an alarming rate today. Air pollution occurs with the addition of harmful chemicals into the Earth's atmosphere. The main pollutants that cause air pollution are carbon monoxide, Chloroflurocarbons, nitrogen oxides, and sulfur dioxide. Water pollution is caused when wastes are released into the water, which contaminates it. Soil can even be contaminated due to various domestic and industrial activities. Noise pollution is also a current environmental issue that causes harm in various ways.

Environmental Management
ISBN 978-0-12-811989-1
http://dx.doi.org/10.1016/B978-0-12-811989-1.00001-4

1.1 WATER POLLUTION

Water pollutants include contamination due to domestic wastes, insecticides and herbicides, food processing waste, pollutants from livestock operations, volatile organic compounds (VOCs), heavy metals, chemical waste, and others. Waterborne diseases caused by polluted drinking water include typhoid, amebiasis, giardiasis, ascariasis, hookworm, etc. Waterborne diseases caused by polluted beach water are rashes, ear ache, pink eye, respiratory infections, hepatitis, encephalitis, gastroenteritis, diarrhea, vomiting, stomach aches, etc.

When water is contaminated with chemicals such as pesticides, hydrocarbons, persistent organic pollutants, or heavy, it could lead to cancer, including prostate cancer and non-Hodgkin lymphoma, hormonal problems that can disrupt reproductive and developmental processes, damage to the nervous system, liver and kidney damage, and damage to the DNA. Specifically, mercury in water can cause abnormal behavior, slower growth and development, reduced reproduction, and death.

Nutrient pollution (nitrogen, phosphates, etc.) causes overgrowth of toxic algae eaten by other aquatic animals, and it may cause death; it can also cause outbreaks of fish diseases. Chemical contamination is known to cause decline in frog biodiversity and tadpole mass. Oil pollution can negatively affect development of marine organisms, increase susceptibility to disease, and affect reproductive processes; it can also cause gastrointestinal irritation, liver and kidney damage, and damage to the nervous system.

Water pollution may disrupt photosynthesis in aquatic plants and thus affects ecosystems that depend on these plants. Terrestrial and aquatic plants may absorb pollutants from water (as their main nutrient source) and pass them up the food chain to consumer animals and humans. Plants may be affected due to large concentrations of sodium chloride in water. They may also be affected by herbicides in water.

Wastewater from the domestic sector is another major contributor to water contamination and subsequent related water-borne diseases. The methodologies for treating potable water are discussed in Chapter 11, and those for treatment of domestic and industrial waste waters are discussed in Chapters 12 and 13.

1.2 AIR POLLUTION

Some of the most important air pollutants are sulfur dioxide, nitrogen dioxide, carbon monoxide, ozone, VOCs, and airborne particles with radioactive pollutants. Combustion of fossil fuels produces extremely high levels of air pollution and is a main area now in focus for control.

The general impacts or effects of air pollution on humans include reduced lung functioning, irritation of eyes, nose, mouth, and throat, asthma attacks, respiratory symptoms such as coughing and wheezing, increased respiratory disease such as bronchitis, reduced energy levels, headaches, and dizziness, disruption of endocrine, reproductive, and immune systems neurobehavioral disorders, and cardiovascular problems, and it can sometimes even lead to cancer and premature death.

The impact of acid rain destroys fish life in lakes and streams. Acid rain can kill trees, destroy the leaves of plants, and can infiltrate soil by making it unsuitable for purposes of nutrition and habitation.

Excessive ultraviolet radiation coming from the sun through the ozone layer in the upper atmosphere, which is eroded by some air pollutants, may cause skin cancer in wildlife. Ozone in the lower atmosphere may damage lung tissues of animals. Ozone holes in the upper atmosphere can allow excessive ultraviolet radiation from the sun to enter the Earth causing damage to trees and plants. Ozone in the lower atmosphere can prevent plant respiration by blocking stomata (openings in leaves) and negatively affecting plants' photosynthesis rates, which will stunt plant growth; ozone can also decay plant cells directly by entering stomata. The details of the technologies that are used to control air pollution are discussed in Chapter 14 of this book.

1.3 LAND POLLUTION

Soil pollution is due to contamination from industrial and domestic pollutants. Some common soil pollutants are hydrocarbons, solvents, and heavy metals. Fossil fuels also contribute to soil contamination and water pollution. Common sources of fossil fuel pollution include power-generating plants, petroleum refineries, petrochemical plants, production and distribution of fossil fuels, road transport (motor vehicles), and shipping and aircraft industries.

Heavy metal contamination in soil is being frequently observed in both developed and developing countries. Lead in soil is especially hazardous for young children, causing developmental damage to the brain, and high levels of mercury can increase the risk of kidney damage; cyclodienes can lead to liver toxicity and cause problems related to depression of the central nervous system as well as causing headaches, nausea, fatigue, eye irritation, and skin rash.

The domestic sector is another significant source of pollution generating municipal solid waste that may end up in landfills or incinerators leading to soil contamination and air pollution. Soil contamination may alter plant metabolism and reduce crop yields. Trees and plants may absorb soil contaminants and pass them up the food chain.

1.4 ENVIRONMENTAL MANAGEMENT

In view of the preceding points, it is necessary to adopt methods to conserve the environment and reduce pollution to the natural systems. Environmental planning and management are strategic challenges confronting businesses in the 21st century. The focus of this book is on environmental management to address the issues related to pollution, discuss the legal position across the globe and in the country, and to help the reader plan and manage the environmental pollution-related problems in a more efficient manner. The book also discusses the technological solutions available for air, water, and hazardous waste management.

CHAPTER TWO

Sustainable Development

2.1 INTRODUCTION

State of environment reports facilitate environmentally related decision-making and contribute significantly to education about the environment and natural resources. This reporting enhances the quality, accessibility, and relevance of data relating to Ecologically Sustainable Development (ESD) and enables progress toward achieving the broader goals of ESD to be monitored.

State of the environment reporting:

- *is scientifically credible*: The information it delivers is objectively based on the best available scientific data and advice,
- *identifies trends*: *important issues and emerging issues* in the environment. This makes it a valuable tool for decision-makers such as managers, industry, or policy developers,
- *assesses efforts to deal with important environmental issues*: This information enables performance evaluation on environmental management and assists strategic planning,
- *is regular*: The frequency of reporting may vary depending upon the issue and the jurisdiction, but the aim is to produce a series of reports that track progress over time.

2.2 SUSTAINABLE DEVELOPMENT

Sustainable development is defined as "Development that meets the needs of the present without compromising the ability of future generations to meet their own needs" (Oxford: Oxford University Press, 1987). Sustainable development depends on how well we balance social, economic, and environmental objectives or needs when making decisions today.

2.2.1 Goals of Sustainability

In 2012, the United Nations Conference on Sustainable Development met to discuss and develop a set of goals to work toward; they grew out of the

Environmental Management
ISBN 978-0-12-811989-1
http://dx.doi.org/10.1016/B978-0-12-811989-1.00002-6

Millennium Development Goals that claimed success in reducing global poverty while acknowledging there was still much more to do. The Sustainable Development Goals (SDG) eventually came up with a list of 17 items **(8)** that included amongst other things:

- the end of poverty and hunger
- better standards of education and healthcare, particularly as it pertains to water quality and better sanitation
- to achieve gender equality
- sustainable economic growth while promoting jobs and stronger economies
- sustainability to include health of the land, air, and sea

Finally, it acknowledged the concept of nature having certain rights, that people have stewardship of the world, and the importance of putting people at the forefront of solving these global issues.

Thus, sustainable development recognizes that growth must be both inclusive and environmentally sound to reduce poverty and build shared prosperity for today's population and to continue to meet the needs of future generations. It is efficient with resources and carefully planned to deliver both immediate and long-term benefits for people, the planet, and prosperity. The three pillars of sustainable development–economic growth, environmental stewardship, and social inclusion (Fig. 2.1)—carry across all sectors of

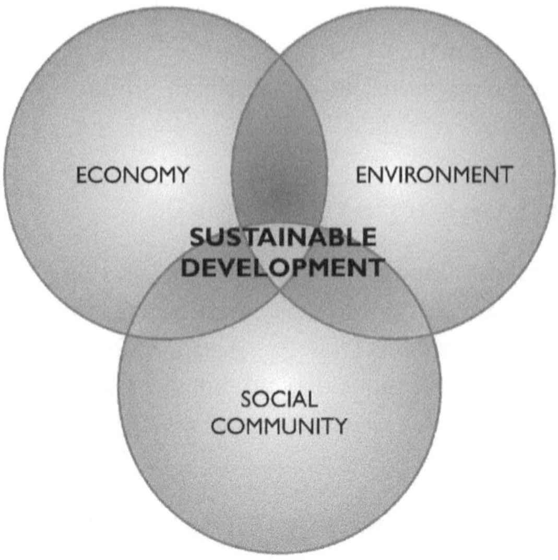

Figure 2.1 Pillars of sustainable development.

development, from cities facing rapid urbanization to agriculture, infrastructure, energy development and use, water availability, and transportation.

Many of these objectives may seem to conflict with each other in the short term. For example, industrial growth might conflict with preserving natural resources. Yet, in the long term, responsible use of natural resources now will help ensure that there are resources available for sustained industrial growth far into the future.

Economic development is about providing incentives for businesses and other organizations to adhere to sustainability guidelines beyond their normal legislative requirements. The supply and demand market is consumerist in nature, and modern life requires a lot of resources every single day; economic development is about giving people what they want without compromising quality of life, especially in the developing world.

Social development is about awareness of and legislation protection of the health of people from pollution and other harmful activities of business. It deals with encouraging people to participate in environmental sustainability and teaching them about the effects of environmental protection as well as warning of the dangers if we cannot achieve our goals.

Environmental protection is the need to protect the environment, whether the concept of 4 Rs (reduce, recycle, recover, and reuse) are being achieved or not. Businesses that are able to keep their carbon emissions low is toward environmental development. Environmental protection is the third pillar and, to many, the primary concern of the future of humanity.

It defines how to protect ecosystems, air quality, integrity, and sustainability of our resources and focuses on the elements that place stress on the environment. It also concerns how technology will drive our greener future; and that developing technology is key to this sustainability and protecting the environment of the future from potential damage that technological advances could potentially bring.

The process of describing indicators helps diverse members of a community reach consensus on what sustainability means. Indicators help put sustainability in concrete terms that demonstrate a new way to measure progress. Concepts like a person's ecological footprint help people understand how their everyday actions relate to issues that seem beyond the reach of a single individual and explain sustainability.

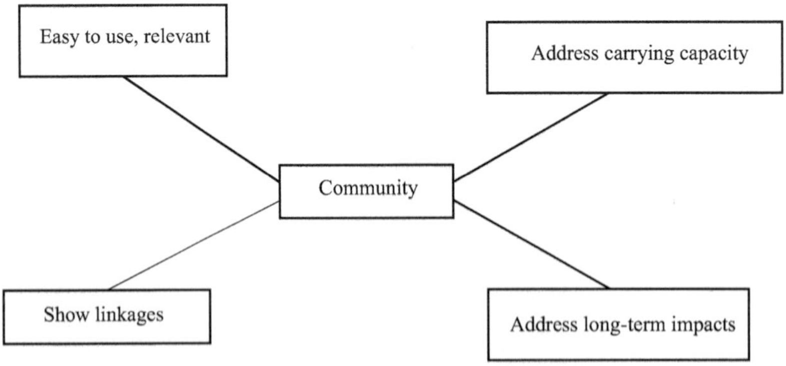

Figure 2.2 Parameters for an indicator.

2.2.2 What Makes an Indicator?

There are a number of characteristics of an indicator (Fig. 2.2) of which some are identified next:

- **address carrying capacity**: An indicator of sustainability needs to address the carrying capacity, i.e., whether the community is using resources at a rate faster than they are being renewed or restored.
- **relevant to community**: An indicator must be relevant, that is, it must fit the purpose for measuring.
- **understandable:** We need to develop indicators that speak to people, so they understand what they personally are doing that is causing problems and what steps, however small, they might be able to take to help solve the problem.
- **useable by the community**: If indicators are not used by the community, they will not have any effect on what people do. Indicators need to help people see how they can change their behavior to have a positive effect on community sustainability.
- **long-term view:** Sustainability is a long-term goal.
- **show linkages**: Traditional indicators tend to be narrowly focused on one aspect of a community.

2.2.2.1 Ecological Footprint

One of the measures or indicators is the ecological footprint. This is a measure of how much of the earth's resources we are using. The amount depends upon consumption and is very different for different countries and for different people living in a particular country. For example, a person who walks or takes public transportation has a smaller footprint than

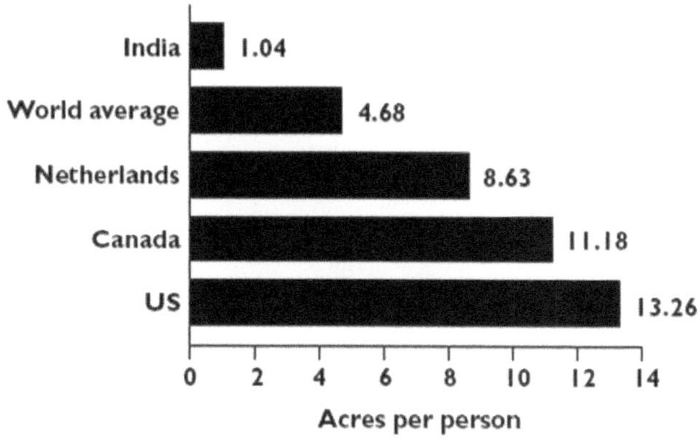

Figure 2.3 The ecological footprints from some countries. *Wackernagal & Rees, Our Ecological Footprint.*

someone who commutes in a private vehicle. Fig. 2.3 gives the ecological footprints from some countries.

2.2.2.2 Pressure–State–Response

Traditionally, organizations tend to measure conditions that exist. This is called the state. For example, an environmental agency measures the condition of the air—the air quality—by measuring how many parts per million of a pollutant are in the air or how many days the air quality is rated "good." The agency may also measure responses related to that state such as what emissions standards have been set for automobiles? These are measures of response to the state of air quality. However, frequently what is not measured is the activity that is causing the state to exist. This activity is called the pressure.

Pressure–state–response discussions are some things that may be a pressure in one context and a state or response in another. For example, if the context is air quality, then the amount of air pollution is the state, and a pressure would be the number of cars being driven. However, if the context is transportation, the state becomes the number of cars driven, and a pressure may by the distance between where people live and where they work. Fig. 2.4 shows the pressure–state–response diagram for air quality as an indicator.

Indicators of sustainability are different from traditional indicators of economic, social, and environmental progress. Traditional indicators–such as stockholder profits, asthma rates, and water quality–measure changes in one part of a community as if they were entirely independent of the other parts.

Pressure	State	Response
Pounds of toxics used	Air quality (ppm)	Number of air permits
Vehicle miles driven	Air quality	Cars inspected
Number of single use / disposable goods purchased	Tons recycled incinerated or landfilled	Number of permitted landfills or incinerators

Figure 2.4 Pressure–state–response.

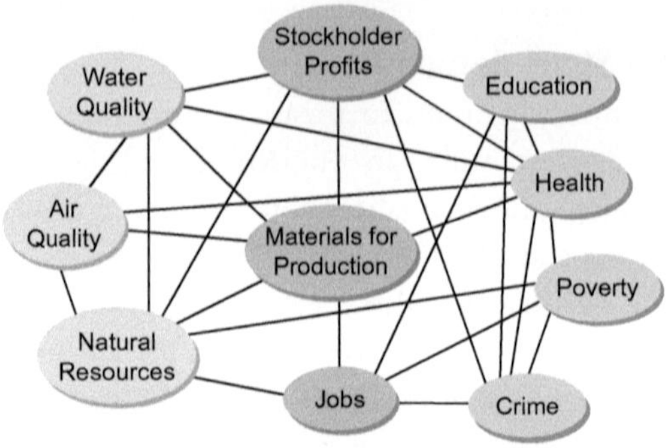

Figure 2.5 Linkages between various sectors.

Sustainability indicators reflect the reality that the three different segments are very tightly interconnected.

2.2.2.3 The Driving forces; Pressures; States; Impacts; Responses (DPSIR) Framework

At present, most indicator reports compile sets of physical, biological, or chemical indicators. They generally reflect a systems analysis view of the relations between the environmental system and the human system (Figs. 2.5 and 2.6).

According to this systems analysis view, social and economic developments exert pressure on the environment and, as a consequence, the state of the environment changes, such as the provision of adequate conditions for

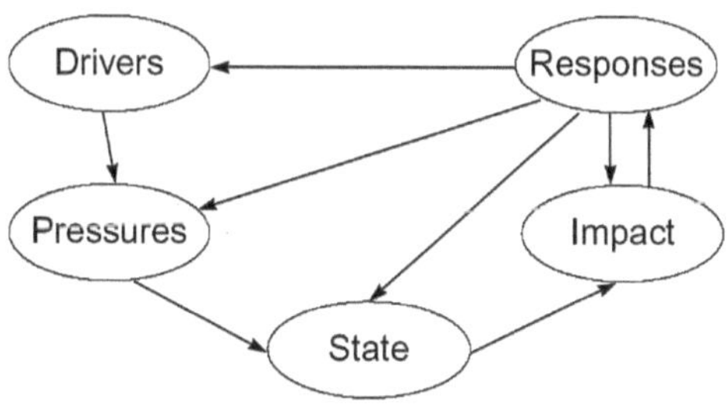

Figure 2.6 DPSIR framework for reporting environmental status.

health, resources availability, and biodiversity. Finally, this leads to impacts on human health, ecosystems, and materials that may elicit a societal response that feeds back on the driving forces, or on the state or impacts directly, through adaptation or curative action. There is arbitrariness in the distinction between the environmental system and the human system. Many of the relationships between the human system and the environmental system are not sufficiently understood or are difficult to capture in a simple framework. Nevertheless, from the policy point of view, there is a need for clear and specific information on the following:

1. driving forces,
2. the resulting environmental pressures the state of the environment,
3. impacts resulting from changes in environmental quality,
4. the societal response to these changes in the environment.

The DPSIR framework (Fig. 2.7) is useful in describing the relationships between the origins and consequences of environmental problems, but to

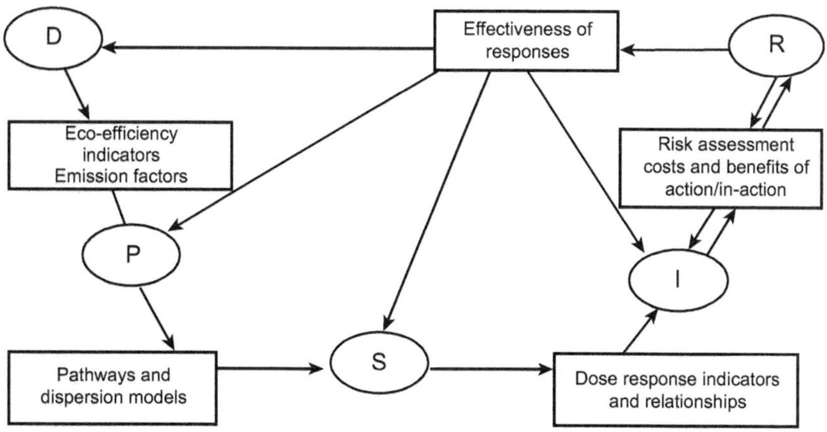

Figure 2.7 Example of the linkages that explain the DPSIR condition.

understand their dynamics it is also useful to focus on the links between DPSIR elements. For instance, the relationship between the "D" and the "P" by economic activities is a function of the eco-efficiency of the technology and related systems in use, with less "P" coming from more "D" if eco-efficiency is improving. Similarly, the relationship between the impacts on humans or ecosystems and the "S" depends on the carrying capacities and thresholds for these systems. Whether society "responds" to impacts depends on how these impacts are perceived and evaluated; and the results of "R" on the "D" depend on the effectiveness of the response.

2.2.3 Environmental Indicators

The environment is complex, and discerning environmental trends can be difficult. Environmental indicators help track changes in the environment by selecting key measures, which may be physical, chemical, biological, or socio-economic, which provide useful information about the whole system. It is possible to evaluate the fundamental condition of the environment using indicators without having to capture the full complexity of the system. Indicators are based on the best scientific understanding currently available so that changes in these simple measures can be related to more complex environmental trends. When time series data for an indicator show a trend, then there is a need to provide some interpretation of the trend and its implications. Therefore an indicator must be backed by a sound theoretical framework so that accurate interpretations can be made. For example, the concentration of ozone depleting substances in the atmosphere is a good example of an indicator. The complex chemistry of stratospheric ozone depletion need not be understood to use this indicator. It is known that increases in the concentration of ozone-depleting substances are harmful to the stratospheric ozone layer, while decreases show that efforts to protect the ozone layer are succeeding.

2.2.4 Why Develop Environmental Indicators?

Environmental indicators simplify state of the environment reporting in two important ways. Trends in the indicators are readily interpreted to yield valuable information about important aspects of the environment. Secondly, environmental indicators can be an aid to communication. They allow information about the environment to be communicated effectively. As users of information about the environment become more familiar with the agreed indicators, they will be able to absorb this information more quickly. Environmental indicators can also help focus and rationalize environmental monitoring programs by drawing attention to the critical measures required to evaluate environmental trends and conditions.

2.2.4.1 *Ozone Depletion*

- Stratospheric ozone depletion trends
- Trends in ultraviolet B (UV-B) measurements
- Phasing out ozone-depleting substances, including level of sales and purchases as an indication of emissions

2.2.4.2 *Climate and Climate Change*

- Greenhouse gas emissions in the area, including per capita, per GDP, per area, and by source and end use
- Estimates of sinks of greenhouse gases

2.2.4.3 *Air Quality*

- Ambient concentrations of ozone, photochemically reactive organic compounds, sulfur dioxide, nitrogen oxides, carbon monoxide, particulate matter, and air toxics
- Concentrations of lead in air
- Precipitation chemistry
- Visibility
- Air pollution indices
- Estimated emissions of various pollutants from industrial, commercial, residential, and mobile sources
- Number and type of complaints to the EPA and other agency telephone information services, indicating level of public concern

2.2.4.4 *Rivers, Estuaries, Lakes, and Wetlands*

- Ecological:
  - extent and condition of riparian vegetation
  - macro invertebrate community characteristics
  - fish community status
  - extent and condition of major wetlands
  - extent and condition of aquatic macrophytes (water plants)
  - occurrence of major cyanobacteria (blue-green algae) blooms
  - algal levels and *chlorophyll-a*
- Physico-chemical water quality:
  - turbidity
  - total phosphorus
  - salinity/electrical conductivity

- Additional water quality:
  - nitrogen
  - dissolved oxygen
  - fecal coliform bacteria
  - pH/acid-alkaline balance
  - pesticides
  - heavy metals
- Water quantity:
  - stream flow characteristics and the impact of water withdrawal

2.2.4.5 *Marine Waters*

- Algal blooms on the NSW open coast, their extent, and recorded frequency
- Beach water quality frequency of compliance with guidelines for bacterial levels in beach bathing waters
- Sewage effluent discharges to marine waters point source loads, nature, and quality of effluent discharged
- Oil spills on the NSW open coast, their extent, and recorded frequency
- Marine water and sediment quality

2.2.4.6 *Ground Water*

- Resource indicators for ground water management areas (GWMAs):
  - bore density
  - water level variations (rises and falls and over what area)
  - water salinity variations (increases and decreases and over what area)
  - ratio of use to entitlements
  - ratio of entitlements to sustainable yield
- Indicators to determine the health of ground water systems:
  - salinity levels
  - nitrate levels
  - pesticide/herbicide levels
- Irrigation water logging, salinization, and dry land salinization

2.2.4.7 *Land*

- Major trends in land use forming a basis for assessment of impacts and sustainability,
- Land degradation by type, e.g., erosion, salinity, scalding, acidity, woody shrub infestation, acid sulfate soils, and soil structure decline. This establishes trends and provides a basis for assessing the causes of such degradation,

- Natural disturbances indicate some of the natural processes that determine the state of the environment and distinguishes between natural processes and human induced degradation,
- Rates of vegetation change, highlighting the causes (e.g., dieback or clearing) of any changes, and indicating the potential impact on the land resources,
- Rates of vegetation regeneration and land rehabilitation; establishing trends in reversing land degradation and defining how this is being achieved.

2.2.4.8 Biological Diversity

- Habitat change may account for the loss or change in composition of both flora and fauna; vegetation change can indicate loss or recovery of forest types and associated vegetation,
- Change in distribution and area of selected ecosystems, such as forest, woodlands, grasslands, rangelands, wetlands, estuaries, riparian zones to indicate trends in area and distribution of major ecosystems,
- Changes in number, size, and level of fragmentation of on-reserve and off-reserve protected areas indicates the level of protection through reserves, the viability of the reserve based on size, and also indicates threats through fragmentation,
- Key indicator species of ecosystem change show clear responses to environmental stresses, indicating potential impacts on environmental systems,
- Number and percentage of naturalized species indicate the threat of displacement to native communities and species,
- Changes in extent and intensity of selected threatening processes indicates trends in the impacts on biodiversity of these processes,
- Number of flora and fauna, compared with number of species known indicates the trends in percentage of known species threatened,
- Number of recovery plans implemented, including number of breeding programs in zoos, botanic gardens, or protected areas indicates a management response to protect species.

2.2.4.9 Public Health

- Sun exposure: incidence of melanomas, cataracts, and pterygium
- Air pollution: number of asthma admissions to hospital, and incidence of asthma related to pollution incidents
- Water drinking water quality
- Food contaminants in food

- Chemical exposure: concentrations of organochlorines in breast milk
- Environmental exposures: asbestos trends in the incidence of mesothelioma, lead trends in blood lead levels in children, and trends in birth defects incidence rates

2.2.4.10 Community Awareness

- Community priorities for the environment in the context of other issues and for specific environmental issues
- The level of concern about the environment
- Knowledge of selected environmental concepts
- Changes in behavior made for environmental reasons; changes in consumption patterns
- Level of skill in environmental areas; number of pollution reports
- Self-reported environmentally damaging behaviors
- Environmental beliefs and values: the new environmental paradigm scale; the environment and economic growth

2.2.4.11 Heritage

- Number of interim and permanent conservation orders made under the Heritage Act
- Number of landscape conservation areas listed by the National Trust
- Number of properties conserved by the Historic Houses Trust
- Number of NSW sites covered by world heritage listing
- Funding programs for the conservation of environmental heritage

2.2.4.12 Urbanization

- Growth of urban areas
- Population and urban density trends: population change by Local Government Area (LGA); and existing density (pop/ha) versus new growth in centers by population size
- Area of open space, green space, urban bushland, and recreational areas
- Water use and pollution
- Urban transport
- Noise

2.2.4.13 Transport

- Infrastructure:
    - trend in space given over to transport roads (motorways and other), rail, ferry, pipeline, shipping (wharves), parking and ancillary space, airports, and heliports

- trends in stock total by type (passenger and goods including rail stock), passenger vehicles per capita/household, age of fleet
- trends in provision of bus lanes, transit lanes, heavy and light rail, cycle routes and facilities, park and ride facilities (including bicycle lockers), and roads converted to pedestrian malls

- Intensity of use of infrastructure:
    - trends in per capita passenger kilometer per mode of transport including walking and air, freight ton kilometer by road, rail, and sea
    - trends in trip time passenger and freight trips, trip lengths
- Urban transport and land use:
    - trends in transport use and land use population density, motor vehicle ownership, mode split for journey to work
- Impacts of transport activity:
    - impacts from emissions from motor vehicles, fuel consumption, exposure to traffic noise, loss of open space

2.2.4.14 Solid Waste

- Quantities and composition of waste generated
- Quantities and composition of waste disposed
- Estimated cost per ton of waste disposed
- Volume of waste and recycled materials imported and exported
- Number, location, capacity, and characteristics of disposal sites

2.2.4.15 Hazardous Waste

- Quantity and composition of hazardous waste generated
- Quantity and composition of hazardous waste disposed
- Quantity and composition of hazardous waste recovered
- Costs and benefits of recycling
- Estimated cost per ton of waste disposed
- Volume of hazardous and recyclable waste imported/exported
- Number, location, capacity, and characteristics of disposal sites

2.2.4.16 Tourism and Recreation

- Rate of growth of tourism in NSW
- Extent to which the tourism industry demonstrates environmental management practices
- Environmental impact, specifically the extent of the following:
    - site degradation/remediation necessary
    - impacts in conservation areas
    - destruction or disturbance of flora, fauna, geology, or topography

- Seasonal effect of tourism activities on natural resources
- Effect of growth in ecotourism on natural resources:
  - total number of visitors
  - destinations
  - environmental impact
- Effect of growth in recreational pursuits on natural resources:
  - participation rates in leisure pursuits
  - participation rates in outdoor recreation
  - environmental impact
  - number of recreational vehicles licensed by coastal councils

2.2.4.17 Fisheries

- Catch per unit effort (recreational and commercial)
- Total catch per species
- Number of species, stocks exploited
- Habitats degraded, pollutants, and changes to fish populations
- Presence of exotic species
- Changes in distribution and diversity of species
- Restoration and enhancement of habitats

2.2.4.18 Agriculture

- Land use consistent with agricultural suitability and agricultural capability attributes:
  - trends in land use
  - agricultural suitability and capability maps

Land inappropriately used for agriculture, or used beyond its capability as agricultural land, is an indicator of unsustainable development.

- Changes in long-term net real farm income;

Attributes:

  - net farm income
  - productivity
  - terms of trade
  - number of farms

Profitable agriculture should not be at the expense of the environment, and it should be recognized that gains might be made by preventing or repairing land degradation. Other factors, such as terms of trade and new technologies, also need to be recognized.

- Changes in the quality of land and water;

Attributes:

  - rate of soil loss per ton of product
  - water balance/efficiency of use
  - weeds and other pests
  - nutrient balance and cycling
  - condition of pasture
  - area of native vegetation and degree of fragmentation

This indicator represents on-site environmental effects. It is possible to define the optimal state of health of the natural resource base (land, soil, and water) to define the critical tolerances for most domesticated plant and animal species.

- Changes in the level of managerial skills of land managers;

Attributes:

  - state of development of land and water management plans
  - land care attitude index
  - catchment management
  - farm planning capacity

The increase in managerial skills and attitudes to stewardship can improve on-site environmental management.

- Changes to product quality, hydrology, and native ecosystems attributable to agricultural practice;

Attributes:

  - ground and surface water quality
  - chemical contaminants of agricultural products
  - length of contact zone with nonagricultural areas

External effects of agriculture must be kept within acceptable bounds so that other ecosystems can be maintained. This indicator focuses on those impacts that will permanently damage other ecosystems, or will be technically or financially difficult to counteract (e.g., siltation of estuaries).

2.2.4.19 Forestry

- Total area of forest by forest type and land tenure, and area available and unavailable for harvesting by logged/unlogged condition
- Environmental impacts of forestry
- Pests and diseases in plantations and the resultant area degraded
- Area and percentage of forest burnt by prescribed burning and wildfire by forest type and tenure

- Area and location of eucalypt and pine plantation forests established
- State timber yield compared with imports
- Progress on National Forest Policy Statement objectives for old growth values and wilderness
- Compliance with statutory provisions

2.2.4.20 *Mining and Quarrying*

- The total area disturbed by mining and ancillary developments as a proportion of the total area of lease
- Land rehabilitation, as a proportion of the total area of lease disturbed
- Derelict sites that have been identified as having environmental impacts, type of mine, nature of the impact, area and type of rehabilitation required, area rehabilitated to date
- Proclaimed mine subsidence districts, and number and cost of successful claims for compensation
- Mined land that has reverted to its pre-mine use or has been converted to an alternative use, compared with the land uses prior to mining
- Off-site impacts of mining
- Number and impact of small quarries, and sand and gravel extraction sites
- Mines on environmentally sensitive land

2.2.4.21 *Energy*

- Primary energy supply
- Primary energy demand and consumption
- Fossil fuel intensity
- Secondary energy supply and consumption
- Energy intensity
- Environmental impact of energy production
- Use of renewable energy sources

2.2.4.22 *Industrial Activity*

- Industrial consumption of energy and resources
- Industrial contribution to air pollution
- Industrial contribution to water pollution
- Industrial contribution to solid waste

Some core environmental indicators for the atmosphere are grouped under four issues: climate variability, enhanced greenhouse effect, stratospheric ozone, and outdoor air quality, and they are discussed briefly.

2.2.4.23 Atmospheric Indicators

1. Climate variability

   Three indicators are suggested for climate variability. They focus mainly on the variability of climate, which is comparatively high. This variability is important in its own right, regardless of any possible anthropogenic effects on the climate.

2. Enhanced greenhouse effect

   Impacts of greenhouse gases to act as a good indicator.

3. Stratospheric ozone

   Three core indicators are suggested for stratospheric ozone: one relating to the concentration of ozone depleting substances in the atmosphere, one to stratospheric ozone concentrations, and one to the destruction of ozone depleting substances.

4. Outdoor air quality

   Measures for ambient air quality relate to the concentration of various pollutants that are potentially harmful to human health. The last core indicator measures emissions of air pollutants.

Motor vehicles are a major source of air pollutants, and emissions are strongly influenced by factors such as the age of the vehicle fleet, driver behavior, and vehicle maintenance. An indicator covering these aspects would be a valuable addition to the core indicator set, but it was not possible to develop an indicator that met the selection criteria. However, part of this information should be covered by the indicator on emissions of air pollutants that includes estimates of emissions from transport-related activities.

2.3 SUMMARY

Since the 1950s, one has been experiencing unprecedented growth including intensive farming, a technological revolution, and a massive increase in our power needs putting great pressure on the Earth's resources. It is now increasingly observed that both natural and human-caused disasters and the effects that these can have on the ecosystems is evident. It is therefore vital that one develops new, cleaner technologies to cope with our demands for natural resources but in a sustainable manner.

CHAPTER THREE

Natural Resource Management and Biodiversity Conservation

3.1 INTRODUCTION

Natural Resource Management (NRM) refers to the sustainable utilization of major natural resources, such as land, water, air, minerals, forests, fisheries, and wild flora and fauna. Together, these resources provide the ecosystem services that provide better quality to human life. Natural resources provide fundamental life support, in the form of both consumptive and public-good services. Ecological processes maintain soil productivity, nutrient recycling, the cleansing of air and water, and climatic cycles.

Biological diversity (biodiversity) is the occurrence of different types of ecosystems, different species of organisms with the whole range of their variants and genes adapted to different climates, and environments along with their interactions and processes. Biodiversity encompasses the variety of all life on earth. India is one of the 17 mega-biodiversity countries of the world. Although India has only 2.5% of land area, it has a large pool and diverse pool of plants and microbes which accounts for 7.8% of recorded species in the world. Genetic diversity describes the variation in the number and type of genes as well as chromosomes present in different species. The magnitude of variation in genes of a species increases with increase in size and environmental parameters of the habitat. Species diversity describes the variety in the number and richness of the spices with in a region. Ecosystem diversity describes the assemblage and interaction of species living together and the physical environment in a given area. It is referred to as landscape diversity because it includes placement and size of various ecosystems.

Countries identified to have mega-biodiversity include Australia, the Congo, Madagascar, South Africa, China, India, Indonesia, Malaysia, Papua New Guinea, Philippines, Brazil, Colombia, Ecuador, Mexico, Peru, United States, and Venezuela.

Environmental Management
ISBN 978-0-12-811989-1
http://dx.doi.org/10.1016/B978-0-12-811989-1.00003-8

3.2 THREE KEY CONCERNS

Three persistent concerns consistently raised regarding NRM include:

- Renewable resources are utilized beyond their regenerative capacity.
- Non-renewable resources are depleted with insufficient savings in man-made, human, or social capital.
- The "sink" capacity of the environment is overburdened by pollution, which in turn damages human health and ecosystem functions.

3.3 NEED FOR PROTECTING NATURAL RESOURCES

Natural habitats consist of land and water areas where the biological communities are formed largely by native plant and animal species, and where human activity has not essentially modified the area's primary ecological functions. Protecting and conserving biodiversity, the variety of life in all its forms, including genetic, species, and ecosystem diversity—and its ability to change and evolve—is fundamental to sustainable development. The components of biodiversity, as defined in the Convention on Biological Diversity (CBD), include ecosystems and habitats, species and communities, and genes and genomes, all of which have social, economic, cultural, and scientific importance.

3.4 THREATS TO BIODIVERSITY

Biodiversity is considered as a reservoir of resources to be used for food, medicine, industrial products, etc. But due to ever-increasing demand, biodiversity is gradually depleting. **The different factors threatening biodiversity include:**

Habitat destruction: The primary cause of loss of biodiversity is habitat loss or destruction which is due to the large industrial and commercial activities associated with mining, irrigation, and large destruction of forests and water bodies, etc.

Habitat fragmentation: With increased population, the habitats are fragmented by urban infrastructure, development, etc. The isolated fragment of habitats restricts the potential of species for dispersal.

Pollution: Loss of biodiversity due to environmental pollution, which includes air pollution, water pollution, industrial pollution, and so on.

The destruction of habitat (natural and modified) is recognized as the major threat to the maintenance of biodiversity. Modified habitats are typically altered natural habitats, often with the introduction of alien species of plants and animals, such as agricultural areas. Both types of habitat can support important biodiversity at all levels, including endemic or threatened species. Within both natural and modified habitats, there may be critical habitats that are required for the survival of critically endangered or endangered species, or for endemic or restricted-range species and migratory species. The intentional or accidental introduction of alien, or non-native, species of flora and fauna into areas where they are not normally found can be a significant threat to biodiversity, since some alien species can become invasive, spreading rapidly and out-competing native species.

Typically, measures to protect biodiversity include:

- **Modified habitats:** Need to minimize any degradation of habitat and identify opportunities to enhance habitat and protect biodiversity as part of operations.
- **Natural habitats:** Need to ensure that habitat is not converted or degraded unless there are no other alternatives, and the overall benefits of keeping operations in the locations outweigh impact to the environment and biodiversity.
- **Legally protected areas:** In cases of legally protected area under national regulations, one needs to comply with the requirements established in the protected area management plans as prescribed.
- **Invasive alien species:** Need to be cautious when introducing any new alien species unless this is appropriately managed within an existing regulatory framework or action plan to determine the potential for invasive behavior.

3.5 BIODIVERSITY HOTSPOTS

Biodiversity hotspots are areas that support natural ecosystems that are largely intact and where native species and communities associated with these ecosystems are well represented. They are also areas with a high diversity of locally endemic species, which are species that are not found, or are rarely found, outside the hotspot. The current, planned, or potential management activities in hotspots place the natural values at risk, and it is likely this risk will increase in the future in the absence of active conservation management. Because the natural values of hotspots are largely intact, undertaking action now to maintain these values has the potential to provide value-for-money in contributing to our efforts in biodiversity conservation.

3.5.1 Biological Diversity Act, [2002]

The CBD is the first attempt by the international community to address biological diversity as a whole in a global legal instrument. It is based on a broad ecosystem approach rather than the sectoral approach (focusing on specific species, ecosystems, or sites) that is characteristic of other international conservation agreements. Indeed, Article 2 of the CBD defines biodiversity as the variability among living organisms from all sources, occurring at three levels: diversity within species (genetic diversity), diversity between species, and diversity of ecosystems. The CBD addresses not only conservation of biodiversity per se, but also related socioeconomic aspects, which makes it a milestone in the field of environment and development. According to Article 1, the CBD has three main objectives:

- conservation of biological diversity;
- sustainable use of its components; and
- fair and equitable sharing of the benefits arising out of the utilization of genetic resources.

This Act provides for setting up of a National Biodiversity Authority (NBA), State Biodiversity Boards (SBB), and Biodiversity Management Committees (BMC) in local bodies.

- All foreign nationals or organizations require prior approval of NBA for obtaining biological resources and/or associated knowledge for any use.
- Similarly, Indian nationals or organizations will be required to give prior intimation to the concerned SBB about any biological resources being imported for commercial use. The SBB may prohibit the import if found to violate the objectives of conservation, sustainable use, and benefit-sharing.
- However, local people and communities of the area, including Vaids and Hakims, will have free access to use biological resources within the country for their own use, medicinal purposes, and research.
- While granting approvals, NBA will impose terms and conditions to secure equitable sharing of benefits.
- There is an enabling provision for setting up a framework for protecting traditional knowledge.

- The monetary benefits, fees, and royalties, as a result of approvals by NBA, are to be deposited in the National Biodiversity Fund, which will be used for conservation and development of areas from where the resource has been accessed, in consultation with local self-government.
- World Wide Fund for Nature (WWF) and World Conservation Union support projects to promote conservation and appropriate development of biosphere reserves.

3.5.2 Nagoya Protocol

The increasing concern about dwindling biological resources due to demographic pressure, habitat destruction, environmental pollution, and other related causes led to the UN-CBD in 1992. CBD provides a road map for the conservation, sustainable and equitable use of biodiversity. It emphasizes that biodiversity occurring within a nation is the sovereign property of its people. The Nagoya Protocol is an international treaty that was adopted in October 2010 by the 193 parties to the CBD. The aim of the Nagoya Protocol on access and benefit-sharing is to give effect to the fair and equitable benefit-sharing provisions of the CBD. Specifically, Article 15 of the CBD recognizes the rights of states to their genetic resources, and Article 8(j) that recognizes the rights of communities to their traditional knowledge. The Nagoya Protocol also ensures that commercial and research utilization of genetic resources and associated traditional knowledge shares the benefits of such utilization with the governments and communities that have conserved such resources and knowledge (adopted October 29, 2010 <http://www.cbd.int/abs/doc/protocol/nagoya-protocol-en.pdf>).

3.6 BIO-PROSPECTING

Biodiversity prospecting refers to the exploration of the commercial value of genetic and biochemical resources. Over the past couple of years, several studies have been devoted to the economic analysis of genetic diversity in the context of the commercial search among genetic codes contained in living organisms in order to develop chemical compounds of industrial and pharmaceutical value in agricultural, industrial, and medical applications. Specifically, biodiversity prospecting refers to "the exploration of biodiversity for commercially valuable genetic and biochemical resources." It should be possible to justify the conservation of biodiversity on the basis of its many pharmaceutical and other commercial applications.

Humans derive many direct and indirect benefits from the living world. Biodiversity is the source of food, medicines, pharmaceutical drugs, fibers, rubber, and timber. The biological resources contain potentially useful resources as well. The diversity of organisms also provides many ecological services free of charge that are responsible for maintaining ecosystem health. Source of food and improved varieties biodiversity is of use to modern agriculture in three ways:

1. as a source of new crops,
2. as a source material for breeding improved varieties, and
3. as a source of new biodegradable pesticides.

Of the several thousand species of edible plants, less than 20 plant species are cultivated to produce about 85% of the world's food. Wheat, corn, and rice, the three major carbohydrate crops, yield nearly two-thirds of the food sustaining the human population.

Fats, oils, fibers, etc. are other uses for which more and more new species need to be investigated. The commercial, domesticated species are cross-bred with their wild relatives to improve their traits. Genes of wild species are used to confer new properties such as disease resistance or improved yield in domesticated species. For example, rice grown in Asia is protected from the four main diseases by genes received from a single wild rice species (*Oryza nivara*) from India.

The Millenium Ecosystem Assessment ("the Assessment") estimates that the current and projected future impact of bio-prospecting on ecosystems is low, because the amount of material that needs to be harvested is normally small. The Assessment also states that there is a strong synergy between biodiversity preservation and bio-prospecting, since the latter benefits from preserving the former. However, it warns that great uncertainty remains about the potential impact of bio-prospecting activities. As the projected impact is minimal, although uncertain, bio-prospecting does not presently implicate provisions of international agreements, which regulate actions likely to have serious adverse environmental impacts in the commons. The legal implications of any potential environmental impacts from bio-prospecting are not explored in detail through this issue brief, but must be a consideration for decision-makers in drafting future laws and policies to regulate this activity and materials between states, particularly for developing nations.

3.6.1 Existing Important Legal Frameworks

- Open-access regime: Refers to the system of law in which no sovereign state controls the resource or area in question, yet they are free to exploit and profit from these resources/areas so long as they adhere to generally accepted principles and obligations of international law.
- The Global Commons: Also known as "the Commons," is a term used to describe domains where common pool resources are found. The Global Commons specifically refers to domains which do not fall within the jurisdiction of any one nation, thus all states have legal access. The High Seas and Antarctica are both considered part of the Global Commons.
- Common Heritage of Mankind: This legal concept dictates that certain resources/areas are the communal property of all humankind. Hence, no person or state has exclusive legal rights to these resources/areas in so far as all uses and benefits must be shared equally for the benefit of current and future generations.
- Access and Benefit Sharing (ABS): Commonly describes legal regimes that seeks to apportion equitable rights to both developed and developing states regarding the exploitation and any derivate benefits from scientific research and commercial development of biological resources.
- The High Seas: The water column and marine environment beyond the territorial waters and exclusive economic zones of coastal states. Activities in the High Seas are broadly regulated by UNCLOS, yet many aspects are left to the jurisdiction of Flag States (states which have vessels carrying their flag and are subsequently responsible for the actions of vessels and crew).
- Precautionary Principle: an approach which dictates that in absence of adequate scientific evidence, decision-makers must err on the side of caution and adopt laws/policies that prevent suspected risks of harm to the environment or human health.

3.6.2 International Intellectual Property Framework: Trade-Related Aspects of Intellectual Property Rights and World Intellectual Property Organization [WIPO]

The agreement on Trade-Related Aspects of Intellectual Property Rights (TRIPS) provides that "patents shall be available for any inventions, whether products or processes, in all fields of technology, provided that they are new, involve an inventive step and are capable of industrial application." Members may exclude from patentability, "plants and animals other than micro-organisms, and essentially biological processes for the production of plants or animals other than non-biological and microbiological processes." These provisions indicate that patents must be available for micro-organisms and microbiological processes which are new, inventive, and capable of industrial application. Genetic materials from bio-prospecting would seem to fit into this category. TRIPS further states that "patents shall be available and patent rights enjoyable without discrimination as to the place of invention." If the location of the source of genetic material is analogous to the place of invention, this provision suggests that members may not decline to patent genetic material because it originates in the commons. Patentability of biological materials focuses on whether they are novel and involve an inventive step.

3.6.3 Merits of Bio-Prospecting

- Bio-prospecting has been an important phenomenon of discovering new drugs since the dawn of civilization. Several millions of people throughout the world have been using more than 8000 species of medicinal plants for healthcare needs. Over 800 medicinal plant species are currently in use by Indian herbal industry alone. Even though pharmaceutical firms and scientists continue to find useful application of components from nature, their search methods and applications have changed.
- With advancement in molecular biology and availability of sophisticated diagnostic tools for screening, it has become pretty effective for pharmaceutical firms to conduct research through bio-prospecting. In high-technology laboratories, extracts from biological specimens undergo rapid and precise screening procedures that allow for the isolation of chemicals displaying a specifically targeted activity.
- Discovery of several life-saving drugs including anti-neoplastic drugs (e.g. Vinblastine, Taxol, topotecan, and etoposide), in recent past has renewed the interest of pharmaceutical industries in bio-prospecting. Efforts are being made to isolate several drugs from plants.
- Bio-prospecting collaborations between pharmaceutical companies and countries supplying the medicinal raw material and knowledge offer not only the revenue source for underdeveloped countries, but also opportunities for society for better education and employment avenues. Many studies have suggested that if the bio-prospecting search is based on the information and knowledge from local people, then the value of bio-prospecting benefits will be higher.

3.6.4 Limitations of Bio-Prospecting

- There is a growing concern that a number of pharmaceutical firms and biotechnology companies are exploring the forests, fields, and waters of the developing world in search of biological riches and indigenous knowledge with sole aim of developing patented and profitable products. In the vast majority of cases, no money has changed hands and no recognition has been given to indigenous communities who selected, maintained, and improved traditional plant varieties for medicine. Pharmaceutical firms are often accused of cheating local people by denying them access to knowledge and financial benefits.
- The multinational companies engaged in bio-prospecting are free to patent bio-materials, but there are no effective guidelines and conditions defined for recognizing and rewarding the contributions of indigenous people and other informal innovators who are responsible for nurturing, using, and developing biodiversity. One of the enduring questions in bio-prospecting has been whether the analysis and identification of active medicinal constituent in biological samples provide the pharmaceutical firms the sole right on ecological habitat in resource rich regions or not.
- Although bio-prospecting agreements are sanctioned by the multilateral CBD, in most cases commercial bio-prospecting agreements cannot be effectively monitored or enforced by source communities, countries or by the convention itself. In several cases, there is no regulation in place to ensure that the source countries of these plants will be adequately compensated.
- Imbalance in ecosystems due to excessive exploitation of material resources is always a possibility. It is a fact that the tropical rainforest regions of the world, which constitute more than 50% of medicinal plants, are disappearing. This is mainly due to multitude of commercial interests including bio-prospecting.

3.7 TRADITIONAL KNOWLEDGE AND BIO-PIRACY

Traditional knowledge has always been an easily accessible treasure, and thus has been susceptible to misappropriation. The traditional knowledge, particularly, related to the treatment of various diseases has provided leads for development of biologically active molecules by the technology rich countries. In other words, traditional knowledge is being exploited for bio-prospecting. Traditional knowledge includes both the codified (documented) as well as non-codified information (not documented but may be orally transmitted). Bio-piracy of codified Indian traditional knowledge continues, since, this information exists in regional languages, and there exists a language barrier due to which the patent offices are unable to search this information as prior art, before granting patents. Formulations used for the treatment of human ailments from traditional knowledge are time-tested since they have been in practice for centuries. The reliability of the traditional medicine systems coupled with the absence of such information with patent offices, provides an easy opportunity for interlopers for getting patents on these therapeutic formulations derived from traditional medicine systems.

While Article 15 of the CBD does not address the issue of traditional knowledge, Article 8(j) of the CBD requires each contracting party, subject to its national legislation, to respect, preserve, and maintain knowledge, innovations, and practices of indigenous and local communities embodying traditional lifestyles relevant for the conservation and sustainable use of biological diversity;

- Promote their wider application with the approval and involvement of the holders of such knowledge, innovations, and practices; and
- Encourage equitable sharing of benefits derived from their utilization.

Traditional Knowledge Digital Library (TKDL) targets Indian systems of medicine, viz., Ayurveda, Unani, Siddha, and Yoga, available in public domain. This is being documented by sifting and collating the information on traditional knowledge from the existing literature existing in local languages such as Sanskrit, Urdu, Arabic, Persian, and Tamil in digitized format, which will be available in five international languages, which are English, German, Spanish, French, and Japanese. Traditional Knowledge Resource Classification, an innovative structured classification system for the purpose of systematic arrangement, dissemination, and retrieval was evolved for about 5000 subgroups against few subgroups available in International Patent Classification (IPC), related to medicinal plants. The information is being structured under section, class, subclass, group and subgroup as per the IPC for the convenience of its use by the international patent examiners.

3.8 ACCESS AND BENEFIT-SHARING

ABS refers to the way in which genetic resources may be accessed, and how the benefits that result from their use are shared between the people or countries using the resources (users) and the people or countries that provide them (providers). The access and benefit-sharing provisions of the CBD are designed to ensure that the physical access to genetic resources is facilitated and that the benefits obtained from their use are shared equitably with the providers. The benefits to be shared can be monetary, such as sharing royalties when the resources are used to create a commercial product, or non-monetary, such as the development of research skills and knowledge. It is vital that both users and providers understand and respect institutional frameworks such as those outlined by the CBD and in the Bonn Guidelines.

3.8.1 Access and Benefit-Sharing Related Obligations and Commitments Under the Convention on Biodiversity

Article 15(1) of the CBD clearly confirms the authority of governments to regulate physical access to genetic resources in areas within its jurisdiction. At the same time, Article 15(1) does not grant the State a property right over these resources [1]. Ownership of genetic resources is not addressed by the CBD at all, but is subject to national and sub-national legislation or law (including common law as well as customary law). The authority of a government to determine access to genetic resources is qualified by Article 15(2) of the CBD, which requires the contracting parties to endeavor to create conditions that:

- facilitate access to their genetic resources for environmentally sound uses by other contracting parties, and
- do not impose restrictions that run counter to the objectives of the CBD.

Article 15(3) of the CBD limits the genetic resources covered by Article 15 (as well as Articles 16 and 19) to those:

- provided by parties that are countries of origin ("country of origin" of genetic resources is defined by Article 2 of the CBD as "the country which possesses those genetic resources in in situ conditions"), or
- provided by parties that have acquired the genetic resources in accordance with the CBD.

Only these two categories of genetic resources entitle a provider to benefits under the CBD.

ABS is based on Prior Informed Consent (PIC) being granted by a provider to a user, and negotiations between both parties to develop Mutually Agreed Terms (MAT) to ensure the fair and equitable sharing of genetic resources and associated benefits.

- PIC: Is the permission given by the competent national authority of a provider country to a user prior to accessing genetic resources, in line with an appropriate national legal and institutional framework.
- MAT: Is an agreement reached between the providers of genetic resources and users on the conditions of access and use of the resources, and the benefits to be shared between both parties. These conditions are required

under Article 15 of the CBD, which was adopted in 1992 and provides a global set of principles for access to genetic resources, as well as the fair and equitable distribution of the benefits that result from their use.

3.8.1.1 Benefits

Article 15(7) of the CBD requires each contracting party to take legislative, administrative, or policy measures, the goal of which is the fair and equitable sharing of benefits with the contracting party providing genetic resources. While the CBD does not give a definition of the term "benefits," it foresees different types (monetary and non-monetary) of benefits to be shared, including:

- research and development results, Article 15(7);
- commercial or other benefits derived from utilizing the genetic resources provided, Article 15(7);
- access to and transfer of technology using the genetic resources, Article 16(3);
- participation in all types of scientific research based on the genetic resources, Article 15(6);
- participation specifically in biotechnological research activities based on the genetic resources, Article 19(1); and
- priority access to the results and benefits arising from biotechnological use of the genetic resources, Article 19(2).

The link between genetic resources and traditional knowledge in the context of ABS is based on the second and third obligations under Article 8(j) of the CBD. CBD acknowledges the value of traditional knowledge to modern society and recognizes that holders of such knowledge, innovations, and practices are to be involved and provide their approval, subject to national laws, and when it gets to the wider application of those knowledge, innovations, and practices.

3.9 SUMMARY

The focus of the biodiversity and conservation is management of all natural resources including biodiversity, heritage, and conservation matters in a manner that ensures equitable and sustainable use, conservation, management, and, where necessary, the restoration of this resource base. It also focuses on mitigating threats to resources as a basis for sustainable and inclusive socioeconomic development, which facilitates sustainable economic growth and development.

REFERENCE

[1] L. Glowka, F. Buihenne- Guilmin, H. Synge, A guide to the convention on Biological Diversity IUCN, Gland and Cammridge, 1994.

FURTHER READING

[1] https://www.cbd.int/ viewed in Jan 2016.
[2] https://www.cbd.int/abs/ viewed in February 2016.
[3] UNU–IAS (United Nations University – Institute of Advanced Studies), Biodiversity Access and Benefit–Sharing Policies for Protected Areas: An Introduction, Yokohama, Japan, 2003.
[4] R. Vernooy, M. Ruiz, Brief Review of Recent ABS Initiatives, in: M. Ruiz, R. Vernooy (Eds.), The Custodians of Biodiversity: Sharing Access to and Benefits of Genetic Resources, Earthscan from Routledge, Abingdon, U.K., 2011.
[5] R. Pavoni, Mutual Supportiveness as a Principle of Interpretation and Law-Making: A Watershed for the 'WTO-and-Competing-Regimes' Debate? European Journal of International Law 21 (3) (2010) 649–679.

CHAPTER FOUR

Environmental Policies and Legislation

4.1 INTRODUCTION CURRENT ENVIRONMENTAL ISSUES

As discussed in Chapter 1, current environmental issues include air and water pollution, global warming, ozone depletion, and loss of natural resources including biodiversity. The issues concerning global problems are governed by international laws, which deal with environmental ethics and conservation of the natural resources between two or more countries.

4.2 GLOBAL WARMING

Global warming is one of the major issues we are being faced with. Global warming has caused change in the climate of the earth, causing temperatures to rise. The greenhouse effect causes the Earth's heat to be trapped in the atmosphere, which results in the increase in temperature. This in turn has an effect on various species dependent on the basic laws of nature. A warmer Earth also causes changes in rainfall patterns, and thus affects humans, plants, and animals as well. Scientists are of the opinion that a further rise in the carbon dioxide levels will aggravate the situation.

4.3 OZONE DEPLETION

Chlorofluorocarbons (CFCs) are considered to be the main cause of ozone depletion. The term ozone depletion implies a decline of the quantity of the ozone in the Earth's stratosphere. The loss of ozone in the lower stratosphere was first recorded in Antarctica in the 1970s. CFCs are used in aerosol sprays as well as air conditioners. When released into the atmosphere, these add to ozone depletion. Due to ozone depletion, humans are faced with various other problems, such as the harmful effects of the UV rays, which in turn affect plants and various other species of animals.

Environmental Management
ISBN 978-0-12-811989-1
http://dx.doi.org/10.1016/B978-0-12-811989-1.00004-X

4.4 LOSS OF NATURAL RESOURCES

With the increase in population, one can see the loss of natural resources. This is caused due to various human activities, and include many reasons. This in turn affects the ecosystem. Forests are being cleared to meet the rising demands for the need of paper, wood, or even land. Mining and the burning of fossil fuels have led to depletion of resources. Under these extremely complex conditions, all nations have a set of policies/laws to preserve the environment. International environmental laws are also laid down to address issues related to the various problems across borders.

4.5 ENVIRONMENTAL ETHICS

Environmental ethics believes in the ethical relationship between human beings and the natural environment. Environmental ethics says that one should base their behavior on a set of ethical values that guide our approach toward the other living beings in nature. Environmental ethics is about including the rights of non-human animals in our ethical and moral values. Even if the human race is considered the primary concern of society, animals and plants are in no way less important. They have a right to get their fair share of existence.

Two levels of ethics are most prevalent – "descriptive ethics" and "prescriptive ethics." Prescriptive ethics deals with moral issues in the conventional sense of that term, that is, with questions of right or wrong, duties and rights, justice and injustice, virtue and wickedness, and so forth. In particular, a new environmental ethic may have to challenge four basic traditions of *anthropocentrism, reductive analysis, egocentric perspective and the fact/value gap.*

4.6 ENVIRONMENTAL SUSTAINABILITY INDEX

In addition to ethics and conservation of the environment, scientists have developed an index to measure the sustainability of the environment. The 2005 Environmental Sustainability Index (ESI) benchmarks the ability of nations to protect the environment over the next several decades. It does so by integrating 76 data sets; tracking natural resource endowments, past and present pollution levels, environmental management efforts, and a society's capacity to improve its environmental performance into 21 indicators of environmental sustainability. The ESI provides a valuable tool for benchmarking environmental stewardship and permits comparative policy analysis. The lack of reliable data to measure performance on a number of issues and across many countries hinders attempts to move toward more data-driven decision-making.

Table 4.1 gives the variables, indicators, and components as identified by Organisation for Economic Co-operation and Development (OECD).

4.6.1 Construct the Environmental Sustainability Index (ESI)

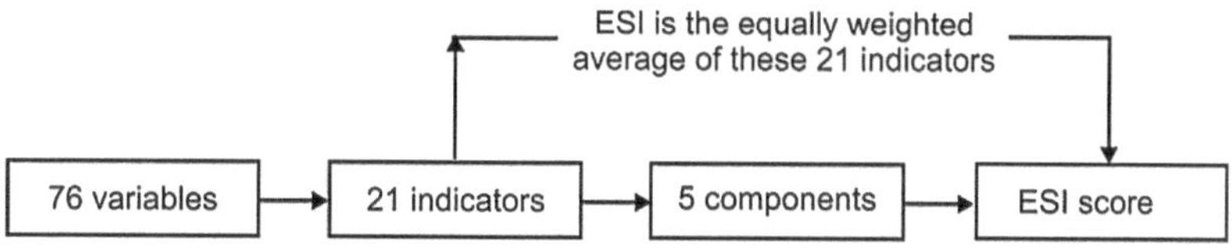

Environmental stewardship depends on both policy efforts and a society's over-arching social, political, and economic systems. While it appears that no country is on a fully sustainable trajectory, at every level of development, some countries are managing their environmental challenges better than others. Measures of governance, including the rigor of regulation and the degree of cooperation with international policy efforts, correlate highly with overall environmental success. Higher ESI scores suggest better environmental stewardship. Environmental stewardship demands attention to a wide range of pollution control and natural resource management issues.

4.7 INTERNATIONAL ENVIRONMENTAL LAW

International environmental law encompasses the body of rules agreed to by countries aimed at protecting various aspects of the global natural environment. Two non-binding instruments adopted by the international community at UN conferences have played important roles in the modern development of international environmental law. The 1972 *Stockholm Declaration of the United Nations Conference on the Human Environment* marked the beginning of increased international action on environmental issues, and focus was reinforced by the 1992 *Rio Declaration on Environment and Development.* As well as contributing to the momentum that has given rise to the many multilateral environmental agreements adopted in subsequent years, the general principles espoused in these non-binding instruments appear throughout the subsequent agreements. The vastness of this area of international law includes the environmental sub-issues of population, biodiversity, global climate change, ozone depletion, preserving the Antarctic regions, movement of toxic and hazardous substances, land or vessel-based pollution, dumping, conservation of marine living resources, transboundary air and water pollution, desertification, and nuclear damage, among others. Some of the most significant environmental agreements are given in the following sections.

Table 4.1 The Variables, Indicators, and Components as Identified by OECD

76 Variables		21 Indicators	5 Components
• Nitrogen dioxide concentration • Sulfur dioxide concentration	• Particulate concentration • Indoor air quality	Air quality	Environmental systems
• Eco-regions at risk • Threatened birds • Threatened mammals	• Threatened amphibians • National biodiversity index	Biodiversity	
• Wilderness area	• Developed area	Land	
• Dissolved oxygen • Electrical conductivity	• Suspended solids • Phosphorus concentration	Water quality	
• Surface water availability	• Groundwater availability	Water quality	
• Coal consumption • Nitrogen oxide emissions • Sulfur dioxide emissions	• Volatile organic compound emissions • Vehicles in use	Reducing air pollution	Reducing environmental stresses
• Forest cover change	• Acidification	Reducing ecosystem stresses	
• Population growth	• Total fertility rate	Reducing population growth	
• Ecological footprint • Waste recycling rates	• Hazardous waste generation	Reducing waste and consumption pressures	
• Industrial organic effluents • Fertilizers consumption	• Pesticide consumption • Area under water stress	Reducing water stress	
• Overfishing • Sustainably managed forests • Market distortions	• Salinization due to irrigation • Agricultural subsidies	Natural resource management	
• Deaths from intestinal infectious diseases • Child mortality rate	• Child mortality due to respiratory infections	Environmental health	Reducing human vulnerability

• Malnutrition	• Safe drinking water supply	Basic human sustenance	
• Casualties due to environmental disasters	• Environmental hazard exposure index	Reducing environment-related natural disaster vulnerability	
• Gasoline price • Corruption • Government effectiveness • Protected land area • Environmental governance • Strength of rule of law • Local agenda 21 initiatives	• Civil and political liberties • Sustainable development data gaps • International environmental engagement • Environmental knowledge creation • Democratic institutions	Environmental governance	Social and institutional capacity
• Energy consumption/GDP	• Renewable energy production	Eco-efficiency	
• Corporate sustainability (Dow Jones) • Corporate sustainability (Innovest) • ISO 14001 certified companies	• ISO 14001 certified companies • Private sector environmental innovation • Participation in responsible care program	Private sector responsiveness	
• Innovation capacity • Digital access index • Female primary education	• University enrollment • Research scientists	Science and technology	
• Intergovernmental environmental activities • Role in international environmental aid	• Participation in international environmental agreements	Participation in international collaborative efforts	Global stewardship
• Greenhouse gas emissions GDP	• Greenhouse gas emissions/capita	Greenhouse gas emissions	
• Transboundary sulfur dioxide spillovers	• Polluting-goods imports	Reducing transboundary environmental pressures	

4.7.1 Biodiversity

Various conventions have been formulated and passed for conservation of biodiversity some of the important ones are listed below:

- 1946 International Convention for the Regulation of Whaling
- 1971 Ramsar Convention on Wetlands of International Importance
- 1972 Convention Concerning the Protection of the World Cultural and Natural Heritage
- 1973 Convention on International Trade in Endangered Species (CITES)
- 1979 Convention on the Conservation of Migratory Species of Wild Animals
- 1992 Convention on Biological Diversity, and its 2000 Cartagena Protocol on Biosafety

4.7.2 Atmosphere

Similar to biodiversity in the area of atmosphere, Kyoto and Montréal protocols are most important.

- 1985 Vienna Convention for the Protection of the Ozone Layer, and its 1987 Montreal Protocol on Substances that Deplete the Ozone Layer
- 1992 United Nations Framework Convention on Climate Change, and its 1997 Kyoto Protocol

4.7.3 Pollution/Hazardous Substances

For prevention of pollution and handling of hazardous substances on land and sea, the various conventions and protocols include:

- 1972 London Convention on the Prevention of Marine Pollution by Dumping of Wastes and Other Matter, and its 1996 Protocol
- 1973 International Convention for the Prevention of Pollution from Ships, as modified by its 1978 Protocol (MARPOL)
- 1982 United Nations Convention on the Law of the Sea (UNCLOS)
- 1989 Basel Convention on the Control of Transboundary Movements of Hazardous Wastes and Their Disposal
- 1992 International Convention on Civil Liability for Oil Pollution Damage
- 1992 International Convention on the Establishment of an International Fund for Compensation for Oil Pollution Damage, and its 2003 Protocol
- 1998 Rotterdam Convention on the Prior Informed Consent Procedure for Certain Hazardous Chemicals and Pesticides in International Trade (PIC)
- 2001 Stockholm Convention on Persistent Organic Pollutants (POPS)

Before 1900, there were few multilateral or bilateral agreements concerning international environmental issues. Relevant international agreements were based on unrestrained national sovereignty over natural resources and focused primarily on boundary waters, navigation, and fishing rights along shared waterways, particularly the Rhine River and other European waterways. They did not address pollution or other ecological issues. In the early 1900s, countries began to conclude agreements to protect commercially valuable species. These agreements include the 1902 Convention for the Protection of Birds Useful to Agriculture, the 1916 Convention for the Protection of Migratory Birds in the United States and Canada, and the Treaty for the Preservation and Protection of Fur Seals signed in 1911. Only one convention focused on wildlife more generally; the 1900 London Convention for the Protection of Wild Animals, Birds and Fish in Africa. By the 1930s and 1940s, states recognized the importance of conserving natural resources and negotiated several agreements to protect fauna and flora generally. These include the 1933 London Convention on Preservation of Fauna and Flora in Their Natural State (focused primarily on Africa), and the 1940 Washington Convention on Nature Protection and Wild Life Preservation (focused on the Western Hemisphere).

During the 1950s and early 1960s, the international community was concerned with nuclear damage from civilian use (a by-product of the Atoms for Peace Proposal), and marine pollution from oil. Thus, countries negotiated agreements governing international liability for nuclear damage and required measures to prevent oil pollution at sea. The scope of international agreements has expanded significantly since 1972, from transboundary pollution agreements to global pollution agreements; from control of direct emissions into lakes to comprehensive river basin system regimes; from preservation of certain species to conservation of ecosystems; from agreements that take effect only at national borders to ones that restrain resource use and control activities within national borders, such as for world heritages, wetlands, and biologically diverse areas.

In 1974, international environmental law was a fledgling field with less than 3 dozen multilateral agreements. Today, international environmental law is arguably setting the pace for cooperation in the international community in the development of international law. There are nearly 900 international legal instruments that are either primarily directed to international environmental issues, or contain important provisions on them. During the period from 1985 to 1992, countries have negotiated a surprisingly large number of global agreements. These include the Vienna Convention on the Protection of the Ozone Layer; the Montreal Protocol on Substances that Deplete the

Ozone Layer with the London Adjustments and Amendments; the Protocol on Environmental Protection to the Antarctic Treaty; the Basel Convention on the Transboundary Movements of Hazardous Wastes and their Disposal; the two International Atomic Energy Agency (IAEA) Conventions on Early Notification of a Nuclear Accident and on Assistance in the Case of a Nuclear Accident or Radiological Emergency; the International Convention on Oil Pollution Preparedness, Response and Co-operation, the Framework Convention on Climate Change; the Convention on Biological Diversity; the principles on forests; the non-binding legal instrument of the Arctic Environmental Protection Strategy; and the London Guidelines for the Exchange of Information on Chemicals in International Trade.

In 1987, Canada and the United States agreed to a protocol to their 1978 Great Lakes Water Quality Agreement, which addresses groundwater contamination affecting the Great Lakes and the airborne transport of toxins into the Great Lakes. In Asia, members of the Association of Southeast Asian Nations (ASEAN) concluded the Convention on the Conservation of Nature, which provides ecosystem protection and controls on trade in endangered species. In Africa, the Bamako Convention on Hazardous Wastes bans the importation of hazardous wastes and creates a strict regimen for moving such wastes within the African continent. In Europe, the Single European Act now provides clear authority for the European Community to act on environmental and natural resource issues [1].

The European Court of Justice has assumed an important role in ensuring that measures adopted by individual nations conform to community directives. At the bilateral level, many international environmental legal instruments have been concluded during this period. In North America, the United States has signed bilateral agreements with Canada and Mexico on the transport of hazardous wastes. In 1991, Canada and the United States concluded an agreement to control acid precipitation. In Latin America, Brazil and Argentina concluded an agreement that provides for consultation in case of nuclear accidents in either country [2].

4.7.3.1 Declaration of the United Nations Conference on the Human Environment

The United Nations Conference on the Human Environment, having met at Stockholm from 5 to 16 June 1972, having considered the need for a common outlook and for common principles to inspire and guide the peoples of the world in the preservation and enhancement of the human environment [3]. It proclaims that:

1. Man is both creature and molder of his environment, which gives him physical sustenance and affords him the opportunity for intellectual, moral, social, and spiritual growth. In the long and tortuous evolution of the human race on this planet, a stage has been reached when, through the rapid acceleration of science and technology, man has acquired the power to transform his environment in countless ways and on an unprecedented scale.
2. The protection and improvement of the human environment is a major issue, which affects the well-being of peoples and economic development throughout the world.
3. Man has constantly to sum up experience and go on discovering, inventing, creating, and advancing. In our time, man's capability to transform his surroundings, if used wisely, can bring to all peoples the benefits of development and the opportunity to enhance the quality of life. Wrongly or heedlessly applied, the same power can do incalculable harm to human beings and the human environment.
4. In developing countries, most environmental problems are caused by under-development. Millions continue to live far below the minimum levels required for a decent human existence, deprived of adequate food and clothing, shelter and education, health and sanitation. Therefore, developing countries must direct their efforts to development, bearing in mind their priorities and the need to safeguard and improve the environment. For the same purpose, industrialized countries should make efforts to reduce the gap themselves and developing countries. In industrialized countries, environmental problems are generally related to industrialization and technological development.
5. The natural growth of population continuously presents problems for the preservation of the environment, and adequate policies and measures should be adopted, as appropriate, to face these problems. It is the people that propel social progress, create social wealth, develop science and technology, and, through their hard work, continuously transform the human environment. Along with social progress and the advance of production, science and technology, the capability of man to improve the environment increases with each passing day.
6. A point has been reached in history when we must shape our actions throughout the world with a more prudent care for their environmental consequences. Through ignorance or indifference, we can do massive

and irreversible harm to the earthly environment on which our life and well-being depend. Conversely, through fuller knowledge and wiser action, we can achieve for ourselves and our posterity, a better life in an environment more in keeping with human needs and hopes. There are broad vistas for the enhancement of environmental quality and the creation of a good life. What is needed is an enthusiastic but calm state of mind and intense but orderly work. For the purpose of attaining freedom in the world of nature, man must use knowledge to build, in collaboration with nature, a better environment.

7. To achieve this environmental goal will demand the acceptance of responsibility by citizens and communities and by enterprises and institutions at every level, all sharing equitably in common efforts. Individuals in all walks of life as well as organizations in many fields, by their values and the sum of their actions, will shape the world environment of the future.

Local and national governments will bear the greatest burden for large-scale environmental policy and action within their jurisdictions. International cooperation is also needed in order to raise resources to support the developing countries in carrying out their responsibilities in this field. A growing class of environmental problems, because they are regional or global in extent or because they affect the common international realm, will require extensive cooperation among nations and action by international organizations in the common interest.

4.7.3.1.1 Principles

States that:

Principle 1: Man has the fundamental right to freedom, equality, and adequate conditions of life, in an environment of a quality that permits a life of dignity and well-being, and he bears a solemn responsibility to protect and improve the environment for present and future generations. In this respect, policies promoting or perpetuating apartheid, racial segregation, discrimination, colonial, and other forms of oppression and foreign domination stand condemned and must be eliminated.

Principle 2: The natural resources of the Earth, including the air, water, land, flora and fauna, and especially representative samples of natural ecosystems, must be safeguarded for the benefit of present and future generations through careful planning or management, as appropriate.

Principle 3: The capacity of the Earth to produce vital renewable resources must be maintained, and, wherever practicable, restored or improved.

Principle 4: Man has a special responsibility to safeguard and wisely manage the heritage of wildlife and its habitat, which are now gravely imperiled by a combination of adverse factors. Nature conservation, including wildlife, must therefore receive importance in planning for economic development.

Principle 5: The non-renewable resources of the Earth must be employed in such a way as to guard against the danger of their future exhaustion, and to ensure that benefits from such employment are shared by all mankind.

Principle 6: The discharge of toxic substances or of other substances and the release of heat, in such quantities or concentrations as to exceed the capacity of the environment to render them harmless, must be halted in order to ensure that serious or irreversible damage is not inflicted upon ecosystems. The just struggle of the peoples of ill countries against pollution should be supported.

Principle 7: States shall take all possible steps to prevent pollution of the seas by substances that are liable to create hazards to human health, to harm living resources and marine life, to damage amenities or to interfere with other legitimate uses of the sea.

Principle 8: Economic and social development is essential for ensuring a favorable living and working environment for man, and for creating conditions on Earth that are necessary for the improvement of the quality of life.

Principle 9: Environmental deficiencies generated by the conditions of under-development and natural disasters pose grave problems, and can best be remedied by accelerated development through the transfer of substantial quantities of financial and technological assistance as a supplement to the domestic effort of the developing countries and such timely assistance as may be required.

Principle 10: For the developing countries, stability of prices and adequate earnings for primary commodities and raw materials are essential to environmental management, since economic factors as well as ecological processes must be taken into account.

Principle 11: The environmental policies of all states should enhance, and not adversely affect, the present or future development potential of developing countries, nor should they hamper the attainment of better living conditions for all, and appropriate steps should be taken by states and international organizations with a view to reaching agreement on meeting the possible national and international economic consequences resulting from the application of environmental measures.

Principle 12: Resources should be made available to preserve and improve the environment, taking into account the circumstances and particular requirements of developing countries and any costs which may emanate- from their incorporating environmental safeguards into their development planning and the need for making available to them, upon their request, additional international technical and financial assistance for this purpose.

Principle 13: In order to achieve a more rational management of resources and thus to improve the environment, states should adopt an integrated and coordinated approach to their development planning so as to ensure that development is compatible with the need to protect and improve environment for the benefit of their population.

Principle 14: Rational planning constitutes an essential tool for reconciling any conflict between the needs of development and the need to protect and improve the environment.

Principle 15: Planning must be applied to human settlements and urbanization with a view to avoiding adverse effects on the environment and obtaining maximum social, economic, and environmental benefits for all. In this respect, projects which are designed for colonialist and racist domination must be abandoned.

Principle 16: Demographic policies which are without prejudice to basic human rights and which are deemed appropriate by governments concerned should be applied in those regions where the rate of population growth or excessive population concentrations are likely to have adverse effects on the environment of the human environment and impede development.

Principle 17: Appropriate national institutions must be entrusted with the task of planning, managing, or controlling the nine environmental resources of states with a view to enhancing environmental quality.

Principle 18: Science and technology, as part of their contribution to economic and social development, must be applied to the identification, avoidance, and control of environmental risks and the solution of environmental problems and for the common good of mankind.

Principle 19: Education in environmental matters, for the younger generation as well as adults, giving due consideration to the underprivileged, is essential in order to broaden the basis for an enlightened opinion and responsible conduct by individuals, enterprises, and communities in protecting and improving the environment in its full human dimension. It is also essential that mass media of communications avoid contributing to the deterioration of the environment, but, on the contrary, disseminates information of an educational nature on the need to project and improve the environment in order to enable man to develop in every respect.

Principle 20: Scientific research and development in the context of environmental problems, both national and multinational, must be promoted in all countries, especially the developing countries. In this connection, the free flow of up-to-date scientific information and transfer of experience must be supported and assisted, to facilitate the solution of environmental problems. Environmental technologies should be made available to developing countries on terms which would encourage their wide dissemination without constituting an economic burden on the developing countries.

Principle 21: States have, in accordance with the Charter of the United Nations and the principles of international law, the sovereign right to exploit their own resources pursuant to their own environmental policies, and the responsibility to ensure that activities within their jurisdiction or control do not cause damage to the environment of other states or of areas beyond the limits of national jurisdiction.

Principle 22: States shall cooperate to develop further the international law regarding liability and compensation for the victims of pollution and other environmental damage caused by activities within the jurisdiction or control of such states to areas beyond their jurisdiction.

Principle 23: Without prejudice to such criteria as may be agreed upon by the international community, or to standards which will have to be determined nationally, it will be essential in all cases to consider the systems of values prevailing in each country, and the extent of the applicability of standards which are valid for the most advanced countries but which may be inappropriate and of unwarranted social cost for the developing countries.

Principle 24: International matters concerning the protection and improvement of the environment should be handled in a cooperative spirit by all countries, big and small, on an equal footing. Cooperation through multilateral or bilateral arrangements or other appropriate means is essential to effectively control, prevent, reduce, and eliminate adverse environmental effects resulting from activities conducted in all spheres, in such a way that due account is taken of the sovereignty and interests of all states.

Principle 25: States shall ensure that international organizations play a coordinated, efficient, and dynamic role for the protection and improvement of the environment.

Principle 26: Man and his environment must be spared the effects of nuclear weapons and all other means of mass destruction.

States must strive to reach prompt agreement, in the relevant international Organizations, on the elimination and complete destruction of such weapons. The United Nations Conference on Environment and Development, having met at Rio de Janeiro from 3 to 14 June 1992.

Reaffirming the Declaration of the United Nations Conference on the Human Environment, adopted at Stockholm on June 16, 1972, and seeking to build upon it:

- With the goal of establishing a new and equitable global partnership through the creation of new levels of cooperation among states, key sectors of societies, and people,
- Working toward international agreements which respect the interests of all and protect the integrity of the global environmental and developmental system,
- Recognizing the integral and interdependent nature of the Earth, our home,

Proclaims That

Principle 1: Human beings are at the center of concerns for sustainable development. They are entitled to a healthy and productive life in harmony with nature.

Principle 2: States have, in accordance with the Charter of the United Nations and the principles of international law, the sovereign right to exploit their own resources pursuant to their own environmental and developmental policies, and the responsibility to ensure that activities within their jurisdiction or control do not cause damage to the environment of other states or of areas beyond the limits of national jurisdiction.

Principle 3: The right to development must be fulfilled so as to equitably meet developmental and environmental needs of present and future generations.

Principle 4: In order to achieve sustainable development, environmental protection shall constitute an integral part of the development process and cannot be considered in isolation from it.

Principle 5: All states and all people shall cooperate in the essential task of eradicating poverty as an indispensable requirement for sustainable development, in order to decrease the disparities in standards of living and better meet the needs of the majority of the people of the world.

Principle 6: The special situation and needs of developing countries, particularly the least developed and those most environmentally vulnerable, shall be given special priority. International actions in the field of environment and development should also address the interests and needs of all countries.

Principle 7: States shall cooperate in a spirit of global partnership to conserve, protect, and restore the health and integrity of the Earth's ecosystem. In view of the different contributions to global environmental degradation, states have common but differentiated responsibilities. The developed countries acknowledge the responsibility that they bear in the international pursuit to sustainable development in view of the pressures their societies place on the global environment, and of the technologies and financial resources they command.

Principle 8: To achieve sustainable development and a higher quality of life for all people, states should reduce and eliminate unsustainable patterns of production and consumption, and promote appropriate demographic policies.

Principle 9: States should cooperate to strengthen endogenous capacity-building for sustainable development by improving scientific understanding through exchanges of scientific and technological knowledge, and by enhancing the development, adaptation, diffusion, and transfer of technologies, including new and innovative technologies.

Principle 10: Environmental issues are best handled with participation of all concerned citizens, at the relevant level. At the national level, each individual shall have appropriate access to information concerning the environment that is held by public authorities, including information on hazardous materials and activities in their communities, and the opportunity to participate in decision-making processes. States shall facilitate and encourage public awareness and participation by making information widely available. Effective access to judicial and administrative proceedings, including redress and remedy, shall be provided.

Principle 11: States shall enact effective environmental legislation. Environmental standards, management objectives, and priorities should reflect the environmental and development context to which they apply. Standards applied by some countries may be inappropriate and of unwarranted economic and social cost to other countries, in particular developing countries.

Principle 12: States should cooperate to promote a supportive and open international economic system that would lead to economic growth and sustainable development in all countries, to better address the problems of environmental degradation. Trade policy measures for environmental purposes should not constitute a means of arbitrary or unjustifiable discrimination or a disguised restriction on international trade.

Unilateral actions to deal with environmental challenges outside the jurisdiction of the importing country should be avoided. Environmental measures addressing transboundary or global environmental problems should, as far as possible, be based on an international consensus.

Principle 13: States shall develop national law regarding liability and compensation for the victims of pollution and other environmental damage. States shall also cooperate in an expeditious and more determined manner to develop further international law regarding liability and compensation for adverse effects of environmental damage caused by activities within their jurisdiction or control to areas beyond their jurisdiction.

Principle 14: States should effectively cooperate to discourage or prevent the relocation and transfer to other states of any activities and substances that cause severe environmental degradation, or are found to be harmful to human health.

Principle 15: In order to protect the environment, the precautionary approach shall be widely applied by states according to their capabilities. Where there are threats of serious or irreversible damage, lack of full scientific certainty shall not be used as a reason for postponing cost-effective measures to prevent environmental degradation.

Principle 16: National authorities should endeavor to promote the internalization of environmental costs and the use of economic instruments, taking into account the approach that the polluter should, in principle, bear the cost of pollution, with due regard to the public interest and without distorting international trade and investment.

Principle 17: Environmental impact assessment, as a national instrument, shall be undertaken for proposed activities that are likely to have a significant adverse impact on the environment and are subject to a decision of a competent national authority.

Principle 18: States shall immediately notify other states of any natural disasters or other emergencies that are likely to produce sudden harmful effects on the environment of those states. Every effort shall be made by the international community to help states so afflicted.

Principle 19: States shall provide prior and timely notification and relevant information to potentially affected states on activities that may have a significant adverse transboundary environmental effect and shall consult with those states at an early stage and in good faith.

Principle 20: Women have a vital role in environmental management and development. Their full participation is therefore essential to achieve sustainable development.

Principle 21: The creativity, ideals, and courage of the youth of the world should be mobilized to forge a global partnership in order to achieve sustainable development and ensure a better future for all.

Principle 22: Indigenous people and their communities and other local communities have a vital role in environmental management and development because of their knowledge and traditional practices. States should recognize and duly support their identity, culture, and interests, and enable their effective participation in the achievement of sustainable development.

Principle 23: The environment and natural resources of people under oppression, domination, and occupation shall be protected.

Principle 24: Warfare is inherently destructive of sustainable development. States shall therefore respect international law providing protection for the environment in times of armed conflict, and cooperate in its further development, as necessary.

Principle 25: Peace, development, and environmental protection are interdependent and indivisible.

Principle 26: States shall resolve all their environmental disputes peacefully and by appropriate means in accordance with the Charter of the United Nations.

Principle 27: States and people shall cooperate in good faith and in a spirit of partnership in the fulfillment of the principles embodied in this declaration and in the further development of international law in the field of sustainable development.

In international law, a distinction is often made between hard and soft law. Hard international law generally refers to agreements or principles that are directly enforceable by a national or international body. Soft international law refers to agreements or principles that are meant to influence individual nations to respect certain norms or incorporate them into national law. If a treaty or convention does not specify an international forum that has subject matter jurisdiction, often the only place to bring a suit with respect to that treaty is in the member state's domestic court system. Only nations are bound by treaties and conventions. In international forums, such as the International Court of Justice (ICJ), countries must consent to being sued. Thus, it is often impossible to sue a country. The final question in the jurisdictional arena is who may bring a suit. Often, only countries may sue countries. Individual citizens and non-governmental organizations (NGOs) cannot. First, the environmental harm must be large and notorious for a country to notice. Second, for a country to have a stake in the outcome of the subject matter, some harm may have to cross the borders of the violating country into the country that is suing. Finally, even if transboundary harm does exist, the issue of causation, especially in the environmental field, is often impossible to prove with any certainty.

International institutions are generally not responsible for directly implementing and enforcing international environmental law, but they often play important monitoring, informational, and diplomatic roles.

The leading norms in the field of international environmental law are addressed in the following:

- Foremost among these norms is Principle 21 of the 1972 Stockholm Declaration on the Human Environment.
- Another widely shared norm is the *duty of a state to notify and consult* with other states when it undertakes an operation that is likely to harm neighboring countries' environments, such as the construction of a power plant, which may impair air or water quality in downwind or downstream states. Over and above the duty to notify and consult, a relatively new norm has emerged whereby states are expected to *monitor and assess* specific environmental conditions domestically, and disclose these conditions in a *report* to an international agency or international executive body created by an international agreement, and authorized by the parties to the agreement to collect and publicize such information.
- Another emerging norm is the guarantee in the domestic constitutions, laws, or executive pronouncements of several states, including India [4], Malaysia, Thailand, Indonesia, Singapore, and the Philippines, that all *citizens have a right to a decent and healthful environment.* In the United States, this fundamental right has been guaranteed by a handful of states but not by the federal government.
- Most industrialized countries subscribe to the *polluter pays principle.* This means polluters should internalize the costs of their pollution, control it at its source, and pay for its effects, including remedial or cleanup costs, rather than forcing other states or future generations to bear such costs. This principle has been recognized by the Indian Supreme Court as a "universal" rule to be applied to domestic polluters as well [5]. Moreover, it has been accepted as a fundamental objective of government policy to abate pollution [6].
- At the 1982 United Nations Conference on the Law of the Sea (UNCLOS) [7], developing countries, led by India, articulated the norm that certain resources, such as the deep seabed, are part of the *common heritage of mankind* and must be shared by all nations.
- The 1992 Rio de Janeiro Earth Summit articulated the norm of *common but different responsibilities.* With regard to global environmental concerns, such as global climate change or stratospheric ozone layer depletion, all nations have a shared responsibility, but richer nations are better able

than poorer nations to take the financial and technological measures necessary to shoulder the responsibility.

More generally, in the quest for environmentally sustainable development, the focus will likely move to considering environmental concerns at the front end of the industrializing process, so as to prevent pollution, minimize environmental degradation, and use resources more efficiently. This should mean an increasing concern with making the whole system of production environmentally sound. If so, international environmental law now reflects its emphasis by focusing on standards and procedures for preventing pollution and minimizing environmental degradation, rather than on liability for damage, and on providing incentives to companies to use environmentally sound processes.

The Copenhagen Consensus Center provides information on which targets will do the most social good relative to their costs. The final decision on choosing goals will definitely rest on a number of factors, not just economics, but knowing the costs and benefits provides an important piece of information. The Post-2015 Consensus brought together renowned experts from the UN, NGO, and private sectors with 60 teams of economists to produce 100+ research papers to establish the most effective targets for the post-2015 development agenda within 22 core issue areas; air pollution, biodiversity, climate change, conflict and violence, data for development, education, energy, food security, gender equality, governance and institutions, health: chronic diseases, health: health systems, health: infant mortality and maternal health, health: infectious diseases, infrastructure, illicit financial flows, nutrition, population and demography, poverty, science and technology, trade, and water and sanitation.

REFERENCES

[1] E.B. Weiss, International environmental law: contemporary issues and the emergence of a new world order, Georgetown Law Journal 81 (675–84) (1993) 702–710.

[2] Reprinted in 31 I.L.M. 818 (1992) and P. Birnie, A. Boyle, Basic Documents on International Law and Environment 390 (1995).

[3] Report of the United Nations Conference on the Human Environment, Stockholm, June 5–16, 1972.

[4] The Fundamental Right to Life Guaranteed under Article 21 of the Indian Constitution Has Been Interpreted by the Supreme Court to Include the Right to a Wholesome Environment. Subhash Kumar V. State of Bihar, AIR 1991 SC 420, 424.

[5] The Bichhri Case (Indian Council for Enviro-legal Action V. Union of India), AIR 1996 SC 1446; and Vellore Citizens' Welfare Forum V. Union of India, AIR 1996 SC 2715.

[6] Ministry of Environment and Forests, Government of India, Policy Statement for Abatement of Pollution para 3.3 (26 February 1992).

[7] Reprinted in 21 I.L.M. 1261 (1982).

CHAPTER FIVE

Life Cycle Assessment

5.1 INTRODUCTION

The concept of conducting a detailed examination of the life cycle of a product or a process is a relatively recent one that emerged in response to increased environmental awareness on the part of the general public, industry, and governments. A number of different terms have been coined to describe the processes. **Life cycle assessment** (also known as **life cycle** analysis, ecobalance, and cradle-to-grave analysis) is a technique to assess environmental impacts associated with all the stages of a product's **life** from cradle-to-grave (i.e., from raw material extraction through materials processing, manufacture, distribution, and use). These better reflect the different stages of the process. The life cycle assessment (LCA) method has a fixed structure and is practiced according to international standards (ISO) 14040.

5.2 STAGES IN LIFE CYCLE ASSESSMENT

Life cycle assessment is a technique for assessing the environmental aspects associated with a product over its life cycle. The most important applications are these:

- analysis of the contribution of the life cycle stages to the overall environmental load, usually with the aim to prioritize improvements on products or processes
- comparison between products for internal use

An LCA study consists of four stages:

Stage 1: Goal and scope aims to define how big a part of product life cycle will be taken in assessment and to what end will assessment be serving. The criteria serving to system comparison and specific times are described in this step.

Stage 2: In this step, inventory analysis gives a description of material and energy flows within the product system and especially its interaction with environment, consumed raw materials, and emissions to the environment. All important processes and subsidiary energy and material flows are described later.

Environmental Management
ISBN 978-0-12-811989-1
http://dx.doi.org/10.1016/B978-0-12-811989-1.00005-1

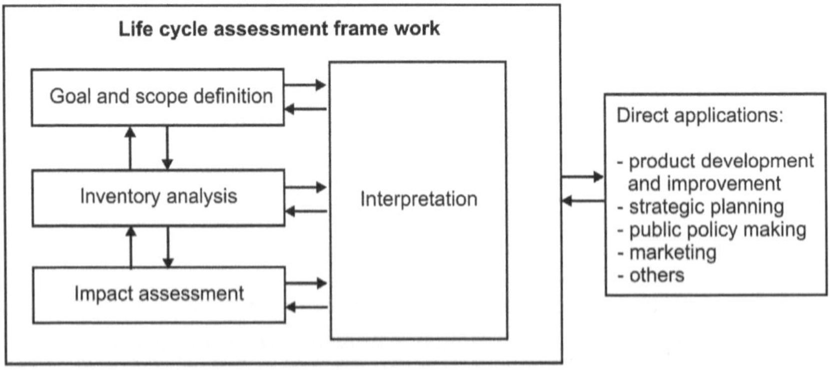

Figure 5.1 Stages of an LCA according EN ISO 14040.

Stage 3: Details from inventory analysis serve for impact assessment. The indicator results of all impact categories are detailed in this step; the importance of every impact category is assessed by normalization and eventually also by weighting.

Stage 4: Interpretation of a life cycle involves critical review, determination of data sensitivity, and result presentation.

Fig. 5.1 gives the four stages under the ISO 14040 guidelines.

When undertaking a life cycle assessment study the following issues need to be addressed:

The burdens imposed on the environment by human activities may be ascertained by accounting for the resources and energy (inputs) consumed at each stage in the life cycle of a product and the resulting pollutants and wastes (outputs) emitted. The inputs and outputs are then assessed for their adverse impacts on long-term sustainability of renewable and nonrenewable resources, human health, and biodiversity, amongst others. Once these are known, measures may be taken to mitigate the impact of the outputs (or inventories) on the environment.

The utilization of LCA method can help in the following:

- searching the most available life cycles, e.g., those with minimal negative impact on environment,
- assuming the decisions in industry, public organizations, or NGOs, which determine direction and priorities in strategic planning, design or design product, or process change,
- choose important indicators of environmental behavior of organization including measurement and assessing techniques, mainly in connection with the assessment of the state of its environment,
- marketing with the link on formulation of environmental declaration or eco-labeling

5.3 LIFE CYCLE ASSESSMENT

5.3.1 Cradle-to-Grave

Cradle-to-grave is the full life cycle assessment from manufacture (cradle) through the use phase to the disposal phase (grave). All inputs and outputs are considered for all the phases of the life cycle.

5.3.2 Cradle-to-Gate

Cradle-to-gate is an assessment of a partial product life cycle from manufacture (cradle) to the factory gate, i.e., before it is transported to the consumer. The use phase and disposal phase of the product are usually omitted. Cradle-to-gate assessments are sometimes the basis for Environmental Product Declarations. The use of biofuel, instead of fossil fuel during transportation, could have an impact on the final evaluation of LCA.

5.3.3 Cradle-to-Cradle

Cradle-to-cradle is a specific kind of cradle-to-grave assessment, where the end-of-life disposal step for the product is a recycling process. From the recycling process originates new, identical products (e.g., aluminum beverage cans from recycled cans), or different products (e.g., glass wool insulation from collected glass bottles).

5.3.4 Life Cycle Energy Analysis

Life cycle energy analysis (LCEA) is an approach in which all energy inputs to a product are accounted for, not only direct energy inputs during manufacture, but also all energy inputs required to produce components, materials, and services needed for the manufacturing process. With LCEA, the total life cycle energy input is established. Also, in this case, it is very important to know the source of energy, whether from fossil fuels or from renewable energies.

Taking the example of the transportation sector, we analyze the society's needs and wants in different spatial scales, as shown in Table 5.1. At the local level, the desire of government for development leads to the construction of rail lines and highways, thus allowing producers ready access to markets and labor supplies. The movement of goods and services is a central focus, and individual transportation becomes less and less central. Transportation as a component of security, and competitiveness assumes interest at the national scale. At the international scale, factors such as the opening of markets and the provision for shipment of large quantities of manufactured goods become elements of transportation planning. A concept for such interactions is true for any activity carried out and is shown in Fig. 5.2.

Table 5.1 Transportation, Evaluated at Different Scales

	Spatial Scale		
Constituency	**Local**	**National**	**International**
Primary producers	Regional development	National security	Trade competitiveness
	Dedicated systems	Dedicated systems	Market diversity, stability of demand
Secondary producers	Labor supply	Product distribution, market access	Exports, market presence
Consumers	Commuting, shopping	Recreation, business	Vacation, business

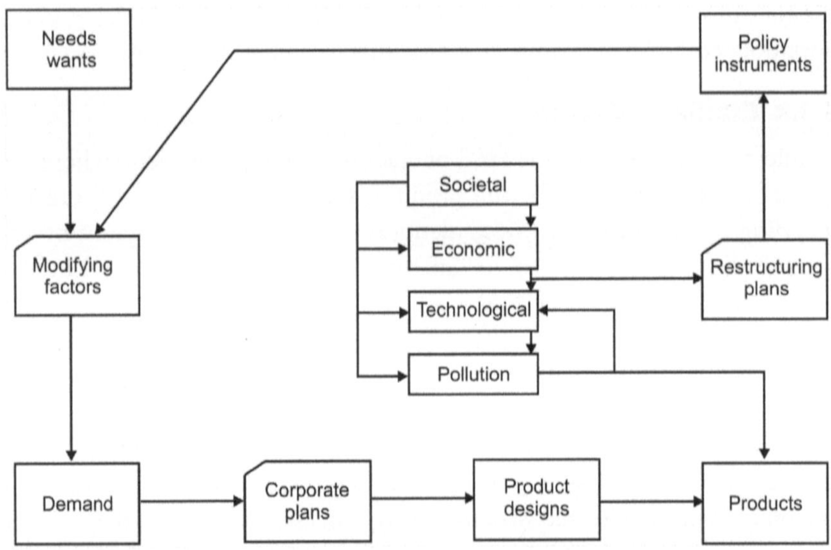

Figure 5.2 Interactions between industrial activities and societal systems.

The flow of information in the figure begins with the needs and wants in the upper left of the diagram. These forces are modified by various societal factors, economic constraints, concerns regarding hazards and environmental impacts, and the state of technology. The result is a demand for specific goods and services.

The concept of life cycle assessment is one that is readily understood and appreciated; its implementation has often proven intractable or at least impractical because of problems related to data needs, time, expense, and uncertainty regarding the defendability of the results. This situation has led to the development of the *streamlined LCA (SLCA)*, which attempts to retain the basic LCA concept while making implementation more efficient and straightforward.

Given either set of guidelines, or any other acceptable set, the relative significance of specific environmental impacts can then be established by consideration of those guidelines in accordance with several related characteristics:

- the spatial scale of the impact,
- the severity of the hazard, i.e., the product of the damage potential of a material, how much material is involved, and the exposed population (highly hazardous substances being of more concern than less highly hazardous substances),
- the degree of exposure (well-sequestered substances being of less concern than readily mobilized substances).

The environmental concerns in the context of LCA are given in Table 5.2, and the target resources and concerns are given in Table 5.3.

Table 5.2 Significant Environmental Concerns

Crucial Environmental Concerns
1. Global climate change
2. Loss of biodiversity
3. Stratospheric ozone depletion
4. Human organism damage
5. Water availability and quality
6. Depletion of fossil fuel resources
Highly Important Environmental Concerns
7. Soil depletion
8. Suboptimal land use
9. Acid deposition
10. Smog
11. Aesthetic degradation
12. Depletion of resources other than fossil fuels
Less Important Environmental Concerns
13. Oil spills
14. Radio nuclides
15. Odor
16. Thermal pollution
17. Landfill exhaustion

Table 5.3 Target Activities in Connection With Environmental Concerns

1. Global climate change	• Fossil fuel combustion (CO_2 emission) • Cement manufacture (CO_2 emission) • Rice cultivation (CH_4 emission) • Coal mining (CH_4 emission) • Ruminant populations (CH_4 emission) • Waste treatment (CH_4 emission) • Biomass burning (CO_2, CH_4 emission) • Emission of CFCs, HFCs, N_2O
2. Loss of biodiversity	• Loss of habitat • Fragmentation of habitat • Herbicide, pesticide use • Discharge of hazardous chemicals to surface waters • Reduction of dissolved oxygen in surface waters • Oil spills • Depletion of water resources • Industrial development in fragile ecosystems
3. Stratospheric ozone depletion	• Emission of CFCs • Emission of HCFCs • Emission of halons • Emission of nitrous oxide
4. Human organism damage	• Emission of hazardous materials to air • Emission of hazardous materials to water • Disposition of hazardous materials in landfills • Depletion of water resources • Physical organism damage
5. Water availability and quality	• Consumptive use of surface water • Use of herbicides and pesticides • Use of agricultural fertilizers • Discharge of hazardous materials to surface or ground waters • Siltation and salinization of surface or ground waters • Depletion of water resources
6. Resource depletion: fossil fuels	• Use of fossil fuels for energy • Use of fossil fuels as feedstocks
7. Soil depletion	• Soil erosion • Discarding or depositing trace metals onto soil • Loss of arable land to development
8. Suboptimal land use	• Loss of arable land to development • Habitat destruction • Abandonment of developed land
9. Acid deposition	• Fossil fuel combustion • Emission of sulfur oxides to air • Emission of nitrogen oxides to air

Table 5.3 Target Activities in Connection With Environmental Concerns—cont'd

10. Smog	• Fossil fuel combustion • Emission of VOCs to air • Emission of nitrogen oxides to air
11. Aesthetic degradation	• Emission of particulate matter to air • Emission of sulfur oxides to air • Incomplete combustion of fossil fuels • Biomass burning • Loss of habitat • Oil spills • Discarding solid residues • Discarding liquid residues
12. Resource depletion other than fossil fuels and soils	• Use of metals in limited supply • Habitat destruction • Use of biomaterials • Discarding solid residues • Discarding liquid residues • Discarding gaseous residues
13. Oil spills	• Transport of petroleum • Refining of petroleum • Distribution of petroleum products
14. Radio nuclides	• Production of nuclear power • Manufacture of products containing radioisotopes
15. Odor	• Odorous industrial emissions • Untreated odorous residues
16. Thermal pollution	• Discharge of heated water to surface waters • Discharge of heated water to groundwater • Discharge of heated air
17. Landfill exhaustion	• Disposition of solid residues in landfills • Disposition of liquid residues in landfills

The LCA covers the environmental and resource impacts of alternative disposal processes, as well as those other processes that are affected by disposal strategies such as different types of collection schemes for recyclables, changed transport patterns, and so on.

The complexity of the task, and the number of assumptions that must be made, is shown in Fig. 5.3.

5.4 THE LIFE CYCLE OF INDUSTRIAL PRODUCTS

The life cycle assessment is an objective process to evaluate the environmental burdens associated with a product, process, or activity by

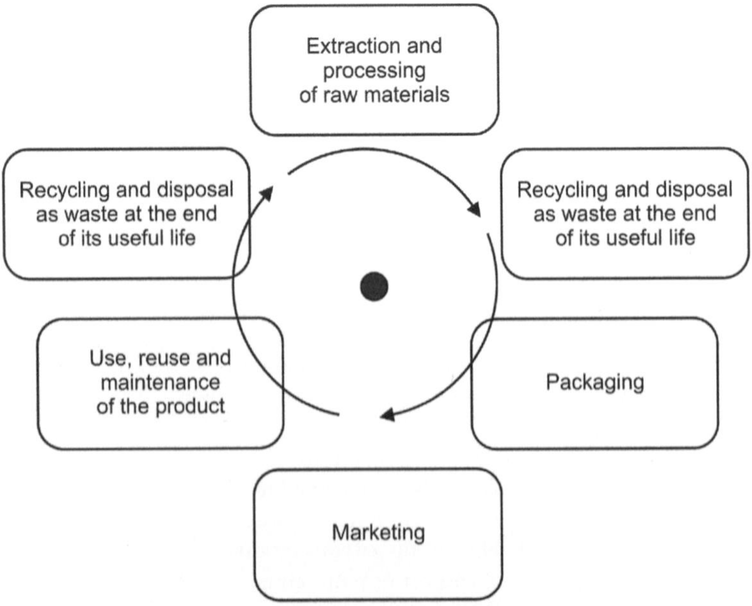

Figure 5.3 Process cycle with reference to LCA.

identifying and quantifying energy and material usage and environmental releases, to assess the impact of those energy and material uses and releases on the environment, and to evaluate and implement opportunities to effect environmental improvements. The assessment includes the entire life cycle of the product, processor activity, encompassing extracting and processing raw materials; manufacturing, transportation, and distribution; use/re-use/maintenance; recycling; and final disposal. An analysis of a typical complex manufactured product is shown schematically in Fig. 5.4.

- Stage 1, pre-manufacturing, is performed by suppliers drawing generally natural resources and producing materials and components,
- Stage 2, the manufacturing operation,
- Stage 3, product delivery; this stage and the previous one are directly under corporate control,
- Stage 4, the customer use stage, is not directly controlled by the manufacturer but is strongly influenced by how products are designed and by the degree of continuing manufacturer interaction,
- Stage 5, a product no longer satisfactory because of obsolescence, component degradation, or changed business or personal circumstances is recycled or discarded.

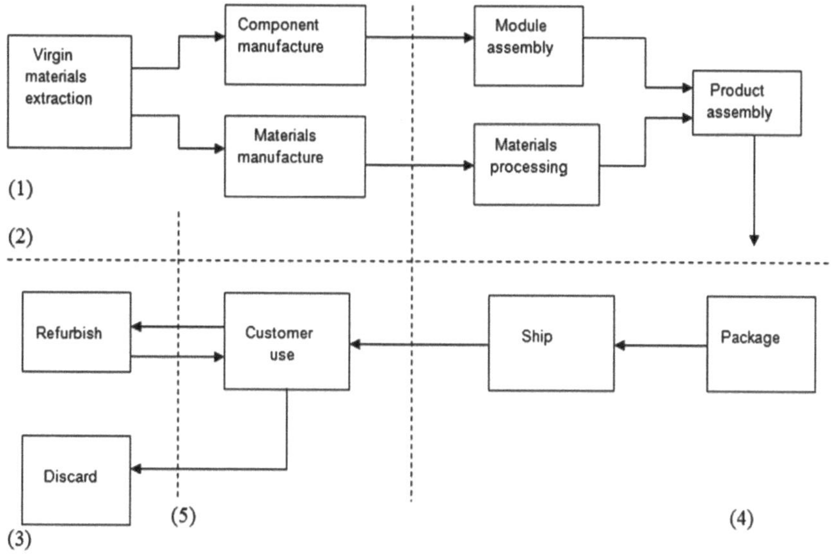

Figure 5.4 Activities in the five life cycle stages.

5.5 THE LCA FRAMEWORK

A life cycle assessment is a large and complex effort, and it has many variations. However, as already mentioned, there is general agreement on the formal structure of LCA, which contains four stages: *goal* and *scope definition*, *inventory analysis*, and *impact analysis*, each stage being followed by *interpretation of results*. LCA normally uses quantitative data to establish the levels and types of energy and materials input to an industrial system and the product output and environmental releases that result, as shown schematically in Fig. 5.5. The main technique used in LCA is that of modeling. In the inventory phase, a model is made of the complex technical system that is used to produce, transport, use, and dispose of a product. This results in a flow sheet or process tree with all the relevant processes. For each process, all the relevant inflows and the outflows are collected. The result is usually a very long list of inflows and outflows that is often difficult to interpret.

5.5.1 Data Collection

The most demanding task in performing LCAs is data collection. Depending on the time and budget available, there are a number of strategies to collect such data. It is useful to distinguish two types of data:

1. Foreground data
2. Background data

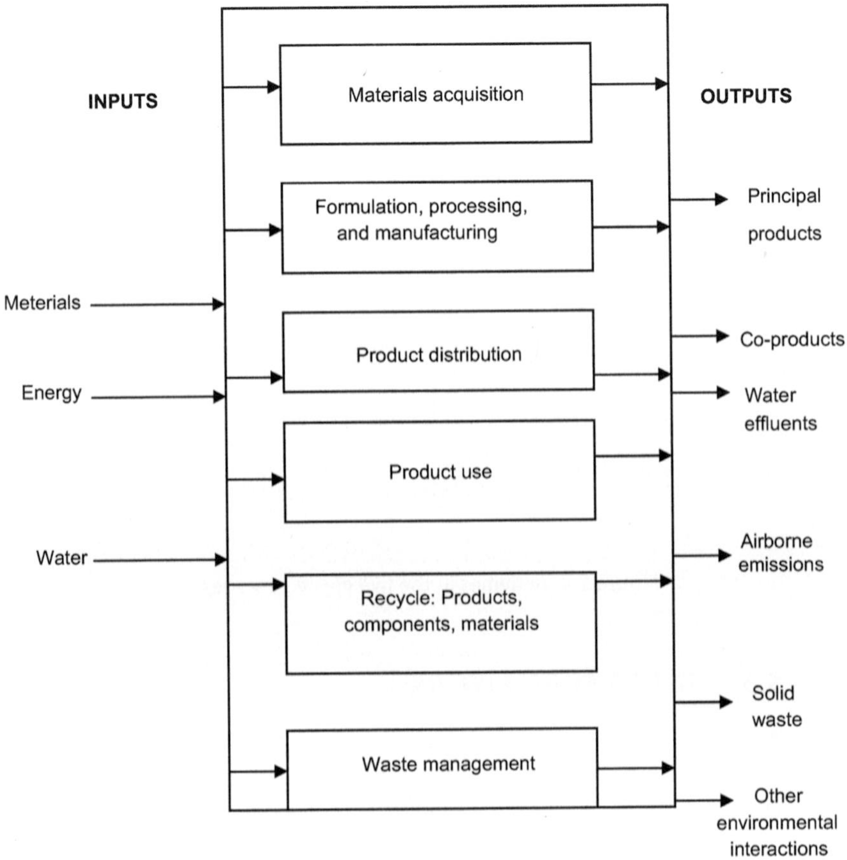

Figure 5.5 The elements of a life cycle inventory analysis. *Adapted from Society of Environmental Toxicology and Chemistry, A Technical Framework for Life Cycle Assessment; Washington, DC 1991.*

Foreground data refers to very specific data needed to model the system. It is typically data that describes a particular product system and particular specialized production system. Background data is data for generic materials, energy, transport, and waste management systems. This is typically data found in databases and literature. Collecting data from other parties is not always easy. It is useful to carefully consider the following points:

- The willingness to supply data is of course determined by the relation you have with these parties. Some parties will be interested as they may have common goals; some will see LCA activities as a threat. In some cases, most of the data collection effort is in the establishing of a good relation, in which parties have trust in each other.

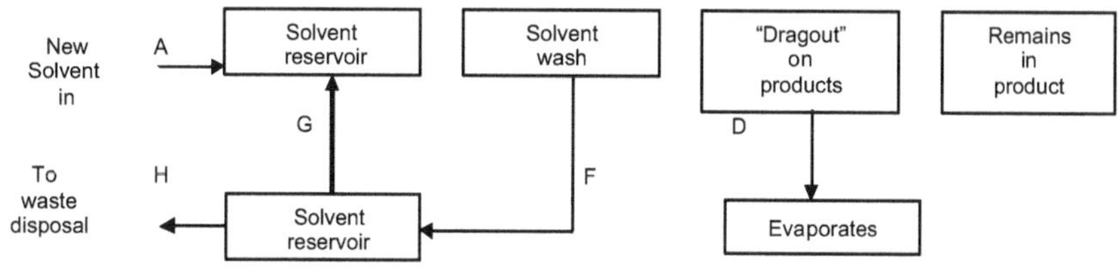

Figure 5.6 Schematic diagram of the flow streams involved in a budget analysis of a chemical solvent in a solvent washing process.

- Confidentiality issues can be very important.
- There are terminology issues in each industry sector; there are different ways of measuring and expressing things. If one develops a questionnaire for a party, it may be applicable within that sector only.
- Questionnaires are often used as a means to collect data. The development of a questionnaire should be done with great care, and it should well connect to the target groups one is addressing.

Once the data is collected and analyzed the budgets and mass balance studies need to be carried out. One of the simple materials budgets is that for a manufacturing process (Fig. 5.6). It shows a chemical process involving the cleaning of a product or product component with a liquid solvent. The process begins with the addition of new solvent to a solvent reservoir, followed by piping or otherwise moving the solvent to the product line where the solvent wash occurs. Most of the solvent eventually enters a recycling (disposal) stream, but a portion (known as "drag out") is retained on the product. Some of the drag out material remains on the product, while a fraction is lost to the atmosphere by evaporation.

5.5.2 Life Cycle Inventory (LCI)

The second phase, "Inventory," involves the modeling of the product system, data collection, as well as the description and verification of data. This implies that data for inputs and outputs for all affected unit processes that compose the product system are available. The inputs and outputs include inputs of materials, energy, chemicals, and "other" and outputs in the form of air emissions, water emissions, or solid waste. Other types of exchanges or interventions such as radiation or land use should also be included if applicable.

The data must be related to the functional unit defined in the goal and scope definition. Data can be presented in tables, and some interpretations can be made at this early stage. The results of the inventory is an LCI that provides information about all inputs and outputs in the form of elementary flow to and from the environment from all the unit processes involved in the study.

Mass balance equations can be set up around any boundary of the system. For example, it is clear that the rate at which the solvent leaves the faculty must be equal to the rate at which it enters: $A = D + E + H$.

A detailed product and process budget is shown in Fig. 5.7.

5.5.3 LAC Impact Analysis

The third phase, "life cycle impact assessment" (LAC), is aimed at evaluating the contribution to impact categories such as global warming, acidification, etc. Characterization is the first step and involves calculation of potential impacts on the basis of the LCI results. The next steps are normalization and weighting, but these are both voluntary according the ISO standard. Normalization provides a basis for comparing different types of environmental impact categories. Weighting implies assigning a weighting factor to each impact category depending on their relative importance. The problem of applying weighting factors is that they distort the scale of values without adding anything to the overall assessment. Impact assessment considerations include thresholds and nonlinearities, temporal scales, spatial scales, and valuation.

The impact assessment can be structured according to the following steps:

- *Classification*: Classification begins with the raw data from the inventory analysis on flows of materials and energy. Given that data, the classification step consists of identifying environmental concerns suggested by the inventory analysis flows. For example, emissions from an industrial process using a petroleum feedstock may be known to include methane, butene, and formaldehyde. Classification assigns the first primarily to global warming, the second to smog formation, and the third to human toxicity.
- *Localization*: Localization is the operation of comparing environmental impacts occurring in different regions with different characteristics. For example, the process of localization attempts to compare the emission of moderately toxic material into a pristine ecosystem with the impact of the same emission into a highly polluted ecosystem. The first consideration is the relationship of emissions from the product or process being assessed relative to all similar emissions in the region. The second is the degree to which the region possesses assimilative capacity for the emitant.

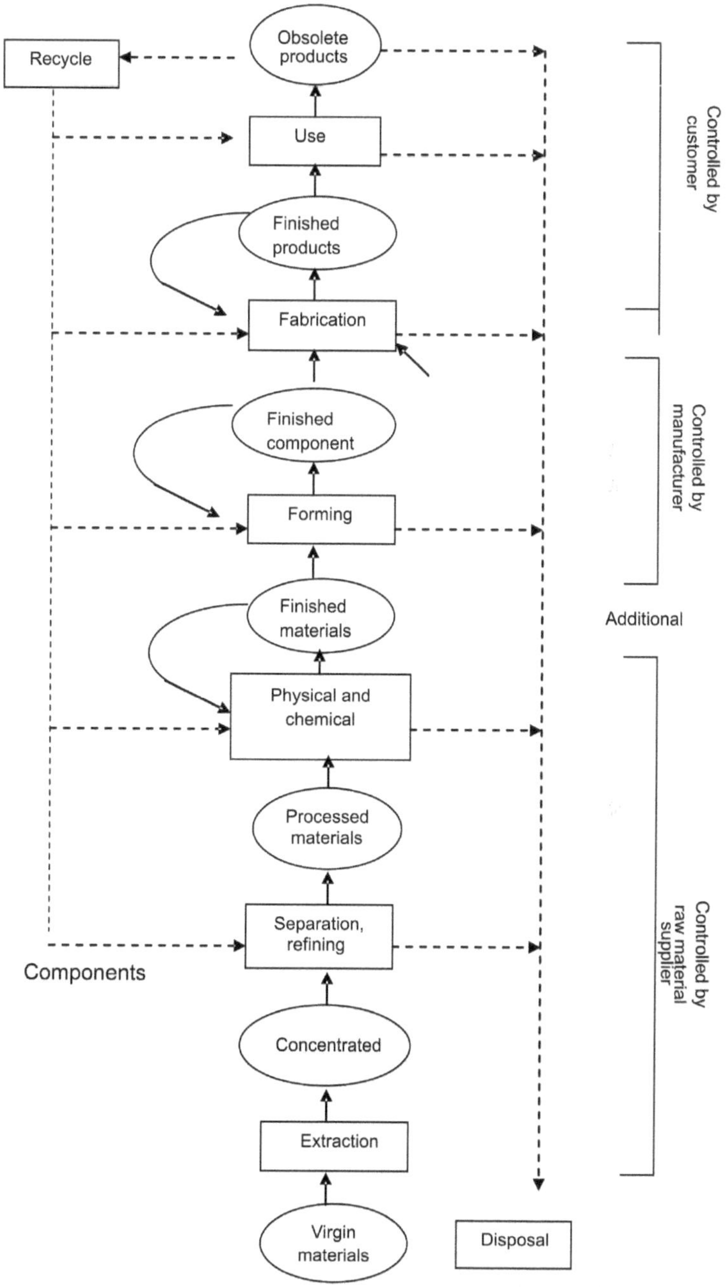

Figure 5.7 Process and product budgets.

- *Valuation*: Valuation is the process of assigning weighting factors to the different impact categories based on their perceived relative importance as set by social consensus. For example, an assessor or some international organization might choose to regard ozone depletion impacts as twice as important as loss of visibility, and apply weighting factors to the normalized impacts accordingly.

5.5.4 Interpretation

The phase stage "interpretation" is the most important. An analysis of major contributions, sensitivity analysis, and uncertainty analysis leads to the conclusion whether the ambitions from the goal and scope can be met. All conclusions are drafted during this phase.

ISO 14040:2006 describes the principles and framework for LCA, including definition of the goal and scope of the LCA, the LCI analysis phase, the life cycle impact assessment phase, the life cycle interpretation phase, reporting and critical review of the LCA, limitations of the LCA, the relationship between the LCA phases, and conditions for use of value choices and optional elements.

5.6 STREAMLINED LIFE CYCLE ASSESSMENT

Techniques that purposely adopt some sort of simplified approach to life cycle assessment, *streamlined life cycle assessments*, form part of a continuum of assessment effort, with the degree of detail and expense generally decreasing as one moves from the left extreme toward the right, as shown in Fig. 5.8. Somewhere within the SLCA region is the ideal

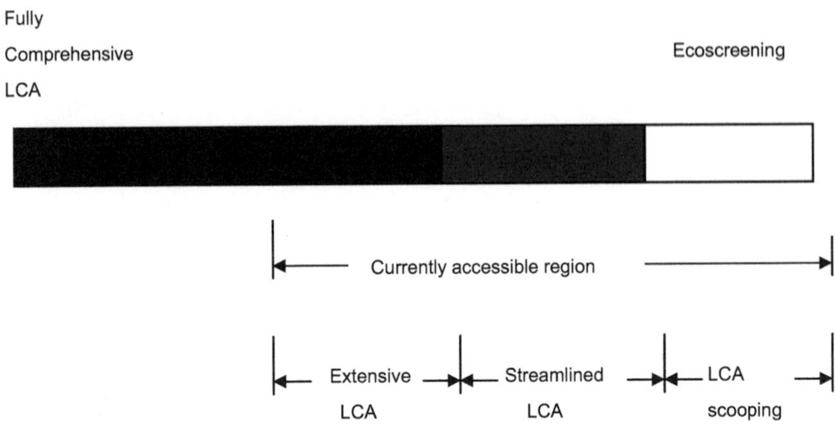

Figure 5.8 The LCA/SLCA continuum.

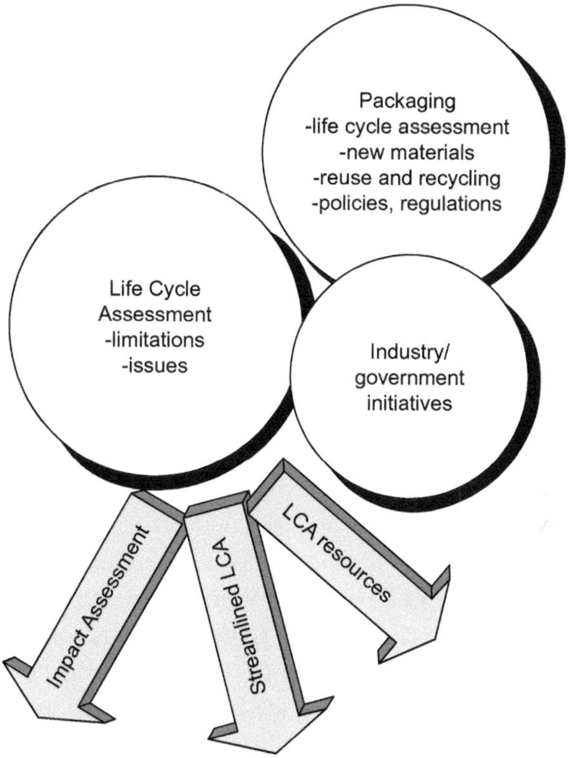

Figure 5.9 Summary of the role of stakeholders and resources in LCA.

point: the assessment is complete and rigorous enough to be a definite guide to industry and an aid to the environment, yet not so detailed as to be difficult or impossible to perform.

Many SLCA approaches have gravitated toward a matrix, one dimension of which is life cycle stages and the other is a list of environmental impacts, potential employee health concerns, or other relevant parameters. The role of the various stakeholders and the resources are shown in Fig. 5.9. The SCLA is an effective tool to assess the impact of the product/activity on the environment. The methods used for the assessment are given next.

5.6.1 Battelle's Pollution Prevention Factors Approach

Battelle has developed what it calls a *P2 approach* to SLCA that also utilizes a matrix tool. The rows in the matrix are 24 items attempting to cover cradle-to-grave aspects of the life cycle (energy use–raw materials, energy use–product assembly, etc.), and the columns are individual components in the products.

5.6.2 Jacobs Engineering's SLCA Approach

Jacobs Engineering has developed a matrix tool utilizing five environmental stresses and seven "risk areas" (global warming, etc.). It has thus far been applied to manufacturing processes, but not to products. The matrix is evaluated for the influence of the process on spatial scales, both at local and global levels. The existing operation is used as a basis from which to evaluate process changes, and the matrix element scores are +1, 0, or −1 depending on whether the alternative proposed is better than, equivalent, or poorer than the base case from an environmental standpoint.

5.6.3 Matrix Calculations

Regardless of how matrix element values are derived, an LCA or SLCA analysis using a matrix-based procedure can be represented mathematically as an exercise in matrix manipulation. To demonstrate, consider the matrix of Table 5.4, which is an SLCA tool devised for ecolabel certification of products. If the matrix elements $f_{m,n}$ are filled with inventory analysis data, the result is a form of an inventory analysis matrix that can be called **F**. Similarly, matrix elements $S_{m,n}$ can be filled with impact assessment data

Table 5.4 The Environmentally Responsible Process Matrix

	Environmental Stressor				
Life Stage	Materials Selection	Energy Use	Solid Residues	Liquid Residues	Gaseous Residues
Resource provisioning	1,1	1,2	1,3	1,4	1,5
Process implementation	2,1	2,2	2,3	2,4	2,5
Primary process operation	3,1	3,2	3,3	3,4	3,5
Complementary process operation	4,1	4,2	4,3	4,4	4,5
Refurbishment, recycling, disposal	5,1	5,2	4,4	5,4	5,5

The numbers are the matrix element indices i,j.

(a one-time operation except for revisions) to give an impact analysis matrix called **S**.

The (S) LCA assessment for a single critical environmental property n is then given by

$$L_n = \sum_{1}^{n} f_{(m,n)} \times S_{(m,n)}$$

In the same fashion, the (S)LCA assessment for a single life stage m is given by

$$L_n = \sum_{1}^{m} f_{(m,n)} \times S_{(m,n)}$$

The overall assessment is given by

$$L_n = \sum_{1}^{m} \sum_{1}^{n} f_{(m,n)} \times S_{(m,n)}$$

As with any matrix, some of the **F** matrix elements may contain zeros. This situation will occur in either of two situations: a null inventory value might be listed for such factors as anticipated soil pollution and degradation during the distribution of a product, or where an inventory value may be deemed unimportant. Similarly, zeros will occur in the **S** matrix if no impact is foreseen from a product or process.

5.7 STAGES IN PROCESS LIFE CYCLES

As with products, industrial processes can be evaluated by SLCA matrix techniques simple enough to permit relatively quick and inexpensive assessments to be made in which all stages of product life cycles and all relevant environmental stressors are encompassed. Fig. 5.10 summarizes the life cycle stages in a process.

Resource provisioning: The first stage in the life cycle of any process is the provisioning of the materials used to produce the consumable resources that are used throughout the life of the product being assessed.

Process implementation: Coincident with resource provisioning is process implementation, which looks at the environmental impacts that result from the activities necessary to make the process happen.

Primary process operation: A process should be designed to be environmentally responsible in operation. Such a process would ideally limit

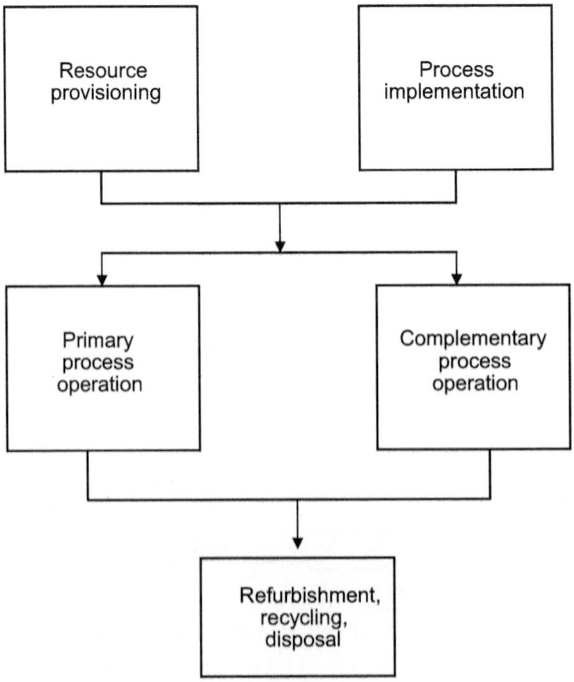

Figure 5.10 The life cycle stages of a process.

the use of hazardous materials, minimize the consumption of energy, avoid or minimize the generation of solid, liquid, or gaseous residues, and ensure that any residues that are produced can be used elsewhere in the economy.

Complementary process operation: It is often the case that several manufacturing processes form a symbiotic relationship, each assuming and depending on the existence of others. Thus, a comprehensive process valuation needs to consider not only the environmental attributes of the primary process itself but also those of the complementary processes that precede and follow.

Refurbishment, recycling, disposal: The process designer must recognize that all process equipment will eventually become obsolete, and it must therefore be designed to optimize disassembly and reuse, either of modules (the preferable option) or materials.

The rating matrix: In arriving at an individual matrix element assessment for processes, or in offering advice to designers seeking to improve the rating of a particular matrix element, the assessor can refer for guidance to underlying checklists and protocols. After an evaluation has been made for each matrix element, the overall environmentally responsible process rating is computed, as in the case of products, as the sum of the matrix element values:

$$R_{ERPS} = \sum \sum Mij$$

Because the process matrix has 25 matrix elements, each with a maximum value of 4, the maximum process assessment rating is 100.

The results from SLCA are often regarded as "approximately correct"; if they even come close to that characterization, carrying out the assessment and implementing their recommendations will be of much value.

SLCA Advantages Over LCA	SLCA Disadvantages When Compared to LCA
SLCA take less time and can be completed in days when compared to months for LCA	SLCA has little capability to track overall material flows
SLCA is less expensive and can use existing staff	It has minimum capacity to compare completely dissimilar approaches to fulfilling a need
Most SLCAs are usable in the early stage of design, when opportunities for change are high	It has minimum capability to track improvements over time
SLCAs are likely to be carried out routinely and applied over a wide variety of products and industrial activities	

In summary, all life cycle analyses collect inventory data on raw material consumption, energy and water use, and waste production. However, a meaningful LCA should contain more than a mere inventory of inputs and outputs: it should also consider the overall contributions and risks to the environment and public health, as well as the social, cultural, and economic impacts of each option.

FURTHER READING

[1] D. Meadows, D. Meadows, J. Randers, Limits to Growth, Universe Books, New York, 1972.

[2] Franklin Associates, Product Life-Cycle Assessment: Guidelines and Principles (EPA Report #68-CO-0003), 1991.

[3] R. Hunt, J. Sellers, W. Franklin, Resource and environmental profile analysis: a life cycle environmental assessment for products and procedures, Environmental Impact Assessment Review (Spring 1992).

[4] U.S. EPA, in: G.A. Keoleian, D. Menerey (Eds.), Risk Reduction Engineering Lab. Life Cycle Design Guidance Manual: Environmental Requirements and the Product System (EPA #600/R-92/226), EPA, Cincinnati, 1993.

CHAPTER SIX

Environmental Impact Assessment and Audit

6.1 INTRODUCTION

One of the main strengths of Environmental Impact Assessment (EIA) is its flexibility. All projects have a planning process in which EIA can be integrated. Given its sensitivity to the social, economic as well as environmental impacts of projects, the EIA process can be used in a project to accomplish many different objectives. EIA can be effectively employed by project managers to compensate for shortcomings in the project planning process. For example, a project that failed to adequately consult the community at the outset can take advantage of the EIA to involve the community in a necessary exchange of ideas and views. The EIA can help to establish and strengthen decision-making and communication mechanisms within a project. It can also pave the way for introducing innovations. While the EIA should not be expected to correct all the weaknesses of a flawed planning process, when properly designed and executed, it can be a valuable tool for project implementation.

Environmental audit was introduced in India for the minimization of generation of wastes and pollution. In this regard a notification was issued which applies to an industry, operation, or process requiring consent to operate under Section 25 of the Water (Prevention and Control of Pollution) Act, 1974, or under Section 21 of the Air (Prevention and Control of Pollution) Act, 1981 (14 of 1981), or both, or authorization under the Environmental Protection Act, 1986 (29 of 1986). The notification requires that an environmental statement for the financial year ending March 31 be submitted to the concerned State Pollution Control Board, on or before September 30 of the same year.

6.2 ENVIRONMENTAL IMPACT ASSESSMENT

Project planning activities can be used to gather necessary information for the EIA process. Each project manager must decide how much importance to accord each planning activity. Across the world where EIA

Environmental Management
ISBN 978-0-12-811989-1
http://dx.doi.org/10.1016/B978-0-12-811989-1.00006-3

studies are undertaken for projects, there are eight guiding principles that govern the entire process, as shown in Fig. 6.1.

1. **Participation**: This is an appropriate and timely access to the process for all interested parties.
2. **Transparency**: All assessment decisions and their basis should be open and accessible.
3. **Certainty**: The process and timing of the assessment should be agreed in advance and followed by all participants.
4. **Accountability**: The decision-makers are responsible to all parties for their action and decisions under the assessment process.
5. **Credibility**: Assessment is undertaken with professionalism and objectivity.
6. **Cost-effectiveness**: The assessment process and its outcomes will ensure environmental protection at the minimum cost to the society.

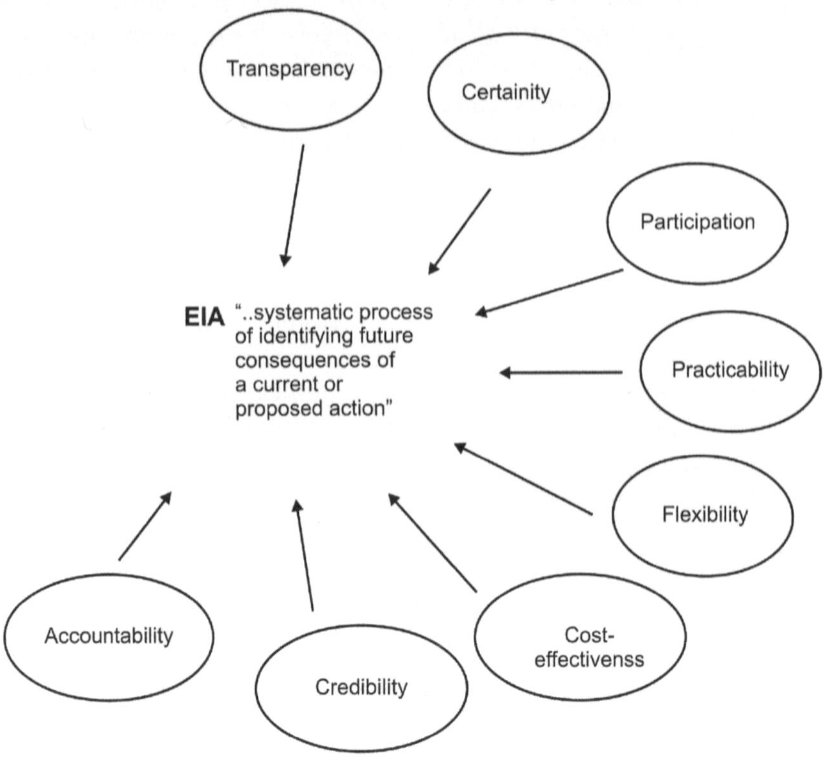

Figure 6.1 Principles of environmental impact assessment process.

7. **Flexibility**: The assessment process should be able to adapt to deal efficiently with any proposal and decision-making situation.
8. **Practicality**: The information and outputs provided by the assessment process are readily usable in decision-making and planning.

A review of environmental practices found the major benefits of the EIA process for project sponsors to be (ESSA Technologies, 1994:16):

- reduced cost and time of project implementation
- cost-saving modifications in project design
- increased project acceptance
- avoided impacts and violations of laws and regulations
- improved project performance
- avoided treatment/clean up costs

The benefits to local communities from taking part in environmental assessments include:

- a healthier local environment (forests, water sources, agricultural potential, recreational potential, aesthetic values, and clean living in urban areas)
- improved human health
- maintenance of biodiversity
- decreased resource use
- fewer conflicts over natural resource use

6.3 ENVIRONMENTAL IMPACT ASSESSMENT IN INDIA

Environmental Impact Notification S.O.1533 (E), of September 14, 2006, as amended in 2009, has made it mandatory to obtain environmental clearance for 36 scheduled development projects. The notification has classified these projects as categories A and B. Category A projects (including expansion and modernization of existing projects) require clearance from Ministry of Environment and Forest (MOEF), Govt. of India (GOI) and for category B, from State Environmental Impact Assessment Authority, constituted by the Govt. of India. The list of project activities identified under the notification for environmental clearance is given in Schedule (Appendix 6). In terms of the EIA notification of the MOEF dated September 14, 2006, the generic structure of the EIA document shall be as follows:

- introduction
- project description
- analysis of alternatives (technology and site)
- description of the environment

- anticipated environmental impact and mitigation measures
- environmental monitoring program
- additional studies
- project benefits
- environmental cost benefit analysis
- environmental management plan
- summary and conclusion

The environmental clearance process for all projects will comprise a maximum of four stages. These four stages in sequential order are as follows:

- **Screening**
  In case of category "B" projects or activities, this stage will entail the scrutiny of an application seeking prior environmental clearance made in Form 1 by the concerned State Expert Appraisal Committee (SEAC) for determining whether the project or activity requires further environmental studies for preparation of an EIA for its appraisal prior to the grant of environmental clearance depending upon the nature and location specificity of the project. The projects requiring an EIA report shall be termed category "B1" and remaining projects shall be termed category "B2" and will not require an EIA report.
- **Scoping**
  "Scoping" refers to the process by which the Expert Appraisal Committee (EAC) in the case of category "A" projects or activities, and SEAC in the case of category "B1" projects or activities, including applications for expansion and/or modernization and/or change in product mix of existing projects or activities, determine detailed and comprehensive terms of reference (TOR) addressing all relevant environmental concerns for the preparation of an EIA report in respect of the project or activity for which prior environmental clearance is sought. The EAC or SEAC concerned shall determine the TOR based on information furnished in the prescribed application Form 1 including TOR proposed by the applicant, a site visit by a subgroup of EAC or SEAC concerned only if considered necessary by the EAC or SEAC concerned and other information that may be available with the EAC or SEAC concerned.
- **Public consultation**
  "Public consultation" refers to the process by which the local concerns by affected persons and others who have plausible stake in the environmental impact of the project or activity are ascertained with a view to taking into account all the material concerns in the project or activity design as

appropriate. All category "A" and category "B1" projects or activities shall undertake public consultation, except wherever explicitly mentioned in the notification. After completion of the public consultation, the applicant shall address all the material environmental concerns expressed during this process, and make appropriate changes in the draft EIA and EMP. The final EIA report, so prepared, shall be submitted by the applicant to the concerned regulatory authority for appraisal. The applicant may alternatively submit a supplementary report to draft EIA and EMP addressing all the concerns expressed during the public consultation.

- **Appraisal**
  Detailed scrutiny by the EAC or SEAC of the application and other documents like the final EIA report, outcome of the public consultations including public hearing proceedings, submitted by the applicant to the regulatory authority concerned for grant of EC.

6.4 ELEMENTS OF AN ENVIRONMENTAL IMPACT ASSESSMENT REPORT

6.4.1 Project Description

Project description in any EIA study carried out across the world may include the following aspects:

- purpose of the project, its goals, and objectives
- overall suitability of the project site and the proposed activity in the light of the existing environmental laws and serious deviations, if any
- significance of the project both at local and national level including background information and overall scenario of the proposed activity
- relevance of the project in the light of the existing development plans of the region, project coverage, outline of master plan, phasing and scope, benefits of the project, etc.
- estimated cost of development of the project, estimated cost of environmental protection both during construction and operation phases of the project, funding agencies
- estimated water budget for the proposed project
- whether project implementation is proposed to be undertaken by the central or state government or through public-private participation or private entrepreneurs
- resources such as construction material, equipment, energy, manpower, time frame, etc., required for project implementation and whether these are available indigenously or to be outsourced

- details of various activities involved both during construction phase and operational phase along with flowcharts duly indicating required resources should be described
- description of the various alternatives like site locations or layouts or modern technologies studied along with the mitigation measures envisaged for each alternative, and the final selection of an alternative should be given.

6.4.2 Baseline Data Analysis

Environmental data to be collected in relation to proposed activity would be (1) land, (2) water, (3) air, (4) biological, (5) noise, (6) socioeconomic, (7) health and safety environment, etc. A map of the study area (core and buffer zone) clearly delineating the location of various monitoring stations (air/water/soil and noise) superimposed with location of habitats is to be shown. Baseline information is required to be collected by field surveys, monitoring, etc. Secondary data with source should be clearly mentioned.

6.4.2.1 Land Environment

Land use/land cover analysis: The changes in the land use/land cover patterns due to the project activity involving changes in land terrain like cutting of high grounds and hillocks, filling of low-lying areas, and reclamation affecting drainage patterns need to be brought out. If land acquisition from either public or private sources is involved, justification for the extent of the area proposed is to be necessarily given. Availability of land for earmarking for the project activity without causing any hardship to local community and their sociocultural and economic aspects is very important. Many a time, acquisitions of large stretches of land and areas being used by the local habitat may be necessitated requiring rehabilitation and resettlement (R & R) measures. It may also become necessary, in some cases, that some of the existing communities and villages may require to be shifted to other areas to earmark the required area for development. The proponent has to undertake required rehabilitation measures in such cases as a part of the project. These aspects should be sufficiently detailed. The communities likely to be affected should be informed well in advance, in consultation with concerned authorities, so they may express their concerns during the public consultation process.

Topography: Landforms, terrain, coastal and inland topography, etc., may be affected due to the project, and its effect on the drainage pattern of the land/terrain has to be studied. Baseline data is to be given on description of existing situation of the land at the proposed project site including description of terrain, viz., hill slopes, coastal and inland topography, coastal features (lowland, beaches, littoral areas, shoal areas in case of beach sand mining), terrain features, slope, and elevation.

Geology: Geology of the area is very important to ascertain mineral resources and recoverable reserves. Baseline data is to be provided on rock types, rock texture and structure, geologic conditions, fractures, fissures, geophysical and morphological details, regional tectonics (intrusives, faults, folds, joints, etc.), hydrology, history of volcanic activity, seismicity, and associated hazards.

Soil quality: Soil data including type, classification, characteristics, soil properties, etc., are important from engineering considerations for design of structures, loading capacities, stockpiles, etc. Changes in parameters of soil also may affect plantation and vegetative growth, which in turn may endanger the health of local community. Baseline data of the soil ascertained by soil investigations is to be carried out as per CPCB norms. Field surveys usually involve a combination of hand auger boring and drilling over the site on a systematic grid pattern, with focus on specific areas of interest. Soil surveys should provide both the physical and engineering properties of the soil.

6.4.2.2 Water Environment

Water environment includes three environmental settings: ground, surface, and marine (in the case of beach sand, mining, and beneficiation). Baseline data with regard to these three environmental settings should be generated.

Groundwater: Groundwater quality is an important parameter, as change in its chemical parameters will affect the water quality. Baseline data of groundwater quality for the season other than monsoon is to be established and other parameters to be decided based on the type of the project.

Surface Waters: Surface water quality is to be monitored at least for one season. The parameters, sampling frequency, and methodology adopted are to be given clearly. The study should indicate locations of monitoring stations with direction and distance from the site. Details of springs, rivers, streams, nullahs, lakes, and reservoirs in the study area need to be identified. Physicochemical characteristics include heavy metals and biological and bacteriological characterization of surface water resources for assessment of water quality. Delineation of watersheds and water drainage pattern in the study area using cadastral/aerial/remote sensing satellite imageries need to studied.

6.4.2.3 Air Environment

The primary sources of dust are construction work and road traffic, crushing, grinding, screening, vehicular traffic, and Diesel Generator (DG) set gaseous pollutants.

Meteorological data: Meteorological data covering the following should be incorporated in the EIA report. The data for at least a 10-year period should be presented from the nearest meteorological station, except for the history of cyclones and floods for which 50-year data is required.

- wind speed and direction
- rainfall
- relative humidity
- temperature
- barometric pressures
- history of cyclones/floods

Wind speed and direction: For preliminary studies, information may be obtained from the available meteorological records of the area. Recording of velocity and direction of wind at the proposed site should be obtained by installing continuous and self-recording anemometers. Seasonal changes and monsoon periods affect the wind direction, intensity, and duration of maximum wind velocity. Obtaining accurate wind data and its interpretation are of importance as wind acts as an agent to convey soot particulate matter, etc., generated both during construction and operational phases of mineral beneficiation from the project area to its neighborhood. The dispersal, however, depends upon the wind direction, intensity, and period as well as the density, size, and shape of the particulate matter.

Rainfall, humidity, temperature: Historical data on other parameters like rainfall, temperature, and humidity at the proposed site area also should be collected. Seasonal changes of climate are associated with the changing monsoons. Data on rainfall and temperature are very important to plan and design safe operating systems, equipment, methods, etc. Data collected should be correlated with data available at places nearest to project site and with recorded data available at the Indian Meteorological Department (IMD) for the region. The length of periods over which data on various meteorological variables should be compiled may vary considerably. The data on annual average, minimum and maximum temperature, rainfall, and relative humidity should be provided in the report. The records on such data for the past may be available with the IMD, Pune, or at the station nearby.

Ambient air quality: Ambient air quality (AAQ) is of utmost importance for the all projects. Describe ambient air parameters, namely, PM2.5,

PM10, oxides of nitrogen (NO_x), sulfur dioxide (SO_2), carbon monoxide (CO), hydrocarbons (HC), heavy metals, and other harmful air pollutants depending upon the ore processed.

Location of AAQ monitoring stations should be presented, monitoring results should be presented, and the values should be compared with National AAQ Standards (Appendix 7).

6.4.2.4 Noise Environment

Construction equipment, crushing and grinding equipment, etc., and road traffic are the major sources of noise and vibration. Baseline data of these parameters at the project area and the neighborhood habitat is to be ascertained and monitored as per CPCB protocol. Noise pollution is generated by road traffic and other activities. Hourly monitoring of noise levels (regs) should be recorded for 24 h by using an noise level meter for 15 min during each hour. Noise standards have been designated for different types of land use, e.g., residential, commercial, industry areas, and silence zones per the Noise Pollution (Regulation and Control) Rules, 2001, notified by the MOEFs, February 14, 2000, in India.

6.4.2.5 Biological Environment

The baseline status for biological environment should be established by studying distribution pattern, community structure, population dynamics, and species composition of flora and fauna. Biological environment like water encompasses both land, coastal, and marine habitat, so field surveys differ widely in the three cases. The information should be collated and given separately.

6.4.2.6 Socioeconomic and Occupational Health

Sociocultural impacts include all kinds of influences on the local community and to people's lifestyle due to relocation of villages, industrialization, population growth, and the formation of slums. The data required for R & R of the affected population per the state norms should be collected and made available. Baseline data of these parameters at the project site and the demography, particularly on human settlements including indigenous people, health status of the communities, existing infrastructure facilities in the proposed area, and distance/area of impact due to the proposed activity should be collected. Present employment and livelihood of these populations and environmental awareness of the population about the proposed activity should be collected. Occupational patterns of people in the area should be presented.

6.4.3 Anticipated Impacts and Mitigation Measures

Mitigation measures with respect to identified anticipated impacts are of major concern. The construction work and operation of a beneficiation plant are likely to affect, significantly, the surrounding environment to varying degrees. The purpose of the EIA is to quantify the impact and ensure that changes to the environment fall within acceptable predefined limits and to give the environment its due place in the decision-making process by clearly evaluating the environmental consequences of the proposed activity. The potential adverse effects of mineral beneficiation include water pollution, air pollution, soil contamination, change in land use and drainage pattern, solid (hazardous) waste, excessive liquid, etc.

By suitable means, including modeling wherever necessary, the impact of all the identified environmental concerns of mineral beneficiation on each facet of the environment should be assessed both during construction phase and operational phase, and suitable mitigation measures against the potential adverse impacts should be considered so an effective EMP can be prepared and its strict implementation adhered to during the project construction and operational phases. Early identification and characterization of critical environmental impacts allow for the environmental acceptability of the proposed developmental project. The impact of the proposed project activities and activities after decommissioning on the concerned environmental attributes must be assessed by standard methods.

Models can be subdivided into three main classes:

1. a scaled-down copy of a physical object such as a ship or tall building;
2. a mathematical representation of a physical or biological process, e.g., of the spread of pollution from a chimney, or the movement of a weather disturbance across a region;
3. an exploratory representation of complex relationships among physical, biological, and socioeconomic factors or indicators, some quantitative, others qualitative.

The third class of mode is often called a *simulation* or a *scenario*. In its simplest form, this kind of representation is extremely useful in the first stages of an EIA, helping to synthesize the widely diverse information reaching the assessor through many specialists. The essential feature of an EIA is the provision of choice between a range of alternatives. Any choice will affect several heterogeneous "elements": physical, ecological, and sociological.

Following are a series of criteria needed for modeling:

- **Volume of data**: The best known feature of computers is their ability to use large amounts of data.
- **Complexity of environmental relations**: The interconnected nature of the elements in the environment poses special problems for impact assessment because the linkages between these elements are often far from simple. If we have two related elements, representing an action and an impact, the simplest assumption to make is that when we alter one element slightly, the other element will change slightly and proportionately and the relationship is "linear." Alternatively, a gradually increasing action may produce negligible change until a point is reached at which dramatic alterations in impact occur. Both of these relationships are technically described as "nonlinear" (Fig. 6.2).
- **Time-dependent relations**: Flows of energy and matter, and changes in these flows, are not only usual but also sometimes necessary for the maintenance of viable ecosystems. Conditions that appear to be static may be slowly changing or may represent only a temporary equilibrium

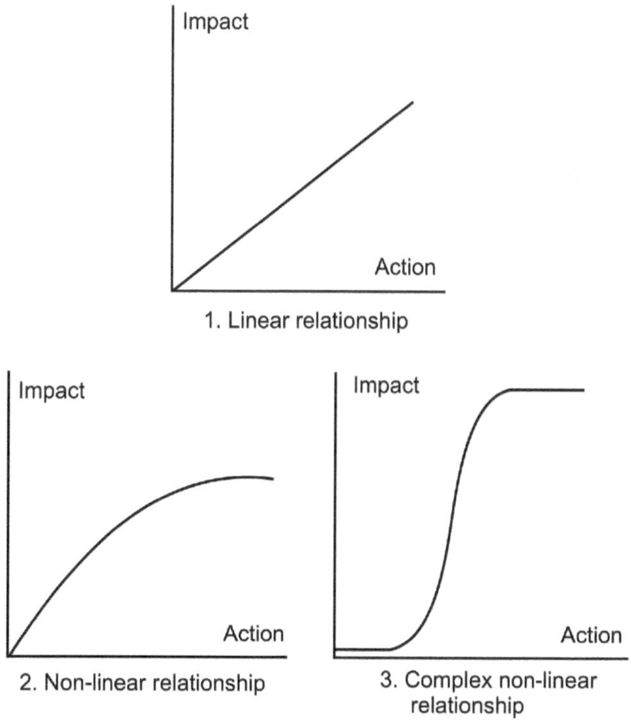

Figure 6.2 Typical forms of relationships between action and impact.

condition among several processes acting in opposite ways. Because man's actions alter these relations, analysis of the time-dependent processes may be necessary to predict the future. Of particular importance is the need to search for possible feedback mechanisms among the various environmental, sociological, and economic processes.

- **Explicit relations**: One apparent disadvantage of a model is that every element and every link must be defined explicitly.
- **Uncertainty**: When the elements and links in a model have been defined, it is likely that very few will have the exactness of simple elements. If the average value of each element is used as a basis for the simulation, then the model will produce only a single, apparently exact, result of the consequences of an environmental change. Wherever feasible, the best approach is to begin with frequency distributions for all of the relevant input variables.
- **Major knowledge gaps**: It is essential that inadequacies in the data or in the assumptions are not conveniently lost within the computer. Facts and values must not become confused. If there is no possibility of defining some of the elements and their relationships in the assessment, there is no point in attempting to include them in a model.
- **Relevance to policy questions**: In the best of all possible worlds, computers and models would be admirably suited to cope with the complexity of impacted environmental systems. It is important to be careful in how we apply any integrated set of modeling techniques and procedures unless they have been designed, from the start, with policy questions as the first consideration. Fig. 6.3 shows the relation between system variables and impact indicators.

The choice will lie between the following classes of models:

a. **Deterministic versus probabilistic models**: In the former, all the relationships are constructed as if they were governed by fixed natural laws; the uncertainties and random fluctuations are not built into the model. In the latter, some or all of the relationships that are defined by statistical probabilities are included explicitly in the model, whose output then directly represents the consequences of those probabilities. This is sometimes called the *Monte Carlo* approach.
b. **Linear versus nonlinear models**: Although it may be convenient to assume that relationships between variables are linear, most practical problems require the more complex assumption of nonlinearity.
c. **Steady-state versus time-dependent models**: Steady-state models compute a fixed final condition based on a fixed pre-action condition, whereas time-dependent models incorporate the way actions affect processes that may eventually produce impacts.

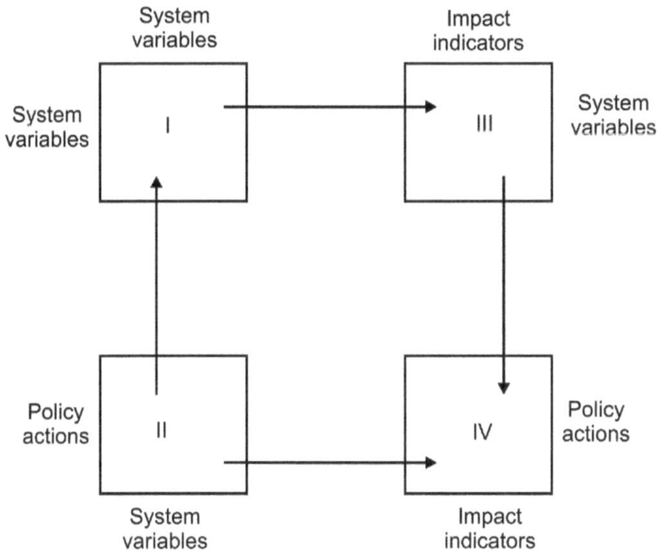

Figure 6.3 Relation between system variables and impact indicators.

d. **Predictive versus decision-making models**: Predictive models enable the consequences of particular decisions to be explored, while decision-making models indicate which of the decisions is "best" in some defined way.

Regardless of whether the model has been adopted for use on a computer, it is at this stage of assessment that computer-aided communication forms (actually communication models) can be of immeasurable value. With a common set of data, a computer system can simultaneously produce a wide variety of specialized displays, including flowcharts, tables, matrices, graphs, response surfaces, maps, and reports in traditional prose form.

6.4.4 Environmental Monitoring Program

It should include the technical aspects of monitoring the effectiveness of mitigation measures (including measurement methodologies, data analysis, reporting schedules, emergency procedures, detailed budget, and procurement schedules). It should also include:

- summary matrix of environmental monitoring, during construction and operation stage
- requirement of monitoring facilities and their onsite installation
- frequency, location, parameters of monitoring
- compilation and analysis of data, comparison with baseline data and compliance to accepted norms and reporting system

6.4.5 Risk Analysis and Disaster Management Plan

In certain projects the risk analysis and disaster management plan forms a very important aspect, and this should be discussed in detailed based on the type of the project undertaken.

6.4.6 Environmental Management Plan (EMP)

The EMP should contain the following:

- administrative and technical set-up for management of environment,
- mechanism of self-monitoring for compliance with environmental regulations,
- institutional arrangements proposed with other organizations/government authorities for effective implementation of proposed environmental management plan,
- integrating in the environmental management plan measures for minimizing use of natural resources, such as water, land, energy, etc., and make provision for reuse and recycle,
- environmental audit of various mitigation measures proposed for different components/sections,
- setting up environmental management cell and formulation of monitoring protocol for various environmental components.

Thus an EIA is a technical exercise to predict environmental impacts, assess their significance, and provide recommendations for their mitigation. The EIA report covers a wide range of technical disciplines and covers areas such as noise and vibration, air quality, ecology, contamination, water quality and hydrology, archaeology and cultural heritage, landscape and visual character, sustainability, and socioeconomics. The EIA report will describe how the project has been improved through the EIA process and what alternatives were considered. It will also present how the conservation of natural resources and minimization of usage of the resources has taken place in the project.

6.5 ENVIRONMENTAL AUDIT

6.5.1 Definition of Environmental Audit

An Environmental audit is a systematic, objective method of reviewing management systems and controls and verifying that environmental standards—regulatory, company, and good industry standards—are being met.

Arthur D. Little.

"Environmental Audit" is a management tool comprising a systematic, documented, periodic, and objective evaluation of how well the pollution control and environmental management systems are performing with the aim of:

- achieving waste prevention through reduction, reuse, and recycle
- assessing compliance with environmental regulatory requirements
- facilitating control of environmental practices by a company's management

The "Environment Audit" tool is proposed to be introduced as a proactive and facilitator tool for the benefit of the industries. The tool provides early warning to the industries on environmental noncompliances to enable them to take timely corrective measures. The tool is perceived to benefit both the "industry" and "environment.

It is a process of systematic, periodic, and objective approaches to investigating business operations from conception to grave. It starts from an analysis of the construction of plant, purchasing of starting building materials and various possible alternatives, processing, recycling, control of waste, storage, distribution, use, repair/recycling, and final disposal of end effluents. A series of questions, i.e., questionnaires and checklists, and their processes are prepared. Where possible, figures are put, and the results are recorded. A report is written, highlighting the both the positive and negative aspects that may need attention. These are then used as the baseline to set objectives and targets against which future audits can be measured. During a typical environmental audit, a team of qualified inspectors, either employees of the organization being audited or contractor personnel, conducts a comprehensive examination of a plant or other facility to determine whether it is complying with environmental laws and regulations. Using checklists and audit protocols and relying on professional judgment and evaluations of site-specific conditions, the team systematically verifies compliance with a management control system.

Environmental audits should answer the following questions:

- What are we doing? In particular, are we in compliance with government regulations, guidelines, codes of practice, and permit conditions?
- Can we do it better? In particular, are there nonregulated areas where operations can be improved to minimize the impact on the environment?
- Can we do it more cheaply? What more should we do?

The "Environmental Audit" covers the following aspects:

- **Manufacturing process and mass balance of various materials used**: process flow, optimal unit operations, materials used (raw materials, water etc.), products and by-products, wastes (gaseous, liquid, solid) generated, unit operations of poor performance.
- **Emissions**: identification of unit operations and processes where emissions are generated, type, characteristics and quantity of emissions and their control measures, compliances with norms, standards, etc., under the Air Act
- **Water balance**: identification of unit operations and processes where water is used, where waste water is generated, its quantity and quality, pretreatment, final treatment, compliance with the norms, standards, etc., under the Water Act
- **Wastes**: identification of unit operations and processes where wastes (hazardous and nonhazardous waste) are generated, their quantity, quality, storage, treatment, disposal, compliances with the HW Management Rules
- **Pollution control measures**: adequacy and performance
- **Potential impact of handling, transportation, and storage** of materials used and products and by-products generated by the industry on the surrounding environment: water bodies, ground water, surrounding population, and agriculture, etc.
- **Compliances with environmental acts, rules, norms, standards** (Air Act, Water Act, HW management as per the Environment Protection Act).

Environmental audits can be carried out for a number of reasons. These could include any of the following:

- to verify compliance
- to review implementation of policies
- to identify liabilities
- to review management systems
- to identify strengths and weaknesses
- to assess environmental performance
- to promote environmental awareness

There are various types of auditing, but all types of auditing include three groups of people and are divided into three stages Table 6.1. The three groups of people are auditor, auditee, and the third party. The various types of environmental audits include liabilities audit, management audit, and activities audit.

Table 6.1 Environmental Audit Types

Liabilities Audit	Management Audit	Activities Audit
Compliance audit	Corporate audit	Site audit
Operational risk audit	Systems audit	Waste audit
Acquisition audit	Policy audit	Product audit
Health and safety audit	Issues audit	Cross-boundary audit

A liabilities audit is often conducted as a prelude to gaining insurance coverage and as a means of demonstrating the regulatory compliance.

- Compliance audit is the most common form of environmental audit that is carried out; it is a verification process whereby the facility establishes the extent to which it is complying with the environmental legislation, regulations, emission limits, etc.
- Operational risk liability audit concentrates on the potential frequency and consequence of environmentally damaging activities in the various functions of the process. Compliance with regulation does not necessarily reduce liability due to operational risks.
- Acquisition audits assess the liabilities due to contaminated land and building remediation costs.
- Health and safety audits normally form part of health, safety, and environment audit and involve assessment of adequacy of personal protective equipment (safety, shoes, goggles, helmets, etc.), emergency preparedness, and disaster management plans.

Management audit pays considerable attention to management systems as they are a guide to how effectively and efficiently the operation runs.

- A corporate audit is initiated by the main board of a parent company and is concerned with the organization structure, roles and responsibilities, policy implementation, awareness, and communication. It is carried out as a reassurance to the main board that their aims and objectives are being implemented throughout the corporate structure.
- Management systems audit are carried out to check the systems against the policy and standards such as British Standard 7750 or ISO 14001.
- Policy audit is carried out to review and reassess the relevance of the policy in the light of developments (legal, technical, financial) within the organization and outside.
- Issues audit is carried out to establish environmental management plan and targets.

Activities audit cover auditing of select technical and management issues.

- Environmental site audit examines all aspects of the facilities performance with respect to the environment. It combines most of the elements of other types of environmental assessment and when undertaken in depth involves considerable time and cost.
- The waste audits are of two types. The first identifies and quantifies waste streams and is a precursor to both waste minimization programs. The second type assesses waste management practice and procedures.
- Product audits cover several aspects of their environmental impacts through design, manufacture, use, and disposal. Such audits are prerequisites for identifying environmentally friendly products for "green labeling".
- Cross-boundary audits assess activities that cut across departments or business units. Transport and supply chain are such examples.

Other types of environmental investigations are frequently conducted with audit-like methodologies such as open inventories and operated-oriented investigations, into prevention of waste and pollution. These studies should be defined separately to avoid mistakes. Thus it is useful to distinguish between three types of audit-like processes:

1. **Environmental reviews**: These involve an unprejudiced investigation of a company's environmental interactions, impacts, and performances, with the aim of establishing an environmental protection system, including organizational and technical measures, in addition to assessment and auditing procedures.
2. **Environmental Audit**: This requires a systematic and objective investigation of a company's environmental activities, management, and equipment against a predetermined set of criteria (legislation, standards, company policy, and objectives, etc.).
3. **Pollution prevention assessments**: These involve a systematic, periodic, and internal investigation of a company's process and operations designed to identify and provide information about opportunities to prevent pollution and waste.

Environmental audits remain an internal process, and results should not be disclosed outside the company for the following reasons:

- to maintain a relationship of trust between auditors and auditees and avoid any hiding of inadequacies,
- because environmental auditing goes beyond compliance, appraisal requires open discussion of confidential business and operational information, and can even involve strategies consideration of ways to improve operations,

- because auditing process is evolving and being improved continuously and would run the risk of becoming "fossilized" if strictly regulated.

6.6 PRINCIPLE ELEMENTS OF AN ENVIRONMENTAL AUDIT

- Internal audit is an audit carried out by the organization to check its own process and progress.
- External audit is an audit carried out by one organization of another organization to check its progress.
- Mandatory audit is an audit carried out by an authority to check the compliance of the process with their requirements, e.g., insurance checks, export, etc.
- Voluntary audit is an audit carried out voluntarily by an organization or an individual of another organization to improve the process and compliance of the process with the laws.

6.7 COMPONENTS OF AUDITING

A systematic examination of performance to ensure compliance includes the following components (International Chamber of Commerce Position paper on environmental auditing adopted November 29, 1988):

- full management commitment
- audit team objectivity
- professional competence
- well-defined and systematic approach
- written reports
- quality assurance
- follow-up

Assessment: It provides expert judgment/opinion on hazards, associated risks, and management and control measures. It helps identify knowable hazards. It also estimates the significance of risks and helps assesses the current practices and capabilities. It also provides the basis for recommendations to improve the organization's management system and environmental performance.

Verification: It determines and documents performance by evaluating the application of, and adherence to, policies and procedures. It certifies the validity of data and reports and evaluates the effectiveness of management systems. It also verifies that regulations and policies are being adhered to and assists in identifying gaps in organizational policies and standards.

6.8 AUDIT PROCESS

Three major steps are involved in environmental auditing:

1. pre-audit activities
2. onsite activities
3. post-audit activities

Pre-audit activities: These comprise scheduling; team selection; logistical arrangements; gathering background information and developing the audit plan. In this phase, information is collected on manufacturing processes along with flow sheets, materials used, waste generated (gaseous, liquid, solid) at each of the productions stage/step, pollution control measures already in place, copies of "Consent to Operate" from pollution control boards, results of monitoring of wastes, before and after respective treatment, drawings of plant layout, drainage, and sewerage. The pre-audit activities also include the finalization of dates for the audit and the arrangement for onsite activities.

Onsite audit process: Key activities include understanding management system; understanding plant process and operating systems; assessing the strengths and weaknesses; gathering audit evidence, evaluating audit findings and reporting audit findings to management. This can include the following:

- discussions with the industry, presentation from industry on the processes used, their operations and performance
- discussions with the management of industries on the preliminary observations
- study of selected manufacturing processes and unit operations
- study of the potential environmental impact on the surrounding area
- study of pollution control measures
- study of various aspects as outlined in the environmental audit format
- feedback from the industries.

Post-audit activities: Objectives of these activities are to ensure the audit results are clearly communicated to the appropriate level of management; ensure all findings and observations are addressed by management; evaluate effectiveness of audit and provide suggestions for improving future audits; share lessons learned during the audit. Tasks are to prepare a draft report (before the team leaves); issue a final report to legal counsel (as appropriate); develop action plans, and follow up.

A management tool comprising a systematic, documented, periodic, and objective evaluation of the performance of the organization, management, and equipment designed to protect the environment with the aim of:

- facilitating management control of environmental practices
- assessing compliance with company policies, including observance of the existing regulatory requirements
- eco-management and audit regulation.

6.9 ENVIRONMENTAL AUDIT (EA) REPORT

The **EA report** needs to communicate the relevant information clearly and concisely and should therefore:

- be presented to make information accessible to the non-specialist, avoiding technical terminology where possible,
- have information presented in summary tables and use good quality maps, charts, diagrams, and other visual aids wherever possible,
- be clearly laid out with a clear table of contents, to allow the reader to find and assimilate information easily and quickly,
- present information without bias and discuss issues with the emphasis appropriate to their importance as in the overall context of the EA.

The report could also be summarized in the form of a checklist, so it would be able to show all the activities that were studied and discussed in the report. A sample checklist is shown in Table 6.2.

6.10 WASTE AUDIT

As an example waste audit is discussed in the next few pages of this chapter. A waste audit is a first step in an on-going program designed to achieve maximum resource optimization and improved process performance. A waste audit enables the management to take a comprehensive look at the site or process to facilitate the understanding of material flows and to focus its attention on areas where waste reduction and therefore cost savings are possible.

The options for environmental management in an industry include end-of-pipe (EOP) technologies and anticipative preventive technologies. Consideration of resource conservation, economic efficiency, and environmental protection warrant the adoption, as far as possible, of a preventive strategy, as EOP control technologies transfer pollutant from one environmental media to another. It is necessary to adopt a strategy of pollution

Table 6.2 Sample CheckList

	Item	Yes	No	Comment
A	Do you audit and record waste in the operations?			
B	Is scrap collected and sold for recycling? • metal off cuts • old vehicle parts • used oil • shrink–wrap			
C	Are there spillages of chemicals, and how it this addressed?			
D	Do you receive complains about odors during your operations?			
E	Is all the air pollution control equipment working properly in the plant?			

A Sample Format for the Environmental Audit for Industry Sector is Shown Next			
(A)	**General**		
1.	Name of the industry		
2.	Location		
3.	Registered office address		
4.	Month and year of establishment		
5.	(a)	Number of workers employed	
	(b)	Male/female	
6.	(a)	Number of electrical connections with service numbers	
	(b)	Total connected load	
	(c)	Electric consumption per ton of product manufactured	
	(d)	Percentage enhancement in energy saving compared to previous year	
7.	Number of D.G. sets and their capacity		
8.	Name/residential address of all directors/partners		
9.	Telephone numbers: residential and industrial, fax number and email of industry and partners/directors		
10.	Number of shifts and timings		
11.	Name and address of the in charge of environment/safety division/cell/unit		

Table 6.2 Sample CheckList—cont'd
A Sample Format for the Environmental Audit for Industry Sector is Shown Next

12.	Number of days during which production activities were in operation during the audit period covered		
13.	Has the industry obtained ISO 9000/ISO 14000/OSHAS 18000/any other EM accreditation/certificate or recognition?		
14.	Has the industry adopted cleaner production/cleaner technology/CDM?		
(B)	**Product details**		
1.	Names of products and capacity with yield/purity per day		
2.	Names of all by-products and their quantity per day		
3.	Date of commencement of production for each product; whether production is as per consented quantity.		
4.	All raw materials required for one kilogram of product(s)		
5.	Whether the manufacturing process is continuous or batch-wise. Indicate the batch capacity. If the process is in batch operation, number of batches/month along with the duration of the completion of each batch		
6.	Detailed manufacturing process with schematic flow diagram, list of unit operations and processes, with all chemical reactions along with the time required (in hrs) for completion of each unit operation/process and the total time for completion of the entire batch Mass balance in respect of the quantity of water, input of raw materials, and waste water generation. (Attached separate sheet)		
(C)	**Water**		
1.	The quantity of water consumed per day as well as per ton of product manufactured, over the last three years Water balance diagram		
2.	The quantity of waste water (trade effluent) generated per ton of each product per day, as well as per batch [a]over the last three years		
3.	The particulars of effluent treatment		
	(i)	Name and size of each unit	
	(ii)	The capacity of the ETP	
	(iii)	Flow diagram and hydraulic diagram of ETP to be submitted	

Continued

Table 6.2 Sample CheckList—cont'd

A Sample Format for the Environmental Audit for Industry Sector is Shown Next

	(iv)	Whether lighting arrangement around ETP is provided	
	(v)	Whether the separate energy meter is installed for effluent treatment plant. If yes, reading of the meter for consumption every month	
	(vi)	Whether flow meters are provided at the inlet and outlet of ETP; please indicate the type of the flow meter	
4.	The method of disposal of effluent and the final point of disposal of effluent (Please attach sketch)		
5.	The quality of trade effluent at the inlet and outlet of ETP and at various stages of treatment.		
6.	The quantity and quality of sewage and its method of treatment and disposal (a) As per norms (b) Total pollution load[a]		
7.	The open area available for disposal of the effluent		
8.	Whether the quality of treated effluent meets the specified norms. If no, the extent of deviation and reason thereof.		
9.	Improvement in effluent quality and quantity since previous environmental audit based on performance evaluation of effluent management systems. If yes, provide details		
10.	Retrofitting undertaken to improve performance of ETP. If yes, provide details.		
11.	Major problems encountered during operation of effluent treatment facilities, if any, and reason thereof.		
12.	The details about the operator/chemist responsible for operation and maintenance of effluent treatment plant		
	(i)	Name of the operators/employees	
	(ii)	Qualification and experience of each operator/employee whether trained in such operation or not	
	(iii)	Salary of operators/employees	
13.	The current status of consent under the water Act-74		
(D)	**Air**		
1.	Number of the flue gas stacks, their height (from ground level), nature, and consumption of fuel		
2.	The details pertaining to the stack monitoring facilities		
3.	Number of process stacks, their height (from ground level), source, expected pollutants, and the details pertaining to the provision of stack monitoring		
4.	The quality of emission from each flue gas stack and the process stack and the extent of deviation from them		

Table 6.2 Sample CheckList—cont'd

A Sample Format for the Environmental Audit for Industry Sector is Shown Next

5.	The ambient air quality within the factory premises, along with the number of ambient air quality monitoring stations outside the industry		
6.	The status of consent under the air Act 81		
7.	The details of air pollution control measures for all process and flue gas stacks		
8.	Improvement in emission quality since previous environmental audit based on performance evaluation of air pollution management systems. If yes, provide details		
9.	Retrofitting undertaken to improve performance emission quality. If yes, provide details		
10.	Major problems encountered during operation of control device, if any and reason thereof		
(E)	**Hazardous (solid) waste**		
1.	The quantity, sources, and composition of hazardous waste/solid waste from each process/sources over the last three years (total sludge generation per ton of product), whether it is as per the consented quantity		
2.	(a) The method of storage, treatment, and disposal of hazardous/solid waste. The details should include area of storage and disposal and whether storage and disposal system is covered and made impervious. The quantity of hazardous waste sent to TSDF. Please also indicate how the quantity of hazardous waste/solid shall be reduced in next three months (b) The data/information about leachate generation, quantity, and characteristics and treatment facility		
3.	The status of authorization under the EPA-86 for solid waste		
4.	Plan, if any to reduce hazardous waste generation or its recycling		
(F)	The site plan showing the location of effluent treatment plant, final point of disposal of effluent, sampling point, drainage line, stacks; solid waste storage; disposal area and green belt (its width)		
(G)	(i)	The details regarding resource recovery including treated effluent for recycle/reuse from environmental pollution control system including ETP	
	(ii)	The details regarding resource recovery/by-product recovery from manufacturing process by using cleaner production technology	
(H)	**Health**		
1.	Whether any hazard is involved in the manufacturing or from the work environment yes/no. If yes, provide details thereof		

Continued

Table 6.2 Sample CheckList—cont'd

A Sample Format for the Environmental Audit for Industry Sector is Shown Next

2.	Whether industry has pre-employment and periodical medical examination facilities, yes/no. If yes, provide details thereof			
3.	Whether health records are maintained regarding adverse effect on the health of workers, yes/no. If yes, details thereof			
4.	Whether industry has appointed a factory medical officer, yes/no. If yes, full time or part time. Include the details about the name, address and qualification of the factory medical officer			
5.	Details of medical facilities available. Please tick (√) correct column: first aid box/dispensary/ambulance/hospital			
6.	Whether sanitary facilities like water closets, urinals, bathroom are provided and are satisfactory			
(I)	The details of accidents in the factory if any and remedial measures taken			
(J)	**Safety measure**			
1.	General environment of the factory. Please tick (√) the appropriate column			
	(a) Housekeeping	Good	Fair	Poor
	(b) Dustiness	High	Medium	Low
	(c) Lighting	Good	Fair	Poor
	(d) Ventilation	Good	Fair	Poor
2.	Whether the following protective appliances are provided to all the persons			
	(a) Goggles		Yes/no	
	(b) Gloves		Yes/no	
	(c) Gumboot		Yes/no	
	(d) Helmet		Yes/No	
	(g) Ear plugs		Yes/utilization level	
	(h) Face masks		Yes/utilization level	
3.	The details of facilities for disaster management/gas leakage			
4.	Whether onsite/offsite emergency plans are prepared and are being implemented/upgraded regularly; please give details			
5.	Whether records of occupational hazards are maintained			
6.	Preventive measures adopted to minimize occupational hazard			

Table 6.2 Sample CheckList—cont'd

A Sample Format for the Environmental Audit for Industry Sector is Shown Next

(K)	**Remedial measures**	
1.	The details of sources; monitoring and measures taken for control of noise pollution in and around the industrial premises	
2.	The measures taken for prevention; treatment and control of odor nuisance in and around the industrial premises	
3.	The details in respect of cases/complaints under the Water Act 1974, the Air Act 1981 and the EPA 1986	
4.	The compliance report with respect to all the conditions of NOC/consent (under all the acts)	
5.	Incidents of spillages, leakages, etc., and remedial measures thereof	
6.	Whether insurance policy obtained under PLI Act Yes/no. If yes, provide details	
(L)	The details regarding payment of the water cess for the previous and the current year	
(M)	The name and address of the consultant engaged by the company/industry	

Total pollution load each for air, water, and hazardous waste should have mentioned along with the quality of effluent, emission, or solid waste as the case may be and whether measures were taken for reduction of pollution load.

[a]Whether production is as per consented quantity.

prevention based on technologies that conserve resources, minimize pollution, and reuse wastes as secondary sources as far as possible.

A good waste audit would involve the following:

(a) define sources, quantities, and types of wastes generated,

(b) collate information on unit operations, raw materials, products, water usage, and waste

(c) highlight process inefficiencies and areas of poor management

(d) help set targets for waste reduction

(e) permit the development of cost-effective waste management strategies

(f) cause awareness in the work force regarding benefits of waste reduction

(g) increase the knowledge of the manufacturing process and help to improve process efficiency

(h) improve operation and maintenance of pollution control equipment

Waste audit as discussed in EA comprises of four distinct phases: (1) formation of audit team, (2) pre-audit phase for audit preparation, (3) data collection phase to derive a material balance, and (4) synthesis phase where findings from the material balance are translated into a waste reduction action plan.

Waste audit is a technical tool meant for waste reduction from all possible sources. Normally the objectives include, (1) pollution prevention and control, (2) solid and hazardous waste reduction, (3) water and energy conservation, and (4) improvement of post-control techniques.

Formation of an audit team is the very basis of the auditing exercise. Typically an audit team should include environmental manager, plant manager, safety officer, chief operators, shift engineers, and laboratory technicians.

In the step of listing unit operations, all the information required for effective industrial waste audit is collected. This information includes (1) site plan, (2) plan of surrounding area indicating topography, water bodies, hydrology, human settlement, etc., (3) process flow diagram, if available, (4) store records giving information on material consumption (5) water use data ,and (6) estimation of quantities of wastewater generated, waste treatment, and disposal costs.

Compilation of this data forms the very basis of the success of the audit. Focus will depend on objectives of waste audit and waste minimization as a whole or concentration on minimization of specific waste or pollutant. Information on (1) raw material, energy losses, (2) wastes th,at cause processing problems, (3) wastes considered to be hazardous, (4) wastes for which disposal costs are high, (5) water conservation, (6) fugitive emissions, (7) thermal pollution and (8) safety or hazard needs to be collected.

For each unit operation, collect information on (1) why water is used in each operation, (2) how often does each action take place, and (3) how much water is used on each action.

Process flow diagrams are constructed based on the information collected. The process flow diagram should represent all the important unit operations along with the raw material inputs, product output, and waste generated per unit operation, and at the same time, it should be as simple as possible, avoiding unnecessary details.

Identification of (1) process layout, piping system, drainage system, (2) transfer points, exhausts, vents, stacks, and (3) material storage, handling and conveying system is important. A properly constructed process flow diagram should communicate all aspects of the production process at a glance. Before proceeding to carry out material balance calculations, attention should be paid to collect information on any obvious waste arising that can be reduced or prevented easily.

Material balance is defined as a precise or quantitative account of the inputs and outputs of a plant, a process, or a unit operation. It is designed to give a better understanding of inputs and outputs of a unit operation such that areas where information is inaccurate or lacking can be identified.

Inputs like raw materials, chemicals, water, air, power to each operation, and process are quantified by examining purchase records. Actual net input to the process is determined by finding out raw material losses at storage and transfer (Table 6.3). A material balance must be prepared at a scale appropriate for the level of detail required in the study. Raw material purchases, storage, and handling losses should be recorded derive at net input to the process.

Energy input to each unit operation has to be noted. It is necessary to study whether waste reduction reduces energy costs. If energy usage is a particularly prominent factor, then a separate energy audit has to be done. Input data should be recoded in a tabular form (Table 6.4).

Some unit operations may receive recycled wastes from other unit operations. These are also inputs. Water is frequently used in industries in process, cooling, gas scrubbing, washing, rinsing, boiler feed, etc. Material balance can become complicated because of the intermittent or batch operations. The audit procedure used to collect evidence fall into three broad categories: (1) inquiry through questionnaires and interviews, (2) observation through inspection, and (3) verification testing.

Table 6.3 Raw Materials Losses

Raw material	Quantity	Quantity Purchased Per Annum	Length of Storage	Losses Expected
1.				
2.				
3.				

Table 6.4 Input Data for Energy Audit

Unit Operation	Raw Material I (m^3or T/ann)	Raw Material 2 (T/ann)	Water (m^3/ann)
1.			
2.			
3.			

Use of water to wash, rinse, and cool should be carefully studied as it is an area where waste reduction is achieved easily and cheaply. The use of water is a factor that should be covered in all waste audits. Information on (1) water source, (2) method of extraction, (3) transport of water and possible loss therein, and (4) water purchased from municipalities or other sources has to be collected. Due consideration should be given to the type of source, method of extraction, and transport and possible loss therein.

To establish water balance and record water usage, water meters should be installed. If it is not possible to provide water meters, the volume of the container in a batch process or recording the number of times the container is filled with water in a day or measuring the discharge and estimating water consumption should be adopted. Reuse and recycling reduces water and raw material requirement. Opportunities for reuse or recycle have to be thoroughly studied. Some wastes that lend themselves to direct reuse of recycle have to be thoroughly studied. Some wastes lend themselves to direct reuse in production like reuse of final rinse in a soft drink plant for bottle washing as initial reuse. Others require some modification before they are suitable for reuse in a process. These reused streams, if any, have to be quantified and suitably represented on the process flow sheet by means of suitable color coding. By the end of this step, all the process inputs should have been quantified.

To calculate the remaining part of the material balance, the outputs from different unit operations and the process as a whole need to be quantified. Outputs include primary product, by-products, wastewater, emissions, hazardous and nonhazardous solid wastes, wastes to be stored or sent offsite for disposal, and reusable or recyclable wastes.

Wastewater flow from each unit operation as well as from the process needs to be quantified, sampled, and analyzed by conducting in-plant survey. It is necessary to understand drainage system in the factory and to design appropriate sampling and a flow measurement program to monitor flow and composition. Samples have to be collected over a range of operating conditions like full production, start up, shut down, etc. Flow measurement, using V-notch, rectangular weir, orifice plate, magnetic flow meter, flow velocity and area method, partial flume, container, and watch method has to be carried out. Based on flow and characteristics of wastewater, pollution loads from each unit operation can be calculated and presented.

Gaseous emissions are not always obvious and are difficult to be measured. It is necessary to estimate the actual and potential gaseous emissions associated with each unit operation from raw material storage to product storage. The audit team should collect information on (1) odors associated with different unit operations, (2) pollution control equipment, (3) gaseous emissions from confined spaces including fugitive emissions vented out, (4) spent scrubber solution disposal, and (5) employee's use of protective equipment. If pollution control equipment is installed, their efficiency in removing pollutants has to be studied.

Wastes that cannot be treated onsite need to be transported offsite for treatment and disposal, which is costly as it involves both transportation and treatment. Hence minimization of wastes that require offsite treatment is of direct benefit from cost point of view. The quality and quantity of wastes needed to be sent for offsite disposal are to be measured and recorded in a tabular form.

Information has to be collected on (1) the source of wastes needing offsite disposal, (2) minimization of waste generation by optimization of manufacturing operations or by use of alternate raw material, and (3) the possibility of recovering any valuable material or separating a particular component, if any, that renders the whole waste hazardous. Hazard identification should start from the conceptual stage and continue through different stages of design. A thorough approach is required in modern chemical industries to systematically identify potential hazards, evaluate risks involved, and take preventive measures. Wastes needing offsite disposal require to be stored safely onsite prior to sending to offsite location.

Material balance for each unit operation and for the whole process has to be made. The total of what goes to a process should equal the total of what comes out. The input information for each unit operation has to be assembled and then a decision has to be made whether all the inputs and outputs need be included in the material balance calculations. The results of material balance for each unit operation can pinpoint the problem areas associated with the production process.

Preliminary material balance for each unit operation has to be made based on data already collected and presented, as shown in Table 6.5.

Table 6.5 Preliminary Material Balance for Each Unit Operation

Inputs:
Raw materials
Waste reuse
Water
Total
Unit process A
Outputs: (Amount in Standard Unit Per Annum)
Raw material, product, by-product storage and handling losses, stored waste, wastewater, gaseous emissions, reused wastes, hazardous liquid waste transported onsite, nonhazardous liquid waste transported onsite, nonhazardous solid, and liquid wastes transported offsite.

Material balance has to be made in weight units. Water balance studies have to be made for all inputs and outputs. Individual material balances have to be summed up to give a balance for the whole process, a production area, or a factory. Once the material balance for each unit operation has been completed for raw materials input and waste output, it is worthwhile to repeat the procedures with reference to each contaminant of concern. The material balance should be used to focus priorities for long-term waste reduction.

If the material imbalance in the data collected is significant, further investigation is needed. If outputs are greater than inputs, it is clear that some inputs are overlooked or there are estimation errors. It may be necessary to examine unit operations to identify where unnoticed uses may be occurring and to repeat collection of data. Preparation of a preliminary material balance and its refinement can be combined into one step in the case of a simple or small manufacturing plant. For more complex manufacturing units, two separate steps, preliminary material balance and its refinement, are more desirable. Inputs should ideally equal output. In case of high-strength hazardous wastes, accurate measurements are required for planning of waste reduction. Based on the data collected a plan on waste minimization has to be developed.

Different waste reduction measures require varying degrees of effort, time, and financial resources. They can be classified into two groups: (1) obvious waste reduction measures including improvements in management techniques and housekeeping procedures that can be implemented cheaply and quickly (i.e., low-cost or no waste measures) and (2) long-term reduction measures involving process modifications and process substitutions to eliminate problem wastes. Increased reuse or recycling to reduce waste is between the immediate and more substantial waste reduction measures.

Some of the waste reduction measures include:

- purchase of raw materials in a form that is easy to handle, e.g., pellets instead of powders
- no over-ordering of materials, particularly when the raw materials or components can be spoiled or are difficult to store
- while receiving materials, making visual inspection and checking sack weights, volumes, composition, and quality of material and demanding quality control from suppliers
- avoiding overflows and spillage in all operations, providing dedicated tanks receiving only one type of material, implementing tank checking procedures, and reducing evaporation losses by using covered closed tanks

- minimizing the number of times materials are moved onsite, checking transfer lines for spills and leaks, plugging leaks, and adopting floor restrictions to reduce excess water consumption
- Informing the employer why actions are taken for waste reduction and how waste reduction improves the process
- Investigating how water can be contained and used again before discharge into drain and how solvents used for cleaning can be used again.

The information gathered for each unit operation should be used to develop better operating practices for all units. Significant waste reductions can often be obtained by improved operation, better handling, and generally taking more care. Simple, quick adjustments should be made to the process to achieve a rapid improvement in the process efficiency.

Material balance has to be used to focus priorities in long-term waste reduction. Factors like poor technology, lack of maintenance, and non-compliance with company procedures leading to waste generation have to be identified. Segregation of weak and strong wastes can offer enhanced opportunities for recycling and reuse with resultant savings in raw material costs. Concentrated simple wastes are more likely to have value compared to dilute or complex wastes. Mixing of strong and weak wastes can enhance of complicate pollution problems.

Waste problems that cannot be solved by simple procedural adjustments or improvements in housekeeping practices will require more substantial long-term changes. Process or production changes that may increase production efficiency and reduce waste generation include:

- changes in the production process: continuous verses batch
- equipment and installation changes
- changes in process control, automation
- changes in process conditions such as retention time, temperature, agitation, pressure, catalysts
- use of dispersants in place of organic solvents where appropriate
- reduction in the quantity or type of raw material used in production
- raw material substitution through the use of wastes as raw materials or the use of different raw materials that produce less waste or less hazardous waste.

It is important to conduct environmental and economic evaluation of waste reduction options. These include environmental economic evaluation. Based on these two the action plan has to be developed and implemented.

Environmental auditing in summary acts as a valuable environmental management tool to:

- provide information for management review (audit findings and recommendations);
- provide competitive advantage by raising corporate profile with respect to environmental issues, especially through ISO 14001 certification;
- save money by preventing incidents due to the proactive nature of auditing; prove corporate environmental claims to meet purchasing guidelines o customers;
- allow checking and corrective action in the light of increasing complexity of environmental legislation;
- help establish due diligence (to avoid personal and corporate liability);
- identify potential environmental problems and current environmental impacts, in addition to gathering basic information about the facility being audited;
- verify compliance with regulations;
- verify conformance with voluntary commitments and contracts;
- provide assurance to management that systems and processes are in place to prevent, detect, and correct nonconformance with audit criteria.

The following is list of the ISO 14000 standards relevant for environmental auditing:

- 14010 Guidelines for Environmental Auditing – General Principles of Environmental Auditing
- 14011 Guidelines for Environmental Auditing – Audit Procedures-Part 1: Auditing of Environmental Management Systems
- 14012 Guidelines for Environmental Auditing – Quantification criteria for environmental auditors
- 14013/15 Guidelines for Environmental Auditing – Audit Programs, Reviews, and Assessments
- 14024 Environmental Labeling – Practitioner Programs – Guiding Principles, Practices, and Certification Procedures of Multiple Criteria Programs
- 14031/32 Guidelines on Environmental Performance Evaluation
- 14040/43 Life Cycle Assessment General Principles and Practices.

FURTHER READING

[1] Benefits to Industry of Environmental Auditing, Centre for Environmental Assurance, Arthur D. Little Inc., Cambridge, Mass., USA, 1983. ISO 14000 Series.

[2] L.W. Canter, Environmental Impact Assessment, second ed., McGraw Hill Publishing Company, Inc., New York, NY, 1996.

[3] B.D. Clark, A. Gilard, R. Bisset, R. Tomlinson, Perspectives on Environmental Impact Assessment, Reidel Publishing Company, Holland, 1984.
[4] S. Dubey, D. Newnes, Green Democracy Peoples Participation in Environmental Decision Making, Environmental Justice Initiative, 2003.
[5] EIA Guide Lines for Planning and Decision Making, U.N. Publications, 1985. ST/ESCAP/351, ES.CAP.
[6] Environmental Auditing – Fundamentals and Techniques, second ed., Centre for Environmental Assurance, Aurthur D. Little Inc., Cambridge, Mass., USA, 1988.
[7] Environmental Auditing, UNEP/IEO, Industry and Environment Review, 1990.
[8] L.B. Cahill (Ed.), Environmental Audits, fifth ed., Government Institutes, Rockville, USA, 1987.
[9] Environmental Statement (As Part of Environmental Audit) Govt. Of India, Ministry of Environment and Forests, New Delhi, 1993.
[10] D.M. McAllister, Evaluation in Environmental Planning, The MIT Press, Cambridge, Mass., 1986, pp. 6–7.
[11] The Environmental Audit, Handbook Series, Executive Enterprises Publication Co., NY, 1988.

CHAPTER SEVEN

Environmental Accounting

7.1 INTRODUCTION

In India, both "Environment Protection" and "Economic Development" are the matters of great importance. However, some sort of tradeoff is needed between the two. For this purpose, environmental accounting is required to measure the environmental impact of economic activities by various sectors. A standard system of this type of accounting is still evolving in India. Environmental accounting needs to work as a tool to measure the economic efficiency of environmental conservation activities and the environmental efficiency of the business as a whole. In many contexts, environmental accounting is taken to mean the identification and reporting of environment-specific costs such as liability cost and waste disposal costs. It is accounting for any costs and benefits that arise from change to a firm's products and processes where the change also involves a change in environmental impact.

"Environmental accounting," sometimes referred to as "green accounting," "resource accounting," or "integrated economic and environmental accounting," refers to modification of the System of National Accounts to incorporate the use or depletion of natural resources. The System of National Accounts (or SNA) is the set of accounts that national governments compile routinely to track the activity of their economies. SNA data are used to calculate major economic indicators including gross domestic product (GDP), gross national product, savings rates, and trade balance figures. The data underlying these aggregate indicators are also used for a wide range of less publicized but equally valuable policy analysis and economic monitoring purposes.

Environmental accounting is a vital tool to assist in the management of the following:

1. Environmental risks
2. Operational costs
   a. Environmental accounting is a set of aggregate data linking the environment to the economy, which will have a long-run impact on both economic and environmental policy-making.

Environmental Management
ISBN 978-0-12-811989-1
http://dx.doi.org/10.1016/B978-0-12-811989-1.00007-5

b. It is not a valuation of environmental goods or services, social cost-benefit analysis of projects affecting the environment, or disaggregated regional or local data about the environment. There are, however, close links between environmental accounting and these three activities, which is why they are frequently discussed together.
c. "Valuation" refers to the process of deriving a monetary value for things that are not sold in a market, for example, fuel wood gathered in the forest, water filtration provided by a wetland, or biodiversity resources that could provide new medicines in the future. Valuation is an essential input into both social cost-benefit analysis and some approaches to environmental accounting. However valuation is only one element in the construction of environmental accounts; it is not the same as the construction of the accounts.

7.2 FORMS OF ENVIRONMENTAL ACCOUNTING

1. *Environmental management accounting*: management accounting with a particular focus on material and energy flow information and environmental cost information. This type of accounting can be further classified in the following subsystems:
   a. Segment environmental accounting: This is an internal environmental accounting tool to select an investment activity, or a project, related to environmental conservation from among all processes of operations, and to evaluate environmental effects for a certain period.
   b. Eco balance environmental accounting: This is an internal environmental accounting tool to support Plan-Do-Check-Act (PDCA) for sustainable environmental management activities.
   c. Corporate environmental accounting: This is a tool to inform the public of relevant information compiled in accordance with the environmental accounting. It should be called corporate environmental reporting. For this purpose the cost and effect (in quantity and monetary value) of its environmental conservation activities are used.
2. *Environmental financial accounting*: financial accounting with a particular focus on reporting environmental liability costs and other significant environmental costs.

3. *Environmental national accounting*: national-level accounting with a particular focus on natural resources stocks and flows, environmental costs, and externality costs, etc.

7.3 NEED OF ENVIRONMENTAL ACCOUNTING

It helps to know whether a corporation has been discharging its responsibilities toward the environment or not. Basically, a company has to fulfill the following environmental responsibilities:

1. Meeting regulatory requirements or exceeding that expectation,
2. Cleaning up pollution that already exists and properly disposing of the hazardous material,
3. Operating in a way that those environmental damages do not occur,
4. Promoting a company having a wide environmental attitude,
5. Control over operational and material efficiency gains driven by the competitive global market,
6. Control over increases in costs for raw materials, waste management, and potential liability.

7.4 SCOPE OF ENVIRONMENT ACCOUNTING

The scope of environmental accounting (EA) is very wide. It includes corporate, national, and international levels. The following aspects are included in EA:

1. From an internal point of view, investment is made by the corporate sector for minimization of losses to environment. It includes investment made into the environment saving equipment/devices.
2. From an external point of view, this includes all types of losses indirectly due to business operation/activities. It mainly includes these:
   a. Degradation and destruction like soil erosion, loss of biodiversity, air pollution, water pollution, noise pollution, problem of solid waste, coastal and marine pollution,
   b. Depletion of nonrenewable natural resources, i.e., loss emerged due to over-exploitation of nonrenewable natural resources like minerals, water, gas, etc.,
   c. Deforestation and land uses: This type of accounting is not easy, as losses to the environment cannot be measured exactly in monetary value. Further, it is very hard to decide how much loss occurred to the environment due to a particular industry.

About 25 countries have experimented with EA over the past 20 years. A few European countries have established physical accounting systems that are routinely compiled and applied to economic and environmental policy-making. Many other countries have undertaken more limited or one-time experiments and case studies with monetary environmental accounts, focused on issues such as forestry, soil erosion, and minerals depletion. A few examples suggest the richness of their experience.

- Norway has compiled physical accounts focused on energy resources and air pollution.
- Indonesia was the first country for which forest depletion was calculated and integrated into a "green GDP."
- Namibia carried out a phased testing and implementation of the system of environmental and economic accounting (SEEA) approach to EA. It is focused on several key natural resource sectors etc.

Conventional accounts address the role of the environment in economic performance in part only. In the conventional accounts, these overlapping flows and stocks consist of the following:

1. Capital formation, that is to say, the acquisition less disposal of fixed assets and changes in inventories,
2. The wear and tear of fixed assets in production, in other words, consumption of fixed capital.

However there is a relationship between the economy and the ecosystem. Fig. 7.1 shows the interrelation between the natural environment and

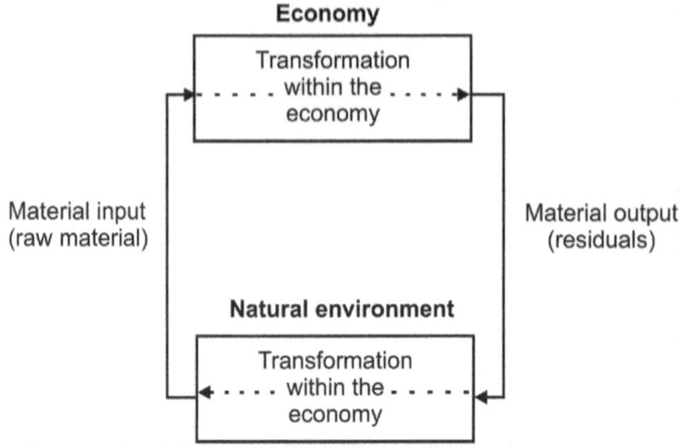

Figure 7.1 Interrelation between the natural environment and the economy. *From United Nations, Integrated Environmental and Economic Accounting (Sales No. E.93. XVII.12), 1993, Figure V.*

the economy. The SEEA incorporates environmental concerns mainly by doing the following:

1. Segregating and elaborating all environment-related flows and stocks that are already included in the conventional accounts. The objective is to present separately environmental protection expenditures;
2. Expanding the asset accounts beyond economic assets to include environmental assets and changes therein;
3. Introducing impacts on natural (economic and environmental) assets, caused by production and consumption activities of industries, households, and government, as environmental costs incurred by these activities.

It is very important to understand the environmental assets that exist. Fig. 7.2 shows the stocks and flows of environmental assets. Environmental assets provide environmental services such as waste absorption, habitat, flood and climate control, and nutrient flows. Environmental asset accounts include the physical accounts of ecosystems. The monetary valuation of stocks or inventories of ecosystems and their components because of the controversial valuation techniques required for determining option or existence values for these environmental assets. It is normal practice to calculate for physical and monetary emission from land, water and air and link it to the production accounts for estimation of contribution of environmental services.

The three main approaches in physical accounting:

1. Natural resource accounts (NRA) describe the stocks and use of different natural resources during the accounting period in a fairly aggregate fashion. They were pioneered by Norway [2] and further developed by France as natural patrimony accounts [3]. NRA are measured in different units (weight, volume, energy equivalent, area) and are largely consistent with the SNA asset accounts. They can be expressed in monetary units, too, and have thus been developed as an integral part of the SEEA.
2. Physical Input–Output Tables can be extended to include material flows from, and back into, the environment, presenting these flows in great sectoral detail [4]. Providing a balance of total material inputs and outputs, these tabulations can also be interpreted as material/energy balances (MEB).
3. Material flow accounts (MFA) attempt to measure the material throughput through the economy as a measure of the sustainability of economic activity in non-monetary terms (usually weight). MFA describe the extraction, production, transformation, consumption, and accumulation

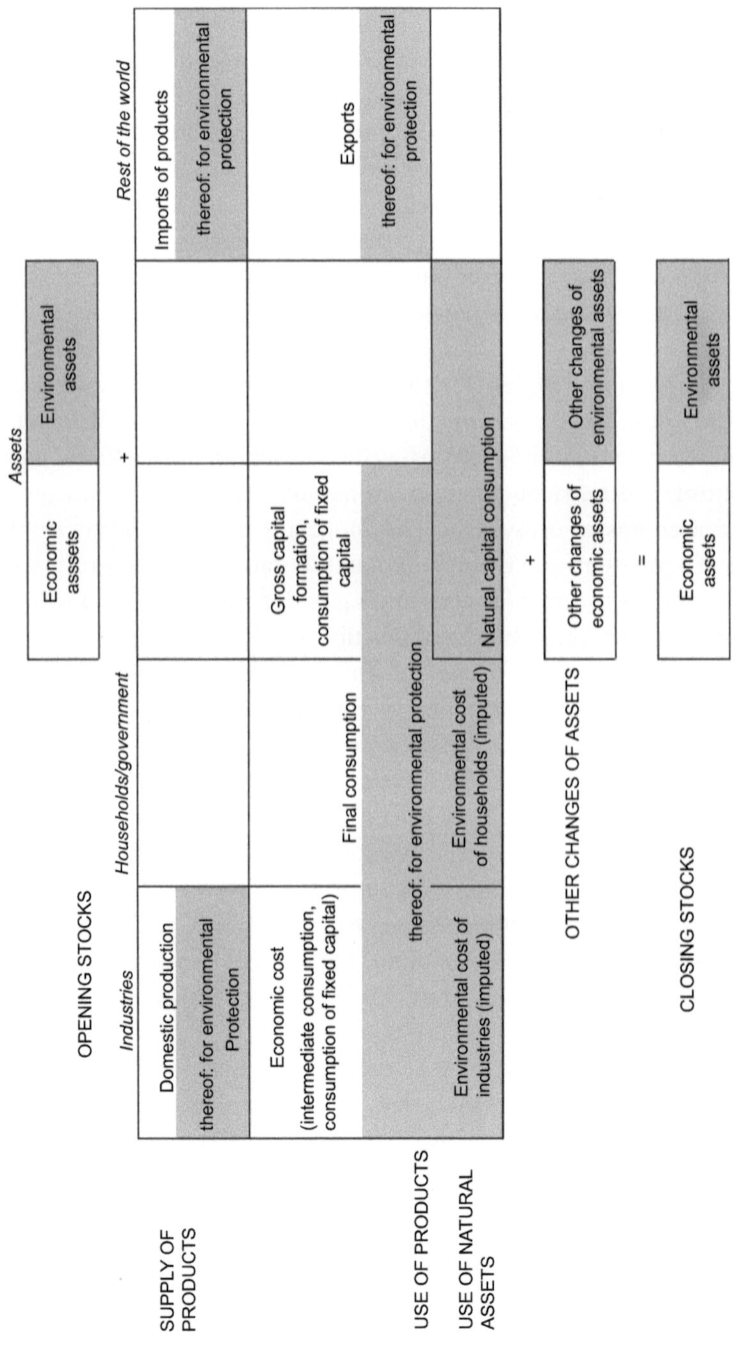

Figure 7.2 Stock and flow of environmental assets.

of chemical elements, raw materials, or products [5]. They may include ecological rucksacks of hidden material flows that are not physically incorporated in a particular output but are required for the production of goods, their use, and the recycling and disposal of wastes (Spangenberg and others, 1999).

The basic and relatively easy to implement modules of the SEEA include these:

- the separate identification of environmental expenditures in the supply, use, and asset accounts of the SNA,
- the accounting of natural resource stocks and use in physical and monetary terms,
- the measurement and valuation of emissions, that is to say, of so-called environmental externalities.

In the SNA, there are those assets of CNFA category 2 (non-produced assets) over which ownership rights are enforced and that provide economic benefits to their owners. Their products are generally valued in the market. Economic non-produced assets are distinguished from environmental ones, not so much because of any scarcity criteria, which apply to environmental assets as well, as because of the following:

- They are clearly different from intangible environmental *services* of waste absorption and life support in general, being material *goods* that enter the economic system for intermediate or final consumption.
- Market values for economic assets are readily available.
- Most of the economic assets are already defined and classified in the conventional accounts.

Accounting Principles for Environmental Resources.

Given that environmental resources are renewable, the change in the environmental resource stock, ΔSt, during a given time, say a year, is estimated as:

$$\Delta St = N(St) - Et + At \tag{7.1}$$

where (St) is the stock of environmental resource (quality of water resources or atmosphere) at time t; N(St) is the natural rate of regeneration of this stock (natural rate of assimilation of pollution loads); (Et) is the rate of depletion of stock (rate of degradation of environmental quality); and (At) is the rate of pollution abatement. The environmentally corrected net national product (ENNP) with the sustainable use of environmental resources could be estimated as:

$$ENNP = Ct + UL\ Lt/pt + \Delta Kt + M'\ \Delta St, \tag{7.2}$$

where (Ct), the consumption; (Lt), the labor employment; (ΔKt), the value of change in manmade capital or investment; (ΔSt), the change in environmental capital; (UL), the marginal disutility of labor; (pt), the price of consumption good; and M(At) and (M′) are, respectively, the pollution abatement cost function and the marginal cost of pollution abatement. The accounting principle, therefore, is to evaluate the changes in environmental resource stocks at the marginal costs of avoiding the depletion of the stocks. The marginal cost of avoiding the depletion is the price of an environmental resource in the context of sustainable development.

7.4.1 Accounting Principles for Exhaustible Natural Resources

Extraction of exhaustible natural resources like fossil fuels, minerals, metals, and ores and their use in various economic activities has implications for sustainable development in two important ways: intergenerational equity and environmental degradation. Given the property right of an exhaustible resource to the present and future generations, sustainable development has to ensure just distribution of benefits from the resource to present and future generations. Exploitation of exhaustible resources contributes to environmental degradation. Mining of resources could result in deforestation, land degradation, and water and air pollution, which could be identified by attempting full life cycle analysis of resource use.

Changes in the resource stocks in this context have to be estimated as follows:

The rate of depletion of the resource stock, ΔR, if fresh discoveries of the resource are not there, is given by:

$$\Delta R_t = -\mathrm{Yt}, \tag{7.3}$$

where Rt and Yt represent, respectively, the stock of exhaustible resource, say coal, and the amount of resource extracted. The rate of depletion of environmental resource stock ΔSt consequent on the extraction of exhaustible resource is given by:

$$\Delta \mathrm{St} = \mathrm{N}\,(\mathrm{St}) \; - \alpha \mathrm{Yt} + \mathrm{At}, \tag{7.4}$$

where At is the rate of pollution abatement and α is the rate of environmental resource degradation per unit extraction of Yt. The ENNP with the sustainable use of exhaustible resources could be estimated as follows:

$$\mathrm{ENNP} = \mathrm{Ct} + \mathrm{UL\ Lt/pt} + \mathrm{Kt} + \mathrm{M/S} + \left((\mathrm{Fy} - \mathrm{vt}) \; - \alpha \mathrm{M}'\right)\mathrm{Rt} \tag{7.5}$$

The term $((Fy - vt) - \alpha M')$ in the equation could be interpreted as the generalized hoteling rent on an exhaustible resource after accounting for the cost of avoiding environmental degradation arising out of use of the resource. By using, say, one ton of fossil fuel, α tons of pollution is generated, and the cost of abatement of it is α M′, which has to be accounted in defining the hoteling rent. It is so because the cost of depletion of environmental quality due to pollution evaluated at the shadow price, the marginal cost of pollution abatement, is already accounted in measuring ENNP through the fourth component in Eqn (7.5).

7.5 VALUATION METHODS

There are two important steps in the development of integrated economic and environmental accounting: first, namely, the description of stocks of natural assets in physical terms and the measurement of changes in stocks during the accounting period; and the second is the valuation of natural assets. Natural assets provide both marketable and non-marketable services, and therefore their valuation requires the use of market and non-market valuation techniques. Several methods for valuation exist; the most important and widely used are discussed next.

7.5.1 Market Valuation of Natural Resources

The market valuation of the SEEA, which measures the depletion of natural resources, in other words, of the economic assets of the SNA, is the closest to conventional accounting. It identifies changes in the values of natural assets that are already accounted for in conventional asset accounts as other changes in volume. These changes in volume include the depletion of natural resources, as well as their degradation from pollution and other degrading activities, to the extent that the underlying environmental impacts are reflected in changed market values of those assets. The SEEA shifts the value of the depletion and degradation of economic natural assets as environmental cost from other volume changes in the asset accounts to the production accounts.

A number of methods to estimate the market value of the stocks of scarce (depletable) natural resources and, by implication, changes in the value of stock have been proposed and applied in practice.

1. **Net present value of natural resources:** The basic principle of market valuation of economic assets for which a price cannot be directly observed in the market rests on using the prices of the goods extracted from, or services provided by, these assets for estimating the

future sales value, reduced by the exploitation costs. If the exploitation is spread over a lengthy period, the flow of future net returns has to be discounted. In some cases, the reserves of depletable natural assets and exploitation rights are marketed. The market prices will then reflect to a high degree the current value of the expected net returns since investors would base their decision of buying an asset on relative present values of future net income streams. This assumption will not hold, however, in those countries where concessions to extract the resource are fixed by the government, and they frequently are done so below market value.

It is also difficult to estimate future returns and costs of natural resource exploitation by industry (agriculture, forestry, mining, construction, and so forth) or type of natural resource used by different industries. Those estimates would require information on the availability of future stocks (reserves), prices, and extraction or harvest costs that are usually available, if at all, only at the microeconomic, rather than the sectoral level.

The method that uses the following to calculate the present value V_0 of the natural resource is the sum of the expected net revenue flows, N_tQ_t, discounted at nominal or real interest rate r:

$$V_o = \sum_{t=0}^{T} \frac{N_tQ_t}{(1+r)^1}$$

where r is assumed to be constant for the lift T of the asset, N_t is defined as the total unit (sales) value of the resource less the production cost (i.e., the cost of extraction, development, and exploration, including a normal return to capital), and Q_t is the quantity exploited during the period t.

2. **Net price method:** The net price valuation neglects future (discounted) losses of net returns from resource depletion. The assumption underlying this simplification is that in long-term equilibrium the net price of the marginal unit extracted will rise at the discount rate, neutralizing the discount factor. The net price has been applied in various studies (for example, Repetto and others, 1989) and country applications of the SEEA. It is defined as the actual market price of the raw material minus its marginal exploitation costs including a "normal" rate of return of the invested produced capital.

The value of a natural resource is then calculated as the product of the quantity of the natural resource stock and the net price. In the case of nonrenewable (mineral) resources, this stock comprises only "proven reserves," which are exploitable under present economic conditions and therefore have a positive net price. The net price method could also be applied to wild biota and water as long as these natural assets are considered economically exploitable assets.

This estimation of stock values can, of course, also be applied for valuing all *changes* of natural assets during the accounting period. In principle, the net price effective at the time of the resource use should be applied. In practice, the cost of depletion is calculated by multiplying the depleted quantities of the natural assets by the *average* net price between the beginning and end of the accounting period.

Assessment Method

a. **Step 1:** Determine the market prices of different natural resource outputs: domestic or export price as applicable; price at the beginning and end of the accounting period and average during the period.
b. **Step 2:** Assess the total production cost per unit of resource output: unit cost at the beginning and end, and average cost during the accounting period.
c. **Step 3:** Assess normal return to the invested capital, used in the exploitation of the resource.
d. **Step 4:** Determine the net operating surplus of the industry exploiting the resource.
e. **Step 5:** Calculate the net price as the difference Step 1 − (Step 2 + Step 3), or as the difference between Step 4 and Step 3.

Net rent method is an alternative to the net price method, avoiding the calculation of a normal return to capital that has to adopt a controversial interest rate. It also prevents the occurrence of negative net rents due to exogenously determined rates of return to produced capital. The idea is to subtract the value of the produced capital stock K from the gross value of the resource stock [6]. Assess the replacement value of the produced capital stock K, at the end of the accounting period.

Calculate total net rent generated during the accounting period as {[(TR − C)/(quantity extracted)] times (total remaining resource stock)} minus K (where TR = total revenue from resource exploitation and C = current extraction costs); Use total net rent for valuing total resource stock directly or as the basis for applying the net present value method.

3. **User cost allowance:** Another method that approximates the net present value for the depletion of natural resources is the user cost valuation. The idea is to convert a time-bound stream of (net) revenues from the sales of an exhaustible natural resource into a permanent income stream by investing a part of the revenues, namely, the "user cost allowance," over the lifetime of the resource; only the remaining amount of the revenues should be considered "true income" [7]. Given particular net revenue for an accounting period, the calculation of the user cost allowance is straightforward, requiring only two additional parameters: the discount rate and the lifespan of the resource at the current extraction rate. The net price and user cost methods differ in their respective objectives of natural capital and income maintenance. The net price method tends to overstate capital consumption, thus representing an upper limit of environmental cost estimates, whereas the user cost allowance assumes full substitutability of natural capital by other production factors; it can thus be considered a lower limit.

   Both valuation methods have been applied in SEEA case studies to assess a range of cost estimates.

   The user cost is the difference between the finite net returns R ($=N_tQ_t$) from the sales of an exhaustible reserve during the accounting period t (expected annually during the lifetime of the resource T) and a "perpetual income stream" X resulting from the investment of the user cost at an interest rate r:

$$R - X = R/(1 + r)^{T+1}$$

   The user cost method is applied to depletion rather than stocks.

7.5.2 Maintenance Valuation of Environmental Assets

Maintenance costs are defined as the costs that one would have had to incur during the accounting period to avoid current and future environmental deterioration from the impacts caused during the accounting period [8]. This is the valuation in addressing the cost that could have been incurred to avoid the impacts of current economic activity, and it does not refer to the actual environmental damage generated by this activity, that is to say, its environmental impacts generated during the accounting period *and* in the future. Maintenance costs are, of course, hypothetical because, in reality, an actual use of the asset that affected the environment *did* take place. The rationale behind this approach is based on the following two criteria:

1. The application of a strong sustainability concept that has gained a central role in the discussion of integrated (environmentally sound) development;
2. The extension of the national accounts concept of replacement cost of the consumption of fixed capital to the use of non-produced natural assets.

In all cases of permanent environmental degradation and destruction, the value of the maintenance costs depends on the avoidance, prevention, or restoration activities chosen. The choice of activities for calculating the imputed maintenance costs of discharging residuals will depend on relative costs and efficiencies, in other words, on the best available technologies. Imputed prevention costs of industries should thus be based on the most efficient methods for preserving environmental assets or meeting environmental standards.

7.5.3 Contingent Valuation of Environmental Services

In assessing the hypothetical costs of keeping environmental assets intact, maintenance costing focuses on the direct impacts of production. In contrast, contingent and related valuations of the demand/benefit side of environmental services attempt to measure the losses of such services, in other words, environmental damage. The well-known problems of applying these valuations in cost-benefit analyses at the project level accumulate at the national level. At least for the time being, such valuations do not seem to be applicable in recurrent national accounting. However, they might be usefully explored in more experimental studies that focus on selected environmental concerns or regions.

7.5.4 Environmentally Adjusted Economic Aggregates

The expansion of the asset boundary of conventional accounts for the inclusion and valuation of natural assets and asset changes permits the calculation of a range of aggregates. The aggregates can be presented as the sum total and elements of conventional accounting identities. These accounting identities are maintained in the SEEA in the following manner:

1. Supply–use identity:

$$O + M = IC + C + CF + X$$

indicating that the supply of goods and services produced (*O*) and imported (*M*) equals their use in intermediate (*IC*) and final consumption (*C*), capital formation (*CF*), and export (*X*);

2. Value-added (environmentally adjusted) identity for industry:

$$EVAi = Oi - ICi - CCi - ECi = NVAi - ECi$$

describing value added as generated by an industry (*EVAi*) as the difference of output and cost, including fixed capital consumption (*CC*) and environmental depletion and degradation costs (*ECi*) or equivalently as the difference of net value added (*NVAi*) and environmental costs (*ECi*);

3. Domestic product identity (environmentally adjusted) for the whole economy:

$$EDP = \sum EVAi - ECh = NDP - EC = C + CF - CC - EC + X - M$$

defining environmentally adjusted net domestic product (*EDP*) as the sum of environmentally adjusted value added of industries (*EVAi*) with a further deduction of environmental costs generated by households (*ECh*).

Therefore the environmentally adjusted accounting indicators would be as follows:

OPENING STOCKS | Economic assets | Environmental assets

+

	DOMESTIC PRODUCTION (industries)	FINAL CONSUMPTION (households, government)	CAPITAL FORMATION	CAPITAL ACCUMULATION	REST OF THE WORLD
SUPPLY OF PRODUCTS	Output (O_i)				Imports (M)
USE OF PRODUCTS	Intermediate consumption (IC_i)	Final consumption (C)	Gross capital formation (CF)		Exports (X)
USE OF FIXED CAPITAL	Fixed capital consumption (CC_i)		Capital consumption (CC)		
Value added (VA)/NDP	$NVA_i = O_i - IC_i - CC_i$ $NDP = \Sigma NVA_i$				
USE OF NATURAL ASSETS (depletion and degradation)	Environmental cost of industries (EC_i)	Environmental cost of households (EC_h)	Natural capital consumption (EC)		
Environmentally adjusted indicators	$EVA_i = NVA_i - EC_i$ $EDP = \Sigma EVA_i - EC_h$		$ECF = (CF - CC) - EC$		

+

Other changes in volume, revaluation

=

CLOSING STOCKS | Economic assets | Environmental assets

7.6 COMPILATION OF PHYSICAL NATURAL RESOURCE ACCOUNTS

The next step is the compilation of the accounts and the records, the stocks and the changes therein during the accounting period in physical measures (square kilometers (km^2), cubic meters (m^3), and metric tons). Opening and closing stocks are measured as the economically exploitable quantity of reserves or stocks available at the beginning and end of the accounting period. Changes in quantity are brought about by the direct economic use/exploitation of the asset, including the extraction of minerals, logging, fish catch, water abstraction, and so forth. For renewable resources, economic use is a gross concept that includes sustainable use and implementation of the SEEA.

A Step-by-step approach is made possible by natural regeneration or replenishment, as well as "depletion," which represents exploitation of the resource beyond long-term sustainable levels or yields. Natural resource depletion is the notion underlying environmental costing. Changes in the quality of natural resources affect their productivity and economic value. Quality changes are thus relevant (physical) aspects of environmental costs, but they are difficult to incorporate into physical *quantitative* asset accounts.

Discussed next are key questions in respect of the compilation of the different resource accounts.

1. **Land and soil accounts:** Opening and closing stocks consist of the land area over which ownership rights are enforced, including land underlying buildings and works, agricultural land, forest and other wooded land, recreational land, and associated surface water, and other open land and areas of artificial watercourses or water impoundment. Soil degradation refers to quantitative changes in the availability of soil and can therefore be treated as depletion of soil. However, from an economic point of view, the main issue with respect to soil degradation is the change in the quality of land due to the loss of topsoil. Soil can be also considered a renewable resource, as it possesses recuperative capacities that allow some erosion to be sustained without productivity losses.
2. **Subsoil asset accounts:** The opening and closing stocks of subsoil assets are proven reserves of mineral deposits, located on or below the Earth's surface, that are economically exploitable, given current technology and relative prices. They consist of coal, oil, natural gas reserves, and metallic and nonmetallic mineral reserves. Probable reserves are defined

as the estimated quantity and grade of a mineralized body for which sufficient information on continuity, extent, grade, operating and capital costs, and so forth is available on the basis of a study indicating an economically viable operation at long-term forecast.

3. **Forest accounts (economic functions):** Economic forest assets include all economic functions such as the provision of timber, bark, fibers, fruits, and other forest products that are commercially exploitable. It is the stock and use of these economic products that are the basis for the application of market valuation discussed later. All the elements of the forest accounts are calculated in volume or weight (cubic meters or tons of biomass). Opening and closing stocks are defined as the total standing volume of timber, including only those trees whose diameter is large enough to allow harvesting. Direct economic use of forests consists of logging or other activities such as land clearance for agriculture. The volume of timber logged, above the long-term net growth of the forest, is considered non-sustainable and termed depletion. Sustainable management of forests aims at maintaining this capacity through practices such as selective cutting. Reforestation and preservation are remedial actions similar in nature to reinvestment of depreciated fixed capital.
4. **Fishery and other biota accounts:** Direct economic use is affected by fish catch. Fish catch refers to the weight/volume of fish caught in the actual fishing site. Sustainable catch is distinguished from depletion and is the amount of fish that can be caught without reducing stock in the long term. It has to be estimated by using models that take into account the current size and age structure of the stock, its reproductive potential, and climatic and environmental variables. Stock data are obtained from direct observation or are inferred from data on catch and catch per unit of effort.
5. **Water accounts:** Measuring the long-term availability of water, under current economic and technical conditions for assessing the (non-) sustainability of water use, is one of the most difficult tasks of natural resource accounting. Water is a cyclic natural resource that moves rapidly, escaping human control and ownership

Changes in stocks are brought about by abstraction of water for industrial or household use, other accumulation, and other volume changes. Natural causes that affect water levels and flows are floods and droughts; their effects are recorded as other volume changes. Detailed water balances have been frequently established without distinguishing between economic and non-economic uses. Those balances are useful for the management of the resource, for example, in a particular drainage basin.

7.7 EXAMPLE FOR FOREST ACCOUNTS

Based on the previous, an example of the forest accounts is given next. A step-by-step approach as applied to forests includes the following steps:

Step 1: Compilation of the supply and use accounts
Step 2: Identification and compilation of environmental protection expenditures related to forests
Step 3: Compilation of produced forest asset accounts
Step 4: Compilation of physical forest accounts
Step 5: Valuation of forests; compiling the monetary accounts
Step 6: Compilation of physical environmental forest accounts
Step 7: Compilation of emissions by economic sector
Step 8: Maintenance costing of environmental degradation

7.7.1 Environmental and Economic Concerns About Forests

Environmental and economic concerns about forests have been expressed in numerous international forums and have given rise to various conventions, notably the International Tropical Timber Agreement; the Tropical Forestry Action Plan; the Non-legally Binding Authoritative Statement of Principles for a Global Consensus on the Management, Conservation and Sustainable Development of All Types of Forests (Forest Principles), [1]; resolution 1, nnex III, adopted at Rio de Janeiro; the Agenda 21 [1]; resolution 1, annex II action program; the Convention on Biological Diversity [9]; the United Nations framework Convention on Climate Change (document A/AC.237/18 (Part II)/Add.1 and Corr.1, annex I); and the United Nations Convention to Combat Desertification in those Countries Experiencing Serious Drought and/or Desertification, particularly in Africa (document A/49/84/Add.2, annex, appendix II).

The main concerns are these:

1. **Economic:** linked to the sustainability of forests as a source of wood and non-wood products and economic activities that take place in the forest. The quantity and quality of forest resources decline because of forest logging, at levels greater than those of regeneration, for timber, fuel, and fodder, and because of forest clearance to provide land for other uses;

2. **Environmental owing to the:**
    a. Role of forests in the carbon cycle and the adverse consequences of large-scale deforestation on regional and global climatic balances;
    b. Role of forests in the water cycle and soil erosion control: forest exploitation and clearance lead to various chains of interlinked problems, notably soil erosion and watershed destabilization;
    c. Role of forests as a habitat and support of biodiversity: prevailing forest management practices contribute to a rapid loss of natural habitat diversity, species diversity, and genetic diversity in forests, both natural and planted;
    d. Degradation of forests due to acidification, fires, inappropriate forestry or logging practices;
    e. Recreational, aesthetic, and cultural functions: the decrease of forests and the increasingly restricted access to forests impact particularly communities that are dependent on forests for their livelihood and traditional cultural activities.

The increasing concern about sustainable forest use, either tropical, temperate, or boreal, has led to the development of various instruments to monitor the health of the forest, changes in land use, and the impact of forest on the national economy. These instruments include integrated environmental and economic accounting for forest, discussed in detail in the present section, and frameworks and lists of environmental statistics and indicators.

7.7.2 Coverage of Forests in the SEEA

The SEEA accounts for forest land and related ecosystems, biological assets (plants, animals, and so forth in the forest), and other assets related to forests.

1. **Land:** The SEEA reorients the SNA_s land classification toward environmental concerns. Wooded land is explicitly distinguished as a land category. Although excluded from SNA_s economic assets, non-exploitable virgin forests are included in the SEEA classification of assets. Cultivated and uncultivated economic forest land is distinguished from non-economic (environmental) forest land.

    Cultivated economic forest land corresponds to land over which ownership rights are enforced, and for which natural growth and/or regeneration of timber and other biological assets is under the direct control, management, and responsibility of institutional units and is likely to produce economic benefits to the owner of the land.

    Uncultivated economic forest land corresponds to land over which ownership rights are enforced (including collective ownership by the government) but for which the natural growth and/or regeneration of timber and other biological assets is not under the direct control, management, and responsibility of institutional units, although growth and regeneration are likely to produce economic benefits to the owner of the land.

Non-economic environmental forest land covers land of both protected and non-exploitable forests. It corresponds to forests that are not exploitable for economic reasons (distance from markets, low productivity and accessibility, and so forth), including virgin forests, and to forests where the exploitation of biological resources is severely restricted by virtue of the protection status.

7.7.3 Biological Assets

Biological assets are related to forest and other wooded land. They consist of animals and plants living in forests: trees and other flora of forests, fauna, and so forth. Produced (cultivated) economic biological assets are distinguished from non-produced (wild or uncultivated) economic biological assets.

They are classified as follows:

1. Fixed assets, when they yield repeat products (for example, trees for cork, gums, berries, and so forth);
2. Work-in-progress inventories, when they yield once-only products (for example, annual plants, trees of timber tracts, and so forth).

Some animals living in cultivated forests can be considered cultivated (for example, livestock grown in forests, game introduced for replenishment in wooded areas for hunting, and so forth). However, in general, fauna of forests is regarded as non-produced. The same applies to flora other than trees. By definition, the natural growth of a produced/cultivated biological asset is the increase in value of an individual specimen during a given year as the result of biological development.

7.7.4 Physical Accounting

1. **Land and land use accounts:** Land accounting is an important aspect in the SEEA, as it is closely linked with several environmental concerns, including soil erosion, changes in land use, and so forth. In the absence of monetary valuation of environmental functions or uses of forests in the more practical versions of the SEEA and indeed the present manual, land accounting provides a measure of the changes in forest from which possible environmental consequences may be derived.

2. **NRAs for forests:** Changes of stocks are due to natural growth, natural losses, and gross removals. These accounts can be subdivided by species (conifers, broadleaves), age classes, or other structural parameters. As far as possible, forest accounts should be compiled by species, type of forest (for example, cultivated, non-cultivated, high forest, coppice, and so forth), and by age.
3. **Commodity balances: uses of wood:** Commodity balances correspond, although in a slightly different format, to the SEEA_s physical accounts for the flows of products and raw materials. They show physical inputs and outputs of wood and wood products in the economy.

7.7.5 Monetary Accounting: Valuation and Aggregation

Monetary accounts are obtained by valuing the physical forest resource accounts, using the following valuation techniques.

1. Valuation of land
2. Valuation of standing timber
3. Valuation of biological non-cultivated assets (other than timber)
4. Segregation of SNA flows

   Such transactions consist of the following:

1. Output related to forest land
2. Costs related to the output: cost of plantations, access roads, felling, and so forth
3. Expenditures for protection of forest
4. Calculation of EDP

   Two types of costs can be considered in the calculation of EDP:

1. Environmental depletion and degradation costs at market value
2. Environmental costs at maintenance cost

REFERENCES

[1] UN, Integrated Environmental and Economic Accounting, United Nations, New York, 1993. Interim version (Sales No. E93 XVII. UN 12).
[2] K.H. Alfsen, T. Bye, L. Lorentsen, Natural Resource Accounting and Analysis, the Norwegian Experience 1978-1986, Central Bureau of Statistics, Oslo, 1987.
[3] J. Theys, Environmental accounting in development policy: the French experience, in: Y.J. Ahmad, S. El Serafy, E. Lutz (Eds.), Environmental Accounting for Sustainable Development, The World Bank, Washington, D.C., 1989.
[4] C. Stahmer, M. Kuhn, N. Braun, Physical Input-Output Tables for Germany, 1990, Eurostat Working Paper No. 2/1998/B/1, Eurostat, Luxembourg, 1998.
[5] A. Steurer, Material flow accounting and analysis: where to go at a European level, in: Eurostat (Ed.), Material Flow Accounting – Experience of Statistical Institutes in Europe, 1997.
[6] A. Born, Accounting for Depletion of Natural Assets in the 1993 SNA, STD/NA/RD 72 (97) (1997).

[7] S. El Serafy, The proper calculation of income from depletable natural resources, in: Y.J. Ahmad, S. ElSerafy, E. Lutz (Eds.), Environmental Accounting for Sustainable Development, The World Bank, Washington, D.C., 1989.
[8] United Nations, Integrated Environmental and Economic Accounting (Sales No. E.93. XVII.12), 1993.
[9] United Nations Environment Program, Section I and II (Social and Economic Dimensions), CONF.151/26 Vol. I (1992).

FURTHER READING

[1] P. Bartelmus, C. Stahmer, J.V. Tongeren, Integrated environmental and economic accounting system: framework for a SNA satellite system, Review of Income and Wealth 37 (2) (1991) 111–148.
[2] C. Brandon, K. Hommann, The Cost of Inaction: Valuating the Economy-wide Cost of Environmental Degradation in India, The World Bank, Asia Environment Division, Washington, D.C., 1995.
[3] K. Chopra, G. Kadekodi, Natural Resource Accounting in the Yamuna Basin: Accounting for Forest Resources, IEG Monograph, New Delhi, 1997.
[4] G.S. Haripriya, Forest resource accounting for the state of Maharashtra in India, Development Policy Review 16 (2) (1998) 131–151.
[5] G.S. Haripriya, Accounting for the forest resources in the national accounts in India, Environmental and Resource Economics 19 (1) (2001) 73–95.
[6] M.N. Murty, S.C. Gulati, A. Banerjee, Hedonic Property Prices and Valuation of Benefits from Reducing Urban Air Pollution in India, Working Paper No. E 237/2003, Institute of Economic Growth, Delhi, 2003.
[7] M.N. Murty, S. Kumar, Environmental and Economic Accounting for Industry, Oxford University Press, New Delhi, 2004.
[8] R.M. Solow, The economics of resources or the resources of economics, American Economic Review 64 (1974) 1–14.

CHAPTER EIGHT

Environmental Risk Assessment

8.1 INTRODUCTION

Risk is broadly defined as the likelihood that a harmful consequence will occur as the result of an action or condition. It involves the combined evaluation of hazards and exposure. Environmental risk assessment (ERA) deals with the interactions of agents or hazards, humans, and ecological resources. It describes human populations, ecological resources, and agents; analyzes agents and exposure potential; characterizes the potential for adverse effects; defines uncertainties; generates options to deal with the risks; and communicates information about the risks to humans and ecosystems. ERA has two components; human health risk assessment and ecological risk assessment. The stages of doing an ERA include: hazard identification and problem formulation, analysis, and risk characterization. The main outputs are the risk management and communication plans.

8.2 USE OF RISK ASSESSMENT IN ENVIRONMENTAL MANAGEMENT

Risk assessment and management approaches to environmental issues are increasingly being used at all levels of policy and regulation. The techniques have a wide range of applications, including:

- The design of regulation; for instance in determining "acceptable" risk levels which may form the basis of environmental standards;
- Providing a basis for site-specific decisions; for instance in land-use planning or siting of hazardous installations;
- Prioritization of environmental risks; for instance in the determination of which chemicals to regulate first;
- Comparison of risks; for instance to enable comparisons to be made between the resources being allocated to the control of different types of risk, or to allow risk substitution decisions to be made.

ERA has traditionally been a function of policy and regulatory agencies and most development has taken place in these fields. ERA is becoming

Environmental Management
ISBN 978-0-12-811989-1
http://dx.doi.org/10.1016/B978-0-12-811989-1.00008-7

more common in industry partly as a result of the use of ERA in regulation. The risks examined in the assessment can be physical, such as radiation; biological, such as a genetically modified organism or pathogen; or chemical, such as an immuno-toxic substance. The target/receptor to be examined in the risk assessment can vary. Many ecological risk assessments can be considered single species, since only a few types of representative organisms are selected as assessment end-points. The end-points can be mortality or morbidity in human health assessments, or other single species assessments. For some ecological risk assessments, end-points may be extinction or total catch.

ERA is used in industry in the following:

- Compliance with legislation
- Product safety
- Financial planning
- Site-specific decision-making
- Prioritization and evaluation of risk reduction measures

8.3 HAZARD AND RISK

A *hazard* represents a chemical, physical, or biological substance that has the potential to produce harm to health if it is present in the environment and comes into contact with people. The hazardous properties of an environmental agent are defined according to the nature and severity of its harmful consequences. Fortunately, many hazards can be either contained or avoided, so not every potential environmental hazard poses an actual health risk.

A *risk* is defined as the likelihood of adverse health effects arising from exposure to a hazard in a human population, which is conceptually expressed as the product of two factors; the probability of exposure and the severity of the consequences. Some risks can be measured directly by observing past and present disease incidence patterns in the human population. Risks can be calculated indirectly, by estimating the theoretical level of human exposure and the potential severity of health effects as predicted by experimental studies.

$$\text{Risk} = \text{probability of exposure} \times \text{severity of consequences}$$

- Three activities, *risk estimation*, *risk evaluation*, and *risk control*, collectively comprise the core of the decision process called *risk management*.

An effective risk management process also requires ongoing consultation among all concerned parties, or *stakeholders*, to resolve questions of policy, science, and societal concerns. The *risk communication* process provides an essential coordinating function, by supporting the activities of information exchange and mutual consultation among various stakeholders. governmental, non-governmental, and private-sector organizations throughout the various phases of the risk management process. Risk assessment p*rovides a valuable tool to inform decision-making* about uncertain future outcomes. One of its strengths is that it *can explicitly* account for uncertainties about future outcomes. In addition, the risk assessment process can be *informed by dialogue* with stakeholders, which can aid decision-making. Risk assessment *can help Strategic Environmental Assessment (SEA)* and sustainability appraisal by providing a framework to evaluate economic, social, and physical outcomes (including impacts on human health and the environment) of proposed policies, plans, and programs.

The typology is shown in Fig. 8.1 and breaks environmental risk assessment into:

- Human health risk assessment
- Ecological risk assessment
- Applied industrial risk assessment

ERA evaluates environmental status to predict future consequences of exposure to hazards and other actions in urban–industrial situations. The chemical industry and other heavy industries make good use of ERA to evaluate the human health and ecological risks that new chemicals and other industrial products may entail if used in various production, commerce, trade, consumption, and disposal, i.e., throughout the chemical product's entire life cycle. Fig. 8.2 shows the components of human health risk and ecological risk.

ERA will include a number of steps:

- Problem formulation
- Hazard identification
- Release assessment
- Exposure assessment
- Consequence assessment

The typology for industrial application for risk assessment is based on use of the method rather than the type of method.

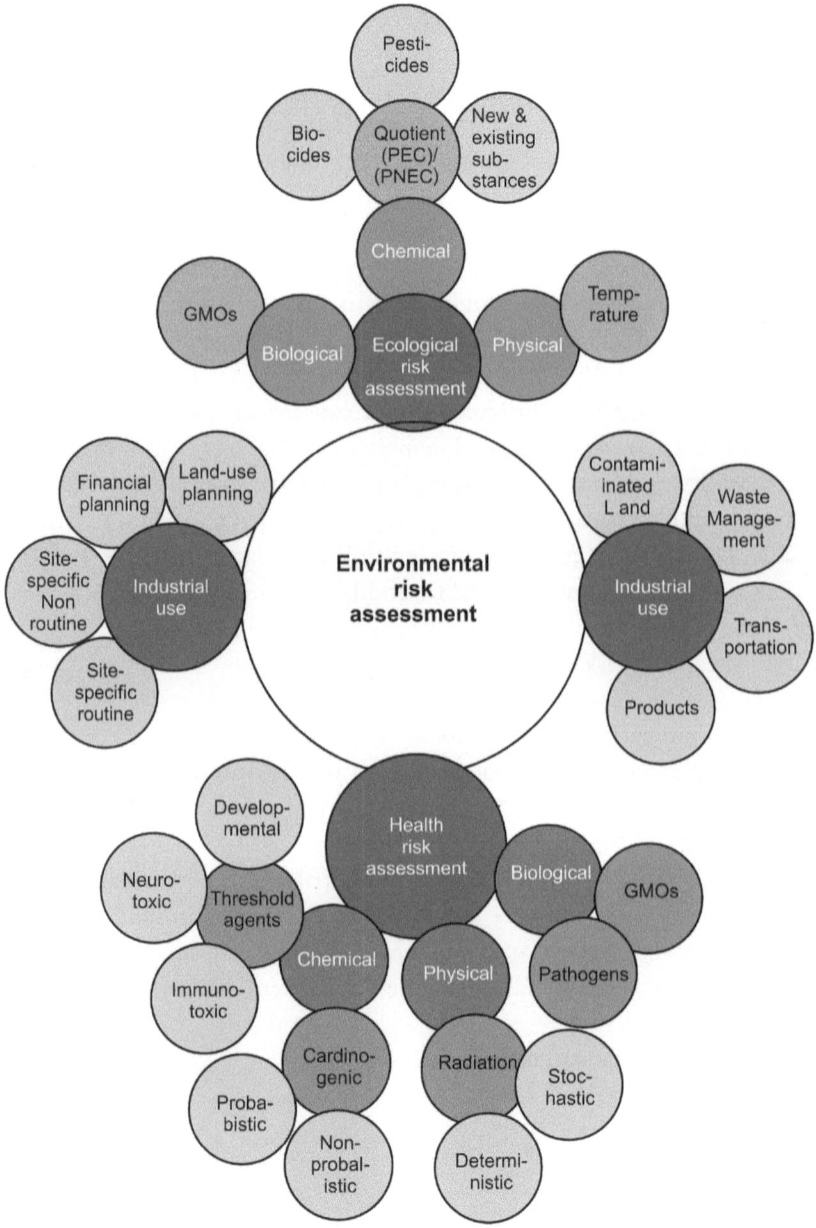

Figure 8.1 A typology of risk assessment. Reproduced from *Fairman, R. and Mead, C. 1996 Approaches to Risk Assessment*.

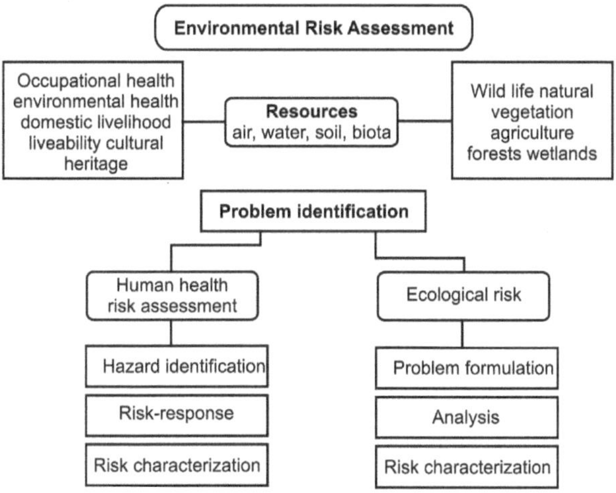

Figure 8.2 Components of human health risk and ecological risk.

8.4 PROCESS OF ENVIRONMENTAL RISK ASSESSMENT AND MANAGEMENT

Environmental risk management is a four-stage process. Risk management decision-making is an iterative process, so stages may need to be revisited as new information comes to light. The flow chart for risk assessment and management is given as follows:

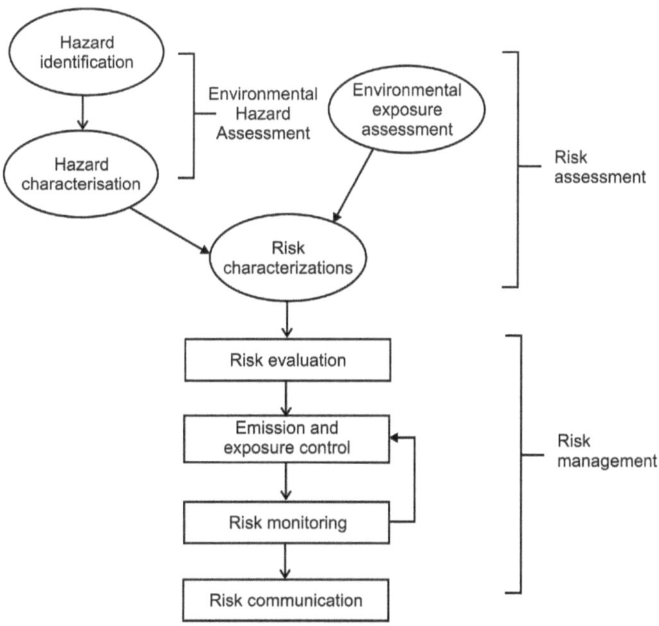

Flow chart for environmental risk assessment and management of chemicals.

8.4.1 Stage 1: Problem Formulation

The problem should be clearly set out along with any constraints on the assessment and the final risk management decision and its implementation. Describing the problem in clear and unambiguous terms will assist in selecting the level or type of risk assessment required (e.g., qualitative/quantitative), and ensure that risk management decisions are as robust as possible.

Problem formulation involves the following:

- Setting out the problem in clear and unambiguous terms
- Defining the spatial and temporal boundaries of the problem
- Identifying all constraints on the assessment, including those imposed by specific legislative requirements
- Starting to document areas of uncertainty and assumptions
- Developing a conceptual understanding ("model") of all sources of hazard, all exposed receptors, and the pathways linking them should be developed at this stage. Without a source–pathway–receptor linkage in place, harm cannot occur.

Hazard identification is defined as, "the identification of the adverse effects which a substance has an inherent capacity to cause" [1]. This involves consultation of any toxicological and epidemiological data.

8.4.1.1 Dose–Response Assessment

Dose–response assessment is the "estimation of the relationship between dose, or level of exposure to a substance, and the incidence and severity of an effect" (CEC, 1993). The dose–response relationship is ascertained from epidemiological and toxicological data.

8.4.1.2 Exposure Assessment

Exposure assessment is described as the "determination of the emissions, pathways and rates of movement of a substance and its transformation and degradation in order to estimate the concentration/doses to which human populations or environmental compartments are or may be exposed" (CEC, 1993). Environmental exposure to chemicals can be direct; as a result of emission to the environment (air, land, water) of a substance through industrial manufacture, use or disposal, or indirect; through drinking water or the food chain. The agents that affect humans and the environment are given in Fig. 8.3.

Identifying and characterizing the inherent properties of chemical substances is basically the first step of ERA. Environmental hazard assessment involves gathering or generating and evaluating data of chemical

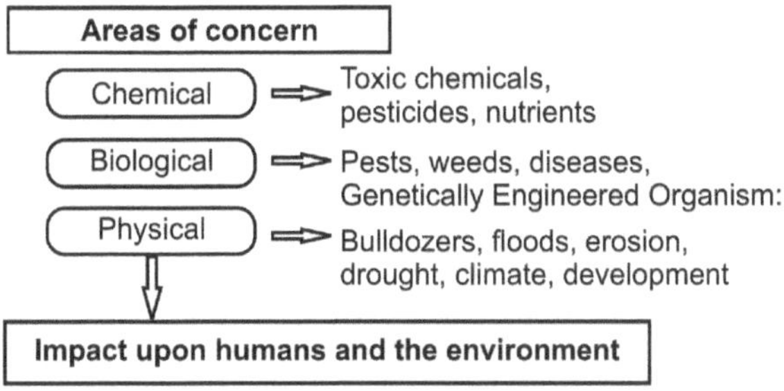

Figure 8.3 Agents that affect humans and the environment.

substances and concluding on their inherent eco-toxicological effects and environmental fate. It does not mean that generating data for every kind of such data is always necessary in the first step of environmental risk assessment.

If hazard information is not available for all relevant end-points, data gaps can be filled through testing or non-testing methods. It is important to note that a data gap is not always a data need, e.g., if a target chemical is insoluble in water, fish toxicity testing may be inappropriate and thus not required.

On the other hand, environmental risk assessment has many problems which need resolution, such as determining the effects at population and community level; selection of end-points; and selection of indicative species.

8.4.2 Stage 2: Risk Assessment

Risk assessments are undertaken to determine the significance of the risk(s) and the need for management action to prevent or limit the risk. Assessments should start at a simple level and can become more sophisticated subject to the nature and complexity of the risks under assessment and the decision-making needs. Using a tiered approach should ensure that the complexity of risk assessment is proportionate to the risk (Fig. 8.4). It is important to be open about uncertainty and, where feasible, define and agree the approaches to be used with relevant stakeholders. Risk assessment takes account of the "probability" of an event and the magnitude of the "consequences" of the event. Risks associated with an event are assessed using a source-pathway-receptor model. Table 8.1 gives some examples of possible source–pathway–receptor linkages for flooding.

The first tier is called scoping, because identification of potential significant issues is stressed with minimal attention to effects. The possible toxicity or possible adverse effects of each hazard identified is only broadly assumed. The scoping tier will determine if the concerns under

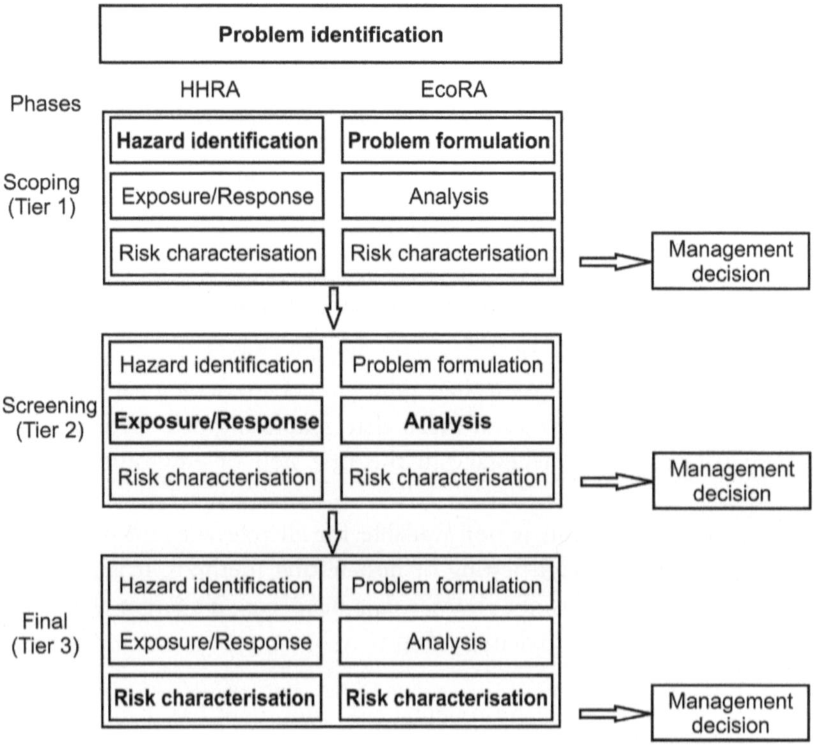

Figure 8.4 The iterative nature of environmental risk management.

Table 8.1 Flood Risk Management Examples of Sources Pathways and Receptors

Sources	Pathways	Receptors
• Rainfall • Sea level • Waves • River flows	• Overtopping and failure of flood defences • Breaching of coastal defences and beaches • Failure of defense components such as barriers and gates • Inundation of floodplains • Inadequate urban drainage	• People • Domestic and commercial property • Infrastructure • Ecosystems

"worst-case" scenario are serious enough. However, if unacceptable risks are indicated, the ERA proceeds to the next tier (called screening), requiring more sophisticated techniques and making more realistic assumptions. The screening tier calls for a revisit of the current hazard information and better estimates of the probable effects. This will now involve more thorough evaluation of the exposure and the hazard, using modeling and extrapolation from additional data sources such as toxicity (eco-toxicological data and dose–response relationships) and environmental pathways.

Consequences can be characterized in a number of ways, including physical "harm," economic, and social impacts. In practice, it is likely that assessments will focus on the spatial and temporal nature of impacts on fauna, flora, and natural resources including factors such as latency and reversibility, and taking account of the sensitivity of receptors. Economic consequences might include impact on resource "value," while social factors such as equity of risk distribution, familiarity, or dread might need to be considered. Risk characterization, is the qualitative, and wherever possible, quantitative determination of the probability of occurrence of the adverse effects of chemicals to the environment under predicted exposure conditions. This process is based on outcomes of the previous steps, i.e., environmental hazard and environmental exposure assessment.

The final stage is the evaluation of the significance of the risk which involves placing it in a context for example with respect to an environmental standard or other criterion defined in legislation, statutory, or good practice guidance.

Illustrative Qualitative Risk Assessment

The probability of the receptors being exposed to the hazard. Example categories:

High
Medium
Low
Very Low

The consequences of a hazard being realized. Example categories:

High
Medium
Low
Very Low

The significance of the risk is determined by combining the probability with the magnitude of the potential consequences. A matrix such as the one that follows can be used to categorize the risk as high, medium, low, or very low.

In each assessment, the meanings of high, medium, low, and very low need to be determined and agreed using a descriptive or numerical scale.

Probability

H	M	M	H	H
M	L	M	M	H
L	L	L	M	M
VL	VL	L	L	M
	VL	**L**	**M**	**H**

Consequences →

The level of risk identified can be used to identify the need for further assessment or appropriate risk management measures.

Pyramid (as follows) of effects using the example of carbon monoxide poisoning, biochemical, and cellular changes at the lowest levels of biological function are invisible, but they may become more visible if higher levels of organization become affected.

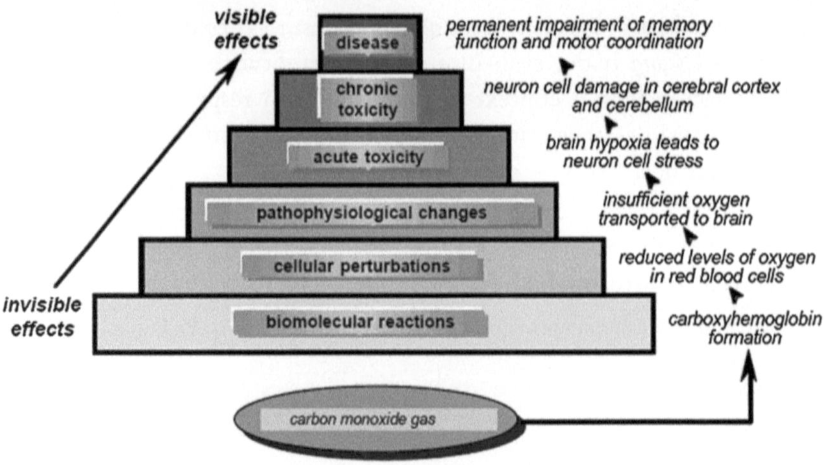

8.4.3 Stage 3: Selection of Preferred Risk Management Technique

The next stage of the process considers the options available for managing the risk, i.e., what measures are to be taken to prevent or control the identified risk. It is likely that more than one possible risk management technique or action is available, including further refinement of the risk assessment options appraisal is used to provide a balanced assessment of each option so that the most appropriate option can be identified.

The options appraisal process involves identifying the advantages and disadvantages of the risk management options (including, environmental, social, economic, technological, and management considerations) and ranking them. Various methods of completing this task are available, including cost–benefit analysis, trade-off analysis and multi-criteria analysis. The output from options appraisal will inform the risk management decision, which must take account of the nature and magnitude of the risks, the level of uncertainty, and any relevant legislative requirements. This stage can inform the assessment and selection of strategic alternatives and mitigation measures by identifying means to address the risks associated with the source–pathway–receptor linkages identified for assessment at the problem formulation stage.

8.4.4 Stage 4: Implementation of Preferred Risk Management Technique

The preferred risk management options are implemented and monitored to ensure that they are effective in reducing risks to acceptable levels. Again, the consideration of how the source–pathway–receptor linkage should be managed is key, e.g., using flooding as an example, it would be possible to:

- Address the source (difficult in this case for rainfall, but action to address climate change may reduce the future severity, catchment management);
- Block or alter pathways (install flood defenses, set up sustainable drainage systems);
- Remove receptors (steer development away from areas that flood, use flood warning and evacuation).

Risk management consists of four elements, i.e., risk evaluation, emission and exposure control, risk monitoring, and risk communication.

8.5 RISK EVALUATION

The first step to risk management, risk evaluation, consists of determining whether the risk(s) identified at the risk assessment stage need to be mitigated. This can be done quantitatively or qualitatively, taking into consideration relevant laws, regulations and policies, societal values, relevant program objectives, and socioeconomic aspects. The objective of this step is to determine whether control measures need to be taken to address specific risks identified at the risk assessment stage. In case of chemical risk to either humans or ecosystems, risks can be ranked to assist decisions and management options may be subjected to any of the three decisions (Fig. 8.5).

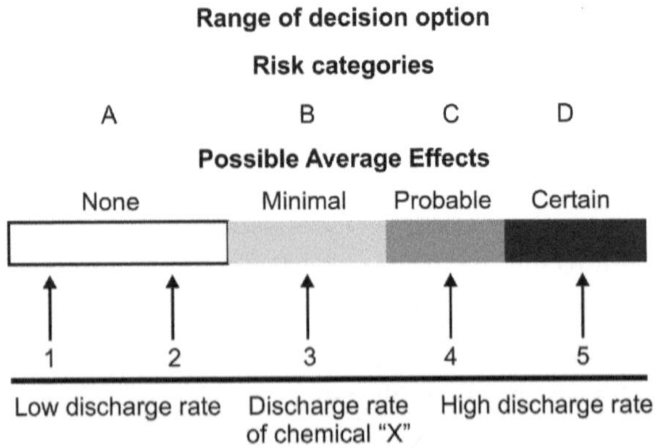

Figure 8.5 Evaluating possible adverse effects.

Two major classes of normative societal issues are considered in the risk evaluation step:

- *Economic evaluation*: Estimating the expected health benefits and anticipated costs of control associated with varying degrees of reduction in risk, using monetary criteria which are amenable to quantitative economic analysis. Among the methods most commonly used are the following analytical techniques:
  - Cost-effectiveness analysis
  - Benefit-cost analysis
  - Risk-benefit analysis
  - Socioeconomic impact analysis
- *Social evaluation*: Characterizing the social issues that reflect value judgments and societal preferences which are not amenable to formal economic analysis, as well as the factors that influence political perceptions of equity and fairness.

8.6 EMISSION AND EXPOSURE CONTROL

Once risk evaluation has been completed and further risk mitigation is necessary, the next step is to take measures to control emission and exposure of chemicals for protecting humans and/or the environment. The process includes identifying and analyzing options for controlling risks to select the most appropriate measure(s) and to implement them.

8.7 RISK MONITORING

Risk monitoring plays an important role in environmental and human risk management with the aim of checking that risk mitigation or reduction has worked effectively. The result of risk monitoring is used as a basis for consideration of further risk mitigation options. The plan should include:

- How and when the risk management strategy will be carried out;
- The roles, responsibilities, and accountabilities of individuals and organizations;
- Plans for communication with involvement of interested and affected parties;
- Criteria that will be used for monitoring and evaluation; and
- Training, staffing, and financing requirements.

Monitoring and review is an essential and integral step in the risk management process. Through monitoring, the actual impacts, benefits, and costs of a risk management strategy can be compared with estimates made earlier in the risk management process. The benefits of a monitoring step in risk management include:

- The identification of new or changing risks;
- The accumulation of evidence to support assumptions and results of analyses;
- The development of a more accurate portrait of the risks; and
- Reduction of costs.

8.8 RISK COMMUNICATION

Risk communication is the interactive exchange of information about risks among risk assessors, managers, news media, interested groups, and the general public. Stakeholders play an important role in risk communication (Fig. 8.5). Risk communication planning must appreciate that decision-making in ERA is difficult, because there are trade-offs to be made among competing objectives and perspectives.

Risk communication is defined as any two-way communication between stakeholders about the existence, nature, form, severity, or acceptability of risks. It is vitally important to understand the basic concepts of risk communication and to ensure that communication among stakeholders (Fig. 8.6) is integral to the risk management process. A stakeholder analysis provides the decision-maker with a profile of potential stakeholders for consideration in decision-making and communication processes.

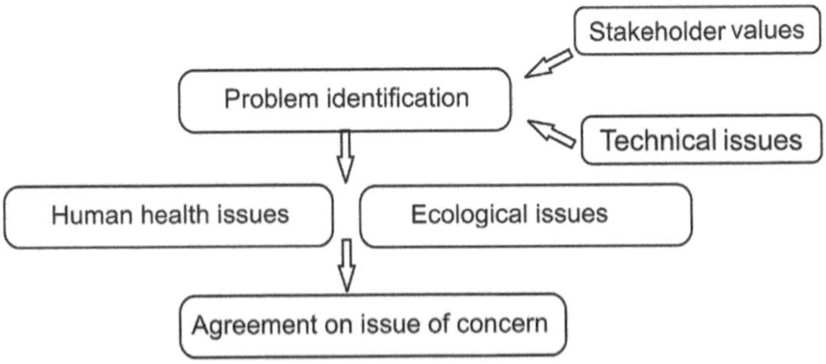

Figure 8.6 Stakeholder involvement.

Risk Communication Tasks in the Risk Management Process

Risk Management Step	Risk Communication Task
Initiation	• Identify stakeholders • Consult with stakeholders in defining scope of issue
Preliminary analysis	• Develop stakeholder analysis for ongoing verification and refinement
Risk estimation	• Discussion of source, exposure issues • Communication of results with stakeholders • Assess changes in knowledge/perception in light of new information
Risk evaluation	• Elicit stakeholder perceptions of the risks and benefits, and the reasons for these, if possible • Assess stakeholder acceptability of the risk
Risk control	• Consult with stakeholders to gain input into identifying and evaluating control options • Inform stakeholders of chosen risk control and financing strategies • Inform stakeholders of benefits, costs, and any new risks associated with proposed control options • Evaluate acceptance of control options and residual risks • Determine if risk trade-offs might be possible
Implementation (Action)	• Communication of risk control decision and implementation
Monitoring	• Ensure implementation of communication strategies • Monitor changes in needs, issues, and concerns of existing or new stakeholders

8.9 DEALING WITH UNCERTAINTY

Uncertainty can arise throughout any risk assessment and risk management process. Types of uncertainty can include:

- Model: Models may not be accurate or complete.
- Environmental: Natural variability may not represent conceptual model assumptions.
- Knowledge: Scientific data may be incomplete.
- Sample: Sample measurements may be inaccurate or the validity may be queried.
- Data: Data may be extrapolated or interpolated from other sources.
- Scenario: Scenarios might not fully describe the problem.

Uncertainty characterization involves the qualitative description of the criteria that lead to the selection or rejection of specific data, scenarios, assumptions, and functional relationships between variables. Although uncertainty cannot be eliminated, it can sometimes be reduced by collecting more information. It is important to acknowledge all uncertainties, including data gaps and assumptions. Evaluating different sources of uncertainty can help with understanding how much confidence can be placed in the risk assessment and aid robust decision-making.

Methods for analyzing uncertainty range from simple to complex. A simple sensitivity analysis can, for example, be undertaken to examine the behavior of a model by measuring the variation in outputs resulting from changes to its inputs.

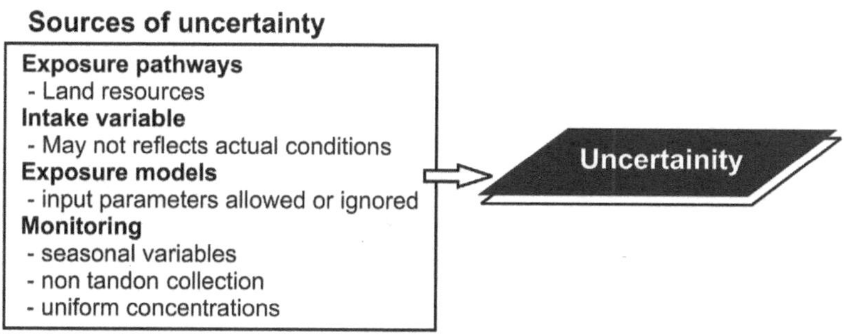

Several risk researchers have identified a system of classification for four major sources of uncertainty in risk estimation:

- Model uncertainty
- Parameter uncertainty
- Decision-rule uncertainty
- Natural variability

Model uncertainty reflects the limited ability of mathematical models to accurately represent the real world. Mis-specification of model form and the functional relationships between critical variables often is the result of lack of sufficient theoretical knowledge to adequately define the structural and operational characteristics of the model.

Parameter uncertainty is attributable both to statistical uncertainty arising in the estimation of model parameters due to measurement error and sampling error, or alternatively from systematic errors arising from biased sampling or flawed experimental design.

Decision-rule uncertainty takes the form of imprecise or inappropriate operational definitions for desired outcome criteria, value parameters, and decision variables. These include the selection of particular types of summary statistics for outcome measures (e.g., lifetime mortality risk versus annual mortality risk), and the choice of variables that express subjective value judgments in the form of utility functions (e.g., the monetary value attributed to loss of life).

Natural variability arises from many inherently random factors that must be considered in a risk assessment. These include demographic factors such as age and sex distributions in the exposed population, or individual variability in terms of susceptibility to health risks.

8.10 DOCUMENTATION

It is important that good documentation is maintained throughout the risk assessment and management process. This will form an audit trail to record assumptions, areas of uncertainty, information sources, and justification for decisions. Good documentation is also essential for communication of risk management decisions, and should take account of the potential audience needs.

8.11 ADVANTAGES AND DISADVANTAGES OF RISK ASSESSMENT

Advantages of Risk Assessment:

- A technique which can weigh-up information that is basically in different "languages"
- A mechanism to aid decision-making
- A basis for effective risk communication
- A method for highlighting and prioritizing research needs

Disadvantages of Risk Assessment:
- Possible over-reliance and over-confidence in results
- Narrow focus on parts of a problem, rather than the whole

8.12 ROLE OF GIS SOFTWARE APPLICATIONS FOR ENVIRONMENTAL RISK MANAGEMENT

Environmental risk management (ERM) needs a multi-disciplinary approach, with input and expertise required from many fields; civil and chemical engineering, physics, life sciences, ecology, geology, hydrology, and statistics being some of them. Wide ranges of simple to complex, spatial as well as non-spatial, and quantitative as well as qualitative input data sets are used in ERA and analysis process. The ERM process involves preparation and use of the processed information derived and presented in various ways, for example, comparative (or relative) risk analysis, cost-benefit analysis, scenario analysis, probabilistic analysis, decision matrix, sensitivity analysis, etc. Due to the need for using and analyzing a large volume of the spatial as well as non-spatial environmental hazards and exposure data in a fast and reasonably accurate way, Geographic Information Systems (GIS)-based software applications, using a variety of modeling techniques, serve as powerful tools for effective ERA and ERM. Such applications can be used for a diverse ERA and analysis purposes. These applications can range from development of databases/inventory systems for simple to complex GIS layers overlays, to complex spatial decision-making systems for study of the impact of air, water, and soil pollution, ecological imbalance, and natural disasters on the natural and man-made environment, including living beings, properties, infrastructure, vegetation, and ecology. These systems could also be interlinked with other related systems, providing online and real-time input data feeds or communication systems, to allow continuous monitoring and tracking of environmental risks in an integrated way.

REFERENCE

[1] CEC 1993 Commission Directive 93/67/EEC, OJ L227/10, 8.9.1993, Commission of the European Communities, Brussels.

FURTHER READING

[1] Environmental Risk Assessment: Dealing with Uncertainty in Environmental Impact Assessment, ADB Environment Paper No. 7, Asian Development Bank, Manila, Philippines, 1990.

[2] P. Calow, C. Streatfield, DIY Environmental Risk Profile, Sheffield Regional Green Business Club, 1995.

[3] E.J. Calabrese, L.A. Baldwin, Performing Ecological Risk Assessments, Lewis Publishers, Michigan, 1993.
[4] C.P.B. Claudio, Risk analysis in developing countries, Guest Editorial, Risk Analysis (1988).
[5] 93/67/EEC on Risk Assessment of New Notified Substances and Commission Regulation (EC) No. 1488/94 on Risk Assessment for Existing Substances. Parts I-IV, Office of Official Publications of the European Communities, 1996.
[6] A. Fairbrother, L.A. Kapustka, B.A. Williams, J. Glicken, Risk assessment in practice: success and failure, Human and Ecological Risk Assessment 1 (1995) 367–375.
[7] GEF/UNDP/IMO, Malacca Straits: Initial Risk Assessment. Regional Programme for the Prevention and Management of Marine Pollution in the East Asian Seas, 1997. Manila, Philippines.
[8] L.A. Kaputska, B.A. Williams, The conceptual basis for assessing ecological risk from incineration facilities, in: Presented at the 84th Air and Waste Management Association Meeting, Vancouver, B.C. Canada, 1991.
[9] Metropolitan Environmental Improvement Program, DENR-World Bank-UNDP, Process Documentation: Community Participation and Advocacy for Ecological Waste Management in Three Barangays and the Public Market in the Paco District of Manila, Department of Environment and Natural Resources, Republic of the Philippines, 1994.
[10] National Research Council-USA. Committee on the Institution of Means for Assessment of Risks to Public Health, Risk Assessment in the Federal Government: Managing the Process, National Academy Press, Washington D.C., 1983.
[11] National Research Council-USA, Science and Judgement in Risk Assessment, National Academy of Sciences, Washington D.C., 1994.
[12] Office of Technology Assessment, Screening and testing chemicals in commerce, in: Congress of the United States, Office of Technology Assessment, Washington D.C., USA, 1995.
[13] K.R. Smith, R.A. Carpenter, M.S. Faulstich, Risk Assessment from Hazardous Chemical Systems in Developing Countries, East-West Environment and Policy Institute Occasional Paper No. 5, East-West Center, Honolulu, 1988.
[14] G.W. Suter, Ecological Risk Assessment, Lewis Publishers, Florida, 1993.
[15] United Nations Centre for Human Settlements, Sustainable Cities: Concepts and Applications of a United Nations Programme. Commission on Sustainable Development, UNCHS (Habitat), Nairobi, 1994.
[16] United Nations Environment Programme - International Environmental Technology Centre, Environmental Risk Assessment for Sustainable Cities. IETC Technical Publication 3, UNEP-IETC, Osaka, 1996.
[17] United States Environmental Protection Agency, Framework for Ecological Risk Assessment. U.S. Environmental Protection Agency. EPA/630/R-92/001, 1992.
[18] United States Environmental Protection Agency, Proposed Guidelines for Ecological Risk Assessment, U.S. Environmental Protection Agency, 1996. EPA/630/R-95/002B.
[19] United States Environmental Protection Agency, Guidance on Cumulative Risk Assessment. Part I. Planning and Scope, U.S. Environmental Protection Agency, Science Policy Council, 1997.

CHAPTER NINE

Energy Management and Audit

9.1 INTRODUCTION

The fundamental goal of energy management is to produce goods and provide services with the least cost and environmental effect. One definition of energy management is: "the judicious and effective use of energy to maximize profits (minimize costs) and enhance competitive positions." Another comprehensive definition is: "the strategy of adjusting and optimizing energy, using systems and procedures so as to reduce energy requirements per unit of output while holding constant or reducing total costs of producing the output from these systems." The objective of energy management is to achieve and maintain optimum energy procurement and utilization throughout the organization, and:

- To minimize energy costs/waste without affecting production and quality.
- To minimize environmental effects.

Energy audit is the key to a systematic approach for decision-making in the area of energy management. It attempts to balance the total energy inputs with its use, and serves to identify all the energy streams in a facility. It quantifies energy usage according to its discrete functions. Industrial energy audit is an effective tool in defining and pursuing comprehensive energy management program. As per the Energy Conservation Act, 2001, energy audit is defined as, "the verification, monitoring and analysis of use of energy including submission of technical report containing recommendations for improving energy efficiency with cost–benefit analysis and an action plan to reduce energy consumption". The energy audit would give a positive orientation to the energy cost reduction, preventive maintenance, and quality control programs, which are vital for production and utility activities. Such an audit program will help to keep focus on variations which occur in the energy costs, availability, and reliability of supply of energy, decide on appropriate energy mix, identify energy conservation technologies, retrofit for energy conservation equipment, etc. In general, energy audit is the translation of conservation ideas into realities, by lending technically feasible

Environmental Management
ISBN 978-0-12-811989-1
http://dx.doi.org/10.1016/B978-0-12-811989-1.00009-9

solutions with economic and other organizational considerations within a specified time frame.

The primary objective of energy audit is to determine ways to reduce energy consumption per unit of product output or to lower operating costs. Energy audit provides a "benchmark" for managing energy in the organization, and also provides the basis for planning a more effective use of energy throughout the organization.

9.2 TYPES OF ENERGY AUDIT

Energy audit can be classified into the following two types:

1. *Preliminary audit*: Preliminary energy audit is a relatively quick exercise to:
   a. Establish energy consumption in the organization
   b. Estimate the scope for saving
   c. Identify the most likely (and the easiest) areas for attention
   d. Identify immediate (especially no or low-cost) improvements/savings
   e. Set a "benchmark"
   f. Identify areas for more detailed study/measurement
   g. Preliminary energy audit uses existing, or easily obtained, data
2. *Detailed audit*: A detailed audit provides a comprehensive energy project implementation plan for a facility, since it evaluates all major energy using systems. This type of audit offers the most accurate estimate of energy savings and cost. It considers the interactive effects of all projects, accounts for the energy use of all major equipment, and includes detailed energy cost saving calculations and project cost. In a detailed audit, one of the key elements is the energy balance. This is based on an inventory of energy using systems, assumptions of current operating conditions, and calculations of energy use. This estimated use is then compared to utility bill charges. Detailed energy auditing is carried out in three phases: phase I, II, and III.

   Phase I: Pre-audit phase
   Phase II: Audit phase
   Phase III: Post-audit phase

The type of energy audit to be performed depends on:

- Function and type of industry
- Depth to which final audit is needed, and
- Potential and magnitude of cost reduction desired

9.3 TEN-STEP METHODOLOGY FOR ENERGY AUDIT

A comprehensive ten-step methodology for conducting an energy audit at field level is given in the following Table 9.1 .

Table 9.1 Methodology for Conducting Energy Audit

Step Number	Plan of Action	Purpose/Results
	Phase I: Pre-audit phase	
Step 1	• Plan and organize • Walk through audit • Informal interview with energy manager, production/ plant manager	• Resource planning, establish/ organize an energy audit team • Organize instruments and time frame • Macro data collection (suitable to type of industry) • Familiarization of process/plant activities • Firsthand observation and assessment of current level operation and practices
Step 2	• Conduct of brief meeting/ awareness program with all divisional heads and persons concerned (2-3 hours)	• Building up cooperation • Issue questionnaire for each department • Orientation, awareness creation
	Phase II: Audit phase	
Step 3	• Primary data gathering, process flow diagram, and energy utility diagram	• Historic data analysis, baseline data collection • Prepare process flow charts • All service utilities system diagram (Example: Single line power distribution diagram, water, compressed air and steam distribution). • Design, operating data and schedule of operation • Annual energy bill and energy consumption pattern (refer manual, log sheet, name plate, interview)
Step 4	• Conduct survey and monitoring	• Measurements: Motor survey, insulation, and lighting survey with portable instruments for collection of more and accurate data Confirm and compare operating data with design data.

Continued

Table 9.1 Methodology for Conducting Energy Audit—cont'd

Step Number	Plan of Action	Purpose/Results
Step 5	• Conduct of detailed trials/ experiments for selected energy guzzlers	• Trials/experiments: • 24-hour power monitoring (MD, PF, kWh etc.) • Load variations trends in pumps, fan compressors etc. • Furnace efficiency trials equipments performance experiments, etc.
Step 6	• Analysis of energy use	• Energy and material balance and energy loss/waste analysis
Step 7	• Identification and development of energy conservation (ENCON) opportunities	• Identification and consolidation of ENCON measures • Conceive, develop, and refine ideas • Review the previous ideas suggested by unit personnel • Review the previous ideas suggested by energy audit, if any • Use brainstorming and value analysis techniques • Contact vendors for new/efficient technology
Step 8	• Cost–benefit analysis	• Assess technical feasibility, economic viability and prioritization of ENCON options for implementation • Select the most promising projects • Prioritize by low, medium, and long-term measures
Step 9	• Reporting and presentation to the top management	• Documentation, report presentation to the top management
	Phase III: Post-audit phase	
Step 10	• Implementation and follow-up	Assist and implement ENCON recommendation measures and monitor the performance • Action plan, schedule for implementation • Follow-up and periodic review

9.4 AUDIT PHASES

9.4.1 Phase I: Pre-Audit Phase Activities

A structured methodology to carry out an energy audit is necessary for efficient working. An initial study of the site should always be carried out, as the planning of the procedures necessary for an audit is most important.

9.4.1.1 Initial Site Visit and Preparation Required for Detailed Auditing

An initial site visit may take one day, and gives the energy auditor/engineer an opportunity to meet the personnel concerned, to familiarize him or her with the site, and to assess the procedures necessary to carry out the energy audit. During the initial site visit, the energy auditor/engineer should carry out the following actions:

- Discuss with the site's senior management the aims of the energy audit.
- Discuss economic guidelines associated with the recommendations of the audit.
- Analyze the major energy consumption data with the relevant personnel.
- Obtain site drawings where available – building layout, steam distribution, compressed air distribution, electricity distribution, etc.
- Tour the site accompanied by engineering/production.

The main aims of this visit are:

- To finalize energy audit team.
- To identify the main energy consuming areas/plant items to be surveyed during the audit.
- To identify any existing instrumentation/additional metering required.
- To decide whether any meters will have to be installed prior to the audit, e.g., kWh, steam, oil, or gas meters.
- To identify the instrumentation required for carrying out the audit.
- To plan with time frame.
- To collect macro data on plant energy resources, major energy consuming centers.
- To create awareness through meetings/program.

9.4.2 Phase II: Detailed Energy Audit Activities

Depending on the nature and complexity of the site, a comprehensive audit can take from several weeks to several months to complete. Detailed studies to establish, and investigate, energy and material balances for specific plant departments or items of process equipment are carried out. Whenever possible, checks of plant operations are carried out over extended periods of time, at night and on weekends as well as during normal daytime working hours, to ensure that nothing is overlooked. The audit report should include a description of energy inputs and product outputs by major department or by major processing function, and needs to evaluate the efficiency of each step of the manufacturing process. Means of improving these efficiencies have to be listed, and at least a preliminary assessment of the cost of the improvements needs to be made to indicate the expected payback on any capital investment needed. The audit report should conclude with specific recommendations for detailed engineering studies and feasibility analyses, which must then be performed to justify the implementation of those conservation measures that require investments.

The information to be collected during the detailed audit includes:

1. Energy consumption by type of energy, by department, by major items of process equipment, and by end-use.
2. Material balance data (raw materials, intermediate and final products, recycled materials, use of scrap or waste products, production of by-products for re-use in other industries, etc.).
3. Energy cost and tariff data.
4. Process and material flow diagrams.
5. Generation and distribution of site services (e.g., compressed air, steam).
6. Sources of energy supply (e.g., electricity from the grid or self-generation).
7. Potential for fuel substitution, process modifications, and the use of co-generation systems (combined heat and power generation).
8. Energy management procedures and energy awareness training programs within the establishment.

Existing baseline information and reports are useful to get consumption pattern, production cost, and productivity levels in terms of product per raw material inputs. The audit team should collect the following baseline data:

- Technology, processes used, and equipment details
- Capacity utilization
- Amount and type of input materials used
- Water consumption
- Fuel consumption
- Electrical energy consumption
- Steam consumption
- Other inputs such as compressed air, cooling water, etc.

- Quantity and type of wastes generated
- Percentage rejection/reprocessing
- Efficiencies/yield

9.4.3 Process Flow Diagram and List Process Steps; Identify Waste Streams and Obvious Energy Wastage

An overview of unit operations, important process steps, areas of material and energy use, and sources of waste generation should be gathered and represented in a flow chart. Existing drawings, records, and shop floor walk-through will help in making this flow chart. Simultaneously, the team should identify the various inputs and output streams at each process step.

9.4.4 Identification of Energy Conservation Opportunities

Fuel substitution: Identifying the appropriate fuel for efficient energy conversion.

Energy generation: Identifying efficiency opportunities in energy conversion equipment/utility such as captive power generation, steam generation in boilers, thermic fluid heating, optimal loading of DG sets, minimum excess air combustion with boilers/thermic fluid heating, optimizing existing efficiencies, efficient energy conversion equipment, biomass gasifiers, co-generation, high efficiency DG sets, etc.

Energy distribution: Identifying efficiency opportunity networks such as transformers, cables, switch gears, and power factor improvement in electrical systems and chilled water, cooling water, hot water, compressed air, etc.

Energy usage by processes: This is where there is major opportunity for improvement. Process analysis is a useful tool for process integration measures.

9.4.4.1 Technical and Economic Feasibility

The technical feasibility should address the following issues:

- Technology availability, space, skilled manpower, reliability, service, etc.
- The impact of energy efficiency measure on safety, quality, production, or process.
- The maintenance requirements and spares availability.

The economic viability often becomes the key parameter for the management acceptance. The economic analysis can be conducted by using a variety of methods. For example; payback method, internal rate of return method, net present value method, etc. For low-investment, short-duration measures, which have attractive economic viability, the simplest of the methods, payback, is usually sufficient. A sample worksheet for assessing economic feasibility is provided as follows:

Sample Worksheet for Economic Feasibility
Name of Energy Efficiency Measure

1. Investment	2. Annual operating costs	3. Annual savings
a. Equipments	• Cost of capital	• Thermal Energy
b. Civil works	• Maintenance	• Electrical Energy
c. Instrumentation	• Manpower	• Raw material
d. Auxiliaries	• Energy	• Waste disposal
	• Depreciation	

Net Savings /Year (Rs./year) = (Annual savings-annual operating costs)	Payback period in months = (Investment/net savings/year) × 12

9.5 CLASSIFICATION OF ENERGY CONSERVATION MEASURES

Based on an energy audit and analyses of the plant, a number of potential energy-saving projects may be identified. These may be classified into three categories:

1. Low cost –high return;
2. Medium cost–medium return;
3. High cost–high return.

Projects relating to energy cascading and process changes almost always involve high costs coupled with high returns, and may require careful scrutiny before funds can be committed. These projects are generally complex and may require long lead times before they can be implemented. Table 9.2 lists the project priority guidelines.

Table 9.2 Project Priority Guidelines

Priority	Economical Feasibility	Technical Feasibility	Risk/Feasibility
A: Good	Well-defined and attractive	Existing technology adequate	No risk/highly feasible
B: Maybe	Well-defined and only marginally acceptable	Existing technology may be updated, lack of confirmation	Minor operating risk/may be feasible
C: Held	Poorly defined and marginally unacceptable	Existing technology is inadequate	Doubtful
D: No	Clearly not attractive	Need major breakthrough	Not feasible

9.6 ENERGY AUDIT REPORTING FORMAT

After successfully carring out an energy audit, the energy manager/ energy auditor should report to the top management for effective communication and implementation. A typical energy audit reporting content and format are given in the following. The following format is applicable for most of the industries. However, the format can be suitably modified for specific requirement applicable for a particular type of industry.

9.6.1 Detailed Energy Audit

Table of contents:

Acknowledgment
Executive summary
Energy audit options at a glance and recommendations
Introduction about the plant
General plant details and descriptions
Energy audit team
Component of production cost (Raw materials, energy, chemicals, manpower, overhead, and others)
Major energy use and areas
Production process description
Brief description of manufacturing process
Process flow diagram and major unit operations
Major raw material inputs, quantity, and costs
Energy and utility system description
List of utilities
Brief description of each utility
Electricity
Steam
Water
Compressed air
Chilled water
Cooling water
Detailed process flow diagram and energy and material balance
Flow chart showing flow rate, temperature, and pressures of all input–output streams
Water balance for entire industry
Energy efficiency in utility and process systems

Specific energy consumption
Boiler efficiency assessment
Thermic fluid heater performance assessment
Furnace efficiency analysis
Cooling water system performance assessment
DG set performance assessment
Refrigeration system performance
Compressed air system performance
Electric motor load analysis
Lighting system
Energy conservation options and recommendations
List of options in terms of no cost/low cost, medium cost, and high investment cost, annual energy and cost savings (CS), and payback
Implementation plan for energy saving measures/projects
Annexure
List of energy audit worksheets
List of instruments
List of vendors and other technical details

Reporting Format for Energy Conservation Recommendations

A: Title of recommendation	Combine Diesel Generator (DG) set cooling tower with main cooling tower
B: Description of existing system and its operation	Main cooling tower is operating with 30% of its capacity. The rated cooling water flow is 5000 m³/h. Two cooling water pumps are in operation continuously with 50% of its rated capacity. A separate cooling tower is also operating for DG set operation continuously.
C: Description of proposed system and its operation	The DG set cooling water flow is only 240 m³/h. By adding this flow into the main cooling tower, will eliminate the need for a separate cooling tower operation for DG set, besides improving the percentage loading of main cooling tower. It is suggested to stop the DG set cooling tower operation.
D: Energy saving calculations	
Capacity of main cooling tower	5000 m³/h
Temperature across cooling tower (design)	8°C
Present capacity	3000 m³/h
Temperature across cooling tower (operating)	4°C

Reporting Format for Energy Conservation Recommendations

% loading of main cooling tower	$(3000\times4)/(5000\times8)=30\%$
Capacity of DG set cooling tower	$240\,m^3/h$
Temperature across the tower	5°C
Heat load $(240\times1000\times1\times5)$	1200,000 kCal/h
Power drawn by the DG set cooling tower	
Number of pumps and its rating	2 nos × 7.5 kW
Number of fans and its rating	2 nos × 22 kW
Power consumption @ 80% load	$(22\times2+7.5\times2)\times80=47\,kW$
Additional power required for main cooling tower for additional water flow of $240\,m^3/h$ (66.67 L/s) with $6\,kg/cm^2$	$(66.67\times6)/(102\times0.55)=7\,kW$
Net energy savings	47–7 = 40 kW
E: Cost benefits	
Annual energy saving potential	40 kW × 8400 h = 3,36,000 Units per year
Annual cost savings	3,36,000 × Rs.4.00 = Rs.13.4 Lakh per year
Investment (only cost of piping)	Rs 1.5 Lakhs
Simple payback period	Less than 2 months

Understanding energy cost is vital factor for awareness creation and savings calculation. In many industries, sufficient meters may not be available to measure all the energy used. In such cases, invoices for fuels and electricity will be useful. The annual company balance sheet is the other sources where fuel cost and power are given with production-related information. Energy invoices can be used for the following purposes:

- They provide a record of energy purchased in a given year, which gives a baseline for future reference.
- Energy invoices may indicate the potential for savings when related to production requirements, etc.
- When electricity is purchased on the basis of maximum demand tariff.
- They can suggest where savings are most likely to be made.
- In later years, invoices can be used to quantify the energy and CS made through energy conservation measures

9.6.2 Fuel Costs

A wide variety of fuels are available for thermal energy supply. A few are listed as follows:

- Fuel oil
- Low sulfur heavy stock (LSHS)

- Light diesel oil (LDO)
- Liquefied petroleum gas (LPG)
- Coal
- Lignite
- Wood, etc.

Understanding fuel cost is fairly simple and it is purchased in tons or kiloliters. Availability, cost, and quality are the main three factors that should be considered while purchasing. The following factors should be taken into account during procurement of fuels for energy efficiency and economics.

- Price at source, transport charge, type of transport
- Quality of fuel (contaminations, moisture, etc.)
- Energy content (calorific value)

9.6.3 Power Costs

Benchmarking of energy consumption internally (historical/trend analysis), and externally are two powerful tools for performance assessment and logical evolution of avenues for improvement. Well-documented historical data helps to bring out energy consumption and cost trends month-wise/day-wise. Trend analysis of energy consumption, cost, relevant production features, and specific energy consumption help in understanding effects of capacity utilization on energy use efficiency and costs on a broader scale. External benchmarking relates to inter-unit comparison across a group of similar units. However, it would be important to ascertain similarities, as otherwise findings can be grossly misleading. A few comparative factors, which need to be looked into while benchmarking externally, are:

- Scale of operation
- Vintage of technology
- Raw material specifications and quality
- Product specifications and quality

Benchmarking energy performance permits:

- Quantification of fixed and variable energy consumption trends vis-à-vis production levels
- Comparison of the industry energy performance with respect to various production levels (capacity utilization)
- Identification of best practices (based on the external benchmarking data)
- Scope and margin available for energy consumption and cost reduction
- Basis for monitoring and target-setting exercises

The benchmark parameters can be:

- Gross production related: [kWh/MT clinker or cement produced (cement plant), kWh/kg yarn produced (textile unit), kWh/MT, kCal/kg, paper produced (paper plant), kCal/kWh Power produced (heat rate of a power plant), million kilocals/MT urea or ammonia (fertilizer plant), kWh/MT of liquid metal output (in a foundry)]
- Equipment/utility related: [kW/ton of refrigeration (on air conditioning plant), % thermal efficiency of a boiler plant, % cooling tower effectiveness in a cooling tower, kWh/NM^3 of compressed air generated, kWh/L in a diesel power generation plant.]

While such benchmarks are referred to, related crucial process parameters need mentioning for meaningful comparison among peers. For instance, in the preceding case:

1. For a cement plant; type of cement, blaine number (fineness), i.e. Portland and process used (wet/dry) are to be reported alongside kWh/MT figure.
2. For a paper plant; paper type, raw material (recycling extent), Grams per Square Meter (GSM) quality are some important factors to be reported along with kWh/MT, and kCal/Kg figure.
3. For a fertilizer plant; capacity utilization (%) and on-stream factor are two inputs worth comparing while mentioning specific energy consumption.
4. For a boiler plant; fuel quality, type, steam pressure, temperature, and flow are useful comparators alongside thermal efficiency, and, more importantly, whether thermal efficiency is on gross calorific value basis or net calorific value basis or whether the computation is by direct method or indirect heat loss method, may mean a lot in a benchmarking exercise for meaningful comparison.

9.7 PLANT ENERGY PERFORMANCE

Plant energy performance (PEP) is the measure of whether a plant is now using more or less energy to manufacture its products than it did in the past; a measure of how well the energy management program is doing. It compares the change in energy consumption from 1 year to the other considering production output. Plant energy performance monitoring compares plant energy use at a reference year with the subsequent years to determine the improvement that has been made. However, a plant production output may vary from year to year, and the output has a significant bearing on plant energy use. For a meaningful comparison, it is necessary to determine the energy that would have been required to produce this year production output, if the plant had operated in the same way as it did during the reference year. This calculated value may then be compared with the actual value to determine the improvement or deterioration that has taken place since the reference year.

9.7.1 Production Factor

Production factor is used to determine the energy that would have been required to produce this year's production output if the plant had operated in the same way as it did in the reference year. It is the ratio of production in the current year to that in the reference year.

$$\text{Production} = \frac{\text{Current year's production}}{\text{Reference year's production}}$$

9.7.2 Reference Year Equivalent Energy Use

The reference year's energy use that would have been used to produce the current year's production output may be called the "reference year energy use equivalent" or "reference year equivalent." The reference year equivalent is obtained by multiplying the reference year energy use by the production factor:

$$\text{Reference year equivalent} = \text{Reference year energy use} \times \text{Production factor}$$

The improvement or deterioration from the reference year is called "energy performance" and is a measure of the plant's energy management progress. It is the reduction or increase in the current year's energy use over the reference, and is calculated by subtracting the current year's energy use from the reference years equivalent. The result is divided by the reference year equivalent and multiplied by 100 to obtain a percentage:

$$\text{Plant energy performance} = \frac{\text{Reference year equivalent} - \text{current year's energy}}{\text{Reference year equation}} \times 100$$

The energy performance is the percentage of energy saved at the current rate of use compared to the reference year rate of use. The greater the improvement, the higher the number will be.

9.7.3 Monthly Energy Performance

Experience, however, has shown that once a plant has started measuring yearly energy performance, management wants more frequent performance information in order to monitor and control energy use on an on-going basis. PEP can just as easily be used for monthly reporting as yearly reporting.

9.8 ENERGY AUDIT INSTRUMENTS

The requirement for an energy audit such as identification and quantification of energy necessitates measurements; these measurements require the use of instruments. These instruments must be portable, durable, easy to operate, and relatively inexpensive. The parameters generally monitored during energy audit may include the following: Basic electrical parameters in AC and DC systems; voltage (V), current (I), power factor, active power (kW), apparent power (demand) (kVA), reactive power (kVAr), energy consumption (kWh), frequency (Hz), harmonics, etc. Parameters of importance other than electrical such as temperature and heat flow, radiation, air and gas flow, liquid flow, revolutions per minute (RPM), air velocity, noise and vibration, dust concentration, total dissolved solids (TDS), pH, moisture content, relative humidity, flue gas analysis; CO_2, O_2, CO, SOx, NOx, combustion efficiency, etc.

9.8.1 Key Instruments for Energy Audit

1. *Electrical measuring instruments*: These are instruments for measuring major electrical parameters such as kVA, kW, PF, Hertz, Amps, and volts. In addition, some of these instruments measure harmonics. These instruments are applied online. Instant measurements can be taken with handheld meters, while more advanced ones facilitate cumulative readings at specified intervals.
2. *Combustion analyzer*: This instrument has built-in chemical cells which measure various gases such as O_2, CO, NO_x and SO_x.
3. *Fuel efficiency monitor*: This measures oxygen and temperature of the flue gas. Calorific values of common fuels are fed into the microprocessor which calculates the combustion efficiency.
4. *Fyrite*: A hand bellow pump draws the flue gas sample into the solution inside the fyrite. A chemical reaction changes the liquid volume revealing the amount of gas. A separate fyrite can be used for O_2 and CO_2 measurement.

5. *Contact thermometer*: These are thermocouples which measure, for example, flue gas, hot air, hot water temperatures by insertion of probe into the stream.
6. *Infrared thermometer*: This is a non-contact type measurement which when directed at a heat source directly gives the temperature read out. This instrument is useful for measuring hot spots in furnaces, surface temperatures, etc.
7. *Pitot tube and manometer*: Air velocity in ducts can be measured using a pitot tube and inclined manometer for further calculation of flows.
8. *Water flow meter*: This non-contact flow-measuring device uses Doppler effect/ultrasonic principle. There is a transmitter and receiver, which are positioned on opposite sides of the pipe. The meter directly gives the flow.
9. *Speed measurements*: In any audit exercise speed measurements are critical as they may change with frequency, belt slip, and loading. A simple tachometer is a contact type instrument which can be used where direct access is possible.
10. *Leak detectors*: Ultrasonic instruments are available which can be used to detect leaks of compressed air and other gases which are normally not possible to detect with human abilities.
11. *Lux meters*: Illumination levels are measured with a lux meter. It consists of a photocell which senses the light output, converts to electrical impulses which are calibrated as lux.

More problems are likely with the accuracy of the energy audits for industrial and manufacturing facilities than for smaller commercial facilities or even large buildings since the equipment and operation of industrial facilities is more complex. However, many of the same problems are discussed here in terms of industrial and manufacturing facilities can occur in audits of large commercial facilities and office buildings.

9.9 CALCULATING ENERGY AND DEMAND BALANCES

The energy and demand balances for a facility are an accounting of the energy flows and power used in the facility. These balances allow the energy analyst to track the energy and power inputs and outputs and see whether they match. Analysts should perform energy and demand balance on a facility before developing and analyzing any energy management recommendations.

Once the complete energy and demand balances are constructed for the facility, we check to see if the cumulative energy/demand for these categories plus the miscellaneous category is substantially larger or smaller than the actual energy usage and demand over the year.

To help solve the problem of overestimating savings from using the average cost of electricity, we divide our energy savings calculations into a demand savings and an energy savings.

In most instances, the energy savings for a particular piece of equipment is calculated by first determining the demand savings for that equipment, then multiplying by the total operating hours of the equipment. To calculate the annual CS, we use the following formula:

$$\begin{aligned} CS = & [\text{Demand Savings} \times \text{Average Monthly Demand Rate} \times 12\text{mos per year}] \\ & + [\text{Energy Savings} \times \text{Average Cost of Electricity without Demand}] \end{aligned}$$

To demonstrate the difference in savings estimates, consider replacing a standard 30 hp motor with a high-efficiency motor. The efficiency of a standard 30 hp motor is 0.901 and a high efficiency motor is 0.931. Assume the motor has a load factor of 40% and operates 8760 h per year (three shifts). Assume also that the average cost of electricity is \$0.068/kWh, the average demand cost is \$3.79/kW/mo, and the average cost of electricity without demand is \$0.053/kWh. The equation for calculating the demand of a motor is:

$$D = HP \times LF \times 0.746 \times 1/Eff$$

The savings on demand (or demand reduction) from installing a high-efficiency motor is:

$$\begin{aligned} DR &= HP \times LF \times 0.746 \times (1/Eff_S - 1/Eff_H) \\ &= 30\text{ hp} \times 0.40 \times 0.746\text{ kW/hp} \times (1/0.901 - 1/0.931) \\ &= 0.32\text{ kW} \end{aligned}$$

The annual energy savings is:

$$\begin{aligned} ES &= DR \times H \\ &= 0.32\text{ kW} \times 8760\text{ h per year} \\ &= 2803.2\text{ kWh per year} \end{aligned}$$

Using the average cost of electricity above, the cost savings (CS_1) is calculated as:

$$\begin{aligned} CS1 &= ES \times (\text{Average cost of electricity}) \\ &= 2803.2\text{ kWh per year} \times \$0.068/\text{kWh} \\ &= \$190.62\text{ per year} \end{aligned}$$

Using the preceding recommended formula:

$$\begin{aligned} CS = & \text{ [Demand Savings} \times \text{Average Monthly Demand Rate} \\ & \times 12 \text{ mos per year } + \text{ [Energy Savings} \times \text{Average Cost of} \\ & \text{Electricity without Demand]} \\ = & \ (0.32 \text{ kW} \times \text{S}3.79/\text{mo} \times 12 \text{ mos per year}) + \\ & (2803.2 \text{ kWh/yr} \times \text{S}0.053/\text{kWh}) \\ = & \ (\text{S}14.55 + \text{S}148.57) \text{ per year} \\ = & \ \text{S}163.12 \text{ per year} \end{aligned}$$

In this example, using the average cost to calculate the energy CS over-estimates the CS by \$27.50 per year, or 17%. Although the actual amount is small for one motor, if this error is repeated for all the motors for the entire facility, as well as all other measures, which only reduce the demand component during the on-peak hours, then the cumulative error in cost savings predictions can be substantial.

9.10 ENERGY AUDIT FOR BUILDINGS

As mentioned earlier, the energy audit may range from a simple walk-through survey at one extreme to one that may span several phases. These phases include a simple walk-through survey, followed by monitoring of energy use in the building services, and then model analysis using computer simulation of building operation. The complexity of the audit is therefore directly related to the stages or degree of sophistication of the energy management program and the cost of the audit exercise.

The first stage is to reduce energy use in areas where energy is wasted and reductions will not cause disruptions to the various functions. The level of service must not be compromised by the reduction in energy consumed. It begins with a detailed, step-by-step analysis of the building's energy use factors and costs, such as insulation values, occupancy schedules, chiller efficiencies, lighting levels, and records of utility and fuel expenditures. It includes the identification of specific Energy Conservation and Commercialization (ECO), along with the cost-effective benefits of each one. The completed study would provide the building owner with a thorough and detailed basis for deciding which ECOs to implement, the magnitude of savings to be expected, and the energy conservation goals to be established and achieved in the energy management program. However, the ECOs may yield modest gains.

The second stage is to improve efficiency of energy conversion equipment and to reduce energy use by proper operations and maintenance. For this reason, it is necessary to reduce the number of operating machines and operating hours according to the demands of the load, and fully optimize equipment operations. Hence, the ECOs would include the following:

- Building equipment operation,
- Building envelope,
- Air-conditioning and mechanical ventilation equipment and systems,
- Lighting systems,
- Power systems, and
- Miscellaneous services.

The first two stages can be can be implemented without remodeling buildings and existing facilities. The third stage would require changes to the underlying functions of buildings by remodeling, rebuilding, or introducing further control upgrades to the building. This requires some investment. The last stage is to carry out large-scale energy reducing measures when existing facilities have past their useful life, or require extensive repairs or replacement because of obsolescence. In this case, higher energy savings may be achieved.

A walk-through survey of a building may reveal several ECOs to the experienced eye of the auditor. The survey could be divided into three parts.

9.10.1 Preliminary Survey

Prior to the walk-through survey, the auditor may need to know the building and the way it is used. The information can be obtained from:

- Architectural blueprints,
- Air-conditioning blueprints,
- Electrical lighting and power blueprints,
- Utility bills and operation logs for the year preceding the audit,
- Air-conditioning manuals and system data, and
- Building and plant operation schedules.

9.10.2 Walk-Through

Thus having familiarized with the building, the walk-through process could be relatively straightforward, if the blueprints and other preliminary information available describe the building and its operation accurately. The process could begin with a walk around the building to study the building envelope. Building features such as building wall color, external sun-shading devices, window screens and tint, and so on are noted as possible ECOs.

The survey inside the building would include confirmation that the air-conditioning system is as indicated on plans. Additions and alterations would be noted. The type and condition of the windows, effectiveness of window seals, typical lighting and power requirements, and occupancy and space usage are noted. This information could be compared against the recommendations in the relevant codes of practices.

System and plant data could be obtained by a visit to the mechanical rooms and plant room. Nameplate data could be compared against those in the building's documents, and spot readings of the current indicating panels for pumps and chillers recorded for estimating the load on the system.

9.10.3 Operator's Input

The auditor may discuss with the building maintenance staff further on the operating schedules and seek clarification on any unusual pattern in the trend of the utility bills. Unusual patterns such as sudden increase or decrease in utility bills could be caused by changes in occupancy in the building, or change in use by existing tenants. It is not uncommon for tenants to expand their computing operations that may increase the energy use significantly.

9.10.4 Report

ECOs could be found in measures such as:

- Reduce system operating hours,
- Adjust space temperature and humidity,
- Reduce building envelope gain,
- Adjust space ventilation rates and building exfiltration,
- Review system air and water distribution,
- Adjust chiller water temperatures, and
- Review chiller operations.

The benefit from adopting each ECO should be compared against cost of implementation. Caution should be exercised in the cost–benefit analysis given the wider range of certainty of the projections made. However, a survey at this level may be sufficient for small buildings.

9.10.5 Measurements

The capability of the energy auditor and the scope of an audit could be extended by the use of in place instrumentation and temporary monitoring equipment. In-place instrumentation refers to existing utility metering, air-conditioning control instrumentation, and energy management systems (EMS). The use of in-place utility metering and temporary monitoring equipment in energy auditing can yield valuable information about the building systems such as:

- Energy signature and end-use consumption analysis,
- Discovery and identification of ECOs,
- Quantification of energy use and misuse,
- Establishing bounds for potential energy reduction, and
- Data acquisition for further calculation and analysis.

9.10.6 Existing Information

Existing instrumentation such as utility meter readings and energy billings could be used to establish energy consumption patterns for the building. The regularity of consumption pattern is an indicator that no significant change in consumption occurred prior to the audit. This can also be used to check the validity of projections based on extrapolated short-term monitored data. Utility data could be used to establish useful indices such as kWh/m^2 per year to compare relative energy performance of buildings.

Air-conditioning control instrumentation such as chilled water temperature probes and water flow meters could be used to estimate cooling load demand and plant operation. For example, chilled water temperature outside the designed range may indicate that cooling coils may be operating under off-design conditions.

9.10.7 Short-Term Monitoring

The building may not be equipped for monitoring energy consumption and it may be necessary to install temporary measurement devices, such as instantaneous recorders (strip chart, data loggers, etc.) and totalizing recorders (kWH meters) to obtain data over the period of a week for the study. Monitored data is also useful for completing the energy model of a building for use in some building energy simulation software. For example, the total building energy consumption would include energy used in the vertical transportation system and potable water pumps which are not modeled in the software.

An estimate for annual consumption is extrapolated from the typical week consumption profile. Regularity of the weekly consumption profile means that the annual consumption could be estimated with confidence and the value used to cross-check with the annual energy bills.

9.10.8 Model Analysis

Building energy consumption in simplest terms is just the product of rate of consumption of a system and the period of operation. In lighting systems, its energy consumption could be determined manually with precision as it does not interact with other consumption variables. Energy consumption of cooling systems, however, is many times more complicated as it is affected by the internal heat gain within a building as well as weather variables, which varies in a complex manner over time.

Building model analysis using computers offers several improvements over manual calculations. These include:

- Precise schedule of building parameters,
- Precise determination of weather impact,
- Specification of part load performance of plant and equipment, and
- Consideration of parameter interactions such as lighting load on air-conditioning consumption.

9.10.9 Software

Some software permit hour-by-hour calculations of building consumption for the entire 8760 h of the year, but require thorough knowledge of the software to carry out accurate and meaningful analysis. Simplified software based on consumption analysis on characteristic days may also be considered. However, the improvements in computational power of the desktop PC has introduced several powerful features and user-friendly graphical interface possible in more recent versions of such software making it more accessible to the practicing engineer.

9.10.10 Analysis

The general procedure for an analysis would be to establish a model giving an annual consumption within 10% of the measured data. This establishes the base model. The impact of ECOs on energy consumption would be compared against the base model. ECOs could be considered singly or in combinations to determine interactions between them. The results of the energy savings in each analysis should not be taken as absolute but rather taken to be relative to the base run so as to give an indication of the order of magnitude of savings. Thus, those ECOs which show significant gains would be implemented.

9.11 SUMMARY

Energy auditing is not an exact science, but a number of opportunities are available for improving the accuracy of the recommendations. Techniques which may be appropriate for small-scale energy audits can introduce significant errors into the analyses for large complex facilities. The audit procedures can be expanded as needed in the various phases of the energy program, with the application of each succeeding phase yielding more information on energy use, and more opportunities for raising energy efficiency.

FURTHER READING

[1] Air-Conditioning & Refrigeration Institute, ARI Unitary Directory, (published every six months).

[2] S. Nadel, M. Shepard, S. Greenberg, G. Katz, A. de Almeida, Energy-Efficient Motor Systems: A Handbook on Technology, Program and Policy Opportunities, American Council for an Energy-Efficient Economy, 1992.

[3] Washington State Energy Office, MotorMaster Electric Motor Selection Software and Database, updated annually.

[4] R.K. Hoshide, Electric motor Do's and Don'ts, Energy Engineering 91 (1) (1994).

[5] W.L. Stebbins, Are you certain you understand the economics for applying ASD systems to centrifugal loads? Energy Engineering 91 (1) (1994).

[6] R.R. Vaillencourt, Simple solutions to VSD pumping measures, Energy Engineering 91 (1) (1994).

[7] G. Kempers, DSM pitfalls for centrifugal pumps and fans, Energy Engineering 92 (2) (1995).

[8] Personal communication with Mr. Mark Webb, Senior Engineer, Virginia Power Company, Roanoke, VA, January 1995. http://www.bdg.nus.edu.sg/BuildingEnergy/publication/papers/BCABEAud.pdf.

CHAPTER TEN

ISO 9000, 14000 Series, and OHSAS 18001

10.1 INTRODUCTION

Organizations are making efforts in adopting environmental management systems (EMS) due to various reasons like:

1. increasing legislation and its stringent application,
2. reduction of cost burden of raw materials, processes, energy, waste disposal, transport, and design for business resulting in increased efficiency,
3. conservation of biodiversity by securing a license to operate, improving stakeholder relationships, appealing to ethical consumers, ensuring sustainable growth, and attracting socially responsible investors,
4. corporate social responsibility (CSR) is the continuing commitment by business to behave ethically and contribute to economic development while improving the quality of life of the workforce and their families as well as of the local community and society at large. Capacity to innovate can be enhanced by both sustainability and CSR. Reputation and the importance of CSR and sustainability are clearly linked, particularly in developed, higher-income markets. Corporate reputation is directly linked as a driver of customer satisfaction, thus establishing a direct link between CSR and customer satisfaction. Key internal and external relationships are widely accepted as a source of competitive advantage that could be enhanced through stakeholder management.

EMS provides the systemic approach for effective environmental management. In the lines of the ISO 9000 series of quality standards, ISO 14001 pertains to environment management systems. "Environment management systems" is registered as an international standard. The British standard BS 7750 for EMS were published in January 1994. These standards do not establish any absolute requirements for environmental performance. However, all of them do require that the system be designed to measure improvement in environmental performance. The thrust is to have curative rather than preventive mechanisms in place.

Environmental Management
ISBN 978-0-12-811989-1
http://dx.doi.org/10.1016/B978-0-12-811989-1.00010-5

10.2 EMS CERTIFICATION

There have been several instances where it has been cited that environmental issues would become barriers to trade. Germany has imposed a ban on imports of textiles containing azo dyes and already has a packaging ordinance that would affect exporters. Developing nations, including India, are going to face increasing trade barriers from the developed countries because of environmental issues. Increasing cases of public litigation, stricter standards and enforcement, and growing public awareness are putting the industry under tremendous pressure. In this background, third-party EMS certification can enable companies to acquire the label of an environmentally sound enterprise, which carries conviction and would also favorably impact the consumer mind.

Reduced pollution means increased efficiency and less waste of resources. Improved health and safety conditions mean a happier, more productive workforce. In short, enterprises become more competitive when they practice good environmental management. EMS helps in responding to such needs of the customer.

Some of the major benefits of EMS that help in gaining a competitive edge over competitors are these:

- continuous improvement in the company's environmental performance
- responds to environmental needs now and for future
- recognizes, reduces, and eliminates the environmental problems before they occur
- brings total employee involvement
- brings a shift from end-of-the-pipe approach to top-of-the-pipe approach
- improves productivity and profitability

10.2.1 Benefits of EMS

Some of the benefits are listed next.

1. cost effectiveness: increased profits
2. increased efficiency
3. reduced wastage
4. increased competitiveness
5. increased access to world market
6. ease of legal compliance
7. improved relation with regulatory bodies

10.3 EMERGING TRENDS IN MANAGEMENT OF ENVIRONMENTAL ISSUES

10.3.1 Life Cycle Assessment

The complex interaction between a product and the environment is dealt with within the life cycle assessment (LCA) method. It is also known as the "life cycle analysis" or eco-balance. LCA systematically describes and assesses all flows to and from nature, from a cradle-to-grave perspective.

10.3.2 Environmental Impact Assessment (EIA)

Environmental impact assessment (EIA) is viewed as both science and art, reflecting the concern both with technical aspects of appraisal and the effects of EIA on the decision-making process. Adopted in many countries, with different degrees of enthusiasm, since its inception in the early 1970s, EIA has been established as a major procedure for assessing the environmental implications of legislation, the implementation of policy and plans, and the initiation of development projects. EIA is increasingly an essential part of environmental management.

10.3.3 Global Compact

The United Nations strongly encourages all vendors to actively participate in the Global Compact. The Global Compact is a voluntary international corporate citizenship network initiated to support the participation of both the private sector and other social actors to advance responsible corporate citizenship and universal social and environmental principles to meet the challenges of globalization. The Principles of the Global Compact includes environmental issues besides human rights and social issues:

Environment (principles 1–6 deal with social issues)

Principle 7: the implementation of a precautionary and effective program to environmental issues;

Principle 8: initiatives that demonstrate environmental responsibility;

Principle 9: the promotion of the diffusion of environmentally friendly technologies.

10.3.4 Triple Bottom Line

Businesses are using a range of approaches to achieve cost savings and improve environmental performance. Among these are cleaner productions, EMS, and "triple bottom line" (TBL) reporting. TBL reporting to stakeholders focuses on the economic, social, and environmental aspects of corporate activities.

10.4 ISO 14000 SERIES

ISO is a network of the national standards institutes of 148 countries, on the basis of one member per country, with a Central Secretariat in Geneva, Switzerland, that coordinates the system. ISO is a nongovernmental organization: its members are not, as is the case in the United Nations system, delegations of national governments. Nevertheless, ISO occupies a special position between the public and private sectors. This is because, on the one hand, many of its member institutes are part of the governmental structure of their countries, or are mandated by their government. On the other hand, other members have their roots uniquely in the private sector, having been set up by national partnerships of industry associations.

The ISO 9000 and ISO 14000 families are among ISOs most widely known standards ever. ISO 9000 and ISO 14000 standards are implemented across various countries.

ISO 9000 has become an international reference for quality management requirements in business-to-business dealings, and ISO 14000 is well on the way to achieving as much, if not more, in enabling organizations to meet their environmental challenges.

The ISO 9000 family is primarily concerned with "quality management."This means what the organization does to fulfill the following:

- the customer's quality requirements,
- applicable regulatory requirements, while aiming to enhance customer satisfaction,
- achieve continual improvement of its performance in pursuit of these objectives.

The ISO 14000 family is primarily concerned with "environmental management."This means what the organization does to:

- minimize harmful effects on the environment caused by its activities,
- achieve continual improvement of its environmental performance.

The vast majority of ISO standards are highly specific to a particular product, material, or process. However, the standards that have earned the ISO 9000 and ISO 14000 families a worldwide reputation are known as "generic management system standards."

"Generic" means that the same standards can be applied:

- to any organization, large or small, whatever its product
- including whether its "product" is actually a service, in any sector of activity,

- whether it is a business enterprise, a public administration, or a government department.

It also signifies that no matter what the organization's scope of activity, if it wants to establish a quality management system or an environmental management system, then such a system has a number of essential features for which the relevant standards of the ISO 9000 or ISO 14000 families provide the requirements.

"Management system" refers to the organization's structure for managing its processes—or activities—that transform inputs of resources into a product or service that meet the organization's objectives, such as satisfying the customer's quality requirements, complying to regulations, or meeting environmental objectives.

ISO 14001 is based on the Deming's model of Plant-Do-Check-Implement. The elements of ISO 14001 are these:

1. environmental policy
2. planning
3. implementation and operation
4. checking and corrective action
5. management review

10.4.1 Environmental Policy

The environmental policy is the driver for implementing and improving the organization's EMS so that it can maintain and potentially improve its environmental performance. The policy should therefore reflect the commitment of top management to compliance with applicable laws and continual improvement and provide the basis for setting of objectives and targets.

ISO 14001 defines an EMS audit as "a systematic, independent and documented process for obtaining audit evidence and evaluating it objectively to determine the extent to which the environmental management system audit criteria set by the organization are fulfilled."

There are different levels of audit:

- **first-party audit** (internal audits),
- **second-party audit** (customers or potential customers auditing EMS suppliers or a company auditing its suppliers or potential suppliers),
- **third-party audit** (independent certification audits by the certification bodies, e.g., Vexil, Lloyds, DnV, etc.).

14000 Guide to Environmental Management Principles, Systems and Supporting Techniques

14001 Environmental Management Systems—Specification with Guidance for Use
14010 Guidelines for Environmental Auditing—General Principles of Environmental Auditing
14011 Guidelines for Environmental Auditing—Audit Procedures—Part 1: Auditing of Environmental Management Systems
14012 Guidelines for Environmental Auditing—Qualification Criteria for Environmental Auditors
(ISO 14010, 14011, and 14012 have been withdrawn and replaced by ISO 19011:2002.)
14013/15 Guidelines for Environmental Auditing—Audit Programs, Reviews, and Assessments
14020/23 Environmental Labeling
14024 Environmental Labeling—Practitioner Programs Guiding Principles, Practices, and Certification Procedures of Multiple Criteria Programs
14031/32 Guidelines on Environmental Performance Evaluation
14040/43 Life Cycle Assessment General Principles and Practices
14050 Glossary
14060 Guide for the Inclusion of Environmental Aspects in Product Standards

10.4.2 ISO 14001: Key Requirements

The ISO 14001 EMS model subscribes to five key principles.

10.4.2.1 Principle 1: Commitment and Policy by Top Management

Initial environmental review (in the absence of existing data)

10.4.2.2 Principle 2: Planning

An organization should formulate a plan to fulfill its environmental policy.

- identification of environmental aspects and evaluation of associated environmental impacts
- legal and other requirements; environmental objectives and targets; environmental management program(s)

10.4.2.3 Principle 3: Implementation and Operation

The successful implementation of an environmental management system calls for the commitment of all employees of the organization. Hence, the responsibility for implementation has to be a line function. Employees and contractors need to be trained on the impacts their work might have on the environment. Documentation sufficient to describe the core elements of the environment management system, which may be integrated with other systems, should be available.

- resources—human, physical, and financial—accountability, and responsibility;
- knowledge skills and training: environmental awareness and motivation;
- environmental management system alignment and integration.

Support action: communication and reporting; environmental management system documentation; operational control; emergency preparedness and response.

10.4.2.4 Principle 4: Checking and Corrective Action

An organization should measure, monitor, and evaluate its environmental performance. The organization has to monitor the key characteristics of its operation and activities that can have a significant impact on the environment and establish records for the same. A periodic audit would help to determine whether the environmental management system conforms to planned arrangements and is being properly implemented and maintained.

- measurement and monitoring (ongoing performance);
- compliance to legal and other requirements;
- corrective and preventive action;
- environmental management system records and information management;
- audit of the environmental management system.

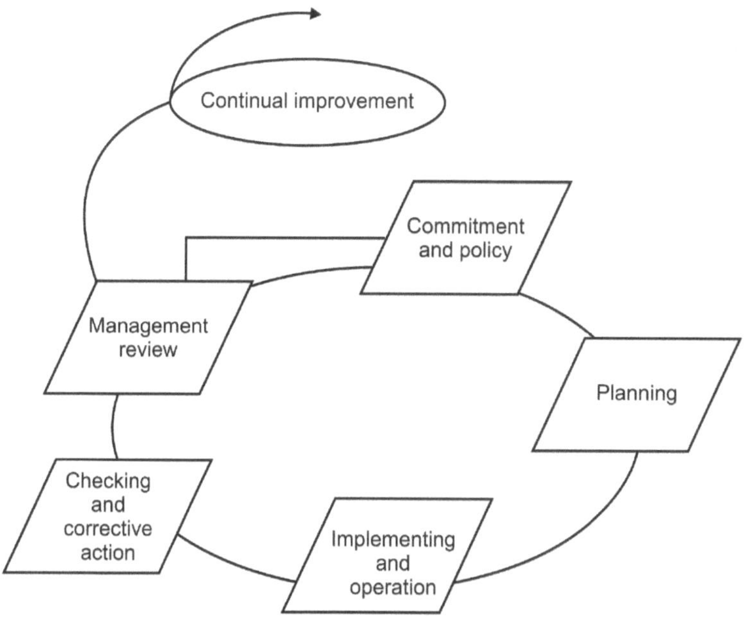

10.4.2.5 Principle 5: Review and Improvement

An organization should review and continually improve its environmental management system, with the objective of improving its overall environmental performance. The organization's top management at intervals reviews the environment system to ensure its continuing suitability, adequacy, and effectiveness. This review addresses the possible need for changes to policy, objectives, and other elements of the environmental management system.

ISO 14001:2004 specifies certain responsibilities of the management representative (MR).

- to establish, implement and maintain the EMS
- to report to top management on system performance as a basis for improvement
- MR also takes up liaison with external agencies, e.g., certificate body, consultants, etc.

The ISO 14000 model is shown next.

The ISO 14000 model

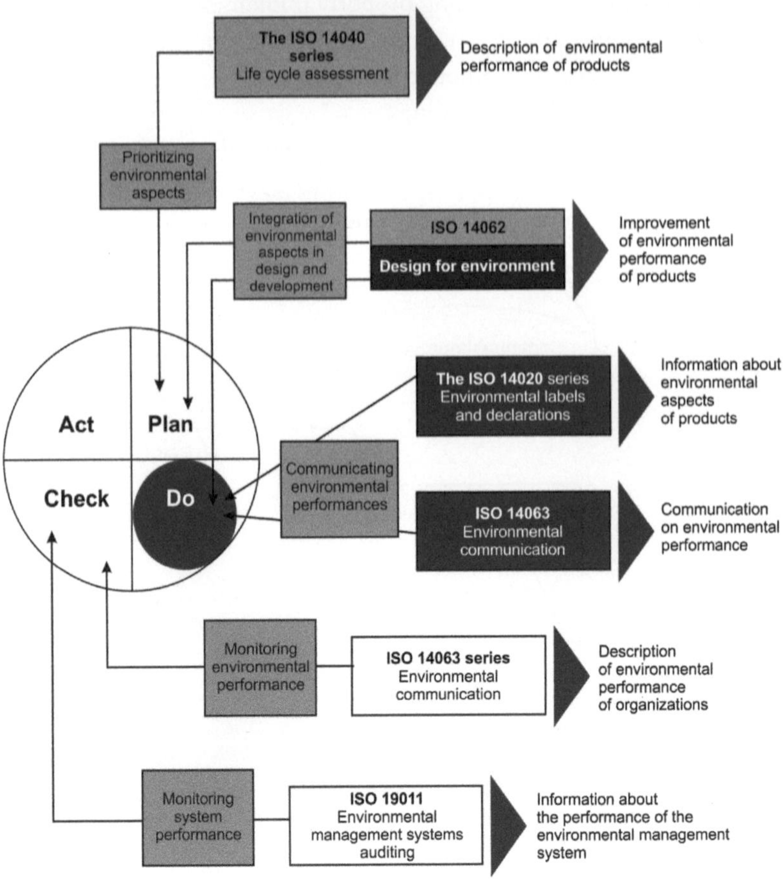

10.4.3 ISO 14001 Requirements

10.4.3.1 Clause 4.1: General Requirements

An organization needs to establish, document, implement, maintain, and continually improve its environmental management system to manage and control its activities, products, and services that do or can give rise to significant environmental impacts. An organization with no existing EMS should, initially, establish its current position with regard to the environment by means of a review. The aim of this review should be to consider all environmental aspects of the organization as a basis for establishing the EMS. In determining how this will be done the organization has to ensure conformance with the remaining clauses in Clause 4.

In addition, the 2004 version of the standard requires that the scope of the environmental management system is defined. Where this is done, it does not say but could be included in a top-level environmental manual. For example, care should be exercised in determining the level of documentation required, e.g., documented procedures, to avoid overly bureaucratic systems. In the definitions of Section 3 of ISO 14001:2004, the point is made that procedures can be documented or not (def.3.19). It is important to recognize the system to deliver improvements in environmental performance.

10.4.3.2 Clause 4.2: Environmental Policy

It is claimed that 90% of environmental problems are a result of a lack of an environmental policy and the lack of management systems to put it into practice. A policy should be good for business. If not, it will disintegrate when faced with commercial or economic pressures.

Basically, an environmental policy is a short written statement explaining your business position on the environment. It is the driver for implementing and improving an organization's EMS so that it can maintain and potentially improve its environmental performance. The contents of the policy should be kept reasonably "high level" to avoid undue complexity. It is therefore important to set a balance between the level of detail required to satisfy stakeholders and provide a framework for the setting of objectives.

Clause 4.2 of ISO 14001 also stipulates that the environmental policy should be sufficiently clear to be capable of being understood by internal and external interested parties and should be periodically reviewed and revised to reflect changing conditions and information.

10.4.3.3 Clause 4.3.1: Environmental Aspects

The relationship between an environmental aspect and an environmental impact is one of cause and effect, with an aspect leading to an impact.

The 2004 version states that this relates to aspects "within the scope of the EMS."

The following factors/subject areas should be taken into account:

- normal and abnormal operation start-up and shut-down situations
- potential emergency situations
- past, current, and likely future activities, products, and services

Although there is no single approach for identifying environmental aspects, the approach selected could, e.g., consider (1) emission to air, (2) releases to water, (3) releases to land, and (4) use of raw material and natural resources.

Although not strictly part of the requirements of the standard, an initial review to establish an organization's position with respect to the environment is recommended in Annex A.1 to ISO 14001. For those organizations without an existing EMS an initial review is essential.

Information on aspects and significant impacts are now required to be documented and kept up to date. The significant aspects are to be considered for establishing EMS.

10.4.3.4 Clause 4.3.2: Legal and Other Requirements

The organization must be capable of demonstrating that it has put in place procedures to identify and have ready access to details of all legislation (particularly environmental legislation and regulations) applicable to the environmental aspects of its activities, products and services. It is vitally important that the organization has systems for identifying, fully understanding, and constantly reviewing and updating its knowledge of regulatory requirements. The organization should establish a list of all laws and regulations relating to its activities, products, and services. These regulations may exist in several forms, including those specific to the organization's activities (e.g., site planning or permissions), those specific to the organization's processes and operations or industrial sector (e.g., operating consents and conditions, permits, etc.), and general national or local authority environmental regulations (including local bylaws). Legislation may be based on local, regional, national, European, or international requirements.

The organization should also include the requirements of other schemes and voluntary agreements to which it subscribes, e.g., codes of conduct, industry best practice schemes, and voluntary agreements with the local community in its procedures.

Internal priorities and criteria should be developed and implemented where external standards or schemes do not meet the needs of the organization or are nonexistent. These will assist the organization in developing its own objectives and targets.

10.4.3.5 Clause 4.3.3: Objectives, Targets, and Programs

When setting objectives and targets, the organization needs to consider a number of issues, namely these:

- the views of interested parties
- its significant impacts
- legal and other requirements
- technological options
- financial, operational, and business requirements

It is worth noting that clause A.3.3 of ISO 14001:2004 indicates that objectives should be specific and targets should be measurable wherever practicable to allow effective auditing of the EMS, and where appropriate should take preventive measures into account. Clause A.3.3 also states that the reference in clause 4.3.3 to the financial requirements of the organization is not intended to imply that they are obliged to use environmental cost-accounting methodologies when setting their objectives and targets.

The environmental management program is the mechanism by which environmental objectives and targets can be met. An organization may choose to develop and implement more than one program to achieve the same objective. Similarly, it may be possible to set up one program, which has a positive effect on a number of environmental aspects or impacts at the same time. For example, should an organization choose to move away from using solvent-based paints in its spraying operations, and use water-based paints with an electro-static system (to attract the new paint to the object being painted and reduce wastage to the air), then it will reduce its emissions of volatile organic compounds (VOCs) and particulate matter, remove the need for VOC authorizations, minimize its paint wastage, and improve the work environment in which its staff operate. An environmental management program should be drawn up to address all the organization's environmental objectives. To be most effective, environmental management planning should be integrated into the organization's overall strategic plan. The program should address issues such as schedules, resources, and responsibilities for achieving the organization's objectives and targets. The program should identify specific actions in order of their priority to the organization.

10.4.3.6 Clause 4.4.1: Resources, Roles, Responsibility, and Authority

Responsibility for the overall effectiveness of the EMS should be assigned to a senior person(s) or functions(s) with sufficient authority, competence, and resources. This person is responsible for reporting on the performance of the environmental management system to top management for review and as the basis for improvement. Operational managers should clearly define the responsibilities of relevant personnel and be responsible and accountable for effective implementation of the EMS and environmental performance. Employees at all levels should be accountable, within the scope of their responsibilities, for environmental performance in support of the overall environmental management system. Organizational infrastructure also includes buildings, communication lines, underground tanks, drainage, Effluent Treatment Plants (ETP), Diesel Generator (D.G.) sets, boilers, etc.

Clause **4.4.1** of **ISO 14004:2004** "Resources—Human, physical and financial" calls for the determination and availability (in a timely and efficient manner) of appropriate human, physical (e.g., facilities, equipment), and the financial resources essential to the implementation of an organization's environmental policies and the achievement of its objectives. In allocating resources, organizations can develop procedures to track the benefits as well as the costs of their environmental activities. Communication of key roles and responsibilities to all persons working for and on behalf of the organization is required.

10.4.3.7 Clause 4.4.2: Competence, Training, and Awareness

Competence of personnel should be judged on the basis of appropriate education, training and/or experience. Clause A.4.2 of ISO 14001:2004 states that the organization should also require that contractors working on its behalf are able to demonstrate that their employees have the requisite training, knowledge, and skills to perform the work in an "environmentally responsible manner." Relevant training should take place at all levels within the organization. At the senior management level, training should focus on raising the awareness of the strategic importance of environmental management to gain top-level commitment to the organization's policies. The competitive advantages of good environmental practices should be stressed.

Employees with very specific environmental responsibilities should receive skill enhancement training to improve performance at an operations level. Employees whose actions can affect compliance should receive training to ensure that regulatory and internal requirements are met.

10.4.3.8 Clause 4.4.3: Communication

The organization should decide whether to communicate about its significant environmental aspects externally and record the decision. Organizations should implement a procedure for receiving, documenting, and responding to relevant information and requests from interested parties. This procedure may include a dialog with interested parties and consideration of their relevant concerns. In some circumstances responses to interested parties concerns may include relevant information about the environmental impacts associated with the organization's operations. These procedures should also address necessary communications with public authorities regarding emergency planning and other relevant issues.

Communication is important to demonstrate management commitment to the environment and to deal with concerns and questions about the environmental issues of the organization's activities, products, or services. Results from EMS monitoring should also be communicated to those people within the organization who are responsible for performance.

Evidence of internal and external communication may be found in the company annual report, newsletters, bulletin boards, media campaigns, staff training programs, and paid advertising.

10.4.3.9 Clause 4.4.4: Documentation

The level of detail of the documentation should be sufficient to describe the core elements of the environmental management system and their interaction, and provide direction on where to obtain more detailed information on the operation of specific parts of the environmental management system. This documentation may be integrated with documentation of other systems implemented by the organization. It does not have to be in the form of a single manual. **These requirements are now very similar to those in ISO 9001:2000.** The mandatory documentation requirements are scope, policy and objectives, core elements of EMS, records recognized by this **ISO** standard, and other records identified by the organization for planning, implementation, and monitoring of effectiveness of EMS.

10.4.3.10 Clause 4.4.5: Document Control

Controls include the procedure to ensure availability of current version of documents at point of use. It is essential that all documentation is legible and readily identifiable, maintained in an orderly manner, and retained for a specified period. Obsolete documents may have to be retained for legal purposes or to provide historical records of the performance of the EMS distribution of external origin documents needs to be controlled.

10.4.3.11 Clause 4.4.6: Operational Control

Implementation is accomplished through the establishment and maintenance of operational procedures and controls to ensure that the organization's environmental policy, objectives, and targets can be met. The organization should consider the different operations and activities contributing to its significant environment impacts when developing or modifying operational controls and procedures. This could include, for example, contractors working on site whose operations give rise to significant impacts. Regulatory requirements contained in documents such as authorizations and permits will stipulate operating criteria. Activities can be divided into three categories:

- activities to prevent pollution and conserve resources in new projects, process changes and resources management, property (acquisitions, divestitures, and property management), and new products and packaging,
- daily management activities to assure conformance to internal and external organizational requirements, and to ensure their efficiency and effectiveness,
- strategic management activities to anticipate and respond to changing environmental requirements.

10.4.3.12 Clause 4.4.7: Emergency Preparedness

Emergency plans and procedures should be established to ensure that there will be an appropriate response to unexpected or accidental incidents. The organization should define and maintain procedures for dealing with environmental incidents and potential emergency situations. The operating procedures and controls should include, where appropriate, consideration of the following:

- accidental emissions to atmosphere,
- accidental discharges to water and land,
- specific environment and ecosystem effects from accidental releases,
- the procedures should take into account incidents arising, or likely to arise, as consequences of abnormal operating conditions and accidents and potential emergency situations.

Clause A.4.7 of ISO 14001:2004 provides a useful list of considerations for emergency preparedness.

10.4.3.13 Clause 4.5.1: Monitoring and Measurement

To determine if the organization is achieving its aims and objectives, it is important that some monitoring activities are carried out. This will also aid identification of improvement opportunities as well as problems.

Some issues to be considered in measuring and monitoring:

- how environmental performance regularly is monitored
- how have specific environmental performance indicators have been established that relate to the organization's objectives and targets and what they are
- what control processes are in place to regularly calibrate or verify the monitoring equipment
- periodical evaluation of compliance to legal requirements

10.4.3.14 Clause 4.5.2: Evaluation of Compliance

Periodical evaluation of compliance to legal (4.5.2.1) and other requirements (4.5.2.2) could be done through internal audits, review of documents and records, site visit, interviews, etc. Frequency of evaluations can be changed depending on the results of previous evaluations.

10.4.3.15 Clause 4.5.3: Nonconformity, Corrective Action, and Preventive Action

In establishing and maintaining procedures for investigating and correcting nonconformance, the organization should include these basic elements:

- apply correction to the nonconformance,
- identifying the cause of the nonconformance,
- identifying and implementing the necessary corrective action,
- implementing or modifying controls necessary to avoid repetition of the nonconformance; also there is a need to take preventive action to address potential nonconformities to prevent their occurrence,
- recording the results of corrective and preventive actions taken and making any necessary changes to the environmental management system documentation resulting from the corrective and preventive action.

10.4.3.16 Clause 4.5.4: Records

Procedures for identification, maintenance, and disposition of records should focus on those records needed for the implementation and operation of the environmental management system, and for recording the extent to which planned objectives and targets have been met. Proper account should be taken of confidential business information. The effective management of these records is essential to the successful implementation of the EMS. The key features of good environmental information management include means of identification, collection, indexing, filing, storage, maintenance, retrieval, retention, and disposal of pertinent EMS documentation and records. Clause A.5.4 of ISO 14001:2004 gives further guidance on types of records that could be covered by this requirement.

10.4.3.17 Clause 4.5.5: EMS Audits

The audit program and procedures should cover these points:

- the activities and areas to be considered in audits
- the frequency of audits
- the responsibilities associated with managing and conducting audits
- the communication of audit results
- auditor competence
- how audits will be conducted

Personnel conducting the audit should be in a position to do so impartially and objectively and should be properly trained. The frequency of audits should be guided by the nature of the operation in terms of its environmental aspects and potential impacts. Also, the results of previous audits should be considered in determining frequency. The EMS audit report should be submitted in accordance with the audit plan.

Consideration should be given to the guidance given on EMS audits in ISO 19011:2002.

10.4.3.18 Clause 4.6: Management Reviews

Reviews need to be carried out at planned intervals to determine if the system remains effective and is suitable for the needs of the organization.

The ISO 14001:2004 now sets out the input requirements for the review:

- results of EMS audits
- communication from external interested parties
- environmental performance of the organization
- the extent to which objectives and targets have been met
- status of corrective and preventive actions
- follow-up actions from previous management reviews

Summary of the Applications of ISO Family

Implementing environmental management systems (EMS)	ISO 14001:1996 ISO 14004:1996 Help an organization to establish a new or improve an existing EMS	ISO 14001:1996 Specifies the requirements for an EMS that may be objectively audited for self-declaration, second or third-party certification/registration purposes	ISO 14004:1996 Provides guidance to help an organization establish and implement an EMS, including guidance that goes beyond the requirements of ISO 14001	ISO/TR 14061:1998 Contains information that assists in the implementation of ISO 14001 and ISO 14004 by forest management organizations and the forest products industry	
Conducting environmental audits and other related investigations	ISO 14010:1996 provides guidance on the general principles common to the conduct of any environmental audit	ISO 14011:1996 provides guidance on the procedures for the conduct of EMS audits, including the criteria for selection and composition of audit teams	ISO 14012:1996 provides guidance on the qualifications of internal or external environmental auditors and lead auditors	ISO 14015:2001 helps an organization to identify and assess the environmental aspects and associated business consequences of sites and organizations to support the transfer of properties, responsibilities and obligations from one party to another	ISO 19011 provides guidance on the principles of auditing; the management of audit programs; the conduct of management system audits as well as on the competence of auditors

Continued

Summary of the Applications of ISO Family

Evaluating environmental performance	ISO 14031:1999 provides guidance on the selection and use of indicators to evaluate an organization's environmental performance	ISO/TR 14032:1999 provides example from real organizations to illustrate the use of the guidance in ISO 14031	Communicating results	ISO/WD 14063 will provide guidance on environmental communication related to an organization's environmental aspects and performance	Understanding terms and definitions	ISO 14050:2002 Helps an organization to understand the terms used in the ISO 14000 series standards
Using environmental declarations and clams	ISO 14020:2000 provides general principles that serve as a basis for the development of ISO guidelines and standard on environmental claims and declarations	ISO 14021:1999 provides guidance on the terminology symbols and testing and verification methodologies an organizations should use for self-declaration of the environmental aspects of its products and services (Type II environmental labeling)	ISO 14024-1999 provides the guiding principles and procedures for third-party environmental labeling certification programs (type I environmental labeling)	ISO/TR 14025:2000 identifies and describes elements and issues for consideration when making declaration of quantified product information based on life cycle inventory data (type III environmental declarations)		

Summary of the Applications of ISO Family

Conducting life cycle assessment (LCA)	ISO 14040:1999 provides the general principles, framework, and methodological requirements for the LCA of products and services	ISO 14041:1998 provides guidance for determining the goal and scope of an LCA study, and for conducting a life cycle inventory	ISO 14042:2000 provides guidance for conducting the life cycle impact assessment phase of an LCA study	ISO/TR 14043:2000 provides guidance for the interpretation of results from an LCA study	ISO 14048:2002 provides information regarding the formatting of data to support life cycle assessment	ISO/TR 14049/14047 provide examples that illustrate how to apply the guidance in ISO 14041 and ISO 14042
Addressing environmental aspects in products and product standards	ISO Guide 64:1997 helps the writers of product standards address environmental aspects in those standards	ISO/TR 14062:2002 provides concepts and current practices relating to integration of environmental aspects into product design and development		Understanding terms and definitions	ISO 14050:2002 helps an organization to understand the terms used in the ISO 14000 series standards	

- changing circumstances such as changing legislation, changing expectations and requirements of interested parties, changes in the products or activities of the organization, advances in science and technology, lessons learned from environmental incidents, market preferences,
- recommendations for improvement.

The review should look at the suitability, adequacy, and effectiveness of the EMS in relation to changing conditions and information.

Decisions and actions related to possible changes to the environmental policy, objectives, targets, and other elements of the environmental management system and improvements should be documented for necessary action.

Continual Improvement

ISO 14001:2004 uses a modified plan, do, check, and act cycle to achieve continual improvements in overall environmental performance.

Depending on the situation, this may be accomplished rapidly and with a minimum of formal planning, or it may be a more complex and long-term activity. The findings, conclusions, and recommendations reached as a result of monitoring, audits, and other reviews of the environmental management system should be documented and the necessary corrective and preventive actions identified. Management should ensure that these corrective and preventive actions have been implemented and that there is systematic follow-up to ensure their effectiveness.

10.5 GUIDELINES FOR AUDITING

ISO 9011:2002 gives guidance on the following areas:

- principles of auditing (clause 4)
- managing an audit program (clause 5)
- conducting an audit (clause 6)
- evaluation and competence of auditors (clause 7)

10.6 TYPES OF AUDITS

10.6.1 Reasons and Benefits to Conduct First-, Second-, or Third-Party Audits

There are different reasons for carrying out audits and they are as follows:

First Party

- as a control mechanism
- a tool for identifying improvement opportunities
- ISO 14001:2004 requires them under clause 4.5.5

Second Party

- as a means of evaluating new suppliers and contractors
- to assist in developing existing suppliers

Third Party

- market recognition
- reduced need for second party audits.

10.6.2 Audit Responsibilities

ISO 19011:2002 is the guidance document for quality and/or EMS auditing. Responsibilities can include the following:

Lead auditor: responsible for management of the entire audit process

- establishing initial contact with the auditee
- confirming the scope and objectives of the audit
- obtaining relevant background information
- conducting document review
- preparing the audit plan
- forming the audit team and communicating the audit plan
- coordinating preparation of working documents
- directing activities of the audit team
- resolving problems
- recognizing when audit objectives become unobtainable and reporting reasons to the client and auditee
- representing the audit team in discussions with the auditee
- notifying auditee of critical audit findings
- determining whether the requirements for an environmental audit have been met
- reporting to the client clearly and conclusively within the time agreed upon
- making recommendations for improvement if agreed within the scope

Auditor: responsible to conduct audit under the directions of the lead auditor

- applying audit principles, procedures, and techniques
- following the directions of and supporting the lead auditor
- planning and carrying out assigned activities objectively, efficiently, and effectively within the scope of the audit
- collecting and analyzing relevant information
- verifying accuracy of collected information
- determine audit findings

- prepare working documents
- documents findings
- safeguarding documents
- assisting in writing the report

Auditee

- informing employees about the scope and objectives of the audit
- providing facilities needed for the audit team
- appointing responsible and competent staff to accompany members of the audit team
- providing access to facilities, personnel, relevant information, and records
- cooperating with the audit team
- taking appropriate corrective/preventive actions on the issues raised by audit team

Auditor attributes and skills include but are not limited to the following:

- ethical
- open-minded
- observant
- competence in expressing concepts and ideas, orally and in writing
- interpersonal skills
- diplomacy, tact, and ability to listen
- maintain independence and objectivity
- methodical and organized
- good judgment
- self-reliant
- industrious
- practical
- professional

10.6.3 Sampling Options to the Auditor

Judgmental sampling: This is used to gather examples of deficiencies or problems to support an auditor's audit of a weak or improper environmental management system. Sampling is directed toward segments of the population where problems are likely to exist. "Judgmental sampling" cannot be used to draw compliance conclusions about an entire population because it focuses on only a portion or subset of that population. Judgmental sampling can be used as a first step to provide the auditor with an indication of whether to pursue probabilistic sampling techniques such as random, block, stratification, or interval sampling.

Random sampling: The objective is to select sample items by a statistically based chance. If properly done, each item in the population should have an equal chance of being selected, and there should be no subjective determinations to bias the sample.

Block sampling: The objective is to analyze certain segments of records or areas of the site. For example, if files were arranged alphabetically, in numerical order, or chronologically, one or more blocks (e.g., all of the "Es," record numbers 51 to 75, of the files for January and June) could be selected. While the block method is easy to use, it neglects entire segments of the population.

Interval sampling: The objective is to select samples at specific intervals (e.g., every nth segment of the population is analyzed) with the first item selected at random. Increased confidence is achieved where several intervals with different random starts are used.

Stratification sampling: The objective is to arrange items by categories (e.g., high versus low effluent volumes; new versus experienced employees; regular versus weekend or off-shift transactions) based on the auditor's judgment of risk. Higher risk categories then receive greater review and testing, but all categories receive some testing. (The extreme of assigning a 0% sampling frequency to some strata is what we would call judgment, not probabilistic, sampling.)

10.6.4 Audit Reports and Follow-Up Action

The audit should provide a balanced view that highlights the good and bad points. Therefore, it is important that definitions are understood:

Conformity: The fulfillment of a specified requirement.

Nonconformity: The absence of, or the failure to implement and maintain, one or more environmental management system requirements, or a situation that would, on the basis of objective evidence, raise significant doubt as to the capability of the EMS to achieve the policy band objectives of the organizations.

Prior to the closing meeting, it is important that all findings have been substantiated. It is important that the auditors are given sufficient time to review their findings and prepare to present them to the management at the closing meeting. This time, usually called the auditors' private meeting allows them to do this. The private meeting is attended by the auditors only, and under the direction of the lead auditor, it is decided what to report and by whom.

The auditors should do the following:

- Review the findings critically.
- Strive for team interaction.
- Play "Devil's advocate."

Integrate and summarize findings:

- Identify common findings.
- Look for patterns or trends.
- Identify underlying causes.

Classify and provide indication of priorities:

- conformity
- nonconformity
- minor/major

Auditing to the root cause is an approach that, if used by the auditors, will add value to the audit and assist the management in taking effective corrective action. It is pointless to undertake a corrective action to address a symptom when the underlying problem remains the same (and is likely to reoccur).

Reports should provide clear insight into the implementation of the various elements of the EMS standard and the effectiveness of the EMS with respect to the achievement of objectives. Nonconformities should be clearly referenced to the applicable standard, to the process/activity under scrutiny, and to any underlying documentation or records.

Remember the auditor normally only recommends certification. It is up to the certification body to decide whether a certificate should be awarded. This may require a head office appraisal of the auditors' report prior to issuing certification. The head office may reject the recommendation for certification and require certain activities to be carried out by the client. Likewise, the client may appeal against the refusal of the head office to issue a certificate if recommended by the auditor.

The key phases to the close out meeting are outlined next:

Open the meeting

- Break the ice.
- Thank the client for their help and time.
- Check the presence of senior management and the management representative (if not available then report the fact).
- State the purpose of the meeting.
- Confirm context, objectives, scope, and limitations.
- Review the reporting process.

- State that the audit is a sample.

Present/agree the audit of findings

- Review each finding succinctly.
- Clarify findings further.
- Note comments made by site personnel.
- Resolve any disagreements.
- Propose/agree corrective action if appropriate.
- Ensure that senior management signs off on the report.

Close the meeting

- Close on a positive note.

10.6.5 A Third-Party Perspective

At this final closing meeting the nonconformances from the whole audit will be presented to the company.

The team leader will only normally make a recommendation as to whether the organization is to be awarded a certificate. It is normally the head office function of the certification body that actually issues the certificate following an evaluation of the audit.

The following points should be addressed in an audit report:

- unique identity
- client name, address, date of audit, etc.
- audit objectives and scope
- EMS standard applied
- names, positions, and qualifications of audit team
- key client names and job titles
- audit program
- names and functions of interviewees
- documentation reviewed
- conformities/nonconformities with respect to compliance with the standard, objectives and procedures
- summary
- approval/recommendation for certification

10.7 RISK AUDIT TECHNIQUES

Risk audit techniques developed in other areas such as FMEA (failure modes effect analysis) and hazard and operability (HAZOP) can be used for environmental impacts. These and some other techniques are described next:

Flowcharting a process(es) can help to identify the inputs and outputs to the environment. It also includes sequencing of the subprocesses in the overall process activity.

The HAZOP method was first introduced by engineers from ICI Chemicals in the United Kingdom in the 1970s. The method entails the investigation of deviations from the design intention for a process facility by a team of individuals with expertise in different areas such as engineering, operations, maintenance, safety, and chemistry. The team is guided in a structured brainstorming process, by a leader who provides structure by using a set of guidewords to examine deviations from normal process conditions at various key points (nodes) throughout the process. The guidewords are applied to the relevant process parameters, e.g., flow, temperature, pressure, composition, to identify the causes and consequences of deviations in these parameters from their intended values. Finally, the identification of unintended (or unacceptable) consequences results in recommendations for improvement of the process. These may comprise design modifications, procedural requirements, modifications in written documentation, further investigations, etc. The basic purpose of a HAZOP study is to identify potentially hazardous scenarios. Therefore, the team should not spend any significant time trying to engineer a solution if a potential problem is uncovered. If a solution to a problem is obvious, the team should document their recommended solution. If a solution is not obvious, they should recommend that someone flows and resolves the problem outside the HAZOP study. Such recommendations will imply the instruction to have specific design items be reviewed by the design or engineering department.

BPEO (best practical environmental option) is the outcome of a systematic and decision-making procedure that emphasizes the protection and conservation of the environment across land, air, and water. The BPEO procedure establishes, for a given set of objectives, the option that provides the most benefit or least damage to the environment as a whole, at acceptable cost, in the long term as well as in the short term.

BAT (best available technique) is the most effective and advanced stage in the development of activities and their methods of operation that indicate the practical suitability of particular techniques for providing, in principle, the basis for emission limit values designed to prevent and, where that is not practicable, generally to reduce emissions and the impact on the environment as a whole.

ALARP (as low as reasonably practical) minimizes any harmful emissions to reduce human exposure to very safe levels.

ALARP, BPM, and BPEO are gradually being displaced by the concept of BAT, in essence an agreed set of *best available techniques*, which will eventually form a uniform level playing field of performance standards for environmental protection across Europe. The growth of BAT is being driven by the need to harmonize differing environmental protection regimes used across the European community's 15 member states. Compliance with relevant good practice alone may be sufficient to demonstrate that risks have been reduced ALARP. For example, recognized standards provide a realistic framework within which equipment designers, manufacturers, and suppliers (including importers) can fulfill their general duties.

FMEA (failure mode and effect analysis) is a risk assessment methodology.

10.8 STATUS OF EMS IN INDIA

Currently, companies cutting across different sectors viz aluminum, pharmaceuticals, refineries, engineering, etc., are already in the process of implementing EMS. Many of them have realized the benefits at the implementing stage itself. It will not be long before there is a surge in the number of companies opting for EMS certification. There are many ways in which the companies' expenses could decrease because of conformances to ISO 14001. As in the case of revenues, it is worthwhile to discuss a few.

1. **Improved process controls and yields**: One of the ways of reducing toxic chemical wastage is to either improve process efficiency (thereby reducing the untreated component discharges) or to substitute them with safer biodegradable constituents (thereby reducing the load and cost of effluent treatment). Another way is to improve material handling, which ultimately reduces spills. Either way, there is an overall reduction in the costs, which ultimately increases competitiveness. ISO 14001 requires objectives and targets to be set and monitored for continuous improvements. Any improvement in efficiency or recycling reduces environmental discharges. So yields are improved, as well as less chemical wastage, and fewer spills, cleaners, and disposals.

   Numerous live examples can be cited from chemical industries where improved yields have automatically also led to reduced effluent loads and simultaneous reduction in effluent treatment costs. Waste minimization is now attracting the attention of small as well as large multinational corporations all over the world. Companies realize that there is money in environmental protection. In Britain, for example, companies like Coca Cola and Schweppes are getting more and more involved in waste minimization projects and learning that the true cost of waste is usually about five times higher than expected.

2. **Reduced litigation costs**: Many companies are finding that the existence of a strong EMS documentation process strengthens the company against liability claims. This is especially true in the United States where liability claims could be varied and heavy.
3. **Reduced multiple assessment**: ISO 14001 certification and implementation of standards promises to reduce the need for multiple assessments. Governmental agencies in the United States and other countries have suggested to do away with site audits to certified companies. Today, customers and suppliers spend a lot of time auditing each other for product and environmental compliances. A certification is generally a good compliance. A certification is generally a good enough proof that the firm has a good management and environment system in place, avoiding costs of multiple assessments.
4. **Reduced accidents and injuries**: A system where process control and waste management is continuously reviewed for improvement greatly reduces the chances of accidents and possible deaths. Obviously, this is a great benefit.
5. **Reduced expenses for ETP and waste handling**: The primary objective of ISO 14001 is to aim for reduction in chemical discharges and solid waste handling. This automatically leads to substantial savings in chemical purchases, handling, and ETP costs. ISO 14001 is a management system standard. It is not a performance or product standard, although the framer developed it with the idea that implementing it would result in improved environmental performance. The standards are not prescriptive since the management determines its own objectives and targets. The standards represent a shift toward holistic proactive management and total employee involvement.

ISO 14001 certification is the beginning of a journey to continual improvement in sustainable growth and quality of life. One very important aspect of this is that it forces organizations to be more proactive. And with it, there is a hope that such a proactive response to anticipated environmental impacts will achieve what two decades of legislation have not.

10.8.1 IT-Enabled Solution

- Design and develop and integrated web-enabled system for all the processes across the organization and with proper authentication and documentation in line with ISO 14001
- Continual improvement by taking possible measures
  - developments in products, services, processes, and facilities
  - enhanced product quality, operational efficiency, and resource utilization

- determine the root cause of incidents and other nonconformances and preparing plans for preventive and corrective actions
- A centralized database is maintained on a remote server that records data with respect to all physical parameters related to environmental components such as, air, water, noise, land, and biological environment (green–belt development), etc.
- EMS-IT-2002 will be extended to all groups of industries in different areas.
- The gaps and deficiencies charts will enable solutions for reengineering and reliability.
- Fuzzy logic solutions are for integrating the process and resulting pollution on spatial and temporal boundary conditions.

All can be made possible by an IT-enabled solutions throughout the operations with different supporting modules right from the project management system to the functional modules, calibration module, module for measuring, and monitoring systems and MIS.

10.9 OHSAS 18001 STANDARD

OHSAS 18000 is an international Occupational Health and Safety Management System specification. It comprises two parts, 18001 and 18002.

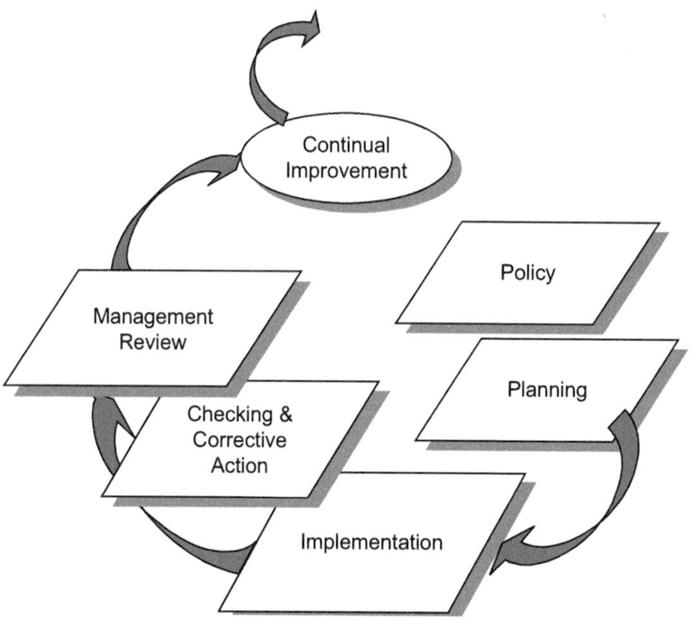

Figure 10.1 OHSAS 18004 flow chart

It helps to minimise risk to employees and improve an existing Occupational health and safety management system.

When OHSAS 18001 is implemented, the organization is expected to:

- establish an OH&S management system to eliminate or minimize risk to employees and other interested parties who may be exposed to OH&S risks
- implement, maintain and continually improve an OH&S management system;
- seek certification/registration of its OH&S management system by an external organization;

As shown in Fig. 10.1, Step 1 the Organisations have to establish and maintain an OH&S management system which meets the requirements of OHSAS18001. The organizations also have to formulate the OH&S policy stating clearly its objectives and its commitment to continual improvement to implement OHSAS. In the planning stage hazard identification, risk assessment and risk control are to be identified. Organisations have to have procedures for risk assessment and risk control and use the outputs from these procedures in setting OH&S objectives. Organisations are required to document OH&S responsibilities and how these responsibilities are structured.

In addition, Organisations must have procedures for communication on OH&S issues and documented arrangements for employee involvement and consultation. There must be control of all documents and data required by the OH&S management system and this control must ensure a number of things, including adequate locating, reviewing and archiving of documents and data as and when required.

When related to the PDCA cycle organizations have to

1. **Plan**

   The planning stage of the process requires the organization to:

   - Devise an OH&S policy
   - Plan for hazard identification, risk assessment and determination of controls
   - Identify relevant legal requirements
   - Plan for emergencies and responses
   - Manage change effectively
   - Devise procedures for performance measuring, monitoring and improvement
   - Provide and ensure the appropriate use of safety equipment

2. **Do**
   - The implementation stage is the easiest part of this process. One must follow the documentation and procedures that have been created.
   - Ensure that the appropriate structure at the organization effectively minimizes risk
   - Start the implementation by breaking the system down into specific elements rather than tackling it as a whole.
   - Concentrating on specific elements in a logical order that creates a solid foundation for the whole system to work efficiently.
   - A matrix should be created showing all groups of personnel, their required competencies, training and status of each.
3. **Check**
   - Conducting internal audits
   - Evaluation of legal compliance
   - Identifying non-conformities and addressing them thorough analysis of incidents and incidental data
   - Measuring performance and monitoring
4. **Act**

   This involves reviewing the suitability, adequacy and effectiveness of the system. It should also include assessing opportunities for improvement and the necessity to change the OH&S policy and the OH&S objectives.
5. **Control of records**
   a. Clause 4.5.4
   b. The organization shall establish and maintain records as necessary to demonstrate conformity to the requirements of its OH&S management system and of this OHSAS standard, and the results achieved.
6. **Internal audit**
   a. Clause 4.5.5
   b. The organization shall ensure that internal audits of the OH&S management system are conducted at planned intervals to:
7. **Management review**
   a. Clause 4.6

Top management shall review the organization's OH&S management system, at planned intervals, to ensure its continuing suitability, adequacy, and effectiveness. Reviews shall include assessing opportunities for improvement and the need for changes to the OH&S management system, including the OH&S policy and OH&S objectives.

FURTHER READING

[1] H. Tilton, The Dawn of ISO 14000, Quality (April 8, 1996) SR5–SR6.
[2] K. Sissell, R. Mullin, Fitting in ISO 14000: a search for synergies, Chemical Week 157 (17) 39–43.
[3] K. Kuryllowicz, ISO 14000: buying into the international environmental standards, Modern Purchasing 38 (3) (1996) 14–17.
[4] International Standards Organization, ISO 9001.
[5] International Standards Organization, ISO 14001.
[6] U.S. Department of Labor, Occupation Health and Safety Assessment Series, OHSAS 18001.
[7] U.S. Department of Labor, Occupational Health and Safety Assessment Series, OHSAS 18002.

CHAPTER ELEVEN

Principles and Design of Water Treatment

11.1 INTRODUCTION

Water is a vital natural resource that forms the basis of all life. Further, water is a key resource in all economic activities ranging from agriculture to industry. With ever increasing pressure of human population, there is severe stress on water resources. The objective of municipal water treatment is to provide a potable supply, one that is chemically and microbiologically safe for human consumption. For domestic uses, treated water must be aesthetically acceptable, free from apparent turbidity, color, odor, and objectionable taste. Groundwater from wells or springs is usually of an acceptable quality due to natural filtration through the ground. However, water from surface sources such as streams, lakes, and ponds will usually require some form of treatment. The quality of this water may vary greatly with the seasons of the year. The treatment of surface water can be an expensive exercise and one that is difficult for communities to sustain without long-term external support. Routine maintenance is essential, and in addition, an ability to vary the method of treatment to respond to changes in water quality is required. Common water sources for municipal supplies are deep wells, shallow wells, rivers, natural lakes, and reservoirs.

11.2 WATER POLLUTION

The demand for water for irrigation and industrial and domestic use is increasing with development. Many developmental projects, industrial/urbanization programs, or policies are likely to have both qualitative and quantitative effects on both the surface and ground water environment (rivers, lakes, estuaries, oceans), which may result in considerable impacts on aquatic faunal or floral species and aquatic ecosystems. The release of complex and diverse industrial wastes impairs the quality of the environment and poses a threat to human health directly or indirectly. The liquid wastes may permeate into the ground or find their way into the surface waters,

Environmental Management
ISBN 978-0-12-811989-1
http://dx.doi.org/10.1016/B978-0-12-811989-1.00011-7

Table 11.1 Non-point and Point Sources

Non-point Pollutants	Point Pollutants
Pollutants from: urban area, industrial area, rural runoff	Specific discharges from: municipalities or industrial complexes
Examples: sediment, pesticides, or nitrates entering a surface water balance of runoff from agricultural farms	Example: organics or metals entering surface water as a result of wastewater discharge from a manufacturing plant

causing severe problems of water pollution due to which the access of safe drinking water in the entire world is tremendously effected. Domestic sewage leads to the spread of waterborne diseases like typhoid, typhoid, cholera, gastroenteritis, dysentery, and hepatitis. The toxic substances present in the industrial wastes can affect the aquatic life, thus disrupting the whole ecosystem. Mercury and organochlorine pesticides are important examples of bioaccumulating substances. A large number of birds like hawks, eagles, and peregrine falcons have also been found to be adversely affected by consuming DDT-containing organisms. Sources of water pollution can be classified as (1) non-point sources, (2) point sources, which are given in Table 11.1, and major types of pollutants, which are given in Table 11.2.

11.2.1 Surface Water Pollution

The effects of pollution sources (Table 11.3) on receiving water quality are manifold and dependent upon the type and concentration of pollutants. Soluble organics, as represented by high biochemical oxygen demand (BOD) wastes, cause depletion of oxygen in the surface water. This can result in fish kills, undesirable aquatic life, and undesirable odors. Even trace quantities of certain organics may cause undesirable taste and odors, and certain organics may be biomagnified in the aquatic food chain. Suspended solids decrease water clarity and hinder photosynthetic processes; if solids settle and form sludge deposits, changes in benthic ecosystems result. Color, turbidity, oils, and floating materials are of concern because of their aesthetic undesirability and possible influence on water clarity and photosynthetic processes. Excessive nitrogen and phosphorous can lead to algal overgrowth, with water treatment problems resulting from algae decay and interference with treatment processes. Chlorides cause a salt taste to be imparted to water, and in sufficient concentration, limitations on water usage can occur. Acids, alkalies, and toxic substances have the potential to cause fish kills and create other imbalances in stream ecosystems.

Table 11.2 The Major Types of Pollutants

Pollutant	Major Sources	Effects
Oxygen-demanding wastes	Sewage effluent; agricultural runoff including animal wastes; some industrial effluents (from paper mills, food-processing, etc.)	Decomposition by aerobic bacteria depletes level of dissolved oxygen in water; flora and fauna perish; further decomposition by anaerobic bacteria produces foul-smelling toxic substances such as hydrogen sulfide
Plant nutrients	Sewage effluent including phosphates from detergents; agricultural runoff, especially nitrates from fertilizers	Algal blooms; death of submerged vegetation; production of large amounts of dead organic matter with subsequent problems of oxygen depletion (see earlier)
Acids	Acid rain; mine drainage; planting of extensive areas of coniferous forests, which acidify the soil	Acidification of natural waters; sharp decline in species richness; fish killed; concomitant increase in level of toxic metals in solution, for example, aluminum
Toxic metals Hg, Pb, Cd, Zn, Sn	Ore mining; associated industries; lead from vehicle exhaust emissions	Biomagnification of toxic metals with each successive stage of food chain; threat to consumers including humans
Oil	Drilling operations; oil tanker spills; natural seepage; waste disposal	Contamination of the aquatic environment, death of birds and mammals
DDT (an organochlorine)	Direct application; agricultural runoff and via aerial crop spraying	Biomagnification; top carnivores (especially birds) at risk; very persistent in the environment
PCBs (a series of organochlorines)	Sewage effluent; waste incineration; toxic dumps; landfill sites	Biomagnification; top carnivores at risk; effects on human health include joint pain, chloracne, and fatigue

Continued

Table 11.2 The Major Types of Pollutants—cont'd

Pollutant	Major Sources	Effects
Radiation	0% from natural sources; 20% from nuclear weapons testing, medical X-rays, nuclear energy industry, etc.	Degree of tissue damage and risk of death depended on exposure; radionuclides can be biomagnified, and some are very persistent in the environment
Heat	Coolant waters from industry, principally the electricity generating industry	Change in species composition usually accompanied by a decrease in species richness; fish may migrate or be killed by suffocation; reproductive cycle of fish and other aquatic organism disrupted

11.2.2 Groundwater Pollution

Groundwater contamination commonly results from human activities where pollutants, susceptible to percolation are stored and spread on or beneath the land surface. Almost every known distance of groundwater contamination has been discovered only after a drinking water supply was affected. Typical pollutant sources are industrial wastewater impoundments, sanitary landfills, storage piles, absorption fields following household septic tanks, improperly constructed wastewater disposal wells, and application of chemicals on agricultural lands. The principal sources and causes of groundwater pollution are under four categories: municipal, industrial, agricultural, and miscellaneous. Most pollution stems from disposal of wastes on or into the ground. Methods of disposal include placing wastes in percolation ponds, on the ground surface (spreading or irrigation), in seepage pits or trenches, in dry streambeds, in landfills, into disposal wells, and into injection wells. Since groundwater in many regions is a high-quality economic source, future demand is expected to increase for domestic use. The amount of water available for infiltration, either from precipitation or the wastewater itself, is a primary factor in carrying pollutants down through a soil profile. Water from the surface passes downward through the unsaturated zone and disperses in an aquifer in a manner depending on site conditions. Dispersion of a contaminant is influenced both physically by soil porosity and hydraulically by the rate of water movement.

Table 11.3 Important Surface Water Contaminants and Their Impacts

Contaminants	Reason for Importance
Color and odor	Aesthetically not acceptable. Laxative effect
Suspended solids	Suspended solids can lead to the development of sludge deposits and anaerobic conditions when untreated wastewater is discharged in the aquatic environment
Biodegradable organics	Composed principally of proteins, carbohydrates, and fats, biodegradable organics are measured most commonly in terms of BOD (biochemical oxygen demand), and Chemical Oxygen Demand(COD) stabilization can lead to the depletion of natural oxygen resources and to the development of septic conditions
Pathogens	Communicable diseases can be transmitted by the pathogenic organisms in wastewater
Nutrients	Both nitrogen and phosphorus, along with carbon, are essential nutrients for growth. When discharged to the aquatic environment, these nutrients can lead to the growth of undesirable aquatic life. When discharged in excessive amounts on land, they can also lead to the pollution of groundwater
Priority pollutants	Organic and inorganic compounds selected on the basis of their known or suspected carcinogenicity, mutagenicity, teratogenicity, or high acute toxicity. Many of these compounds are found in wastewater
Refractory organics	These organics tend to resist conventional methods of wastewater treatment. Typical examples include surfactants, phenols, and agricultural pesticides
Heavy metals	Heavy metals are usually added to wastewater from commercial and industrial activities and may have to be removed if the wastewater is to be reused
Dissolved inorganics	Inorganic constituents such as calcium, sodium, and sulfate are added to the original domestic water supply as a result of water use and may have to be removed if the wastewater is to be reused
Gases	Decomposition of domestic wastes. Domestic water supply surface water infiltration. Decomposition by aerobic bacteria depletes level of dissolved oxygen in water; flora and fauna perish; further decomposition by anaerobic bacteria produces foul-smelling toxic substances such as hydrogen sulfide

Reproduced from Metcalf and Eddy, Waste Water Engineering, 1991.

Changes in water quality resulting from withdrawal of groundwater or channelization of rivers, although induced by human activities, are often viewed as natural contamination. Under normal undisturbed conditions, groundwater moves toward a river, lake, or the sea. Pumping of nearby wells can reverse this flow, causing migration of polluted surface waters into the aquifer. An important aspect of groundwater pollution is the fact that it may persist underground for years, decades, or even centuries. This is in marked contrast to surface water pollution. Reclaiming polluted groundwater is usually much more difficult, time consuming, and expensive than reclaiming polluted surface water. Underground pollution control is achieved primarily by regulating the pollution source and secondarily by physically entrapping and, when feasible, removing the polluted water from the underground.

The principal natural chemicals found in groundwater are dissolved salts, iron and manganese, fluoride, arsenic, radionuclides, and trace metals. Both geologic and climatic conditions influence mineral composition. In arid regions with limited water recharge, slow percolation results in mineralized poor-quality water high in sodium chloride.

In humid climates, weathering of sedimentary rock leaches calcium and magnesium, creating excessive hardness and often dissolved iron and manganese. Fluoride is a constituent of mineral fluorite found in sedimentary, igneous, and metamorphic rocks. In some regions, high concentrations of fluoride in groundwater result in fluorosis (mottling of teeth) and, in extreme cases, bone damage. Arsenic can be a significant problem in aquifers of volcanic deposits where concentrations are higher than the maximum contaminant level. Small amounts of metals, such as selenium, cadmium, lead, copper, and zinc, are found in rocks and unconsolidated deposits. Despite this, groundwater generally contains only traces of these metallic elements, and their presence is rarely a water quality problem. For control of groundwater pollution the point sources and diffuse sources (Table 11.4) have to be monitored and regulated properly.

11.3 WATER QUALITY MANAGEMENT

Many uses of water are restricted within narrow ranges of water quality, such as the public and industrial water supply. Therefore, control of quality is required to ensure that the best employment of the water is not prevented by indiscriminate use of watercourses for disposition of wastes. Surface waters are classified according to intended uses that dictate the specific physical, chemical, and biological quality standards, thus ensuring the most beneficial uses will not be deterred by pollution. Criteria defining quality are dissolved oxygen, solids, coliform bacteria, toxins, pH, temperature, and other parameters as necessary.

Table 11.4 Point Sources and Diffuse Sources for Groundwater Pollution Monitoring

Point Sources	**Diffuse Sources**
***Prevention* (Groundwater Protection)**	
Siting, design, construction, and operation	Site preparation and control of chemical application
***Monitoring* (Early Warning)**	
Testing of groundwater from wells installed around the site	Testing of groundwater from observation, irrigation, and water supply wells
***Abatement* (Elimination of Source)**	
Reconstruction of site or modifying or abandoning operation (Examples: wastewater ponds, landfills, refuse piles, buried storage tanks and deep injection wells)	Changing method of application or restricting or banning chemical usage (Examples: application of pesticides or fertilizers on agricultural land and land application of wastewater)

As water travels through the hydrologic cycle, it changes from pure salt-free moisture suspended in the troposphere as clouds to the brine of the sea. Through the cycle, it progressively picks up salts by trickling through the atmosphere, flowing on the earth's surface, percolating through the soil medium and the unsaturated zone, and moving in the saturated zone (Fig. 11.1). Different human activities like agriculture (Fig. 11.2), industry (Fig. 11.3), and land use changes (Fig. 11.4) contribute significantly for the deterioration of water quality, which if managed properly can help in its conservation.

11.3.1 Agricultural Uses

Irrigation, in particular, degrades the quality of water due to the following:
1. Concentration of salts in irrigation return water as the result of evapotranspiration,
2. Pick-up of fertilizers and other soluble additives applied to the soil.

11.3.2 Industrial Uses

The major applications of water in industrial plants are (1) cooling, (2) boiler feed, (3) processing, and (4) sanitation. The quality and chemical composition of the wastewater varies with respect to the type of industry and water use. Pollution effects of industrial waste water are well recognized.

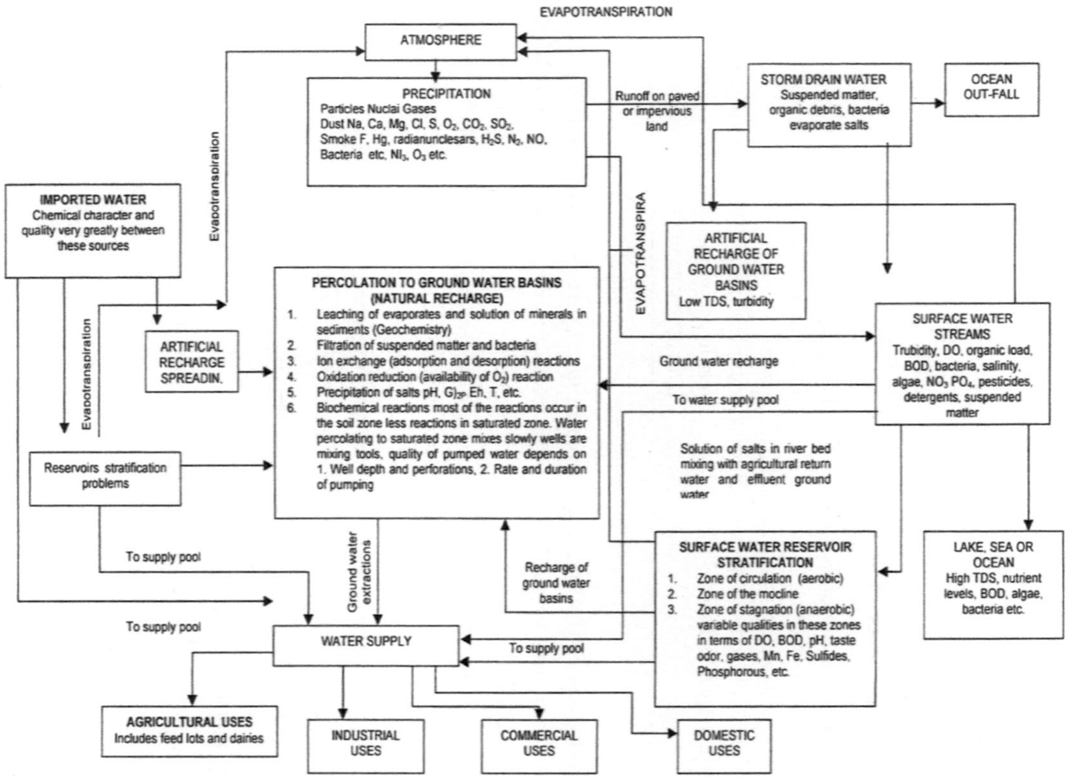

Figure 11.1 Water quality cycle, sources, uses, and effects on water quality. Reproduced from *A. A. Hassan, Water quality cycle, reflectance of activities on nature and man, Ground Water 12 (1) (1974) 16–21.*

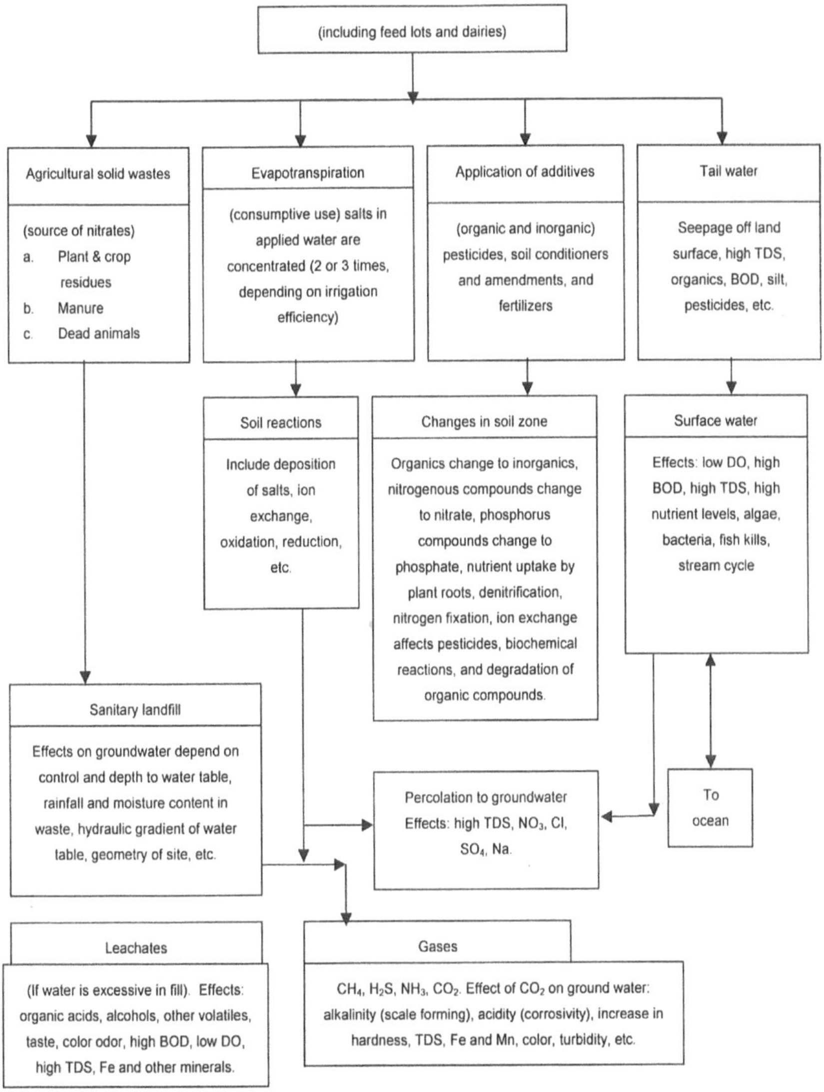

Figure 11.2 Agricultural uses of water and their effects on water quality. Reproduced from: *A.A. Hassan, Water quality cycle, reflectance of activities on nature and man, Ground Water 12 (1) (1974) 16–21.*

11.3.3 Land Use Changes

Changes in land use due to industrialization and unplanned urbanization also contribute significantly for deterioration of environmental quality.

In view of the large-scale pollution occurring in the surface and ground water, it becomes essential to treat the water so that it is potable. The next section deals with the treatment technologies for drinking purposes.

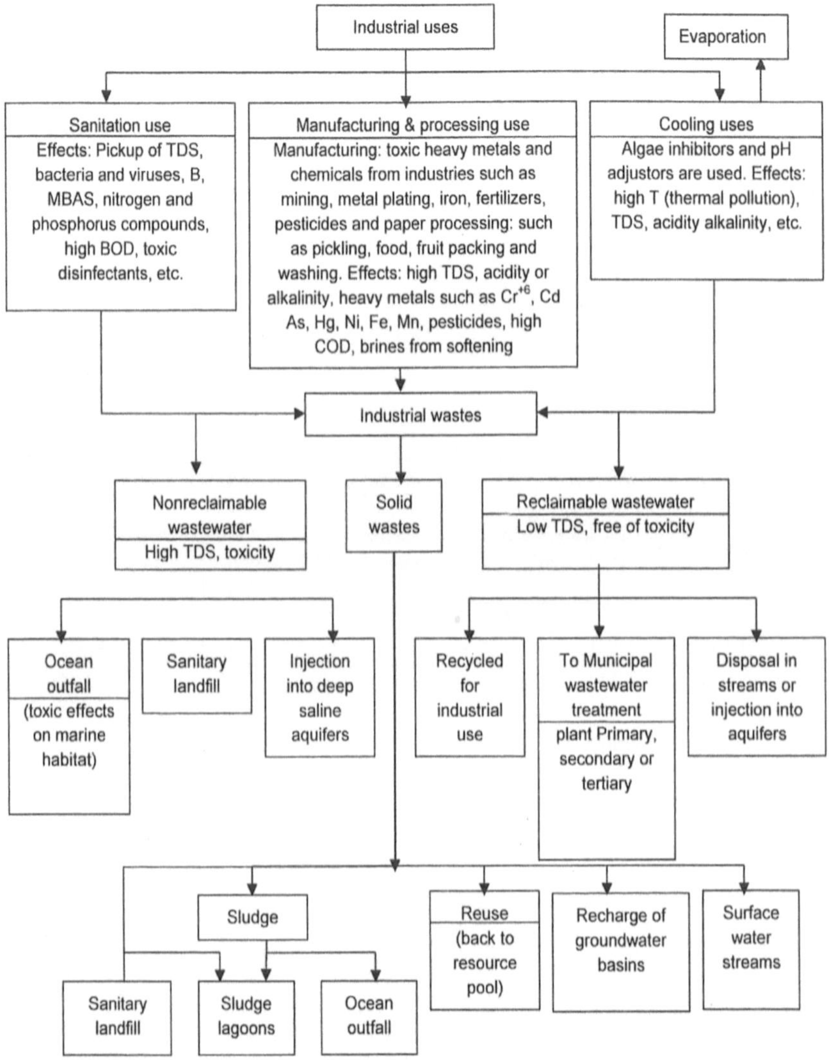

Figure 11.3 Industrial uses of water and their effects on water quality.

11.4 TREATMENT TECHNOLOGIES

The primary process in surface water treatment is chemical clarification by coagulation, sedimentation, and filtration, as illustrated in Fig. 11.5. Lake and reservoir water has a more uniform year-round quality and requires a lesser degree of treatment than river water. Natural purification results in

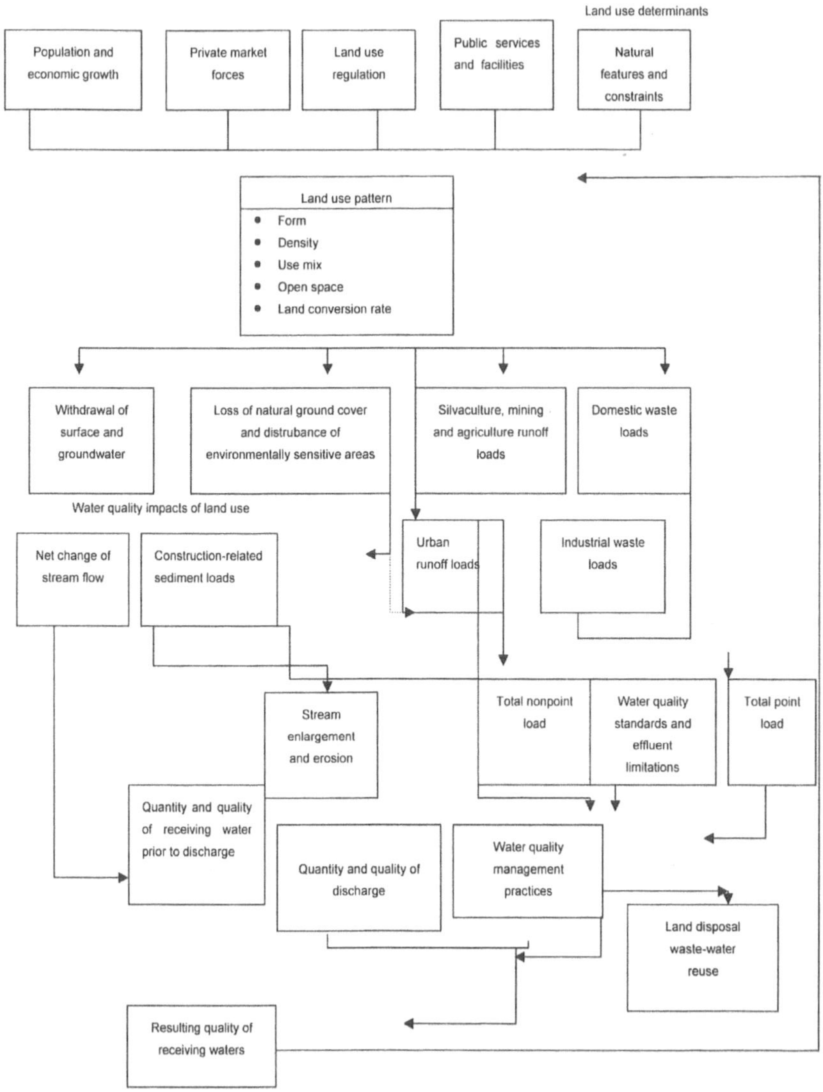

Figure 11.4 Schematic diagram of the land use–water quality relationship.

reduction of turbidity, coliform bacteria, color, and elimination of day-do-day variations. On the other hand, growths of algae cause increased turbidity and may produce difficult-to-remove tastes and odors during the summer and fall. The specific chemicals applied in coagulation for turbidity removal depend on the character of the water and economic considerations.

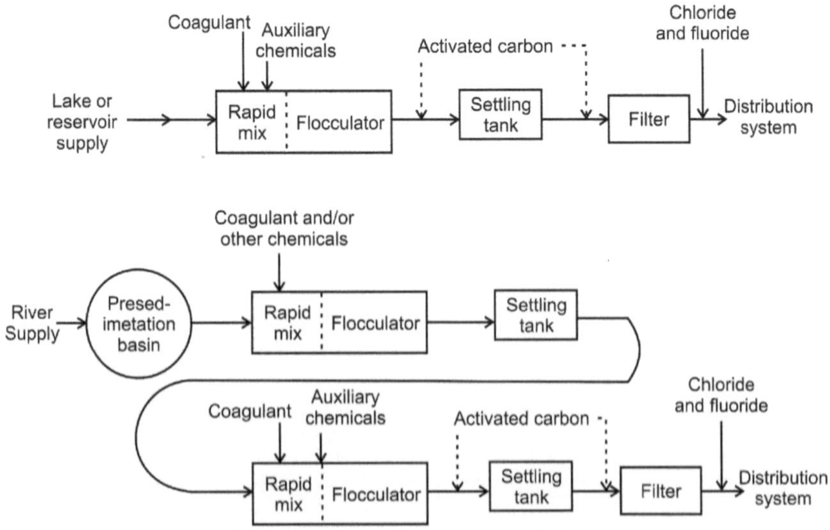

Figure 11.5 Schematic patterns of typical surface water treatment systems.

Well supplies normally yield cool, uncontaminated water of uniform quality that is easily processed for municipal use. Processing may be required to remove dissolved gases and undesirable minerals. The simplest treatment illustrated in Fig. 11.6A is disinfection and fluoridation. Deep-well supplies may be chlorinated to provide residual protection against potential contamination in the water distribution system. Fluoride is added to reduce the incidence of dental caries. Dissolved iron and manganese in well water oxidize when contacted with air, forming tiny rust particles that discolor the water. Removal is performed by oxidizing the iron and manganese with chlorine or potassium permanganate and removing the precipitates by filtration (Fig. 11.6B) and softening methods given in Fig. 11.6C.

11.4.1 Sedimentation

Sedimentation, or clarification, is the removal of particulate matter, chemical floc, and precipitates from suspension through gravity settling. The common criteria for sizing settling basins are detention time, overflow rate, weir loading, and with rectangular tanks, horizontal velocity. Detention time, expressed in hours, is calculated by dividing the basin volume by average daily flow, Eq. (11.1):

$$t = \frac{V \times 24}{Q} \tag{11.1}$$

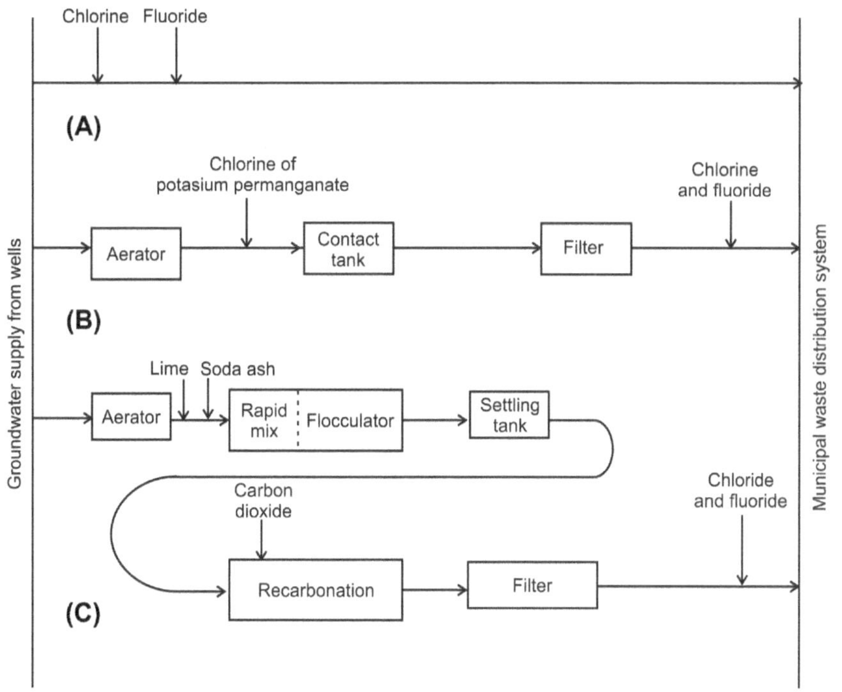

Figure 11.6 Flow diagrams of typical groundwater treatment systems: (A) disinfection and fluoridation, (B) iron and manganese removal, and (C) precipitation softening.

where t, detention time, hours; V, basin volume, million gallons (cubic meters); Q, average daily flow, million gallons per day (cubic meters per day); 24, number of hours per day.

The overflow rate (surface loading) is equal to the average daily flow divided by total surface area of the settling basin, expressed in units of gallons per day per square foot, Eq. (11.2).

$$V_o = \frac{Q}{A} \tag{11.2}$$

where V_o, overflow rate (surface loading), gallons per day per square foot (cubic meters per square meter per day); Q, average daily flow, gallons per day (cubic meters per day); A, total surface of basin, square feet (square meters).

Most settling basins in water treatment are essentially upflow clarifiers where the water rises vertically for discharge through effluent channels (Fig. 11.7). Water flows horizontally through the basin and then rises vertically, overflowing the weir of a discharge channel at the tank surface.

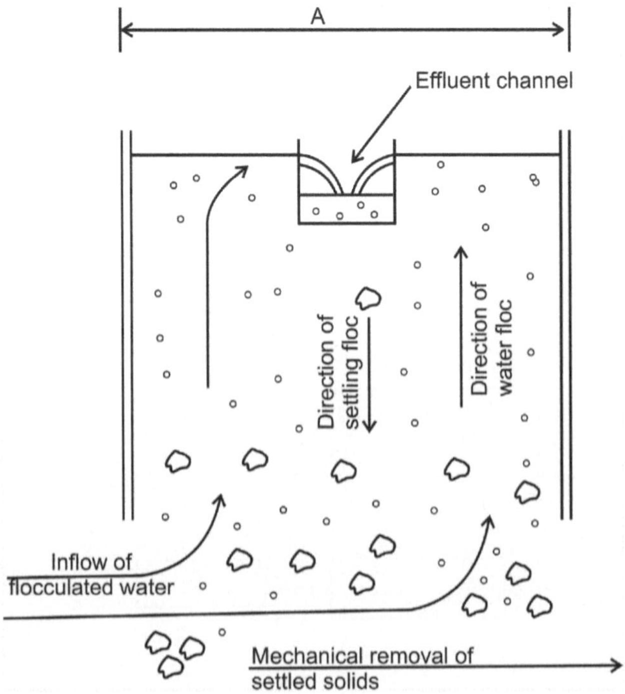

Figure 11.7 Sedimentation tank as defined by Eqs. (11.1) and (11.2).

Flocculated particles settle downward, in a direction opposite to the upflow of water, and are removed from the bottom by a continuous mechanical sludge removal apparatus. The particles with a settling velocity v greater than the overflow rate Q/A are removed, while lighter flocs, with settling velocities less than the overflow rate, are carried out in the basin effluent.

Weir loading is computed by dividing the average daily quantity of flow by the total effluent weir length and expressing the results in gallons per day per foot (cubic meters per meter per day).

Sedimentation basins, either circular or rectangular, are designed for slow uniform water movement with a minimum of short-circuiting.

A design of settlement tank is shown in Fig. 11.8. The length is usually about three times the width, and a practical depth is about 2 m. A capacity between 2- and 4-h retention at maximum flow should be sufficient to remove most sand and silt.

On small installations, it may be better to fill the basin with stone or gravel to prevent the incoming flow from disturbing the settled solids. The sediment can then be washed out with a hose pipe.

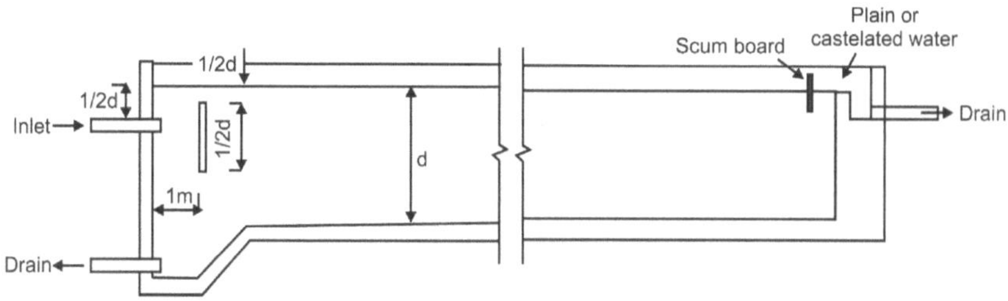

Figure 11.8 Section through a sedimentation tank.

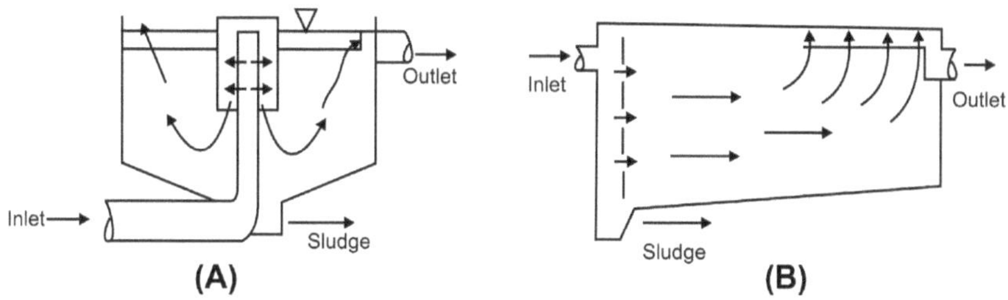

Figure 11.9A Rectangular sedimentation tank.

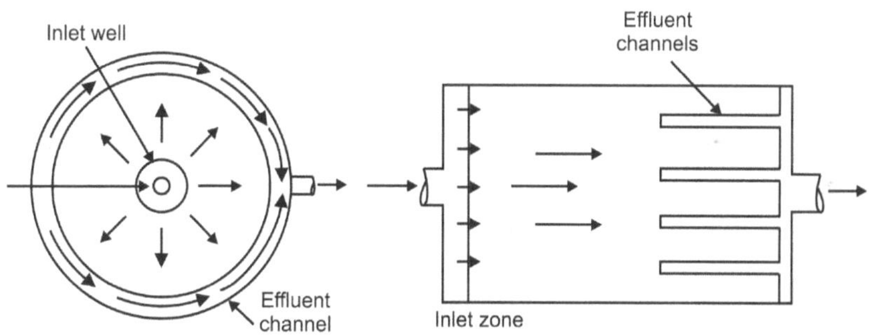

Figure 11.9B Circular tank.

In water treatment, sedimentation basins are more commonly rectangular (Fig. 11.9A). The influent is spread across the end of the basin by a baffle structure to dissipate velocity and uniformly distribute the flow. The water passes horizontally to effluent overflow channels with V-notch weirs mounted along the edges.

In a circular tank (Fig. 11.9B), the influent enters through a vertical riser pipe in the center with outlet ports discharging behind a circular inlet well. This baffle dissipates the horizontal velocity by directing the flow downward. Radial flow from the center is collected into the effluent channel attached to the outside wall by overflowing a V-notch weir mounted along the edge of the channel. The settled solids that accumulate on the bottom are gently pushed by scraper blades into a central hopper for discharge. The sludge scraper can be driven by a central turntable supported by a bridge or pier. In a warm climate without ice and snow, it can be driven around by a wheel running on the peripheral wall of the tank.

11.4.2 Clariflocculators

Flocculator-clarifiers, also referred to as solids contact units or upflow tanks, combine the processes of mixing, flocculation, and sedimentation in a single compartmented tank.

11.4.2.1 Coagulants

Commonly used metal coagulants in water treatment are these:

1. those based on aluminum, such as aluminum sulfate, sodium aluminate, potash alum, and ammonia alum;
2. those based on iron, such as ferric sulfate, ferrous sulfate, chlorinated ferrous sulfate, and ferric chloride.

Aluminum sulfate, $Al_2(SO_4)_3 \cdot 14.3H_2O$, is by far the most widely used coagulant; the commercial product is commonly known as alum, filter alum, or alumina sulfate. Filter alum is a grayish-white crystallized solid containing approximately 17% water-soluble Al_2O_3 and is available in lump, ground, or powdered forms as well as concentrated solution. Ground alum is commonly measured by a gravimetric-type feeder into a solution tank from which it is transmitted to the point of application by pumping. Amber-cooled liquid aluminum sulfate contains about 8% available Al_2O_3.

The hydrolysis of aluminum ions in solution is complex and is not fully defined. In pure water at low pH, the bulk of aluminum appears as Al^{+++}, while in an alkaline solution complex, species such as $Al(OH)_4^{e}$ and $Al(OH)_5^{]}$ have been shown to exist. In the hypothetical coagulation equations, aluminum floc is written as $Al(OH)_3$. This is the predominant form found in a dilute solution near neutral pH in the absence of complexing anions other than hydroxide. The reaction between aluminum and natural alkalinity is given in Eq. (11.i). If lime or soda ash is added to the water with the coagulant, the theoretical reactions are as shown in Eq. (11.ii) and (11.iii).

$$Al_2(SO_4)_3 \cdot 14.3H_2O + 3Ca(HCO_3)_2 = 2Al(OH)_3\downarrow + 3CaSo_4 + 14.3\ H_2O + 6CO_2 \qquad (11.i)$$

$$Al_2(SO_4)_3{\cdot}14.3H_2O + 3Ca(HCO_3)_2 = 2Al(OH)_3\downarrow + 3CaSo_4 + 14.3\ H_2O \quad (11.ii)$$

$$Al_2(SO_4)_3{\cdot}14.3H_2O + 3Na_2CO_3 + 3H_2O = 2Al(OH)_3\downarrow + 3Na_2SO_4 + 3\ CO_2 + 14.3\ H_2O \quad (11.iii)$$

Based on these reactions, 1.0 mg/L of alum with a molecular weight of 600 reacts with 0.50 mg/L natural alkalinity, expressed as $CaCO_3$; 0.39 mg/L of 95% hydrated lime as $Ca(OH)_2$, or 0.33 mg/L of 95% hydrated lime as $Ca(OH)_2$, or 0.33 mg/L 85% quicklime as CaO; or 0.53 mg/L soda ash as Na_2CO_3. When lime or soda ash is reacted with the aluminum sulfate, the natural alkalinity of the water is unchanged. Sulfate ions added with the alum remain in the finished water. In the case of natural alkalinity and soda ash, carbon dioxide is produced. The dosages of alum used in water treatment are in the range of 5–50 mg/L, with the higher concentrations needed to clarify turbid surface waters. Alum coagulation is generally effective within the pH limits of 5.5–8.0.

Sodium aluminate, $NaAlO_2$, may be used as a coagulant in special cases. The commercial grade has a purity of approximately 88% and may be purchased either as a solid or as a solution. Because of its high cost, sodium aluminate is generally used to aid a coagulation reaction instead of being the primary coagulant. It has been found effective for secondary coagulation of highly colored surface waters and as a coagulant in the lime–soda ash softening process to improve settleability of the precipitate.

11.4.2.2 Coagulant Aids

Difficulties with coagulation often occur because of slow-settling precipitates or fragile flocs that are easily fragmented under hydraulic forces in basins and filters. Coagulant aids benefit flocculation by improving settling and toughness of flocs. The most widely used materials are polymers; others are activated silica, adsorbent weighting agents, and oxidants.

Synthetic polymers are long-chain, high-molecular weight, organic chemicals commercially available under a wide variety of trade names. Polymers are classified according to the type of charge on the polymer chain. Those possessing negative charges are called anionic, those positively charged are called cationic, and those carrying no electric charge are nonionic. Anionic or nonionic are often used with metal coagulants to provide bridging between colloids to develop larger and tougher floc growth.

A variety of alternative processing methods are available: however, because of unique characteristics of each plant's waste, no specific process can be universally applied. Fig. 11.10 is a typical system for dewatering alum sludge. Filtered backwash water is discharged to a clarifier-surge tank. Overflow is recycled to the raw water inlet of the treatment plant while settled solids are discharged, along with waste sludge from the settling tank to a clarifier thickener. Supernatant from this unit may be recycled to the head of the plant or may be discharged to a watercourse. Thickened sludge is mechanically dewatered, usually by centrifugation or filtration. Dewatered solids may be processed for recovery of chemicals or may be discharged to drying beds or to land burial.

11.4.3 Filtration

Several types of filtration systems have been used extensively in developing countries. The design and application of different types of filters depend on the volume, flow rate, and quality of the inflowing water; the desired degree of water purification; and the use of the filtered water. The availabilities of filtering materials and skilled personnel are also factors to be considered in the selection of an appropriate filtration system. Normally, the quality of the product water can be improved by mechanical straining through a porous

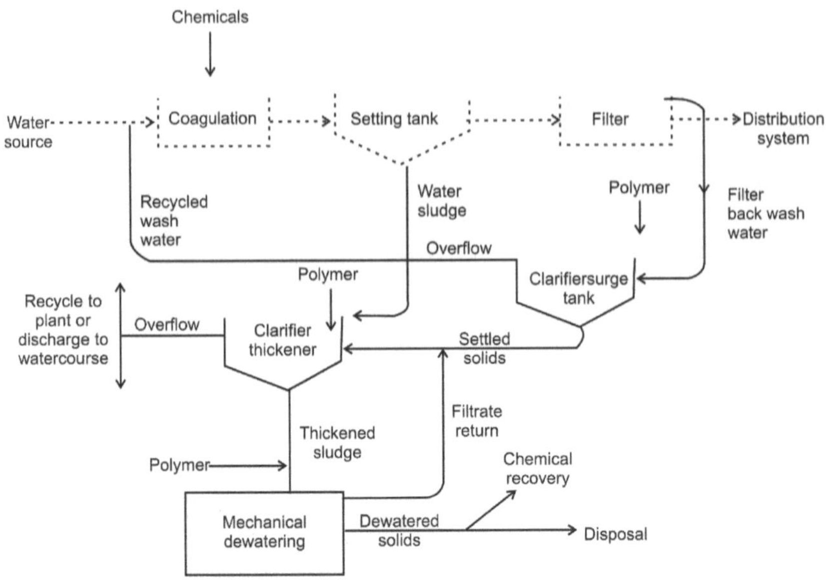

Figure 11.10 Dewatering for alum sludge.

material, such as sand or gravel. Depending on the size of the pores and the nature of the filter material, straining, or filtering, may remove a significant portion of the undesirable contents of the feedwater: suspended and colloidal matter, bacteria and other microorganisms, and, sometimes, certain chemicals. The filter material may be any porous, chemically stable material, but sand (silica and garnet) is used most often. Sand is cheap, inert, durable, and widely available. It has been extensively tested and has been found to give excellent results.

The granular media gravity filter is the most common type used in water treatment to remove non-settleable floc remaining after chemical coagulation and sedimentation. A typical filter bed is placed in a concrete box with a depth of about 9 ft. The granular media, about 2 ft deep, are supported by a graded gravel layer over underdrains. During filtration, water passes downward through the filter bed by a combination of water pressure from above and suction from the bottom. Filters are cleaned by backwashing (reversing the flow) upward through the bed. Wash troughs suspended above the filter surface collect the backwash water and carry it out of the filter box.

11.4.3.1 Roughing Filters

Roughing filters are usually vertical flow filters where the sediment is deposited on the filter media as the water flows down through it. This filter media can vary in size but 4–20 mm is the range to be considered, depending on the sediment to be removed. The filter media can be cleaned by washing it down, or by backwashing under pressure, with a supply of clean water. The ultimate roughing filter is a rapid gravity filter with sand of size 1–4 mm as the filtering media. Such a filter needs careful design and must have the facility for backwashing under pressure with water at a high rate of flow. The filter rate for this type of filter is usually about 5 m/h, but this can safely be exceeded in many situations.

11.4.3.2 Slow Sand Filters

Slow sand filters function by forming a film of bacteria and algae on the surface of the sand as the water passes through it. The rate of flow must be controlled to 2.5 m^3/m^2/day, or a vertical flow rate of 0.1 m/h, and the filter must be cleaned periodically as the flow rate drops, by removing a skin of sand (20 mm). The incoming water must be of a reasonable quality, or must receive pretreatment, to prevent the slow sand filter from blocking too quickly. It is usually necessary to have two units in parallel,

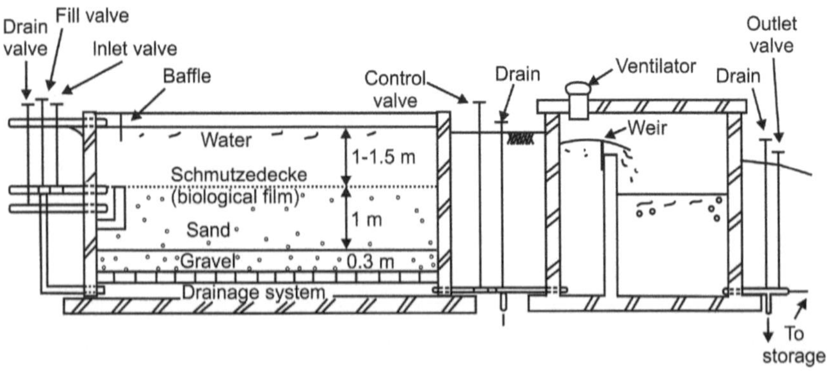

Figure 11.11 Slow sand filters.

so some supply can be maintained when one unit is out of commission for cleaning.

A typical layout is shown in Fig. 11.11.

- The open tank should be about 3 m deep.
- The filter media is 1 m deep with clean sand of one size, between 0.15 and 0.35 mm.
- The filter media is supported on gravel, varying between 2 and 10 mm.
- An underfloor drainage system is required, which is constructed of bricks, blocks, or pre-cast slabs.
- The baffled inlet should be about 1 m above the sand.
- The outlet flow needs to be controlled by a weir and outlet valve.

Many different media have been used for the underdrain system. Bricks, stone, and even bamboo have been used for this purpose; bamboo, however, requires frequent replacement because it is organic and subject to decomposition. The effective size of the sand used in slow sand filters is about 0.2 mm, and it may range between 0.15 and 0.35 mm, with a coefficient of uniformity of between 1.5 and 3.0. In a mature bed, a layer of algae, plankton, and bacteria forms on the surface of the sand. The walls of the filter can be made of concrete or stone. Sloping walls, dug into the earth and supported or protected by chicken wire reinforcement and a sand or sand-bitumen coating, could be a cost-effective alternative to concrete. Lateral pipes range from 2 to 8 in., while the bottom drains are normally between 10 and 30 in. Bottom drains consist of a system of manifold and lateral pipes.

The successful performance of a slow sand filter depends mainly on the retention of inorganic suspended matter by the straining action of the sand. Filtration rates usually employed range between 2.5 and

$6.0\,m^3/m^2/day$. Higher rates may be used after a series of tests demonstrate that the effluents are of good quality. The system should be designed for flexibility, and it should consist of a number of separate units to enable maintenance to be performed without interruption of the water service. The suggested number of units for a given population size ranges from two units for a population of 2000 up to six units for a population of 200,000.

Slow sand filters are very effective in removing solids and turbidity when the raw water has low turbidity and color (turbidity up to 50 NTU and color up to 30 Pt units). Taste and odor are also improved. However, if the raw water quality is poor, filtration is often less effective. In such situations, roughing filters, or pre-filters, are often used before the feedwater enters the slow sand filters. The slow sand filters are very effective in removing bacteria; in general, their effectiveness in removing bacteriological contaminants ranges between 80% and 99%, depending on the initial level of contaminants and the number and design of the filtration units.

11.4.3.2 Rapid Sand Filters

Rapid sand filters differ from slow sand filters in the size of the media employed. Media in rapid sand filters may range in size from 0.35 to 1.0 mm, with a coefficient of uniformity of 1.2–1.7. A typical size might be 0.5 mm, with an effective size of 1.3–1.7 mm. This range of media size has demonstrated the ability to handle turbidities in the range of 5–10 NTU at rates of up to $4.88\,m^3/m^2/h$. Filtration rates for rapid filters may be as high as $100–300\,m^3/m^2/day$, or about 50 times the rate of a slow sand filter. The number of filters used for a specific plant ranges from 3 filters for a plant capacity of 50 L/s to 10 filters for a plant capacity of 1500 L/s. A typical rapid sand filter consists of an open watertight basin containing a layer of sand 60–80 cm thick, supported on a layer of gravel. The gravel, in turn, is supported by an underdrain system. In contrast to a slow sand filter, the sand is graded in a rapid rate filter configuration. The sand is regraded each time the filter is backwashed, with the finest sand at the top of the bed. The underdrain system, in addition to performing the same functions served in the slow rate filter, serves to distribute the backwash water uniformly to the bed. The underdrain system may be made of perforated pipes, a pipe and strainer, vitrified tile blocks with orifices, porous plates, etc. A clear well is usually located beneath the filters (or in a separate structure) to provide consistent output quantity. The minimum number of filter units in a system is two. The surface area of a unit is normally less than $150\,m^2$. The ratio of length to width is 1.25–1.35.

Rapid sand filtration plants are complicated to operate, requiring operator training for the plant to produce a product water of consistent quality and quantity. The filters require frequent backwashing to maintain satisfactory operating heads in the system (filter runs may vary from only a few hours to as many as 24–72 h, depending on the suspended solids in the influent). Backwashing rates are typically 0.6 m^3/min or higher, for a period of several minutes. In addition, the initial production following backwashing is channeled to waste for several minutes. Thus, the water backwashing uses can be as much as 10–15% of the total plant output. On the other hand, rapid sand filtration plants (including chemical treatment) can effectively treat higher solids loadings and produce higher outputs than slow sand filters. The land area requirements are significantly lower. Rapid sand filters are more complex to operate than slow sand filters, but they are widely used, especially in areas with high turbidity and where land requirements may be an important design consideration.

11.4.3.3 Dual-Media or Multimedia Filters

Dual-media filtration uses two layers, a top one of anthracite and a bottom one of sand, to remove the residual biological floc contained in settled, secondary-treated wastewater effluents and residual chemical-biological floc after alum, iron, or lime precipitation in potable water treatment plants. Gravity filters operate by using either the available head from the previous treatment unit or the head developed by pumping the feedwater to a flow cell above the filter cells. A filter unit consists of an open watertight basin; filter media; structures to support the media; distribution and collection devices for influent, effluent, and backwash water flows; supplemental cleaning devices; and the necessary controls to sequence water flows, levels, and backwashing.

Dual or multimedia filters are limited to developing countries that can inexpensively acquire anthracite. The higher skill level and energy requirements for the operation of these high rate systems may limit their application.

11.4.3.4 Upflow Solids Contact Filter

These units eliminate the need for separate flocculators and settling tanks since they perform liquid–solid separation, filtration, and sludge removal in a single unit process. Coagulation and flocculation are performed in a granular medium (such as a layer of gravel under a sand bed). The use of flocculent aids improves filtration results. This process should be restricted to raw waters of low turbidity (up to 50 JTU) and no more than 150 mg/L of suspended solids. These filters are designed for rates of filtration between 120 and 150 m^3/m^2/day. Upflow solids contact filters, because of their simplicity and low cost, could be an effective technology in many developing countries.

A number of factors affect the operation and maintenance of filters. During filtration, impurities are deposited in and on the surface layer of the sand bed, and the loss of head increases. At a predetermined limit the filter is taken out of service and cleaned. The period between cleaning is dependent on the type of filter. Dual-media and rapid sand filters are cleaned by hydraulic backwashing (upflow) with potable water. Thorough cleaning of the bed makes it advisable in the case of single medium filters, and mandatory in the case of dual- or mixed-media filters, to use auxiliary scour or so-called surface wash devices before or during the backwash cycle. In dual-media and mixed-media beds, such additional effort is needed to remove accumulated floc, which is stored throughout the bed depth to within a few inches of the bottom of the fine media. Backwashing is generally carried out every 24–72 h. The optimum rate of wash water application is a direct function of water temperature, as expansion of the bed varies inversely with the viscosity of the wash water. For example, a backwash rate of 18 gpm/ft^2 at 20°C equates to 15.7 gpm/ft^2 at 5°C, and to 20 gpm/ft^2 at 35°C. The time required for backwashing varies from 3 to 15 min. After the washing process, water should be discharged to waste until the turbidity drops to an acceptable value.

11.4.3.5 Flow Control

A filter is cleaned by backwashing when the measured head loss through the media is approximately 8 ft. As shown in Fig. 11.12. This valving arrangement drains away the water from above the filter bed down to the elevation of the outlet channel wall. If air agitation is used, compressed air is introduced through the underdrain to loosen accumulated impurities from the grains of media by rubbing them against each other in the turbulence caused by the air mixing. If rotary washer units are employed, they are placed in operation prior to starting the water backwash.

11.4.3.6 Filter Media

The action that takes place in a granular media filter is extremely complex, consisting of straining, flocculation, and sedimentation. Gravity filters do not function properly unless the applied water has been chemically treated and, if necessary, settled to remove the large floc. Coagulant carryover is

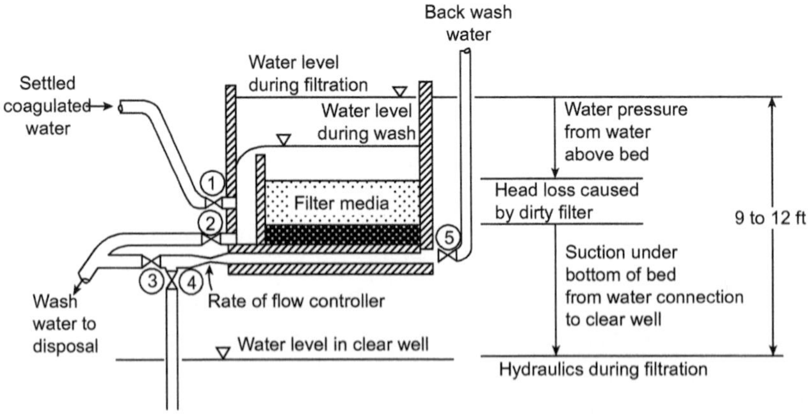

Figure 11.12 Filter media.

essential in removing microscopic particulate matter that would otherwise pass through the bed. If an excessive quantity of large floc overflows the settling basin, a heavy mat forms on the filter surface by straining action and clogs the bed.

11.4.3.7 Advantages

- Filtration systems have a low construction cost, especially when built using manual labor.
- These systems are simple to design, install, operate, and maintain, which makes them ideal for use in areas where skilled personnel are few.
- No chemicals are required, although flocculent aids are sometimes used in conjunction with large-scale filtration systems; supplies of sand can usually be found locally.
- Use of filtration to pretreat water and wastewaters results in fewer sludge disposal problems because fewer contaminants are left to be removed during the treatment process.

11.4.4 Disinfection by Boiling and Chlorination

Boiling and chlorination are the most common water and wastewater disinfection processes in use throughout the world. Boiling is primarily used in rural areas in developing countries to eliminate living organisms, especially bacteria, present in the water. It is also used in emergencies when other, more sophisticated methods of disinfection are not available. Prior to the development of chlorination, boiling was the principal method used to kill pathogenic organisms.

11.4.4.1 Boiling

Boiling is a very simple method of water disinfection. Heating water to a high temperature, 100°C, kills most of the pathogenic organisms, particularly viruses and bacteria causing waterborne diseases. For boiling to be most effective, the water must boil for at least 20 min. Since boiling requires a source of heat, rudimentary or nonconventional methods of heat generation may be needed in areas where electricity or fossil fuels are not available.

11.4.4.2 Chlorination

The most common application of chlorination is disinfection of drinking water to destroy microorganisms that cause diseases in humans. The disinfecting action of chlorine results from a chemical reaction between HOCl and the microbial cell structure, inactivating required life processes. Rate of disinfection depends on the concentration and form of available chlorine residual, time of contact, pH, temperature, and other factors. Hypochlorous acid is more effective than hypochlorite ions; therefore, the power of free chlorine residual decreases with increasing pH. The disinfecting action of combined available chlorine is significantly less than that of free chlorine residual.

As an oxidant, it is used in iron and manganese removal, destruction of taste and odor compounds, and elimination of hydrogen sulfide. One volume of liquid chlorine confined in a container under pressure yields about 450 volumes of gas. It is a strong oxidizing agent reacting with most elements and compounds. Most chlorine is extremely corrosive; consequently, conduits and feeder parts in contact with chlorine are either special alloys or nonmetal. The vapor is a respiratory irritant that can cause serious injury if exposure to a high concentration occurs.

Hypochlorites are salts of hypochlorous acid (HOCl). Calcium hypochlorite [$Ca(OCl)_2$] is the predominant dry form used in the United States. High-test calcium hypochlorites, available commercially in granular, powdered, or tablet forms, readily dissolve in water and contain about 70% available chlorine. Sodium hypochlorite (NaOCl) is commercially available in liquid form at concentrations between 5% and 15% available chlorine. Most water treatment plants in the United States use liquid chlorine since it is less expensive than hypochlorites. The latter are used in swimming pools, small waterworks, and emergencies.

Chlorine combines with water forming hypochlorous acid, which, in turn, can ionize to the hypochlorite ion. Below pH 7, the bulk of the HOCl remains unionized, while above pH 8, the majority is in the form of OCL^-, Eq. (11.iv).

$$Cl_2 + H_2O \rightarrow HCl + HOCl \underset{PH<7}{\overset{PH>8}{\Leftrightarrow}} H^+ + OCL^- \quad (11.iv)$$

Hypochlorites added to water yield the hypochlorite ion directly, Eq. (11.v).

$$Ca\ (OCL)_2 + H_2O = Ca^- + 2OCl^- + H_2O \quad (11.v)$$

Chlorine existing in water as hypochlorous acid and hypochlorite ion is defined as free available chlorine.

Chlorine readily reacts with ammonia in water to form chloramines as follows:

$$HOCl + NH_3 = H_2O = NH_2Cl \text{ (monochloramine)} \quad (11.vi)$$

$$HOCl + NH_2Cl = H_2O + NHCl_2 \text{ (dichloramine)} \quad (11.vii)$$

$$HOCl + NHCl_2 = H_2O + NCl_3 \text{ (trichloramine)} \quad (11.viii)$$

The reaction products formed depend on pH, temperature, time, and initial chlorine-to-ammonia ratio. Monochloramine and dichloramine are formed in the pH range of 4.5–8.5. Above pH 8.5, monochloramine generally exists alone, but below pH 4.4, trichloramine is produced. Chlorine existing in chemical combination with ammonia–nitrogen or organic nitrogen compounds is defined as combined available chlorine. When chlorine is added to water containing ammonia, the residuals that develop yield a curve similar to that shown in Fig. 11.13. The straight line from the origin is the concentration of chlorine applied, or the residual chlorine if all of that applied appeared as residual. The curved line represents chlorine residuals, corresponding to various dosages, remaining after a specified contact time, such as 20 min. Chlorine demand at a given dosage is measured by the vertical distance between the applied and residual lines. This represents the amount of chlorine reduced in chemical reactions and, therefore, the amount that is no longer available. With molar chlorine to ammonia–nitrogen ratios less than 1 to 1, monochloramine and dichloramine are formed with the relative amounts depending on pH and other factors. Higher dosages of chlorine increase the chlorine-to-nitrogen ratio and result in oxidation

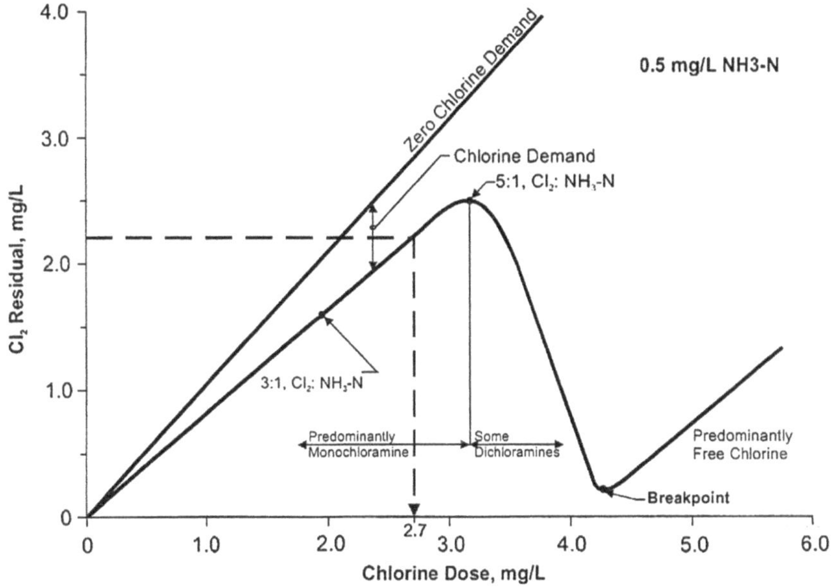

Figure 11.13 Typical breakpoint chlorination curve.

of the ammonia and reduction of the chlorine. Theoretically, 3 mol of chlorine react with 2 mol of ammonia to drive nitrogen off as gas and to reduce chlorine to the chloride ion:

$$2NH_3 + 3Cl_2 = N_2 + 6HCl \tag{11.ix}$$

Chloramine residuals decline to a minimum value referred to as the breakpoint. Dosages in excess of this breakpoint produce free chlorine residuals. The breakpoint curve is unique for each water tested since chlorine demand depends on the concentration of ammonia, presence of other reducing agents, contact time between chlorine application and residual testing, and other factors.

Chlorine is a much stronger oxidizing agent for manganese than is dissolved oxygen. Therefore, one of the treatment processes for iron and manganese removal uses chlorine to remove these metals from solution.

Hydrogen sulfide present in groundwater can be rapidly converted to the sulfate ion using chlorine.

$$H_2S + 4Cl_2 + 4H_2O = H_2SO_4 + 8HCl \tag{11.x}$$

Breakpoint chlorination in the treatment of surface waters may be used to destroy objectionable tastes and odors and to eliminate bacteria, minimizing biological growths on filters and after growths in the distribution system. Breakpoint chlorination of polluted surface waters, however, can result in formation of trihalomethanes.

11.4.5 Defluoridation

When water supplies contain excess fluorides, the teeth of most consumers over a period of several years become mottled with a permanent brown to gray discoloration of the enamel. Children who have been drinking water containing 5 mg/L develop fluorosis to the extent that the enamel is severely pitted, resulting in loss of teeth.

Treatment methods for defluoridation use either activated alumina or bone char. Water is percolated through insoluble, granular media to remove the fluorides. The media are periodically regenerated by chemical treatment after becoming saturated with fluoride ion. Regeneration of bone char consists of backwashing with a 1% solution of caustic soda and then rinsing the bed. Reactivation of alumina also involves backwashing with a caustic solution. Removal of excess fluoride from public water supplies is a sound economic investment when related to the increased cost of dental care and loss of teeth. Despite the obvious need, some communities have not installed defluoridation units, allegedly because of excessive costs of construction and operation.

11.4.6 Precipitation Softening

Hardness in water is caused by calcium and magnesium ions resulting from water coming in contact with geological formations. Public acceptance of hardness varies, although many customers object to water harder than 150 mg/L. The maximum level considered or public supply is 300–500 mg/L. A moderately hard water is generally defined as 60–120 mg/L. Hardness interferes with laundering by causing excessive soap consumption and may produce scale in hot water heaters and pipes. Precipitation softening uses lime (CaO) and soda ash (Na_2CO_3) to remove calcium and magnesium from solution. In addition, lime treatment has the incidental benefits of bactericidal action, removal of iron, and aid in clarification of turbid surface waters. Carbon dioxide can be applied for re-carbonation after lime treatment to lower the pH by converting the excess hydroxide ion and carbonate ion to bicarbonate ion.

Lime is sold commercially in the forms of quicklime and hydrated lime. Quicklime, available in granular form, is a minimum of 90% CaO with magnesium oxide being the primary impurity. A slaker is used to prepare quicklime for feeding in a slurry containing approximately 5% calcium hydroxide. Powdered, hydrated lime contains approximately 68% CaO and may be prepared by fluidizing in a tank containing a turbine mixer. Lime slurry is written as $Ca(OH)_2$ in chemical equations. Soda ash is a grayish-white powder containing at least 98% sodium carbonate.

The chemical reactions in precipitation softening are these:

$$CO_2 + Ca(OH)_2 = CaCO_3\downarrow + H_2O$$

$$Ca(HCO_3)_2 + Ca(OH)_2 = 2CaCO_3\downarrow + 2H_2O$$

$$Mg(HCO_3)_2 + Ca(OH)_2 = CaCO_3\downarrow + MgCO_3 + 2H_2O$$

$$MgCO_3 + Ca(OH)_2 = CaCO_3\downarrow + Mg(OH)_2\downarrow$$

$$Mg(HCO_3)_2 + 2Ca(OH)_2 = 2CaCO_3\downarrow + Mg(OH)_2\downarrow + 2H_2O$$

$$MgSO_4 + Ca(OH)_2 = Mg(OH)_2\downarrow + CaSO_4$$

$$CaSO_4 + Na_2CO_3 = CaCO_3\downarrow + Na_2SO_4$$

11.5 DRINKING WATER QUALITY MONITORING

11.5.1 Microbiological Quality of Drinking Water

Drinking water must be free of all pathogenic microorganisms. The viruses, bacteria, protozoa, and helminths are most likely to be transmitted by water. Testing water for this broad diversity of pathogens is not feasible because of the difficulty in performing laboratory analyses and their poor quantitative reproducibility. Therefore, the microbial quality of drinking water is controlled by specified treatment techniques and monitoring for the presence of coliform bacteria. The disinfection of surface waters is defined by treatment techniques for removal of protozoal cysts by chemical coagulation and granular media filtration, such as giardia and cryptosporidium resistant to chlorine residual. Inactivation of any remaining cysts and enteric viruses is by chemical treatment, commonly chlorination. Effective coagulation and filtration of the treated water is determined by a turbidity equal to or less than 0.5 NTU in at least 95% of the measurements. Effective chemical disinfection of the water prior to entering the distribution system is determined by C-t product, which is the disinfectant concentration multiplied by the time of contact.

11.5.2 Chemical Quality of Drinking Water

Drinking water standards are likely to remain flexible with continuous adjustment to accommodate changing chemical nature in the environment. Chloride, sulfate, and total dissolved solids have taste and laxative properties, and highly mineralized water affects the quality of coffee and tea. The quality of water can be quantified by water quality index (WQI).

11.5.3 Water Quality Index

For calculation of WQI, selection of parameters has great importance. Since selection of too many parameters might widen the water quality index and importance of various parameters depends on the intended use of water, eight physicochemical parameters, namely, pH, TDS, total alkalinity, total hardness, chloride, sulfate, Dissolved Oxygen (DO), and BOD, are normally used to calculate WQI. The calculation of WQI can be made using a weighted arithmetic index method in the following steps.

Calculation of subindex or quality rating ($\mathbf{q}_\eta$)

Let there be η water quality parameters, and the quality rating or subindex ($\mathbf{q}_\eta$) corresponding to the ηth parameter is a number reflecting the relative value of this parameter in the polluted water with respect to its standard permissible value. The $\mathbf{q}_\eta$ is calculated using the following expression:

$$\mathbf{q}_\eta = 100(V_\eta - V10)/S - V$$

where $\mathbf{q}_\eta$, quality rating for the ηth water quality parameter; V_η, estimated value of the ηth parameter at a given sampling station; S_η, standard permissible value of ηth parameter; V_{10}, ideal value of ηth parameter in pure water.

All the ideal values (V) are to be taken as zero for the drinking water except pH = 7.0 and dissolved oxygen = 14.6 mg/L.

Calculation of unit weight (W_n)

The unit weights (W_n) for various water quality parameters are inversely proportional to the recommended standards for the corresponding parameters:

$$W_n = \frac{K}{S_n}$$

where W_n, unit weight for parameters; S_n, standard value for parameters; K, constant for proportionality.

Calculation of WQI[1]

WQI is calculated from the following equation:

$$WQI = \sum_{\eta-1}^{\eta} \mathbf{q_n}\,\mathbf{W_n} / \sum_{\eta-1}^{\eta} \mathbf{W_n}$$

[1] T.N. Tiwari, M. Mishra, IJEP, 5(4) (1985) 276–279.

11.6 REMOVAL OF DISSOLVED SALTS: DESALINATION TECHNOLOGIES

Processes for separating salts from water in potable water treatment include distillation, reverse osmosis, and electrodialysis. Distillation and reverse osmosis are the common methods for desalinization of seawater. Reverse osmosis and electrodialysis are used to desalt brackish groundwater or to reduce the concentration of contaminants that are hazardous to human health, such as nitrate, fluoride, and radionuclides. In addition to high energy consumption, a significant problem at inland desalting plants is the disposal of reject brine. Common methods are evaporating ponds, deep-well injection, and piping to the ocean depending on the volume of reject brine, site geography, and climate. There are essentially five basic techniques to desalt water: distillation, reverse osmosis, electrodialysis, ion exchange, and freezing processes. Distillation and freezing remove fresh water from saline, leaving behind a more concentrated brine. Reverse osmosis and electrodialysis are processes in which membranes are used to separate salts from fresh water. Ion exchange involves passing saline water over resins that exchange more desirable ions for less desirable dissolved ions.

Distillation involves boiling the saline water at atmospheric or reduced pressure and condensing the vapor as fresh water, leaving behind a more concentrated brine solution. Even though distillation chambers are run in series to conserve energy (i.e., the incoming water to one unit is preheated by using it to cool the vapor in another unit), the energy consumption of distillation methods is still relatively high compared to other methods. Because distillation involves vaporizing water from the salty feedwater, the energy required for distillation does not increase appreciably with increasing salinity. Thus, distillation plants have commonly been used for desalting seawater, although membrane systems are competing in this area. Solar distillation has also been developed, but even though the energy source is free, the conversion rate is fairly low.

11.6.1 Membrane Processes

A process of dialysis and osmosis occurs in the body. Membranes are used in two commercially important desalting processes: electrodialysis (ED) and reverse osmosis (RO). Each process uses the ability of the membranes to differentiate and selectively separate salts and water. However, membranes are used differently in each of these processes.

Electrodialysis depends on the ability of electrically charged ions in saline water to migrate to positive or negative poles in an electrolytic cell. Two different types of ion-selective membranes are used: one that allows passage of positive ions and one that allows negative ions to pass between the electrodes of the cell. When an electric current is applied to drive the ions, fresh water is left between the membranes. The amount of electricity required for electrodialysis, and therefore its cost, increases with increasing salinity of feedwater. Thus, electrodialysis is less economically competitive for desalting seawater.

ED is a voltage-driven process and uses an electrical potential to move salts selectively through a membrane, leaving fresh water behind as product water. RO is a pressure-driven process, with the pressure used for separation by allowing fresh water to move through a membrane, leaving the salts behind. The development of ED provided a cost-effective way to desalt brackish water for producing potable water for municipal use. ED depends on the following general principles:

1. Most salts dissolved in water are ionic, being positively or negatively charged.
2. Membranes can allow passage of either cation or anions.

The dissolved ionic constituents in a saline solution, such as chloride, sodium, calcium, and carbonate, are dispersed in water, effectively neutralizing their individual ions. When the electrodes are connected to an outside source of direct current like a battery and placed in a container of saline water, individual membranes that will allow either cations or anions to pass are placed between a pair of electrodes. These membranes are arranged alternately, with an anion selective membrane followed by a cation selective membrane. A spacer sheet that permits water to flow along the face of the membrane is placed between each pair of membranes.

One spacer provides a channel that carries feed (and product) water, while the next carries brine. As the electrodes are charged and saline feedwater flows along the product, water spacers are attracted and diverted through the membrane toward the positive electrode. This dilutes the salt content of the water in the product water channel. The anions pass through the anion selective membrane, but they cannot pass any further than the cation selective membrane, which blocks their path and traps the anions in the brine stream. Similarly, cations (such as chloride or carbonate) under the influence of the negative electrode move in the opposite direction through the cation selective membrane to the concentrate. A basic ED unit consists of the following:

- pretreatment train
- membrane stack
- low-pressure circulating pump

- power supply for direct current (a rectifier)
- posttreatment

ED has the following characteristics that make it suitable for a number of applications:

- capacity for high recovery
- energy usage that is proportional to the salts removed
- ability to treat feedwater with a higher level of suspended solids than RO
- unaffected by nonionic substances such as silica
- low chemical usage for pretreatment ED units is normally used to desalinate brackish water.

RO is a membrane process that relies on the tendency for fresh water to diffuse through a semipermeable membrane into a salt solution, thereby diluting the more saline water. The fresh water migrates through the membrane as though was were pressure on it, and the effective driving force is called osmotic pressure. By applying pressure to saline water on one side of a semipermeable membrane, fresh water can be driven through in the direction opposite to the osmotic flow. This process is called reverse osmosis. Although energy intensive, one of the major advantages of RO is lower energy consumption than distillation, particularly for brackish water, although RO is used to desalt seawater. The major energy required for desalting is for pressurizing the feedwater. Similarly, an RO system is made up of the following basic components:

- pretreatment
- high-pressure pump
- membrane assembly
- posttreatment

Pretreatment is important in RO because the membrane surfaces must remain clean. Therefore, suspended solids must be removed and the water pretreated so that salt precipitation or microbial growth does not occur on the membranes. Two developments have helped to reduce the operating cost of RO plants during the past decade: the development of more cost-effective membranes and the use of energy recovery devices.

RO uses semipermeable membranes for removing organic and inorganic pollutants. It is directly opposite to the classical osmosis process that occurs in nature. In RO, pure water is forced out of saline water by applying a pressure exceeding the osmotic pressure. This pressure is applied on the side of the feed solution, and the feed solution is in contact with the semipermeable membrane. This membrane is made of an assortment of fibers. The process can remove over 95% of the hardness, more than 90% of the inorganic solids, and almost all of the bacteria, virus, and spores.

During operation, the pressure used to force the clean water through the membrane ranges from 200 to 1000 psig. Thus, in actual operation, salt-free water flows from one end of the system, and concentrated brine is removed from another. If seawater containing 35,000 parts per million of salt is subjected to a single-stage RO operation, the water effluent will contain 3500 ppm of salts, which is still brackish. However, if it is subjected to another stage of RO, the salt concentration will be brought down from 3500 to 350 ppm, which is in the range of potable water; operational cost, in this case, will be approximately doubled.

Ion exchange resins substitute hydrogen and hydroxide ions for salt ions. For example, cation exchange resins are commonly used in home water softeners to remove calcium and magnesium from "hard" water. A number of municipalities use ion exchange for water softening, and industries requiring extremely pure water commonly use ion exchange resins as a final treatment following RO or electrodialysis. The primary cost associated with ion exchange is in regenerating or replacing the resins. The higher the concentration of dissolved salts in the water, the more often the resins will need to be renewed. In general, ion exchange is rarely used for salt removal on a large scale.

In ion exchange, dissolved salts from aqueous solutions are removed by exchanging the unwanted ions with the desired ions, or hydroxyl ions, which in turn give water. Let R be the symbol for the cation exchanger or radical. The reactions involved are these:

$$\begin{matrix}Ca\\Mg\end{matrix}\left\{\begin{matrix}(HCO_3)_2\\SO_4\\Cl_2\end{matrix}\right\} + 2NaR \rightarrow \left\{\begin{matrix}Ca\\Mg\end{matrix}\right\} R_2 + Na_2 \left\{\begin{matrix}(HCO_3)_2\\SO_4\\Cl_2\end{matrix}\right\}$$

(soluble from Water feed) — Sodium cation exchanger (insoluble) — cation exchanger (insoluble) — (soluble)

$$\left\{\begin{matrix}Ca\\Mg\\Na_2\end{matrix}\right\}(HCO_3)_2 + 2\,HR \rightarrow \left\{\begin{matrix}Ca\\Mg\\Na_2\end{matrix}\right\} R_2 + 2H_2O + 2CO_2$$

(soluble) — Hydrogen cation exchanger (insoluble) — (insoluble)

$$\left\{\begin{matrix}Ca\\Mg\\Na_2\end{matrix}\right\}\left\{\begin{matrix}SO_4\\Cl_2\end{matrix}\right\} + 2\,HR \left\{\begin{matrix}Ca\\Mg\\Na_2\end{matrix}\right\} R_2 + H_2 \left\{\begin{matrix}SO_4\\Cl_2\end{matrix}\right\}$$

(soluble) — Hydrogen cation exchanger (insoluble) — (insoluble) — (soluble)

Thus the calcium and magnesium, which cause hardness in water, react with the radical (R) and give an insoluble substance that is removed as waste, and the sodium gives some soluble substance that is removed as waste, and the sodium gives some soluble substance, which is not desirable.

The hydrogen cation exchanger behaves similarly. The ability of the cation exchanger bed to produce softened water decreases with time, and the exchanger must be regenerated after a certain period. Regeneration is accomplished by using sodium chloride or sulfuric acid, which removes the calcium and magnesium, respectively, as soluble chlorides or sulfates, and regenerates the exchanger to its original state. Now the unit is ready again to do its job of softening the water. The regeneration reactions are these:

$$\left\{\begin{matrix} Ca \\ Mg \end{matrix}\right\} R_2 + 2\,NaCl \rightarrow 2NaR + \left\{\begin{matrix} Ca \\ Mg \end{matrix}\right\} Cl_2$$

$$\left\{\begin{matrix} CA \\ Mg \\ Na_2 \end{matrix}\right\} R_2 + H_2SO_4 \rightarrow 2HR + \left\{\begin{matrix} Ca \\ Mg \\ Na_2 \end{matrix}\right\} SO_4$$

The freezing processes involve three basic steps: partial freezing of the feedwater in which ice crystals of fresh water form ice-brine slurry, separating the ice crystals from the brine, and melting the ice. Freezing has some inherent advantages over distillation in that less energy is required, and there are a minimum of corrosion and scale formation problems because of the low temperatures involved. Freezing processes have the potential to concentrate waste streams to higher concentration than other processes, and the energy requirements are comparable to RO. While the feasibility of freeze desalination has been demonstrated, further research and development remains before the technology will be widely available.

11.6.2 Various Desalination Technologies Options for Production of Fresh Waters From Sea

Many countries are now considering desalination as one important source of water supply to meet the fresh water needs of drinking and agriculture. Desalination of seawater has been practiced regularly for over 50 years and is a well-established means of water supply in many countries. It is now feasible, technically and economically, to produce large quantities of water of excellent quantity from desalination processes.

The main raw material and product of desalination process is presented as follows:

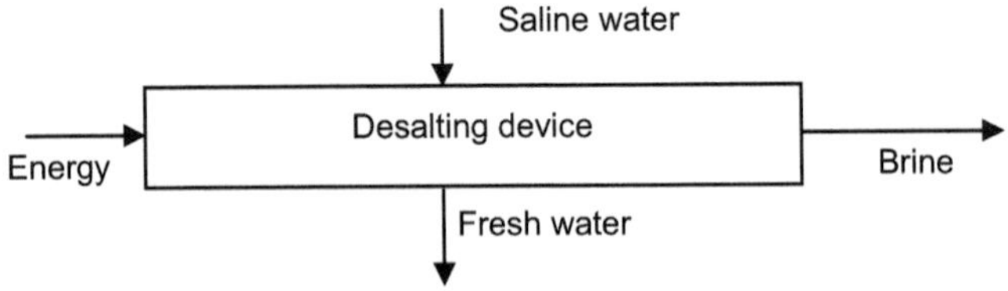

Two main directions survived the evolution of desalination technology, namely, evaporation and membrane techniques. A variety of desalting technologies have been developed over the years, and based on their commercial success, they can be classified into major and minor desalting processes and are given next:

Major Processes

Thermal

- Multistage flash distillation
- Multiple effect distillation
- Vapor condensation

Membrane

- Electrodialysis
- Reverse osmosis

Minor Processes

Freezing
Membrane distillation
Solar humidification

The continual growth of desalination has been monitored over the years, and the desalting equipment is now used in over 100 countries. According to reports, 10 countries have about 75% of all the capacity. Almost half of this desalting capacity is used to desalt seawater in the Middle East and North Africa. Saudi Arabia ranks first in total capacity of the world.

11.6.3 Thermal Processes

About half of the world's desalted water is produced with heat to distill fresh water from seawater. The distillation process mimics the natural water cycle in that salt water is heated, producing water vapor that is in turn condensed to form fresh water. The process that accounts for the most desalting capacity for seawater is the multistage flash distillation (MSF) process.

In the MSF process, seawater is heated in a vessel called the brine heater. This is generally done by condensing steam on a bank of tubes that carry seawater that passes through the vessel. This heated seawater then flows into another vessel, called a stage, where the ambient pressure is lower, causing the water to immediately boil. The sudden introduction of the heated water into the chamber causes it to boil rapidly, almost exploding or flashing steam. Generally, only a small percentage of this water is converted into steam, depending on the pressure maintained in this stage, since boiling will continue only until the water cools to the boiling point.

11.6.4 Multi-effect Distillation

The multi-effect distillation (MED) process has been used for industrial distillation for a long time. Traditional uses for this process are the evaporation of juice from sugar cane in the production of sugar and the production of salt with the evaporative process. Some of the early water distillation plants used the MED process, but the MSF units, because of a better resistance against scaling, displaced this process. Diverse designs have been or are being used for the heat exchange area, such as vertical tubes with falling brine film or rising liquids, horizontal tubes with falling film, or plates with a falling brine film. By far the most common heat exchanger consists of horizontal tubes with a falling film.

11.6.5 Vapor Compression Distillation

The vapor compression distillation process is generally used in combination with other processes and by itself for small- and medium-scale seawater desalting applications. The heat for evaporating the water comes from the compression of vapor rather than the direct exchange of heat from steam produced in a boiler. The plants that use this process are also designed to take advantage of the principle of reducing the boiling point temperature by reducing the pressure. Vapor compression units have been built in a variety of configurations to promote the exchange of heat to evaporate the seawater.

11.7 DUAL WATER DISTRIBUTION

As the name implies, dual distribution systems involve the use of water supplies from two different sources in two separate distribution networks. The two systems work independently of each other within the same service area. Dual distribution systems are usually used to supply potable water through one distribution network and non-potable water through the other. The systems would be used to augment public water supplies by providing untreated, or poorly treated, water for purposes other than drinking. Such purposes could include fire-fighting, sanitary flushing, street cleaning, or irrigation of ornamental gardens or lawns.

The systems are designed as two separate pipe networks: a potable water distribution system and a system capable of distributing seawater or other non-potable waters. The system includes distribution pipes, valves, hydrants, standpipes, and a pumping system, if required. Pipes in the systems are generally cast iron or ductile iron, although fiberglass has also been used. The technology is, however, suitable only in areas where a supply of raw water is available. This type of system is generally used near the coast where seawater is abundant, or in places where wastewater is readily available as a source of supply. It can also be utilized in areas that have rivers, streams, or other water sources but lack of treatment facilities.

11.7.1 Potential Risks

A dual distribution system requires that two distribution systems have to be installed, at essentially double the cost of a single system. Having non-potable water in a distribution system creates a potential to cross-contaminate the potable water system. Use of untreated seawater or wastewater to irrigate leafy vegetables could also threaten human health. Seawater can be highly corrosive to metal pipes, fittings, and appurtenances.

11.8 SOURCES OF WASTES IN WATER TREATMENT

Wastes originating from water treatment in approximate order of abundance are residues from chemical coagulation, precipitates from softening, filter backwash water, settled solids from pre-sedimentation, oxides from iron and manganese removal, and spent brines from regeneration of ion exchange units. These wastes vary widely in composition, containing the concentrated materials removed from raw water and chemicals added in treatment. They are produced continuously and discharged intermittently. Settled floc is allowed to accumulate in clarifiers over relatively long periods of time, while backwashing of filters produces a high flow of wastewater for a few minutes, usually once a day for each filter.

11.8.1 Coagulation Wastes

The chief constituent in coagulation sludge is either hydrated aluminum from alum or iron oxides from iron coagulants. Small quantities of activated carbon and coagulant aids, such as polymers and activated silica, may be included. Particulate matter entrained in the floc is mostly inorganic in nature, being principally silt and clay. Since the organic fraction is small, the sludge does not undergo active biological decomposition. Aluminum hydroxide sludges are gelatinous in consistency, which makes them difficult to dewater. Settled sludges have low solids concentrations, usually between 0.2 and 2.0%. Iron precipitates are slightly denser than alum sludges. The total solids produced in alum coagulation of a surface water can be estimated by using the following relationship:

$$\text{Sludge solids, lb/mil gal} = 8.34\,(0.44 \times \text{alum dosage} + 0.74 \times \text{turbidity})$$

11.8.2 Filter Wash Water

Backwashing of filters produces a relatively large volume of wastewater with low solids concentration in the range of 0.01–0.1% (100–1000 mg/L). The total solids content depends on efficiency of prior coagulation and sedimentation and may be a substantial fraction, say 30%, of the residue resulting from treatment. Two to three percent of all water processed is used for filter washing; the exact amount is contingent on the type of treatment system and the filter backwashing technique. Wash water may be discharged to a recovery basin and recycled for processing with the raw water. In the case of a lime softening plant that is treating groundwater, backwash maybe collected, mixed, and returned to the inlet of the plant without solids removal. However, in surface water plants, this often creates a buildup of undesirable solids, for example, algae, that keep cycling through the system. Here, the suspended solids are allowed to settle, often after the addition of a polymer to improve flocculation, and only the overflow is returned for reprocessing.

FURTHER READING

[1] H.T. Man, D. Williamson, Water Treatment and Sanitation: Simple Methods for Rural Areas, Intermedia Technology Publications, London, 1986.

[2] [a] J.M. Montgomery, Consulting Engineers, Water Treatment Principles and Design, Walnut Grove, California, 1985.
[b] PAHO, Evaluation of the utilization of new technology in water treatment in Latin America, in: Paper presented for the Seventeenth Meeting of the PAHO Advisory Committee on Medical Research, Lima, Peru. Washington, D.C, 1978.

[3] W.-P. Schmidt, S. Cairncross, Household water treatment in poor populations: is there enough evidence for scaling up now, Environmental Science Technology (June 2009).

[4] D.M. Norton, M. Rahman, A.L. Shane, Z. Hossain, R.M. Kulick, M.I. Bhuiyan, et al., Flocculant-disinfectant point-of-use water treatment for reducing arsenic exposure in rural Bangladesh, International Journal of Environmental Health Research 19 (1) (February 2009) 17–29 Centers for Disease Control and Prevention, Atlanta, GA 30333, USA.

[5] R. Ainsworth (Ed.), Safe Piped Water: Managing Microbial Water Quality in Piped Distribution Systems, IWA Publishing, London, 2004 (for the World Health Organization, Geneva).
[6] O. Schmoll, et al., Protecting Groundwater for Health: Managing the Quality of Drinking-water Sources, IWA Publishing, on behalf of the World Health Organization, London, 2006. Available at: http://www.who.int/water_sanitation_health/publications/protecting_groundwater/en/index.html.
[7] M. Sobsey, Managing Water in the Home: Accelerated Health Gains from Improved Water Supply, World Health Organization, Geneva, 2002 (WHO/SDE/WSH/02.07).

CHAPTER TWELVE

Wastewater Treatment Technologies

12.1 INTRODUCTION

The term "wastewater" includes liquids and waterborne solids from domestic, industrial, or commercial uses as well as other waters that have been used in man's activities, whose quality has been degraded, and which are discharged to a sewage system. The term "sewage" has been used for many years and generally refers to waters containing only sanitary wastes. However, "sewage" technically denotes any wastewaters that pass through a sewer.

Two general categories of wastewaters, not entirely separable, are recognized: domestic wastewaters and industrial wastewaters. Domestic wastewaters originate principally from domestic, household activities but will usually include waters discharged from commercial and business buildings and institutions as well as groundwater. Surface and storm waters may also be present. Domestic wastewaters are usually of a predictable quality and quantity.

Industrial wastewaters, on the other hand, originate from manufacturing processes and are usually of a more variable character. They are often more difficult to treat than domestic wastes. While domestic wastewaters can be dealt with in general terms with respect to character and treatment, industrial wastewaters must be examined on an industry-by-industry basis. This chapter deals with the treatment of domestic wastewaters.

12.2 COLLECTION OF WASTES

A network of pipes, pumps, and pump stations collect and transport wastewater to the treatment plant. The length of time required for the wastes to reach a treatment facility is very important and can affect treatment plant efficiency. A velocity of at least 2 feet per second (2 fps) should be maintained within the collection system to prevent any settling of solids, which tend to clog pipes and cause odors. Manholes should be located

Environmental Management
ISBN 978-0-12-811989-1
http://dx.doi.org/10.1016/B978-0-12-811989-1.00012-9

every 300–500 ft to allow for inspection and cleaning of the sewer. When lowland areas and areas a great distance away from the treatment facility must be sewered, pump and lift stations are normally installed. These pump stations lift the wastewater to a higher elevation where it again can flow by gravity or may be pumped under pressure to the treatment facility. Even though pump stations convey the wastewater to the treatment plant, they can cause operational problems throughout the treatment units. A pump station located just ahead of the plant can cause problems by periodically sending large volumes of flow to the plant one minute and virtually nothing the next minute.

In most wastewater systems the sewer coming into the treatment plant that carries wastes from households, commercial establishments, and industry is called a sanitary sewer. While a storm sewer carries storm water runoff from street catch basins, land, roofs of buildings, etc., a system that conveys both sanitary wastes and storm runoff is called a combined sewer.

Combined sewers can cause operational problems at a treatment plant. Unfortunately, most plants with combined sewers are not designed to handle the increased flow loads during storms and usually cause a decrease in plant efficiency. During high flow periods, detention times are decreased, solids may be washed out of the secondary system and large amounts of grit, sand, and silt may be washed into the plant.

12.3 SOURCES AND TYPES OF WASTEWATER

Domestic wastewaters consist primarily of liquid discharges resulting from sanitary facilities, bathing, laundering, and cooking activities as well as from other sources. The principal sources of domestic and commercial wastewaters are shown next:

Domestic and Commercial Wastewater Sources

- human wastes
  - urine
  - feces
- household wastes
  - laundry
  - bathing
  - kitchen
- storm flows/street washings
  - sand, grit, etc.
  - animal wastes

Domestic and Commercial Wastewater Sources

- groundwater infiltration
  - leaky pipes, manholes
- industrial wastes
  - manufacturing process waste
  - equipment cleaning
  - cooling waters

12.4 COMPOSITION OF WASTEWATER

Wastewaters consist of water in which solids exist as settleable particles, dispersed as colloids, which are materials that do not settle readily, or solids in a dissolved state. The wastewater mixture will contain large numbers of microscopic organisms, mostly bacteria that are capable of consuming the organic component (fats, proteins, and carbohydrates) of the mixture and bringing about rapid changes in the wastewater. Since the sources of wastewater as well as the inputs are highly variable and since there is also an active microbial component, the composition of all wastewaters is constantly changing. Prior to entering a wastewater treatment plant, it is called raw sewage. The solid components of wastewaters actually represent a very small part of most discharges, usually less than 0.1% by weight. However, it is this small component of the wastewater that presents the major challenges in wastewater treatment, operation, and disposal. Essentially, the water component, the other 99.9%, can be viewed as providing the volume and the vehicle for transporting the solid and microbial component of the wastewater. The amount of solid component in wastewater is expressed as a concentration in milligrams per liter or parts per million. Considered chemically, wastewater is a very complex mixture of components that would be difficult to completely define. In broad terms, it consists of an organic and an inorganic component. Probably the most often measured characteristics of wastewater are suspended solids and biological oxygen demand (BOD).

12.4.1 Solids in Wastewater

Since solids are classified in a variety of ways, they should be discussed with regard to the various categorizations that are used as well as with respect to their chemical make-up. There will, of course, be some overlap in the classification method.

12.4.2 Organic Solids

In domestic wastewater, solids are about 50% organic. This fraction is generally of animal or vegetable life, dead animal matter, plant tissue, or organisms, but it may also include synthetic (artificial) organic compounds. These are substances that contain carbon, hydrogen, and oxygen, some of which may be combined with nitrogen, sulfur, or phosphorous. The principal organic compounds present in domestic wastewater are proteins, carbohydrates, and fats together with the products of their decomposition. These compounds are subject to decay or decomposition through the activity of bacteria and other living organisms and are combustible; that is, they can be ignited or burned. Since the organic fraction can be driven off at high temperatures, they are sometimes called volatile solids.

12.4.3 Inorganic Solids

Inorganic solids are substances that are inert and not subject to decay. Exceptions to this characteristic are certain mineral compounds or salts—such as sulfates—which under certain conditions can be broken down. Inorganic solids are frequently called mineral substances and include sand, gravel, and silt as well as the mineral salts in the water supply that produce the hardness and mineral content of the water. In general, they are noncombustible. The amount of these solids, both organic and inorganic, gives to wastewater the characteristic termed as "strength." Actually, the amount or concentration of the organic solids and their capacity to undergo decay or decomposition is the most important part of this strength. The greater the concentration of organic or volatile solids is, the stronger is the wastewater. A "strong" wastewater can be defined as one containing a large amount of solids, particularly organic solids. Solids can also be grouped depending on their physical state as suspended solids, colloidal solids, and dissolved solids, each of which can include both organic and inorganic solids.

12.4.4 Suspended Solids

Suspended solids are those that are visible and in suspension in the water. They are the solids that can be removed from the wastewater by physical or mechanical means, such as sedimentation or filtration. More precisely, they are the solids that are retained on the filter mat or glass fiber pad in a Gooch crucible. Suspended solids will include the larger floating particles and consist of sand, grit, clay, fecal solids, paper, pieces of wood, particles of food and garbage, and similar materials. Suspended solids are approximately 70% organic solids and 30% inorganic solids, the latter being principally sand and grit. The suspended solids portion consists of settleable solids and colloidal solids.

12.4.5 Settleable Solids

Settleable solids are that portion of the suspended solids that are of sufficient size and weight to settle in a given period of time, usually 1 h. These will settle in an Imhoff Cone in 1 h. The results are reported as milliliters of settled solids per liter of wastewater. Settleable solids are approximately 75% organic and 25% inorganic.

12.4.6 Colloidal Suspended Solids

Colloidal suspended solids are solids that are not truly dissolved and yet do not settle readily. These are somewhat loosely defined as the differences between the total suspended solids and the settleable solids. There is, at present, no simple or standard laboratory test to specifically determine colloidal matter. Most colloids will not settle out even after long quiescent periods of settling. They constitute that portion of the total suspended solids (about 40%) that are not readily removed by physical or mechanical treatment facilities but may be filtered out in a Gooch crucible. Colloidal solids are about 65% organic, 35% inorganic, subject to rapid decay, and are an important factor in the treatment and disposal of wastewater.

12.4.7 Dissolved Solids

Dissolved solids are smaller in size than suspended and colloidal solids. As used, the term means all of the solids that pass through the filter pad of a Gooch crucible. Of the total dissolved solids, about 90% are in true solution and about 10% colloidal. Dissolved solids, as a whole, are about 40% organic and 60% inorganic in nature.

12.4.8 Total Solids

Total solids, as the term implies, includes all of the solid constituents of a wastewater. Total solids are the total of the organic and inorganic solids or the total of the suspended and dissolved solids. In average domestic wastewater, total solids are about half organic and half inorganic, and about two-thirds in solution (dissolved) and one-third in suspension. The organic solids, which are subject to decay, constitute the main problem in wastewater treatment.

12.4.9 Solids Determinations

The solid components of domestic wastewater can be classified in a number of ways. For example, wastewater solids can be categorized on the basis of several operational procedures used in the wastewater treatment laboratory. Total solids may be determined by driving off the water fraction, and suspended solids may be determined by filtering out the solid fraction on a porous pad and drying. Settleable solids may be determined by permitting a sample to settle in a special Imhoff cone apparatus. The categories used most often in the wastewater treatment field are suspended solids and total solids. The colloidal fraction of domestic wastewaters comprises about 20% of the solid component of an "average" wastewater. As stated, this component is characterized by being non-settleable; that is, usually long periods of time would be required for them to settle by gravity alone. Any estimate of wastewater composition can give only an average composition. The amounts of solids indicated cannot be applied equally to all wastewaters at all times.

12.4.10 Characteristics of Settleable Solids

The settleable solids to be removed from wastewater in primary or secondary settling tanks after grit removal are mainly organic and flocculent in nature, either dispersed or flocculated. Basically, four categories of settling occur depending on the tendency of particles to interact and the concentration of solids. These settling types are (1) discrete settling, (2) flocculent settling, (3) hindered or zone settling, and (4) compression.

12.4.10.1 Discrete Settling

Discrete particles do not change their size, shape, or mass during settling. Grit in wastewater behaves like discrete particles. The settling velocity of discrete particles is determinable using Stokes or Transition law. Organic solids in raw wastewater and bioflocs in biologically treated wastewaters cannot be considered as discrete particles, and hence, Stoke's law is not applicable for these particles.

12.4.10.2 Flocculent Settling

Flocculent particles coalesce during settling, increasing the mass of particles that settle faster. Flocculent settling refers to settling of flocculent particles of low concentration, usually less than 1000 mg/L. The degree of flocculation depends on the contact opportunities, which in turn are affected by the surface overflow rate, the depth of the basin, the concentration of the particles, the range of particle sizes, and the velocity gradients in the system. No adequate mathematical equation exists to describe flocculent settling, and therefore, overflow rates to achieve a given removal efficiency are determined using data obtained from settling column studies.

12.4.10.3 Hindered or Zone Settling

When the concentration of flocculent particles is in the intermediate range, they are close enough that their velocity fields overlap, causing hindered settling. The settling of particles results in significant upward displacement of water. The particles maintain their relative positions with respect to each other and the whole mass of particles settles as a unit or zone. This type of settling is applicable to concentrated suspensions such as are found in secondary settling basins following activated sludge units. In the hindered settling zone the concentration of particles increases from top to bottom, leading to thickening of sludge. Such secondary clarifiers where zone settling occurs are designed on the basis of solid flux or solids loading and checked for surface overflow rate, both of which can be determined by conducting settling column analysis.

12.4.10.4 Compression

In compression zone, the concentration of particles becomes so high that particles are in physical contact with each other, the lower layers supporting the weight of upper layers. Consequently, any further settling results from compression of the whole structure of particles and accompanied by squeezing out of water from the pores between the solid particles. This settling phenomenon occurs at the bottom of deep sludge mass, such as in the bottom of secondary biological treatment by trickling filters and activated sludge process and in tanks used for thickening of sludge.

12.5 CLASSIFICATION OF TREATMENT PROCESSES

Sewage, before being disposed of either in river streams or on land, has generally to be treated to make it safe. The degree of treatment required, however, depends upon the characteristics of the source of disposal, as discussed in the previous chapter.

Sewage can be treated in different ways. Treatment processes are often classified as follows:

1. preliminary treatment/physical
2. primary treatment/chemical

3. secondary/biological treatment
4. complete final treatment

Physical

- sedimentation (clarification)
- screening
- aeration
- filtration
- flotation and skimming
- degasification
- equalization

Chemical

- chlorination
- ozonation
- neutralization
- coagulation
- adsorption
- ion exchange

Biological

- *aerobic*
  - activated sludge treatment methods
  - trickling filtration
  - oxidation ponds
  - lagoons
  - aerobic digestion
- *anaerobic*
  - anaerobic digestion
  - septic tanks
  - lagoons

12.5.1 Physical Methods of Treatment

Physical methods include processes where no gross chemical or biological changes are carried out and strictly physical phenomena are used to improve or treat the wastewater. Examples would be coarse screening to remove larger entrained objects and sedimentation (or clarification). In the process of sedimentation, physical phenomena relating to the settling of solids by gravity are allowed to operate. Usually, this consists of simply holding a wastewater for a short period of time in a tank under quiescent conditions, allowing the heavier solids to settle, and removing the "clarified" effluent. Sedimentation for solids separation is a very common process operation and is routinely employed at the beginning and end of wastewater treatment operations. While sedimentation is one of the most common physical treatment processes that is used to achieve treatment, another physical treatment process consists of aeration usually to provide oxygen to the wastewater. Still, other physical phenomena used in treatment consist of filtration. Here, wastewater is passed through a filter medium to separate solids. An example would be the use of sand filters to further remove entrained solids from a treated wastewater. Certain phenomena will occur during the sedimentation process and can be advantageously used to further improve water quality. Permitting greases or oils, for example, to float to the surface and skimming or physically removing them from the wastewaters is often carried out as part of the overall treatment process.

12.5.2 Chemical Methods of Treatment

Chemical treatment consists of using chemical reactions to improve the water quality. The most commonly used chemical process is chlorination. Chlorine, a strong oxidizing chemical, is used to kill bacteria and to slow down the rate of decomposition of the wastewater. Another strong oxidizing agent that has also been used as an oxidizing disinfectant is ozone.

Coagulation consists of the addition of a chemical that, through a chemical reaction, forms an insoluble end product that serves to remove substances from the wastewater. Polyvalent metals are commonly used as coagulating chemicals in wastewater treatment and typical coagulants would include lime (that can also be used in neutralization), certain iron containing compounds (such as ferric chloride or ferric sulfate), and alum (aluminum sulfate). Certain processes may actually be physical and chemical in nature. The use of activated carbon to "adsorb" or remove organics, for example, involves both chemical and physical processes. Processes such as ion exchange, which involves exchanging certain ions for others, are not used to any great extent in wastewater treatment.

12.5.3 Biological Methods of Treatment

Biological treatment methods use microorganisms, mostly bacteria, in the biochemical decomposition of wastewaters to stable end products. More microorganisms, or sludges, are formed, and a portion of the waste is converted to carbon dioxide, water, and other end products. Generally, biological treatment methods can be divided into aerobic and anaerobic methods, based on availability of dissolved oxygen. The purpose of wastewater treatment is generally to remove from the wastewater enough solids to permit the remainder to be discharged to receiving water without interfering with its best or proper use. The solids that are removed are primarily organic but may also include inorganic solids. Treatment must also be provided for the solids and liquids that are removed as sludge. Finally, treatment to control odors, to retard biological activity, or destroy pathogenic organisms may also be needed.

12.6 PROCESS DETAILS

While the devices used in wastewater treatment are numerous and will probably combine physical, chemical, and biological methods, they may all be generally grouped under six methods:
- preliminary treatment
- primary treatment
- secondary treatment
- disinfection
- sludge treatment
- tertiary treatment

Degrees of treatment are sometimes indicated by use of the terms primary, secondary, and tertiary treatment. Tertiary treatment, properly, would be any treatment added onto or following secondary treatment. A typical flow diagram incorporating some of the units is shown in Fig. 12.1. The units shown cover pretreatment, clarification, filtration, adsorption, and filtration, and they may use reverse osmosis, ion exchange, electrodialysis, and evaporation. The final step is disinfection by chlorination. Of course, other unit processes may be added, depending on the need, the type of waste, and the pollutants in it. The pretreatment is intended for removing floating debris, and for settling grit and sand, along with other sludge deposits.

12.6.1 Preliminary Treatment

At most plants, preliminary treatment is used to protect pumping equipment and facilitate subsequent treatment processes. Preliminary devices are designed to remove or cut up the larger suspended and floating solids, to remove the heavy inorganic solids, and to remove excessive amounts of oils or greases. To effect the objectives of preliminary treatment, the following devices are commonly used:
1. screens: rack, bar, or fine
2. Comminuting devices: grinders, cutters, shredders
3. grit chambers

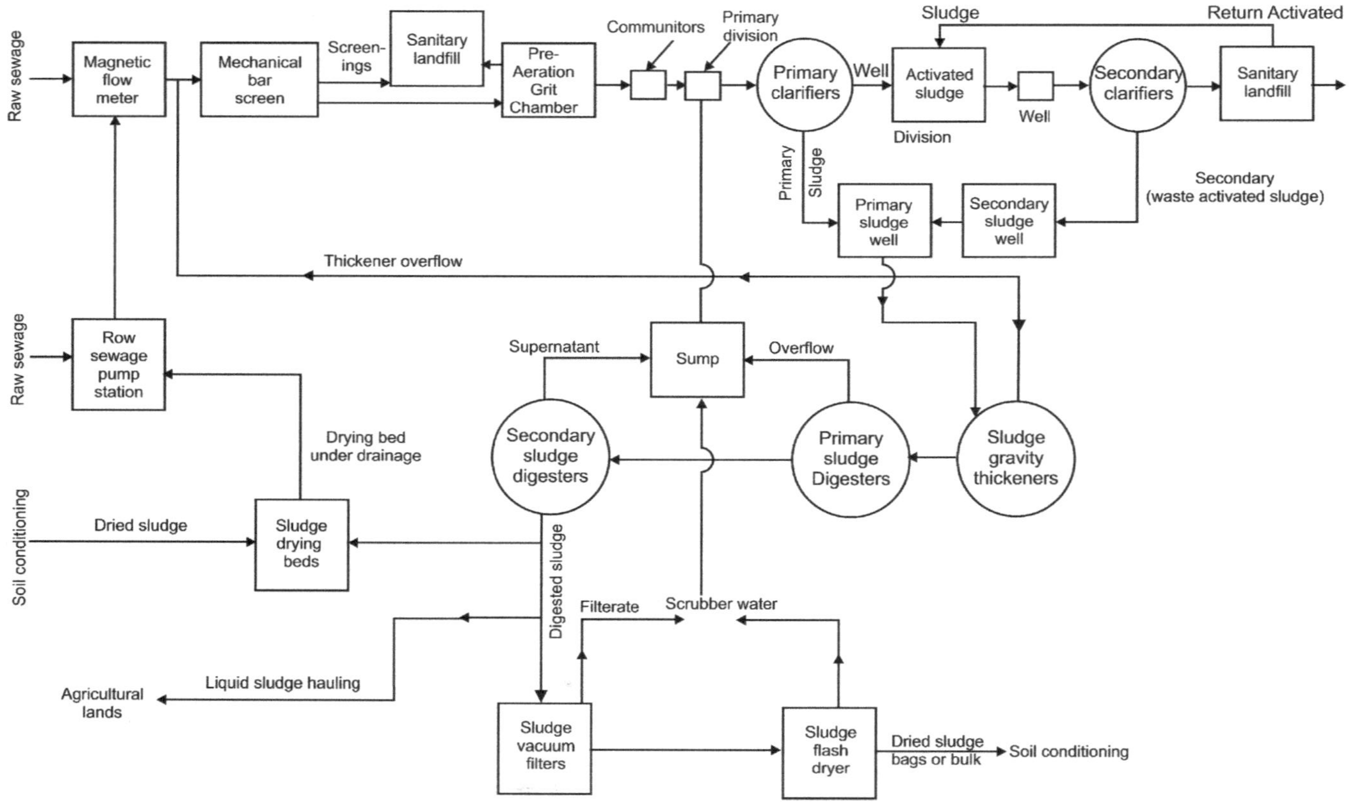

Figure 12.1 Wastewater treatment plant floc diagram. Reproduced from *Principles of Design and Operations of Wastewater Treatment Pond Systems for Plant Operators, Engineers, and Managers; EPA/600/R-11/088/ August 2011 / www.epa.gov /nrmr.*

4. pre-aeration tanks

In addition to the aforementioned, chlorination may be used in preliminary treatment. Since chlorination may be used at all stages in treatment, it is considered to be a method by itself. Preliminary treatment devices require careful design and operation.

12.6.1.1 Racks and Bar Screens

These consist of bars usually spaced 0.75–6 in. Those most commonly used provide clear openings of 1–2 in. Although large screens are sometimes set vertically, screens are usually set at an angle of 45–60 degrees with the vertical. The incoming wastewater is passed through the bars or screens, and periodically the accumulated material is removed. The racks or screens may be cleaned either manually or by means of automatically operated rakes. The solids removed by these units can be disposed of by burial or incineration.

12.6.1.2 Comminuting Devices

Grinders, *cutters*, and *shredders* are devices to break or cut up solids to such size that they can be returned to the wastewater without danger of clogging pumps or piping or affecting subsequent treatment devices. They may be separate devices to grind solids removed by screens or a combination of screen and cutters installed within the wastewater flow channel in such a manner that the objective is accomplished without actually removing these larger solids from the wastewater. These latter devices are made by a number of manufacturers under various trade names and, in most cases, consist of fixed, rotating, or oscillating teeth or blades, acting together to reduce the solids to a size that will pass through fixed or rotating screens or grids having openings of about 0.25 in. This is usually witnessed in the aeration system of activated sludge plants. These shredded solids tend to clog diffusers and cling to the impeller blades of mechanical aerators.

12.6.1.3 Grit Chambers

Wastewater usually contains a relatively large amount of inorganic solids such as sand, cinders, and gravel, which are collectively called grit. The amount present in a particular wastewater depends primarily on whether the collecting sewer system is of the sanitary or combined type. Grit will damage pumps by abrasion and cause serious operational difficulties in sedimentation tanks and sludge digesters by accumulation and plugging of outlets and pump suctions. Consequently, it is common practice to remove this material by grit chambers. Grit chambers are usually located ahead of pumps or comminuting devices, and if mechanically cleaned, they should be preceded by coarse bar rack screens. Grit chambers are generally designed as long channels. In these channels the velocity is reduced sufficiently to deposit heavy inorganic solids but to retain organic material in suspension. Channel type chambers should be designed to provide controlled velocities as close as possible to 1.0 fps. Velocities substantially greater than 1.0 fps cause excessive organic materials to settle out with the grit. The detention period is usually between 20 s and 1.0 min. This is attained by providing several chambers to accommodate variation in flow or by proportional weirs at the end of the chamber or other flow control devices that permit regulation of flow velocity. One development is the injection of air several feet above the floor of a tank type unit. The action of the air keeps the lighter organic matter in suspension and allows the grit relatively free from organic matter to be deposited in the quiescent zone beneath the zone of air diffusion. Excessive quantities of air can cause the roll velocity to be too high, resulting in poor grit removal. Insufficient quantities of air results in low roll velocities, and excessive organic matter will settle with the grit. These grit chambers are usually called aerated grit chambers.

Cleaning the Grit Chamber

Grit chambers are designed to be cleaned manually or by mechanically operated devices. If cleaned manually, storage space for the deposited grit is usually provided. Grit chambers for plants treating wastes from combined sewers should have at least two hand-cleaned units or a mechanically cleaned unit with bypass. Mechanically cleaned grit chambers are recommended. Single, hand-cleaned chambers with bypass are acceptable for small wastewater treatment plants serving sanitary sewer systems. Chambers other than channel type are acceptable, if provided with adequate and flexible controls for agitation and/or air supply devices and with grit removal equipment. There are a number of mechanical cleaning units available that remove grit scrapers or buckets while the grit chamber is in normal operation. These require much less grit storage space than manually operated units.

Washing Grit

Grit always contains some organic matter that decomposes and creates odors. To facilitate economical disposal of grit without causing nuisance, the organic matter is sometimes washed from the grit and returned to the wastewater. Special equipment is available to wash grit. Mechanical cleaning equipment generally provides for washing grit with wastewater as it is removed from the chamber.

Quantity of Grit

This depends on the type of sewer system, the condition of the sewer lines, and other factors. Strictly domestic wastewater collected in well-constructed sewers will contain little grit, while combined wastewater will carry large volumes of grit, reaching a peak at times of severe storms.

Disposal of Screenings and Grit

Screenings decompose rapidly with foul odors. They should be kept covered in cans at the screens and removed at least daily for disposal by burial or incineration. The walls and platforms of the screen chamber and screen itself should be hosed down and kept clean. Grit containing much organic matter may have to be buried to prevent odor nuisances.

12.6.1.4 Pre-Aeration Tanks

Pre-aeration of wastewater, that, is aeration before primary treatment, is sometimes provided for the following purposes:

1. to obtain a greater removal of suspended solids in sedimentation tanks,
2. to assist in the removal of grease and oil carried in the wastewater,
3. to freshen up septic wastewater prior to further treatment,
4. BOD reduction.

Pre-aeration is accomplished by introducing air into the wastewater for a period of 20–30 min at the design flow. This may be accomplished by forcing compressed air into the wastewater at a rate of about 0.10 cubic feet per gallon of wastewater when 30 min of aeration is provided or by mechanical agitation whereby the wastewater is stirred or agitated, so new surfaces are continually brought into contact with the atmosphere for absorption of air. To ensure proper agitation when compressed air is forced into the wastewater, air is usually supplied at the rate of 1.0–4.0 cubic feet per minute per linear foot of tank or channel. When air for mechanical agitation is used for the additional purpose of obtaining increased reduction in BOD, the detention period should be at least 45 min at design flow. The agitation of wastewater in the presence of air tends to collect or flocculate lighter suspended solids into heavier masses that settle more readily in the sedimentation tanks. Pre-aeration also helps to separate grease and oil from the wastewater and carry them to the surface. By the addition of air, aerobic conditions are also restored in septic wastewater to improve subsequent treatment. The devices and equipment for introducing the air into the wastewater are the same or similar to those used in the activated sludge process.

12.6.2 Primary Treatment

In this treatment, most of the settleable solids are separated or removed from the wastewater by the physical process of sedimentation. When certain chemicals are used with primary sedimentation tanks, some of the colloidal solids are also removed. Biological activity of the wastewater in primary treatment is of negligible importance. The purpose of primary treatment is to reduce the velocity of the wastewater sufficiently to permit solids to settle and floatable material to surface. Therefore, primary devices may consist of settling tanks, clarifiers, or sedimentation tanks. Because of variations in design, operation, and application, settling tanks can be divided into four general groups:

1. septic tanks
2. two story tanks: Imhoff and several proprietary or patented units
3. plain sedimentation tank with mechanical sludge removal
4. upward flow clarifiers with mechanical sludge removal

When chemicals are used, other auxiliary units are employed:

1. chemical feed units
2. mixing devices
3. flocculators

The results obtained by primary treatment, together with anaerobic sludge digestion as described later, are such that they can be compared with the zone of degradation in stream self-purification. Primary treatment consists of removing large suspended organic solids. This is usually accomplished by sedimentation in **settling basins**. The liquid effluent from primary treatment often contains a large amount of suspended organic material, and it has a high BOD (about 60% of original). Sometimes, the preliminary as well as primary treatments are classified together under primary treatment. The organic solids, which are separated out in the sedimentation tanks, are often stabilized by anaerobic decomposition in a digestion tank or are incinerated. The residue is used for landfills or soil conditioners.

12.6.2.1 Septic Tanks

The septic tank was one of the earliest treatment devices developed. Septic tanks provide wastewater treatment for small populations, such as individual residences, small institutions, schools, etc. Currently, this practice has been dispensed with. They are designed to hold wastewater at low velocity, under anaerobic conditions for a minimum detention time of 36 h. During this period, a high removal of settleable solids is achieved.

These solids decompose in the bottom of the tank with the formation of gas, which entrained in the solids causes them to rise through the wastewater to the surface and lie as a scum layer until the gas has escaped, after which the solids settle again. This continual flotation and resettling of solids carries some of them in a current toward the outlet to be discharged with the effluent. The final effluent disposal occurs by subsurface methods. The effectiveness of this method is dependent on the leaching ability of the soil. These primary type units require a minimum of attention, which involves an annual inspection and the periodic (3–5 years) removal of sludge and scum accumulations.

Recommended sizes of septic tanks for individual households (up to 20 users) and for housing colonies (up to 300 users) are given next in Tables 12.1 and 12.2, respectively:

Table 12.1 Recommended Sizes of Septic Tank up to 20 Users

			Liquid Depth (Cleaning Interval of)	
No. of Users	**Length (m)**	**Breadth (m)**	**2 years**	**3 years**
5	1.5	0.75	1.0	1.05
10	2.0	0.90	1.0	1.40
15	2.0	0.90	1.3	2.00
20	2.3	1.10	1.3	1.80

Note 1: The capacities are recommended on the assumption that discharges from only Water Closet(WC) will be treated in the septic tank.
Note 2: A provision of 300 mm should be made for free broad.
Note 3: The sizes of septic tank are based on certain assumption on peak discharges, as estimated in IS:2470 (part 1) 1985, and while choosing the size of septic tank, exact calculations shall be made.

Table 12.2 Recommended Sizes of Septic Tank for Residential Colonies

			Liquid Depth (Cleaning Interval of)	
No. of Users	**Length (m)**	**Breadth (m)**	**2 years**	**3 years**
50	5.0	2.00	1.0	1.24
100	7.5	2.65	1.0	1.24
150	10.0	3.00	1.0	1.24
200	12.0	3.30	1.0	1.24
300	15.0	4.00	1.0	1.24

Note 1: A provision of 300 mm should be made for free board.
Note 2: The sizes of septic tank are based on certain assumptions on peak discharges, as estimated in IS:2470 (Part 1) 1985, and while choosing the size of septic tank, exact calculations shall be made.
Note 3: For population over 100, the tank may be divided into independent parallel chambers of maintenance and cleaning.

12.6.2.2 Plain Sedimentation Tank

The purpose of sedimentation of sewage is to separate the settleable solids so the settled wastewater, if discharged into water courses, does not form sludge banks and when used for land disposal does not lead to clogging of soil pores and excessive organic loading. Sedimentation is used in wastewater treatment to remove:

1. organic and residual inorganic solids
2. bioflocculated solids
3. chemical flocs

Several factors such as flow variations, density currents, solids concentration, solids loading area, detention time, and overflow rate influence the design and performance of sedimentation tanks. Sedimentation tanks are designed for average flow conditions. For primary sedimentation tanks, surface overflow rate and detention period are important design criteria as the solids to be settled are flocculent in nature and undergo flocculation. The important design parameters are mentioned in Table 12.3.

Circular tanks are more common than rectangular or square tanks. Upflow tanks have been used for sewage sedimentation, but horizontal flow types are more popular. Rectangular tanks need less space than circular tanks and could be more economically designed where multiple units are to be constructed in a large plant. They can form a more compact layout with the rectangular secondary treatment units such as aeration tanks in the activated sludge system.

For rectangular tanks, maximum length and widths of 90 and 30 m, respectively, with length to width ratios of 1.5–7.5 and length to depth ratios of 5–25 are recommended. A minimum depth of 2.5 m in case of primary settling tanks and 3.5 m in case of secondary settling tanks for activated sludge should be provided. Bottom slopes of 1% are normally adopted. Peak velocities that are greater than 1.5 mph should generally be avoided.

Diameters of circular tanks vary widely from 3 to 60 m, although the most common range is 12–30 m. Diameters and depths could be chosen at the discretion of the designer in conformity with the manufactured sizes of scraper mechanisms in the country. The water depth varies from 2 m for primary to 3.5 for secondary settling tanks. Floors are sloped from periphery to center at a rate of 6.5–10%. The inlet to the tank is generally at the center and outlet is a peripheral weir, the flows begin radial and horizontal from center to the periphery of the tank. Multiple units are arranged in pairs with feed from a central control chamber.

Table 12.3 Design Parameters for Settling Tanks

Types of Settling	Overflow Rate, $m^3/m^2\cdot day$ Average	Overflow Rate, $m^3/m^2\cdot day$ Peak	Solid Loading, $kg/m^2\cdot day$ Average	Solid Loading, $kg/m^2\cdot day$ Peak	Depth, m	Detention Time, h
A. Primary Settling						
1. Primary settling only	25–30	50–60			2.5–3.5	2.0–2.5
2. Primary settling followed by secondary treatment	35–50	80–120			2.5–3.5	—
3. Primary settling with activated sludge return	25–35	50–60			3.5–4.5	—
B. Secondary Settling						
4. Secondary settling for trickling filter	15–25	40–50	70–120	190	2.5–3.5	1.5–2.0
5. Secondary settling for activated sludge (excluding extended aeration)	15–35	40–50	70–140	210	3.5–4.5	—
6. Secondary settling for extended aeration	8–15	25–35	25–120	170	3.5–4.5	—

12.6.2.3 Chemical-Aided Sedimentation

Chemical-aided sedimentation of sewage or industrial wastewater is analogous to chemical coagulation, flocculation, and sedimentation in water treatment. The colloidal and finely dispersed solids that cannot be removed by plain primary sedimentation alone, as they possess extremely low settling velocities, are aggregated into settleable particles by addition of chemicals in chemical-aided sedimentation. Commonly used chemicals are trivalent or divalent metallic salts such as aluminum sulfate, ferric and ferrous salts, and lime. Polyelectrolytes and polymers, cationic, anionic, and nonionic, have also been used both as primary coagulants as well as coagulant-aids. The colloidal particles in domestic wastewaters are usually negatively charged and therefore do not agglomerate naturally into settleable mass. Addition of chemical coagulants results in destabilization, aggregation, and binding together by any one or more than one mechanisms of (1) ionic layer compression, (2) adsorption and charge neutralization, (3) enmeshment in precipitate, and (4) interparticle bridging. When aluminum and iron salts are added to wastewaters in quantities sufficient to exceed the solubility limits of metal hydroxide, polymers of hydroxometal $\left[\text{e.g., } Al_{13}(OH)_{34}^{5+}\right]$ complexes are formed, which are adsorbed on colloidal particles and neutralize their charge or form bridging between colloidal particles if the polymers are of the same charge as that on colloidal particles. Further, colloidal particles can get entrapped within the metal hydroxide precipitates, which are heavy and settleable. Chemical-aided sedimentation products intermediate results between plain sedimentation and secondary biological treatment. With proper dosages of chemicals, this treatment process may be expected to remove 60–80% of suspended solids and 45–65% of BOD when it is not preceded by any plain sedimentation.

Chemical-aided sedimentation involves the unit processes and operations of chemical coagulation, flocculation, and sedimentation. Therefore, it will not remove dissolved solids. On the contrary, addition of chemicals that are soluble may add to the total dissolved solids concentration of wastewater.

As compared to secondary biological treatment methods such as trickling filter or activated sludge, chemical treatment methods will be less efficient and will work out to be uneconomical and are therefore recommended only in the following situations:

1. Plants are operated seasonally or variations in strength and volume of sewage are high.
2. Intermediate treatment between plain sedimentation and secondary biological treatment is adequate.
3. Sludge conditioning for dewatering is needed.

12.6.2.4 Chemicals Used

The most commonly used chemicals are ferrous sulfate, ferric chloride, ferric sulfate, chlorinated copperas, alum, aluminum chloride, lime, and sodium carbonate. Choice of chemical and its dosage depend on cost of chemical, degree of treatment required, and the characteristics of the waste, pH being one of the more important factors. Optimum dosage is determined by conducting a jar test in the laboratory.

Iron Salts

Ferric salts are better coagulants than ferrous salts because of their higher valency and their efficiency over a wider pH range. Ferric salts are effective at approximate pH values above 3, the efficiency increasing in pH, while the useful pH range of ferrous salts is above 10. But when wastewaters are highly alkaline due to presence of trade wastes, it may be cheaper to use larger dosage of ferrous salts as they are relatively cheaper. Chlorinated copperas, which is an equimolar mixture of ferric sulfate and ferric chloride formed by the addition of chlorine to ferrous sulfate, is also used in place of ferric salts.

Aluminum Salts

Aluminum chloride and sulfate of alumina (filter alum) are the commonly used aluminum salts. Where alum is used, the sludge produced is greater in volume and also bulkier than with iron salts, making it less easily settleable.

Lime and Sodium Carbonate

These are used for pH adjustment to favorable ranges of coagulants, especially when sewage is highly acidic. Lime is sometimes used independently as precipitant, particularly when iron pickling liquors are present in sewage. The action may be due to formation of calcium carbonate floc or reactions with small amounts of aluminum or iron salts present in sewage. Lime incidentally helps in grit separation, oil and grease removal, and is perhaps the cheapest chemical used in chemical precipitation.

Flocculation

The principle of flocculation in sewage is similar to flocculation in water purification. The floccules that are formed after flash mixing with chemicals are made to coalesce into bigger sizes by either air flocculation or mechanical flocculation. Both diffused air and mechanical vertical draft tubes are used for air flocculation. The revolving paddle type is the most common of the mechanical flocculators. The tanks are usually in duplicate with a detention period of 30–90 min depending upon results required and the type of sewage treated. However, the dose of chemical required as well as the flocculation period is best determined by laboratory test followed by pilot plant studies for optimum results. The paddles are mounted either on a horizontal or vertical shaft. The peripheral speed of the paddles is kept in the range of 0.3–0.45 mps. The flow-through velocity through the flocculator should be in the range of 15–25 cm/s to prevent sedimentation.

12.6.3 Secondary Treatment

Secondary treatment depends primarily upon aerobic organisms, which biochemically decompose the organic solids to inorganic or stable organic solids. It is comparable to the zone of recovery in the self-purification of a stream.

The treatment reactors, in which the organic matter is decomposed (oxidized) by aerobic bacteria, are known as *aerobic biological units*, and they may consist of (1) **Filters** (*intermittent sand filters* as well as *trickling filters*); (2) **Aeration tanks**, with the feed of *recycled* **activated sludge** (i.e., the sludge, which is settled in a secondary sedimentation tank, receiving effluents from the aeration tank); and (3) **Oxidation ponds and aerated lagoons**. Since all these aerobic units generally make use of primary settled sewage, they are easily classified as secondary units.

The devices used in secondary treatment may be divided into four groups:

1. trickling filters with secondary settling tanks
2. activated sludge and modifications with final settling tanks
3. intermittent sand filters
4. stabilization ponds

The key step in secondary treatment is aeration. By aerating the wastewater, the number of aerobic bacteria is increased tremendously. This large number of bacteria consumes organic matter that is in the waste, and when the food or organic matter has been consumed, the bacteria will die off at a rapid rate. The aeration is usually done by bubbling compressed air through the wastewater. After about 5 to 6 h of aeration, the waste is sent to a final settling tank; some of the settled substance, called activated sludge, is recycled back to the aeration tank because this activated sludge has a large number of bacteria in it, and it will help in consuming organic matter and also in seeding bacteria. Thus, it will complement the aeration process by increasing the number of bacteria for consuming the waste food. Final settling tanks are used to settle the bacteria out. The treated water is then chlorinated and discharged. When this sludge is removed from the sedimentation tank and placed in digesters for some time, it may be disposed of, whether by burning or by drying it in small ponds. The dried substance may be used as a fertilizer. During the secondary treatment process, almost 90% of the organic matter is removed through the use of millions of bacteria. After the addition of chlorine, the waste is sterilized and is no longer heavily polluted. It is safe for industrial reuse.

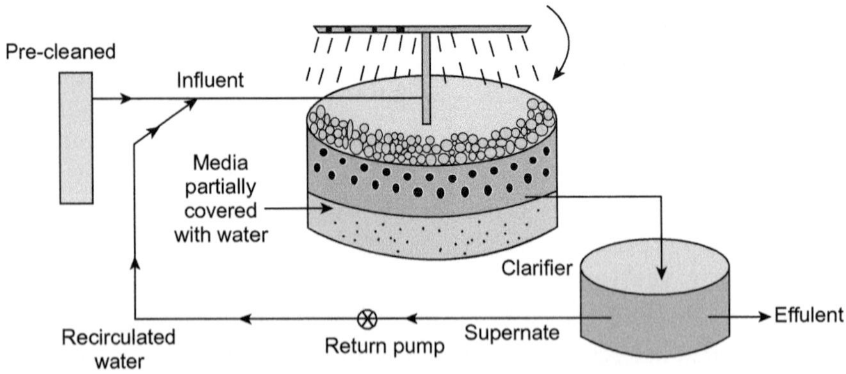

Figure 12.2 Schematic diagram of trickling filter.

12.6.3.1 Trickling Filters

This is the most common secondary treatment in use today; it is a circular tank filled with 3–5″ stones. The wastewater is sprayed or sprinkled over the top of the stones, trickling down through them; the water is then removed at the bottom of the tank (Fig. 12.2). The stones constitute a surface upon which the aerobic bacteria grow, consume, and remove organic matter. The flow from the trickling filter is usually followed by a settling tank that helps in removing sludges that may have been dislodged from the trickling filter. The effluent is finally disinfected by chlorination and discharged safely into a water stream. Trickling filters may be enclosed or not, depending on weather conditions.

Trickling filters are used to remove organic matter from wastewater. The trickling filter is an aerobic treatment system that utilizes microorganisms attached to a medium to remove organic matter from wastewater. This type of system is common to a number of technologies such as rotating biological contractors and packed bed reactors (bio-towers). These systems are known as attached growth processes. By this, the systems in which microorganisms are sustained in a liquid are known as suspended growth processes. Trickling filters enable organic material in the wastewater to be absorbed by a population or microorganisms (aerobic, anaerobic, and facultative bacteria; fungi, protozoa, and algae) attached to the medium as a biological film or slime layer (ranging from 0.1 to 0.2 mm thick). As the wastewater flows over the medium, microorganisms already in the water gradually attach themselves to the rock, slag, or plastic surface and form a film. The organic material is then degraded by aerobic microorganisms in the outer part of the slime layer. As the layer thickens through microbial growth, oxygen

Table 12.4 Design Features for Trickling Filters

S.No.	Design Feature	Low Rate Filter	High Rate Filter	Super Rate or Roughing Filter
1	Hydraulic loading, $m^3/m^2 \cdot day$	1.4	10–40 (including recirculation)	40–200 (including recirculation)
2	Organic loading, kg BOD_5/ $m^3 \cdot day$	0.08–0.32	0.32–1.0 (excluding recirculation)	0.8–6.0 (excluding recirculation)
3	Depth, m	1.8–3.0	0.9–2.5	4.5–12
4	Recirculation ratio	0	0.5–3.0 (domestic wastewater) up to 8 or more for strong industrial wastes	1–4
5	Filter media	Rock, gravel, slag etc.	Rock, slag, synthetic materials	Plastic media

cannot penetrate the medium face, and anaerobic organisms develop. As the biological film continues to grow, the microorganisms near the surface lose their ability to cling to the medium, and a portion of the slime layer falls off the filter. This process is known as sloughing. The sloughed solids are picked up by the underdrain system and transported to a clarifier for removal from the wastewater.

12.6.3.2 Types of Filters

Trickling filters may be categorized as low rate, high rate, and super rate, primarily based on hydraulic and organic loading rates. Although there is no well-demarcated practice, some important distinguishing design criteria and features for the three types of filters are presented in Table 12.4.

The hydraulic loading rate is the total flow including recirculation applied on unit area of the filter in a day, while the organic loading rate is the 5-day 20°C BOD, excluding the BOD of the recirculant, applied per unit volume in a day. Much higher organic loadings than indicated here have been used in roughing filters. Recirculation is not generally adopted in low rate filters and media depths for low rate filters range from 1.8 to 3.0 m. They require larger media volumes than high rate filters. However, they are easy to operate and give consistently good quality of effluent and are preferred when plant capacities are small, as in the case for institutions. In contrast to the low rate filters, in high and super rate filters a part of the settled or filter effluent is recycled through the filter. Recirculation has the advantage of bringing the organic matter in the waste in contact with the biological slime more than once, thus increasing the efficiency of the filters. It enables higher hydraulic loading and thereby reduces filter clogging and aids uniform distribution of organic load over the filter surface. It also helps to dampen the variations in the strength and the flow of sewage applied on the filter. The ratio of the sewage flow is known as the recirculation ratio. Recirculation ratios usually range from 0.5 to 3, and values exceeding 3 are considered to be uneconomical in the case of domestic sewage, but ratios of 8 and above have been used with industrial wastes and super-high rate filters, which may be single stage or two stage. Media depths of 0.9–2.5 m have been used for high rate filters with an optimum range of 1.5–2.0 m for the first stage and 1–2 m for the second-stage filters. Single-stage units consist of a primary setting tank, the filter, secondary settling tank, and facilities for recirculation of the effluent.

Two-stage filters consist of two filters in series with a primary settling tank, an intermediate settling tank, which may be omitted in certain cases, and a final settling tank. Recirculation facilities are provided for each stage. The effluent from the first-stage filter is applied on the second-stage filter either after settlement or without settlement. An intermediate clarifier is used for settling the first-stage effluent before it is applied to the second-stage filter, and the recirculation is only through the settling tanks. The intermediate settling is omitted, and their circulation flows are settled. In the series-parallel system, part of the settled raw sewage is applied directly to the second-stage filter, increasing the efficiency of that stage. Two-stage filtration will provide a higher degree of treatment than the single stage for the same total volume of media. Two-stage units are used for strong sewage when the effluent BOD has to be less than 30 mg/L.

A well-operated low rate trickling filter in combination with secondary sedimentation tank may remove 75–90% BOD and produce highly nitrified effluent. It is suitable for treatment of low to medium strength domestic wastewaters. The high rate trickling filter, single stage, and two stage are recommended for medium to relatively high-strength domestic and industrial wastewaters. The BOD removal efficiency is around 75–90%, but the effluent is only partially nitrified. The super rate or roughing filters find application for high-strength wastewaters. They have also been used as roughing filters to reduce the BOD of high-strength wastewaters for further treatment. The effluent from these filters may be partially nitrified only when low organic loadings are employed.

12.6.3.3 *Rotating Biological Contractor*

Rotating biological contractor (RBC) is one of the relatively recent addition to the family of biological treatment devices. This is a relatively simple attached growth system operating on the principle of moving media. The RBC units can be adopted for small and medium towns.

The advantages claimed for RBC include (1) low food to microorganism ratio resulting in higher efficiencies of organic matter removal, (2) low hydraulic retention periods minimizing tank volume and capital costs, (3) low head loss and lower power requirements, (4) inherent simplicity and low operational and maintenance cost, (5) ability to resist shock loads, and (6) ability to lend itself to modular fabrication to suit required effluent quality.

Process Description

The RBC unit (Fig. 12.3) consists of a series of closely spaced vertical discs mounted on a horizontal shaft rotating at slow speeds, normally less than 10 rpm, the movement of the discs being perpendicular to the wastewater movement in a cylindrical vessel. The discs, also called biodiscs, support biomass and are partially (40–60%) submerged in the wastewater. The rotation of the biodiscs causes the biomass to be alternatively submerged in wastewater to absorb food and to pick up a thin layer of wastewater and then are raised out of the liquid into the air to oxidize the absorbed substrate and to allow the wastewater film to slide down the biomass. It has been suggested that the orthogonal rotational motion of biomass on discs relative to horizontal liquid movement results in ideal shear and turbulence conditions at the solid–liquid interface to cause exceptionally high transfer of substrate and oxygen into biological slime and waste products from it.

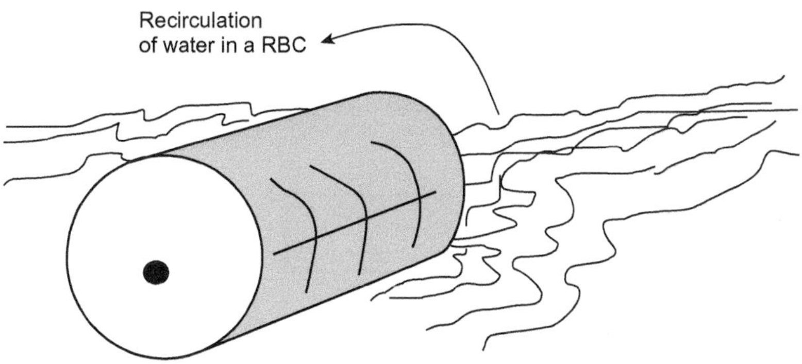

Figure 12.3 Rotating biological contractor.

Excess biomass growing on the disc surfaces is sheared off and sloughed biomass is kept in suspension by the mixing action of the discs and carried out of the cylindrical tank along with the effluent. Both the substrate utilization within the microbial film and the sloughing of excess biomass are continuous processes that help in maintaining a constant thickness of microbial film on the discs. Thickness of biofilm may reach up to 2–4 mm depending upon the strength of wastewater and rotational speed of the discs.

The basic process flow sheet of wastewater treatment system may consist of primary sedimentation following screening and grit removal, aerobic biological treatment in RBC unit, and secondary settling for solid–liquid separation of sloughed film from treated wastewater. The settled sludge from primary and secondary sedimentation has to be suitably treated and disposed.

The RBC unit essentially consists of the following:

1. cylindrical bottomed horizontal flow tank usually divided into an appropriate number of stages that are hydraulically connected; the tank may be constructed of steel, fiber, glass, concrete, or masonry,
2. circular discs of PVC, asbestos cement, or any inert light material of high durability mounted on a shaft of sufficient rigidity; the disc diameters usually vary between 1 and 4 m and thickness up to 10 mm,
3. a driving mechanism comprised of a motor and a reduction gear.

A reactor module consists of a tank with circular discs mounted on a shaft driven by a motor through reduction ear. Several modules may be arranged in parallel and/or in series to meet the flow and effluent quality requirements.

Activated Sludge Process

The activated sludge process (Fig 12.4) uses microorganisms to feed organic contaminants that are in wastewater to produce a high-quality effluent. The basic principle behind all activated sludge processes is that as microorganisms

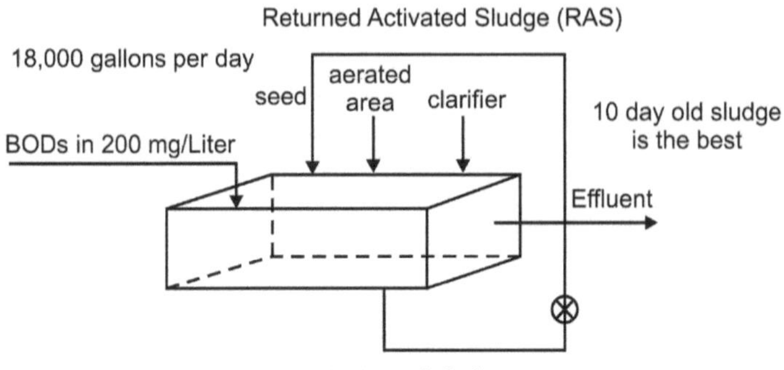

Figure 12.4 Activated sludge process.

grow, they form particles that clump together. These particles, which are referred to as floc, are allowed to settle to the bottom of the tank, which results in a relatively clear liquid free organic material and suspended solids. The screened wastewater is mixed with varying amounts of recycled liquid that contains a high proportion of organisms that are taken from a secondary tank, and it becomes a product that is called mixed liquor. The next step for the mixture is to stir and inject it with large quantities of air to provide oxygen and keep the solids in suspension. After a period of time, the mixed liquor flows to a clarifier where it is allowed to settle. During this settling a portion of the bacteria is removed and the partially cleaned water flows on for additional treatment. The settled solids that resulted, the activated sludge, are then returned to the first tank to begin the process again. The basic activated sludge process consists of several interrelated components:

- The aeration tank where the biological reactions occur
- An aeration source that provides oxygen and mixing
- A tank, known as the clarifier, where solids settle and are separated from treated wastewater
- A collecting means for the solids either to return them to the aeration tank, returned activated sludge, or to remove them from the process (waste-activated sludge)

Aerobic suspended growth systems are of two basic types, those that employ sludge recirculation, viz., conventional activated sludge process and its modifications, and those that do not have sludge recycle, viz., aerated lagoons. In both cases sewage containing waste organic matter is aerated in an aeration basin in which microorganisms metabolize the soluble and suspended organic matter. The suspended solid concentration in the aeration tank liquor, also called mixed liquor suspended solids (MLSS), is generally taken as in index of the mass of active microorganisms in the aeration tank. However, the MLSS will contain not only active microorganisms but also dead cells as well as inert organic and inorganic matter derived from the influent sewage. The mixed liquor volatile suspended solids (MLVSS) value is also used and mostly preferred to MLSS as it eliminates the effect of inorganic matter.

Activated Sludge Process Variables

An activated sludge plant essentially consists of the following: (1) aeration tank containing microorganisms in suspension in which the reaction takes place, (2) activated sludge recirculation system, (3) Excess sludge wasting and disposal facilities, (4) aeration systems to transfer oxygen, and (5) secondary sedimentation take to separate and thicken activated sludge.

The main variables of the activated sludge process are the loading rate, the mixing regime, and the flow scheme.

12.6.3.4 Loading Rate

The loading rate expresses the rate at which the sewage is applied in the aeration tank. A loading parameter that has been developed empirically over the years is the hydraulic retention time (HRT), Θ, day:

$$\Theta = \frac{V}{Q} \tag{12.1}$$

where V, volume of aeration tank, m^3, and Q, sewage inflow, m^3/day.

Another empirical loading parameter is volumetric organic loading, which is defined as the BOD applied per unit volume of aeration tank, per day.

A rational loading parameter, which has found wider acceptance and is preferred, is specific substrate utilization rate, per day, which is defined as:

$$U = Q\frac{(S_o - S)}{VX} \tag{12.2}$$

A similar loading parameter is mean call residence time or sludge retention time (SRT), Θ_c, day:

$$\Theta_c = \frac{VX}{Q_w X_s} \tag{12.3}$$

where S_o and S are influent and effluent organic matter concentrations, respectively, conventionally measured as BOD_5 (g/m^3), X and X_s are MLSS concentration in aeration tank and waste-activated sludge from secondary settling tank under flow (g/m^3), respectively, and Q_w, waste-activated sludge rate (m^3/day).

Under steady-state operation the mass of waste-activated sludge is given by:

$$Q_w X_s = YQ\,(S_o - S)\; - k_d XV \tag{12.4}$$

where Y, maximum yield coefficient (microbial mass synthesized/mass of substrate utilized), and k_d, endogenous respiration rate constant (day^{-1}).

From these equations, the following is seen:

$$1/\Theta_c = YU - k_d \tag{12.5}$$

Since both Y and k_d are constants for a given waste, it is, therefore, necessary to define either Θ_c or U Eq. (12.5) is plotted for typical values of Y = 0.5 and k_d = 0.06/day for municipal wastewaters. If the value of S is small compared to S_o, which is often the case for activated sludge systems treating municipal wastewater, U may also be expressed as food applied to microorganism ratio, F/M:

$$F/M = QS_o/XV \tag{12.6}$$

The Θ_c value adopted for design controls the effluent quality, settleability, and drainability of biomass. Other operational parameters that are affected by the choice of Θ_c values are oxygen requirement and quantity of waste-activated sludge.

Secondary Settling

Secondary settling assumes considerable importance in the activated sludge process as the efficient separation of the biological sludge is necessary not only for ensuring final effluent quality but also for return of adequate sludge to maintain the MLSS level in the aeration tank. The secondary settling tank of the activated sludge process is particularly sensitive to fluctuations in flow rate, and on this account, it is recommended that the units be designed not only for average overflow rate but also for peak overflow rates. The high concentrations of suspended solids in the effluent require that the solids loading rate should also be considered.

Sludge Recycle

The MLSS concentration in the aeration tank is controlled by the sludge recirculation rate and the sludge settleability and thickening in the secondary sedimentation tank:

$$\frac{Q_R}{Q} = \frac{X}{X_s - X} \tag{12.7}$$

where Q_R, sludge recirculation rate, m^3/day.

The sludge settleability is determined by sludge volume index (SV1) defined as volume occupied in ml by 1 g of solids in the mixed liquor after settling for 30 min and is determined experimentally. If it is assumed that sedimentation of suspended solids in the laboratory is similar to that in sedimentation tank, then X_s, 10^6/SVI. Values of SVI between 100 and 150 mL/g indicate good settling of suspended solids. The X_s value may not be taken more than 10,000 g/m^3 unless separate thickeners are provided to concentrate the settled solids or a secondary sedimentation tank is designed to yield a higher value. Using the aforementioned value for X_s and 5000 mg/L for X in Eq. (12.7), the sludge recirculation ratio comes out to be 1.0. The return sludge is always to be pumped, and the recirculation ratio should be limited to the values suggested in Table 12.5.

Table 12.5 Characteristics and Design Parameters of Activated Sludge Systems for Municipal Wastewaters

Process Type	Flow Regime	MSLL mg/L	MLVSS/ MLSS	F/M, kg BOD_5/kg MLSS day	HRT, h	Θ_c, day	Q_R/Q	Eff. %	kg O_2/ kg BOD_5 Removed
Conventional	Plug flow	1500–3000	0.8	0.3–0.4	4–6	5–8	0.25–0.5	85–92	0.8–1.0
Completely mixed	Completely mixed	3000–4000	0.8	0.3–0.5	4–5	5–8	0.25–0.8	85–92	0.8–1.0
Extended aeration	Completely mixed	3000–5000	0.6	0.1–0.18	12–24	10–25	0.5–1.0	95–98	1.0–1.2

Aerated Lagoons

Aerated lagoons are generally provided in the form of simple earthen basins with an inlet at one end and outlet at the other to enable the wastewater to flow through while aeration is usually provided by mechanical means to stabilize the organic matter. The major difference between activated sludge systems and aerated lagoons is that in the latter settling tanks and sludge recirculation are absent.

Aerated lagoons are of two principal types depending on how the microbial mass of solids may leave with the effluent stream, and some settle down in the lagoon since aeration power input is just enough for oxygenation and not for keeping all solids in suspension. As the lower part of such lagoons may be anoxic or anaerobic while the upper layers are aerobic, the term facultative is used. The characteristics are shown in Table 12.6.

Stabilization or Oxidation Ponds

Stabilization or oxidation ponds are large open flow-through earthen basins specifically designed and constructed to treat sewage and biodegradable industrial wastes. Sewage is pumped to the ponds, which may be as large as 1000 ft by 1000 ft, and there may be a number of them. The sewage enters the first pond, and then it progressively moves to the other ponds in the

Table 12.6 Some Characteristics of Aerated Lagoons

S.No.	Characteristics	Facultative Aerated Lagoons	Fully Aerobic	Extended Aeration System (for Comparison)
1.	Detention time, days	3–5	2–3	0.5–1.0
2.	Dept, m	2.5–5.0	2.5–4.0	2.5–4.0
3.	Land required, square meter/person	0.15–0.30	0.10–0.20	—
4.	BOD removal efficiency %	80–90	50–60	95–98
5.	Overall BOD removal rate, K, per day 20°C (soluble only)	0.6–0.8	1–1.5	20–30
6.	Suspended solids (SS) in unit, mg/L	40–150	150–350	3000–5000
7.	VSS/SS	0.6	0.8	0.6
8.	Desirable power level watts/cubic meter of lagoon volume	0.75	2.75–6.0	15–18
9.	Power requirement, kWh/person/year	12–15	12–14	16–20

series. They are scientifically built, taking advantage of the effects of sunlight, algae, and oxygen to improve the quality of the wastewater. Algae use carbon dioxide resulting from the decomposition of organic matter, and they release oxygen. Aerobic bacteria are multiplied extensively by this oxygen release; they digest the organic waste. The sunlight penetration provides for the life and growth of algae. The light penetration to the lagoons may reach a depth of 3 ft, which helps in this process of stabilization. Cleaning the settled substance at the bottom of a lagoon is accomplished at certain intervals. The wastewater depth may vary from 3–5 ft. The detention time is normally about 10 days. In this type of treatment, the effluent is sometimes discharged without any further treatment, and it bears a green color due to the presence of algae. It should be mentioned here that the BOD is reduced from 125–5 mg/L for the filtered effluent. Ammonia is also reduced from 25 mg/L at the inlet to a range of 10–15 mg/L at the outlet. Also, the most probable number of estimated bacteria is reduced from 1 million or more to 100 bacteria per milliliter.

Aerobic Aerobic ponds are designed to maintain completely aerobic conditions. The ponds are kept shallow with a depth less than 0.5 and BOD loadings of 40–120 kg/ha day. The pond contents may be periodically mixed. Such ponds develop intense algal growth and have been used on an experimental basis only.

Anaerobic Completely anaerobic ponds are used as pretreatment for high-strength industrial wastes and sometimes for municipal wastewaters. They are also used for digestion of municipal sludges, depending on temperature and waste characteristics. A BOD load of 400–3000 kg/ha day and 5–50 day detention period would result in 50–85% BOD reduction. Such ponds are constructed with a depth of 2.5–5 m to conserve heat and minimize land area requirement. Usually, they have an odor problem.

Facultative The facultative pond functions aerobically at the surface, while anaerobic conditions prevail at the bottom. The aerobic layer acts as a good check against odor evolution from the pond. The treatment effected by this type of pond is comparable to that of conventional secondary treatment process. The facultative pond is hence best suited and most commonly used for treatment of sewage.

Anaerobic Treatment of Wastewaters Anaerobic treatment of wastewaters has a number of advantages over aerobic treatment process, namely, the energy input of the system is low, as no energy is required for oxygenation, lower production of excess sludge (biological synthesis) per unit mass of organic matter stabilized, lower nutrient requirement due to lower biological synthesis, and the degradation of waste organic material leads to the production of biogas, which is a valuable source of energy.

Anaerobic digestion as a unit process in municipal wastewater treatment has been in use for many years now. It is employed for stabilization of sludge solids from primary and secondary sedimentation tanks either in closed digesters or open lagoons. Anaerobic lagoons are also used for treatment of industrial wastes. Conventionally the anaerobic process is considered a slow process requiring digesters of large HRT.

Anaerobic Filter In anaerobic filters, microbial cells are both entrapped as clumps of cells in the interstices between packing material and as biofilm attached to the surface of the packing material. The packing or filter media is usually of naturally crushed rock of 15–25 mm size or consisting of plastic or ceramic material. The filter media should have high specific surface and porosity to allow for maximum possible film growth and retention of biomass. The reactor is operated as up flow submerged packed bed reactor. A number of such filters have been constructed for treatment of low strength wastes such as municipal wastewater.

Anaerobic Fixed Films Reactor In an anaerobic fixed film reactor, the microbial mass is immobilized on fixed surfaces in the reactor. It is operated in downflow mode to prevent accumulation of refractory particulates contained in the influent and sloughed biofilm. The sloughed biofilm is also discharged with the effluent. The reactor may be operated in either submerged or unsubmerged condition. The reactor is usually of modular construction consisting of plastic sheets proving a high void ratio. Such reactors have been constructed to treat high-strength wastes.

Fluidized and Expanded Bed Reactor The fluidized bed reactor, incorporates an upflow reactor partly filled with sand or a low density carrier such as coal or plastic beads. A very large surface area is provided by the carrier material for growth of biofilm. The system readily allows passage of particulates, which could plug a packed bed, but it requires energy for fluidization. Expanded bed reactors do not aim at complete fluidization and use a lower upflow velocity, resulting in a lower energy requirement. These reactors can be used for treatment of municipal wastewater as well.

Upflow Anaerobic Sludge Blanket Reactor The upflow anaerobic sludge blanket reactor maintains a high concentration of biomass through formation of highly settleable microbial aggregates. The wastewater flows upward through a layer of sludge. At the top of the reactor, phase separation between gas–solid–liquid takes place. Any biomass leaving the reaction zone is directly recirculated from the settling zone. The process is suitable for both soluble wastes and those containing particulate matter. The process has been used for treatment of municipal wastewater at few locations, and hence, limited performance data and experience is available presently.

12.6.4 Chlorination

This is a method of treatment that has been employed for many purposes in all stages in wastewater treatment, and even prior to preliminary treatment. It involves the application of chlorine to the wastewater for the following purposes:

1. disinfection or destruction of pathogenic organisms
2. prevention of wastewater decomposition
   a. odor control
   b. protection of plant structures

12.7 SLUDGE TREATMENT

The solids removed from wastewater in both primary and secondary treatment units, together with the water removed with them, constitute wastewater sludge. It is generally necessary to subject sludge to some treatment to prepare or condition it for ultimate disposal. Such treatment has two objectives: the removal of part or all of the water in the sludge to reduce its volume and the decomposition of the putrescible organic solids to mineral solids or to relatively stable organic solids. This is accomplished by a combination of two or more of the following methods:

1. thickening
2. digestion with or without heat
3. drying on sand bed: open or covered
4. conditioning with chemicals
5. vacuum filtration
6. heat drying
7. incineration
8. wet oxidation
9. centrifuging

The organic solids/sludge, separated out in the primary as well as in the secondary settling tanks, will be disposed of by stabilizing them under anaerobic process in a sludge digestion tank.

The principle purposes of sludge digestion are to reduce its putrescibility and pathogenic contents and to improve its dewatering characteristics. This is mainly achieved through anaerobic or aerobic digestion.

Sludge is usually disposed of on land as manure to soil, as a soil conditioner, or barged into sea. Burial is generally resorted to small quantities of putrescible sludge. The most common method is to utilize it as fertilizer. Ash from the incinerated sludge is used as a landfill.

12.7.1 Sludge Pumping

Pumping is important in handling sludge because sludge produced in the different units of a sewage treatment plant has to be moved from point to point. The selection of a pump depends upon the type of sludge to be handled, viz., whether the sludge is primary, secondary, return, elutriated, or thickened and concentrated. The sludge may be watery, thick, or occasionally sum. Sludge is more viscous than water. An important characteristic of the different types of sludges is the percentage content of the suspended solids, as summarized in Table 12.7.

Sludge pumping may be intermittent or continuous, depending upon the type and design of the waste-treatment processes and of the sludge-handling and treatment units.Pumping of sludge is required in the following situations:

1. for transfer of the sludge from the sedimentation tanks to thickeners and/or digesters
2. for recirculation of secondary sludge

Table 12.7 Solids in Different Types of Sludges

Type of Sludge	% of Solids
Raw primary sludge	4–8
Secondary sludge	1–5
Raw primary and secondary sludge	3–8
Digested sludge	6–10
Chemical sludge	4–12
Alum and ferric sludge	2–6
Chemical slurries	1–30
Incinerator as slurries	5–20

3. for transfer of excess sludge from secondary biological treatment units to thickeners and/or digester or to primary settling tanks
4. for carrying sludge from extended aeration system directly to drying beds
5. for disposal of sludge into lagoons or on land

12.7.2 Sludge Pumps

Sludge pumps have to be resistant to abrasion, as sludges quite often contain sand and grit. The sludge pumps should be slow-speed machines to contain the rate of wear and tear.

Since a sludge pump may have to run intermittently or continuously, a sludge pump has to be dependable in respect of satisfactory, trouble-free operation, whether under the fatigue of the intermittent operation or with the endurance desired for long, continuous operation.

The types of pumps used for pumping sludges are these:

1. centrifugal pumps
2. air lift pumps
3. screw pumps
4. reciprocating pumps of the plunger type or of diaphragm type

Table 12.8 shows the typical applications of pumps of these different types and the types of sludges handled by them.

There are specific considerations to be borne in mind in the use of the different types of pumps for handling sludge.

Table 12.8 Typical Applications of Sludge Pumps

Type of Pump	Max. Suction Lift (m)	Max. % Solids Generally Handled	Typical Applications
Centrifugal			
1. Non-clog	4.5	2	Primary settled sludge, secondary settled sludge, chemically treated sludge, incinerator slurries
2. Vortex flow	4.5	6	Sludge recirculation
Air lift	0	6	Return sludge
Screw pump	0	6	Return sludge
Positive displacement, plunger, or diaphragm pump	6.5	10	Primary settled sludge, thickened sludge, digested sludge, incinerated sludge, heat-conditioned sludge, chemically treated sludge, slurries

12.8 DISINFECTION OF WASTEWATER

Primary, secondary, and even tertiary treatment cannot be expected to remove 100% of the incoming waste load, and as a result, many organisms still remain in the waste stream. To prevent the spread of waterborne diseases and also to minimize public health problems, regulatory agencies may require the destruction of pathogenic organisms in wastewaters. While most of these microorganisms are not pathogens, pathogens must be assumed to be potentially present. Thus, whenever wastewater effluents are discharged to receiving waters that may be used for the water supply, swimming, or fishing, the reduction of bacterial numbers to minimize health hazards is the desirable goal. Disinfection is treatment of the effluent for the destruction of all pathogens. Another term that is sometimes also used in describing the destruction of microorganisms is *sterilization*. Sterilization is the destruction of *all* microorganisms. While disinfection indicates the destruction of all disease causing microorganisms, no attempt is made in wastewater treatment to obtain sterilization. However, disinfection procedures applied to wastewaters will result in a substantial reduction of all microbes, so bacterial numbers are reduced to a safe level. In general, disinfection can be achieved by any method that destroys pathogens. A variety of physical or chemical methods are capable of destroying microorganisms under certain conditions. Physical methods might include, for example, heating to boiling or incineration or irradiation with X-rays or ultraviolet rays. Chemical methods might theoretically include the use of strong acids, alcohols, or a variety of oxidizing chemicals or surface active agents (such as special detergents). However, the treatment of wastewaters for the destruction of pathogens demands the use of practical measures that can be used economically and efficiently at all times on large quantities of wastewaters that have been treated to various degrees. In the past, wastewater treatment practices have principally relied on the use of chlorine for disinfection. The prevalent use of chlorine has come about because chlorine is an excellent disinfecting chemical and, until recently, has been available at a reasonable cost. However, the rising cost of chlorine coupled with the fact that chlorine even at low concentrations is toxic to fish and other biota as well as the possibility that potentially harmful chlorinated hydrocarbons may be formed has made chlorination less favored as the disinfectant of choice in wastewater treatment. As a result, the increased use of ozone (ozonation) or ultraviolet light as a disinfectant in the future is a distinct possibility in wastewater disinfection. Both ozone and ultraviolet light, as well as being an effective disinfecting agent, leave no toxic residual. Ozone will additionally raise the dissolved oxygen level of the water. However, ozone must be generated and has only recently begun to compete favorably with chlorination in terms of economics. Ultraviolet light has recently undergone studies to determine its effectiveness and cost when used at large wastewater treatment plants. The use of both chlorine and ozone as chemical disinfectants and their disinfecting properties and actions will be considered individually. However, chlorine continues to be used extensively as a disinfectant.

12.9 TERTIARY AND ADVANCED WASTEWATER TREATMENT

Tertiary treatment is supplementary to primary and secondary treatment for the purpose of removing the residual organic and inorganic substances and in some cases the refractory and dissolved substances to the degree necessary. Tertiary treatment of sewage is increasingly being adopted in India. Some of the reuse could be:

- industrial reuse of the reclaimed water in cooling systems, boiler feed, process water etc.,
- reuse in agriculture, horticulture, watering of lawns, golf courses, and such purposes,
- ground water recharge for augmenting groundwater resources for downstream users or for preventing saline water intrusion in coastal areas.

The treatment processes are mainly physicochemical in nature and include disinfection, oxidation, chemical dosing for water quality correction, chemically aided settling, filtration, softening, activated carbon treatment, ion exchange, and reverse osmosis.

12.9.1 Treatment Plant Operation and Maintenance

Maintenance comprises those operations that are well-planned systematic programs of maintaining the machinery by taking appropriate steps to prevent breakdown well in advance before it causes major damage. This prevents wastage of time and production loss and prolongs the life of machine. This maintains better efficiency in the system and economics, the running cost of the plant. It can be classified as follows:

1. preventive maintenance that constitutes works and precautions to be taken to prevent breakdown
2. corrective maintenance that involves carrying out repairs after breakdown.

Preventive maintenance is more economical than corrective maintenance and provides uninterrupted service that is essential to achieve the basic objectives of nuisance.

The various units of the plant are designed for maximum efficiency within a certain flow range and sewage quality. Close control and coordination of operation of different units are, therefore, required within the limits of design to achieve maximum efficiency. Hence, accurate measurements of flow of raw and settled sewage, air, recirculated sludge/effluent, sludges, and final effluent are required. For this purpose, flow measuring devices and meters, preferably of the indicating and recording types, are provided to guide the operator in his supervision and obtain data for progressive improvement. The broad list of operation troubles are given in Table 12.9.

12.9.2 Safety in the Plant

The work of an operator in a sewage treatment plant presents many hazards that must be guarded against. Common types of accidents include injuries from falls, deaths from drowning, and asphyxiation. Narrow walks or steps over tanks (particularly in darkness, rains, and wind), ladders, and spiral staircases are potential danger spots where the operator should be alert; overexertion during operation of valves, moving weights, and performing other arduous tasks should be avoided. All open tanks should be provided with guard rails to prevent accidental falls. Glass parts as well as moving parts should be protected by screen or guards. Adequate lighting within the plant and around the plant should be provided, which gives a better working facility, reducing accidents due to slipping, etc. Honeycomb grating be provided on open channels to avoid accidents due to falling or drowning. The staff should be trained and compelled to use helmets, gumboots, hand gloves, etc. Wherever necessary, precautionary boards/danger boards/sign boards should be displayed in the plant wherever necessary, drawing attention to the potential danger spots.

Gas poisoning, asphyxiation, and gas explosion are other hazards. Hence smoking or carrying open flames in and around digesters should be prohibited. Covered tanks, wet wells, or pits should be well ventilated. Before entering, they should be kept open for sufficient time or preferably forced ventilated, as they present problems of asphyxiation. Entry into them should be permitted only after ensuring the safety by testing for the presence of hazardous gases.

Gas masks should be stored in a location where no possibility of contamination by gas exists and should be easily accessible. A first aid kit should

Table 12.9 Operation Troubles in Sewage Treatment Plant

Signs and Symptoms (1)	Possible Causes (2)	Suggested Action (3)
Pretreatment		
Unusual or excessive screenings	Increase in domestic sewage or industrial waste	Clean screens more often
Excessive grit	Road washings, ashes, or material from building site	Report and get them diverted
Excessive organic matter in grit	Velocity is too low and detention period too long	Reduce the cross-sectional area of the channel occupied by flowing sewage Install planks, bricks, or tile along sides of channel or reshape or repair outlet weir to proportionally reduce depth of flow for all normal present flow rates, decrease the number of channels used, or reduce length of channel by moving outlet weir
Carryover of grit	Velocity is too high and detention too short	Remove grit more frequently or increase number of channels or increase cross-sectional area of channels
Sedimentation tank		
Floating sludge in all tanks	Accumulated sludge decomposing in the tank and buoyed to the surface	Remove sludge more completely and more often
Floating sludge-not in all tanks	Affected tanks receiving too much sewage	Reduce flow to affected tanks

Bubbles rising in tanks	Septic conditions	Reduce flow to affected tanks
Bubbles rising in tanks	Septic conditions	Report and empty tank completely as soon as possible
Contents black and odorous	Septic sewage or strong digester supernatant	Take action to eliminate septicity by improving hydraulics of sewer system, pre-aeration of organic industrial wastes admitted to the system etc., or improve digester operation to have improved quality supernatant, or reduce flow into settling tank or bypass completely supernatant to lagoons, etc., till situation improves
Excessive settling in inlet channels	Velocity too low	Reduce cross-sectional area by installing inner wall of suitable material along one wall of channel, or agitate with air or water otherwise to prevent deposition
Excessive suspended matter in effluent, all tanks	Accumulates sludge flow through tanks too fast (overloading)	Clean tanks more often; report and get the loading reduced
	Humus sludge or under drainage returned too fast	Reduce pumping rate
Not all tanks	Some tanks receiving too much sewage	Reduce flow to affected tanks
Excessive floating matter in the effluent	Detective scum boards or none	Repair scum boards or install new ones
Sludge pipes choke	Sludge too thick	Clean more often
	Sludge contains grit	Clean more often
		Clean grit chamber more often; if chokage persists report; change sludge piping if necessary

Continued

Table 12.9 Operation Troubles in Sewage Treatment Plant—cont'd

Signs and Symptoms (1)	Possible Causes (2)	Suggested Action (3)
Intermittent surging of flow	High intermittent pumping rates	Adjust pumping rates to keep close to rates of flow or install or adjust baffling to reduce inlet velocity and to have effective flow distribution across the width of tank
Sludge hard to remove from hopper	High content of grit and/or clay Low velocity in withdrawal line	Reduce grit content, reduce clay-content, or rod the clogged lines Pump sludge more often, or change sludge piping
Trickling filters Filter ponding	Rock or other media too small or not sufficiently uniform in size Organic loading excessive	Rake or fork the rocks or film surface with light equipment; wash the filter surface with a stream of water under high pressure; or stop the distributor over the ponded area and allow continuous flow of sewage; or dose the filter with heavy application of chlorine (5 mg/L Cl_2 in filter influent) for several hours at weekly intervals or take the filter out of use for one day or longer to allow it to dry out or replace filter media if above methods do not succeed
Filter files	Develop most frequently in an alternate wet and dry environment	Dose filter continuously not intermittently; or remove excessive biological growth; or flood the filter for 24 h at weekly or biweekly intervals (it should be done at intervals frequent enough to prevent the fly completing its life cycle between floodings); or wash vigorously the inside of the exposed filter walls; or chlorinate the sewage (3–5 mg/L) for several hours at frequent intervals of 1–2 weeks or apply DDT or other insecticides
Odors	Anaerobic decomposition of sewage sludge or biological growths	Maintain aerobic conditions in all units including sewer system or reduce accumulation of slime and biological growth; or chlorinate filter influent for short periods when flow is low; or reduce unusually heavy organic loadings as from milk wastes

Icing of filter surface	Air temperature at or below 0°C; or progressive lowering of temperature of applied sewage by recirculation or uneven distribution of sewage on filter	Decrease number of times sewage is recirculated; or where two-stage filters are used; operate filters in parallel with little or no recirculation; or adjust orifices, etc., to improve uniformity of distribution over filter and to reduce spray effect; or erect a wind screen at the filter in the path of prevailing winds; or breakup to remove ice frequently
Activate sludge		
Change in sludge volume index	High soluble organic loads in sewage	Decrease aeration liquor suspended solids; or bulking of activated sludge may be controlled by proper application of chlorine to return sludge; or control sludge index by converting digested sludge to activated sludge
Rising sludge (in settling tanks)	Due to excessive nitrification	Increase the rate of return activated sludge from the final settling tank; or decrease the rate of flow of aeration liquor into the tank; or increase the speed of sludge collecting mechanism in the final settling tank to increase the rate of removal of sludge; or decrease nitrification by reducing aeration or lowering the detention period
Frothing	Synthetic detergents cause, frothing. The froth increases with decrease in aeration liquor suspended solids or increase in aeration; or increase in degree of purification of sewage; or increase in atmospheric temperature	Use water, effluent or clarified sewage sprays in the frothing areas; or apply defoamants in small quantities to tank surface (repeated dosing is necessary); or increase aeration liquor SS concentration
Sludge digestion	Fluctuation in sludge temperature	Pump large quantities of thin sludge at high rate for cooling
Temperature drops in unit with hot water coils.	Sludge solids adhering to coils forming a thick insulting layer preventing heat transfer to digester	Clean the surfaces of coils or replace this form of heating with an external heat exchanger

Continued

Table 12.9 Operation Troubles in Sewage Treatment Plant—cont'd

Signs and Symptoms (1)	Possible Causes (2)	Suggested Action (3)
Temperature constant gas production drops	Increase in scum accumulation; or increase in grit accumulation; or excessive acid production or acid condition due to 1. organic over loading 2. acid wastes 3. toxic metals, Cu, Ni, Cr, and Zn	Control scum; or control grit; or prevent excessive acid conditions by reducing organic overloads; or reduce acid wastes by pretreatment; or eliminate toxic metals or add lime to keep pH between 6.8 and 7.2 or proper quantity of over digested sludge should be withdrawn from digester
Foaming	Insufficient amount of well-buffered sludge in the digester; or excessive additions of raw sludge (with high volatile content); or poor mixing of digester contents; or temperature too low for prolonged periods followed by rise in temperature of digester contents; or withdrawal of too much digested sludge; or excessive scum or grit accumulations	Temporarily reduce or stop raw sludge additions; or add lime to keep pH between 6.8 and 7.2 while other corrective measures are undertaken; or restore good mixing within digester; or raise temperature to normal range; or breakup and remove excessive scum layer; or if large quantities of oil or grit are present, empty digester
Sludge drying beds	Sludge layer too thick	Put on less sludge
	Second dose applied too late	Do not apply second dose if first has started to dry off
Sludge dries more slowly than usual	Standing water	Decant water
	Bed surface clogged	Rake over skim if necessary and redress the surface
	Broken or clogged drains	Set them right

be available readily at hand. Fire extinguishers of the proper type should be located at strategic points and maintained in good operating condition at all times by testing them. All staff should be trained in rendering first aid and operating fire extinguishing equipment.

An adequate number of toilets and bathing facilities, drinking water facilities, and locker should be provided for the convenience of operating staff and protection from risk of infection. Eating facilities and canteen should be maintained hygienically.

All workers should be compelled to observe personal, hygiene such as washing with soap after work as well as washing before taking food. The use of antiseptics along with washing should be emphasized. The employees should be medically checked after every 6 months especially for eye sight, hearing, indigestion, mental capability, Tuberculosis (T.B.), diabetes, heart troubles, etc.

FURTHER READING

[1] G. Lettinga, et al., High rate anaeróbio waste water treatment using the UASB reactor under a wide range of temperature conditions, Biotechnology and Genetic Engineering Review 2 (1989).

[2] S.C. Reed, E.J. Middlebrooks, R.W. Crites, Natural Systems for Waste Management and Treatment, McGraw-Hill, New York, 1988.

[3] USEPA., Process Design Manual: Onsite Wastewater Treatment and Disposal Systems. Cincinnati, Ohio (EPA Report No. EPA-625/1-80-012), 1980.

[4] USEPA, Innovative and Alternative Technology Assessment Manual. Washington, DC (Report No. PA-430/9-78-009), 1980.

[5] USEPA, Planning Wastewater Management Facilities for Small Communities. Cincinnati, Ohio (Report No. EPA-600/8-80-030), 1980.

[6] USEPA, Process Design Manual: Land Treatment of Municipal Wastewater. Cincinnati, Ohio (Report No. EPA-625/1-81-013), 1981.

[7] USEPA, Process Design Manual: Municipal Wastewater Stabilization Ponds. Cincinnati, Ohio (Report No. EPA-625/1-83-015), 1983.

[8] USEPA, Process Design Manual: Constructed Wetlands and Aquatic Plant Systems. Cincinnati, Ohio (Report No. EPA-625/1-88-022), 1988.

[9] USEPA, State Design Criteria for Wastewater Treatment Systems. Washington, DC (Report No. EPA-430/9-90-014), 1990.

[10] USEPA, Process Design Manual: Wastewater Treatment/Disposal for Small Communities. Cincinnati, Ohio (Report No. EPA-625/R-92/005), 1992.

CHAPTER THIRTEEN

Industrial Wastewater Treatment Technologies, Recycling, and Reuse

13.1 INTRODUCTION

While a huge amount of water is required for different industrial processes, only a small fraction of the same is incorporated in their products and lost by evaporation; the rest finds its way into the water courses as waste water. Thus the industries join the municipalities to contribute to the "pollution" of the natural bodies of water.

The industrial wastes either join the streams or other natural water bodies directly, or are emptied into the municipal sewers. Thus these wastes affect in some way or other the normal life of a stream or the normal functioning of sewerage and sewage treatment plants. Streams can assimilate certain amount of wastes before they are "polluted," and the municipal sewage treatment plants can be designed to handle any kind of industrial wastes. Three alternatives for the disposal of the industrial wastes exist:

1. The direct disposal of waste into the streams without any treatment,
2. Discharge of the wastes into the municipal sewers for combined treatment,
3. Separate treatment of the industrial wastes before discharging the same into the water bodies.

The selection of a particular process depends on various factors like these:

- self-purification capacity of the streams,
- permissible limits of the pollutants in the water bodies, as established,
- technical advantages, if any, in mixing the industrial wastes with domestic sewage.

If it is decided to treat the industrial waste either independently or along with the domestic sewage, the treatment plants are to be designed after the following:

1. A thorough investigation of the characteristics of the wastes,
2. A cost study for the final choice of the particular method of treatment.

Environmental Management
ISBN 978-0-12-811989-1
http://dx.doi.org/10.1016/B978-0-12-811989-1.00013-0

13.2 TREATMENT OF INDUSTRIAL WASTES

The treatment of industrial waste water can be done in part or as a whole either by the biological processes, as done in the case of sanitary sewage, or by processes very special for the industrial waste water only. The important factors that affect the planning for industrial waste water treatment plants are these:

- discontinuous and sometimes seasonally discharged wastes,
- high concentration of the waste,
- nonbiodegradability and toxicity of some wastes.

Depending upon the mode of discharge of the waste, and the nature of the constituents present in it, the treatment may consist of any one or more of the following processes:

- equalization
- neutralization
- physical treatment
- chemical treatment
- biological treatment

When the characteristics of the waste vary in a day and also the discharge rate is not uniform or continuous, the waste may require equalization before it is subjected to the treatment. Equalization consists of holding the waste for some predetermined time in a continuously mixed basin, which produces an effluent of fairly uniform characteristics. When the waste contains an excessive amount of acid or alkali (particularly acid), the waste requires neutralization in the neutralization tank. Neutralization may be carried out in the equalization tank, when the conditions permit.

When the industrial waste is treated along with the municipal sewage or discharged into a stream, the waste may be subjected to another prior unit operation, known as proportioning. Proportioning consists of the control of the discharge of the waste into the receiving sewer or stream, in a fixed proportion to the flow of domestic swage or the steam. This helps not only in protecting the treatment device from the shock load but also in improving the sanitary quality of the treated effluent.

Before an industrial waste is subjected to a chemical or biological treatment, or both, it may be required to separate the suspended matter by physical operations like sedimentation and flotation. Sedimentation tanks are to be provided only when the waste contains a high percentage of settleable solids. Flotation is employed to separate fine particles with very low settling characteristics. Flotation consists of creation of fine air bubbles in the waste body by the introduction of air to the system. The rising air bubbles attach themselves to the suspended particles and thereby increase the buoyancy of the particles. The particles thus lifted to the liquid surface are removed by skimming. The domestic/industrial wastewater treatment processes and pathways are summarized in Fig. 13.1.

Some of the industrial wastes, amenable to biological treatment, may require prior chemical treatment; some requires only chemical treatment without any biological treatment. The important contaminants of concern in wastewater treatment are listed in Table 13.1. Secondary treatment standards for wastewater are concerned with the removal of biodegradable organics, suspended solids, and pathogens. Each of the categories of solids may be further classified on the basis of their volatility at 550 + 50°C. The organic fraction will oxidize and will be driven off as gas at this temperature, and the inorganic fraction remains behind as ash.

Other important physical characteristics include odor, temperature, color, and turbidity.

In cases where the industrial effluents are being mixed with the waste waters, Biological Chemical Demand (BOD) plays a very important role in the type of treatment processes to be used. Fig. 13.2 shows the BOD with time for sewage, combined wastes, and industrial wastes. It illustrates one possible effect of a given industrial wastewater on a sewage plant. In this instance the industrial wastewater, with its constant rate of degradation, tends to smooth out the rate of decomposition of the sewage so that the result shows less upsurge due to nitrogenation. Also, the rate of decomposition of the industrial wastewater tends to slow down the initial rapid rate of domestic sewage.

Some of the important parameters to be studied for specific industries have been identified and listed in Table 13.2. The table is indicative, and certain parameters may need to be studied in addition depending on the production processes, raw material, etc.

As in wastewater, industrial wastes also are treated by physical, chemical, and biological means. The individual methods usually are classified as physical unit operations, chemical unit processes, and biological unit processes.

13.2.1 Physical Unit Operations

These are treatment methods in which the application of physical forces predominates are known as physical unit operations. Screening, mixing, flocculation, sedimentation, flotation, filtration, and gas transfer are typical unit operations.

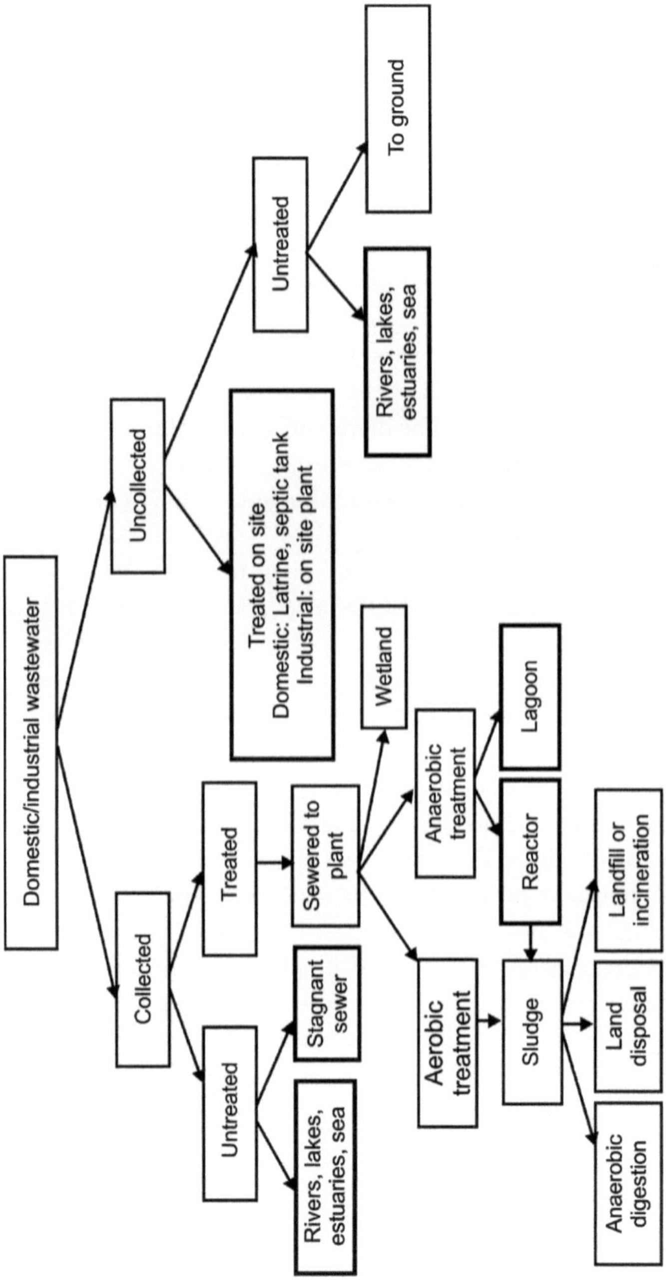

Figure 13.1 Flow chart showing steps in wastewater treatment processes.

Table 13.1 Important Contaminants in Wastewater Treatment

Contaminants	Reason for Importance
Suspended solids	Suspended solids can lead to the development of sludge deposits and anaerobic conditions when untreated wastewater is discharged in the aquatic environment.
Nutrients	Both nitrogen and phosphate, along with carbon, are essential nutrients for growth. When discharged to the aquatic environment, these nutrients can lead to the growth of undesirable aquatic life. When discharged in excessive amounts on land, they can also lead to the pollution of groundwater.
Priority pollutants	Organic and inorganic compounds selected on the basis of their known or suspected carcinogenicity, mutagenicity, teratogenicity, or high acute toxicity. Many of these compounds are found in wastewater.
Refractory organics	These organics tend to resist conventional methods of wastewater treatment. Typical examples include surfactants, phenols, and agricultural pesticides.
Heavy metals	Heavy metals are usually discharged to wastewater from commercial and industrial activities and have to be removed if the wastewater is to be reused.
Dissolved inorganics	Inorganic constituents such as calcium, sodium, and sulfate are added to the original domestic water supply as a result of water use and may have to be removed if the wastewater is to be reused.

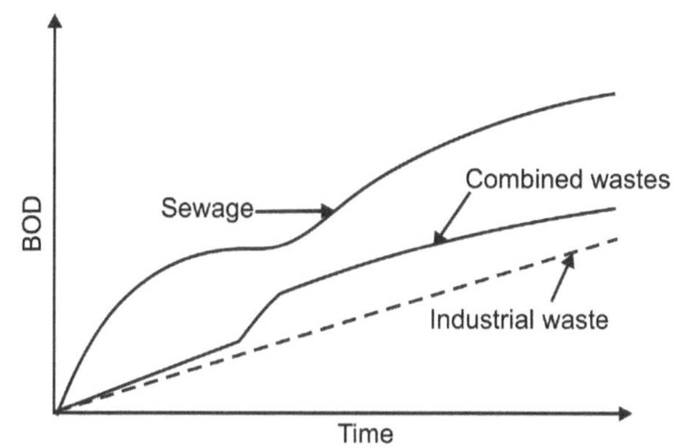

Figure 13.2 BOD for sewage, combined wastes, and industrial wastes.

Table 13.2 Important Parameters to Be Checked in Various Industries

Para Meter	Industry													
	Automobile	Beverage	Caning	Fertilizer	Inorganic chemicals	Organic chemicals	Meat products	Metal finishing	Plastics and synthetics	Pulp and paper	Petroleum refining	Steel	Textiles	Dairy
BOD_5	x	x	x		x	x	x		x	x	x		x	x
COD	x		x	X	x	x		x	x	x	x		x	x
TOC			x			x				x	x			x
TOD					x									
pH		x	x	X	x	x	x		x	x	x	x	x	x
Total solids	x				x									
Suspended solids	x	x	x	X	x	x	x	x	x	x	x	x	x	x
Settable solids		x					x							
Total dissolved solids		x	x	X	x	x	x		x	x	x		x	
Volatile suspended solids											x			
Oil and grease	x	x		X		x	x	x	x	x	x	x	x	
Heavy metals, general						x		x		x			x	
Chromium	x			X	x						x	x	x	
Copper											x			
Nickel	x													
Iron	x			X	x						x	x		
Zinc	x			x					x		x	x		
Arsenic					x									
Mercury				X	x									
Lead	x				x						x			
Tin	x											x		
Cadmium	x													
Calcium				X										
Fluoride				X	x									
Cyanide					x	x		x	x		x	x		
Chloride	x			X	x	x					x	x		x
Sulfate	x			X	x				x		x	x		

Table 13.2 Important Parameters to Be Checked in Various Industries—cont'd

Para Meter	Industry													
	Automobile	Beverage	Caning	Fertilizer	Inorganic chemicals	Organic chemicals	Meat products	Metal finishing	Plastics and synthetics	Pulp and paper	Petroleum refining	Steel	Textiles	Dairy
Ammonia	x			X		x	x		x	x	x	x		
Sodium				X										
Silicates					x									
Sulfite										x				
Nitrate	x			X		x			x	x	x			x
Phosphorus			x	X	x	x	x		x	x	x			x
Urea of organic nitrogen				X										
color		x	x				x			x	x		x	x
Total coliforms		x					x			x				
Fecal coliforms			x							x				
Toxic materials		x					x		x	x	x		x	x
Temperature		x	x	X	x						x	x	x	x
Turbidity		x					x			x	x			x
Foam	x													
Phenols	x				x	x			x	x	x	x	x	
Chlorinated benezoids and polvnclear aromatics					x				x					
Mercaptans / sulfide									x		x		x	

13.2.2 Chemical Unit Processes

Treatment methods in which the removal or conversion of contaminants is brought about by the addition of chemicals or by other chemical reactions are known as chemical unit processes. Precipitation, adsorption, and disinfection are the most common examples used in wastewater treatment. In chemical precipitation, treatment is accomplished by producing a chemical precipitate that will settle. In most cases, the settled precipitate will contain both the constituents that may have reacted with the added chemicals and the constituents that were swept out of the wastewater as the precipitate settled. Adsorption involves the removal of specific compounds from the wastewater on solid surfaces using the forces of attraction between bodies.

13.2.3 Biological Unit Processes

Treatment methods in which the removal of contaminants is brought about by biological activity are known as biological unit processes. Biological treatment is used primarily to remove the biodegradable organic substances (colloidal or dissolved) from industrial wastewaters. Basically, these substances are converted into gases that can escape to the atmosphere and into biological cell tissue that can be removed by settling.

The major biological processes used for wastewater treatment are five major groups: aerobic processes, anoxic process, anaerobic process, combined aerobic anoxic, and anaerobic/aerobic processes. The aerobic biological processes are further subdivided depending on whether treatment is accomplished in suspended growth systems, attached growth systems, or combinations thereof. All the biological processes used for the treatment of wastewater are derived from processes occurring in nature.

Aerobic suspended growth:

- activated sludge processes
  - plug flow with recycle
    - aerated lagoons

Aerobic attached growth:

- trickling filter
  - roughing filter rotating biological contactor
    - fixed film nitrification reactor

The anaerobic process has been developed for the treatment of sludge and high-strength organic load. Some selected industrial wastes, their major characteristics, and disposal methods are summarized in Table 13.3.

One of the most important chemical and physicochemical processes, employed in the industrial wastes treatment, for the removal of dissolved inorganic materials is membrane separation technique.

13.3 MEMBRANE SEPARATION

Membrane separation processes play an important role in the reduction and/or recycling of wastes. These processes include reverse osmosis (RO), ultrafiltration (UF), hyperfiltration (HF), and electrodialysis, each of which

Table 13.3 Major Characteristics and Disposal Methods for Industrial Wastes

Industrial Producing Wastes	Major Characteristics	Major Treatment and Disposal Methods
Textile	Highly alkaline, colored, COD, temperature, high suspended solids	Neutralization, chemical precipitation, biological treatment, aeration, and/or trickling filtration
Leather goods	High total solids, hardness, salt sulfides, chromium, pH, precipitated lime, and BOD_5	Equalization, sedimentation, and biological treatment
Laundry trades	High turbidity, alkalinity, and organic solids	Screening, chemical precipitation, flotation, and adsorption
Canned goods	High in suspended solids, colloidal products, and dissolved organic matter	Screening, lagooning, soil absorption, or spray irrigation
Dairy	High in dissolved organic matter, mainly protein, fat and lactose	Acidification, flotation biological treatment, aeration trickling filtration, activated sludge
Meat and poultry products	High in dissolved and suspended organic matter, blood, other proteins, and fats	Screening, setting and/or flotation, trickling filtration
Brewed and distilled beverages	High in dissolved organic solids, containing nitrogen and fermented starches or their products	Recovery, concentration by centrifugation and evaporation, trickling filtration; use in feeds; digestion of slops
Beet sugar	High in dissolved and suspended organic matter, containing sugar and protein	Reuse of wastes, coagulation, and lagooning
Pharmaceutical products	High in suspended and dissolved organic matter	Activated sludge
Yeast	High in solids (mainly organic) and BOD_5	Anaerobic digestion, trickling filtration
Pickles	Variable pH, high suspended solids, color, and organic matter	Good housekeeping, screening, equalization
Coffee	High BOD_5 and S.S.	Screening, settling, and trickling filtration
Fish	Very high BOD_5, total organic solids, O&G, and odor	Oil removal, biological treatment
Glass	Red color, alkaline nonsettleable suspended solids	Calcium chloride precipitation
Fuel oil use	High in emulsified and dissolved oils	Leak and spill prevention, flotation
Rubber	High BOD_5 and odor, high suspended solid, variable pH, high chlorides	Aeration chlorination, sulfonation, biological treatment

Continued

Table 13.3 Major Characteristics and Disposal Methods for Industrial Wastes—cont'd

Industrial Producing Wastes	Major Characteristics	Major Treatment and Disposal Methods
Cane sugar	Variable pH, should organic matter with relatively high BOD_5 of carbonaceous nature	Neutralization, calculation, chemical treatment, some selected aerobic oxidation
Palm oil	High BOD_5, COD, solids and total fats and low pH	Neutralization, coagulation, flotation, filtration
Pulp and paper	High or low pH, color, high suspended, colloidal, and dissolved solids, inorganic filters	Settling lagooning, biological treatment, aeration, recovery of by-products using flotation
Photographic	Alkaline, containing various organic and inorganic reducing agents	Recovery of silver; discharge
Steel	Low pH, acids, cyanogen, phenol, ore, coke, limestone, alkali, oils, mill scale, and fine suspended solids	Neutralization, recovery and reuse, chemical coagulation
Metal-plated	Acid, metals, toxic, low volume, mainly mineral matter	Alkaline chlorination of cyanide, reduction and precipitation of chromium, lime precipitation on other metals
Oil fields and refineries	High dissolved salts from field; high BOD_5, odor, phenol, and sulfur compounds from refinery	Recovery of salts; acidification burning of alkaline sludge
Petrochemical	High COD, TDS, metals, COD/ BOD_5 ratio	Recovery and reuse, equalization and neutralization, chemical coagulation, settling or flotation, biological oxidation
Cement	Heated cooling water, suspended solids, some inorganic salts	Segregation of dust-contact streams, neutralization, and sedimentation
Asbestos	Suspended asbestos and mineral solids	Detention in ponds, neutralization, and land filling
Paint and inks	Contain organic solids from dyes, resins, oils, solvents, etc.	Settling ponds for detention of paints, lime coagulation of printing inks
Pesticides	High organic matter, benzene ring structure, toxic to bacteria and fish, acid	Activated carbon adsorption, alkaline chlorination
Organic	Varied types of organic chemicals	Biological degradation plant control, process modification

separates a contaminant (solute) from a liquid phase (solvent, typically water). In addition, newer membrane separating technologies, such as pervaporation, are now commercially available. Membrane separation processes can function in several ways: volume reduction, recovery and/or purification of the liquid phase, and concentration and/or recovery of the contaminant or solute.

The following basic characteristics should be taken into account:

1. *UF* is primarily used to separate organic components from water according to the size (molecular weight) of the organic molecules. Membranes are manufactured with the capability to remove contaminants with molecular weights between 500 and 1,000,000.
2. *HF* separates ionic or organic components from water by limiting the size of membrane pores through which a contaminant can pass. It is typically used to remove contaminants having a molecular weight of 100–500 from water.
3. *RO* is primarily used to separate water from a feed stream containing inorganic ions. The purity of the recovered water is relatively high, and the water is generally suitable for recycling. The maximum achievable concentration of salt in the reject stream is usually about 100,000 mg/L because of osmotic pressure considerations.
4. *Electrodialysis* is used to remove ionic components from water. It produces moderate-quality product water (i.e., several hundred mg/L salt) and is capable of producing concentrate streams containing up to 20% salt.
5. *Pervaporation* removes volatile organic compounds (VOCs) from contaminated water. Depending on the contaminant, concentration factors of 5- to 200-fold are achievable.

13.3.1 Ultrafiltration and Hyperfiltration

UF and *HF* utilize pressure and a semipermeable membrane to separate nonionic materials from a solvent (such as water). These membrane separation techniques are particularly effective for the removal of suspended solids, oil and grease, large organic molecules, and complexed heavy metals from wastewater streams.

In UF and HF systems, the membrane retains materials based solely on size, shape, and molecule flexibility. As the feed solution is pumped through a membrane module, the membrane acts as a sieve to retain dissolved and suspended materials that are physically too large to pass through its pores. The retained materials then exit the module separately from the purified solvent or permeate.

The major difference between HF and UF is that HF typically removes species having a molecular weight of 100–500; UF removes species having a molecular weight greater than 500. The two membrane separation methods utilize identical operating principles.

UF and HF membranes have an asymmetric structure designed to maximize productivity per unit surface area. They are composed of a thin (0.1–1.0 μM), selective, surface layer supported by a porous, spongy layer about 100 μM thick. Membrane pore sizes typically range from 10 to 1000 A. Two common membrane materials are polysulfone and cellulose acetate. Polysulfone is the most versatile because it can tolerate temperatures between 0 and 79°C (32 and 175°F) and pH from <1 to 13. It also can be cleaned with a wide array of cleaning agents. Cellulose acetate is also a popular membrane material; however, it can only be used at pH 2.5 to 7 and temperatures from 0–50°C (32–122°F). UF membranes are available that retain molecular weights from 500 to 1,000,000.

UF and HF are used in various pollution control applications to do the following:

1. Remove complexed toxic metals from metal-finishing wastewater.
2. Concentrate oily wastes from metal-finishing rinse waters, aluminum and steel-coil cleaning rinse waters, and detergent rinse waters from automotive and aircraft-chassis cleaning.
3. Concentrate electrodeposition paint baths to remove water and contaminants.
4. Recover oil from waste-oil paint baths to remove water and contaminants.
5. Remove dyes from textile industry effluents.

13.3.2 Reverse Osmosis

When a semipermeable membrane separates two solutions of different dissolved solids concentrations, pure water will flow through the membrane into the concentrated solution, while ions are retained behind the membrane. This process is known as *osmosis*. During *RO*, pressure is applied to the more concentrated solution to reverse the normal osmotic flow, and pure water is forced through the semipermeable membrane into the less concentrated solution. The purified stream that passes through the membrane is called *permeate*; the concentrated stream retained by the membrane is known as *concentrate*.

Conceptually, a RO system has several advantages for treating wastes:

- Both the recovered solvent and the concentrated solute in some cases can be recycled to a manufacturing process, rather than requiring treatment and disposal.
- The separation process does not require an energy-intensive phase change such as is required for distillation or evaporation.

- RO equipment does not require a large amount of space. Also, because RO is a straightforward mechanical process, it requires a low degree of operational skill.

One of the major applications of RO has been in the electroplating industry.

13.3.3 Electrodialysis

Electrodialysis relies on ion-exchange membranes in a direct-current electrical field to separate ionic species from solution. Like RO membranes, electrodialysis membranes are sensitive to fouling. This sensitivity has limited waste treatment applications of the technology, although the development of the electrodialysis reversal process has significantly reduced complications due to fouling.

Since electrodialysis is suited only to the removal or concentration of ionic species, the metal-finishing and electroplating industry is probably the greatest potential market for these systems.

13.4 CHEMICAL OXIDATION

Electrolytic recovery techniques are used primarily for the recovery of metals from process streams or rinse waters. These metals must be removed or recovered from the effluent streams before discharge to recover the metals for their economic value. The mining industry has used electrolytic techniques to refine ores for several years. The process is now finding wide application with electroplaters, rolling mills, manufacturers of printed circuit boards, and metal coating firms.

Electrolytic recovery techniques are based on the oxidation–reduction reaction that takes place at the surface of conductive electrodes (cathode and anode). The electrodes are immersed in a chemical medium under the influence of an applied potential. At the cathode, the metal ion is reduced to its elemental form. Simultaneously, gaseous products such as oxygen, hydrogen, or nitrogen may evolve at the anode. The gases produced at the anode depend on the chemical composition of the medium. Dissolved species such as cyanide are generally oxidized at the anode.

The major process equipment consists of these:

1. The electrochemical reactor containing the electrodes
2. A venting system for gas
3. Recirculation pumps
4. A power supply

After the metal coating or deposition at the cathode reaches the desired thickness, the metal can be removed and generally reused or sold. In the electroplating industry, for example, either the recovered metal, which is essentially pure, is returned to the plating tank, or the metal-plated cathode can now be used for an anode in the plating bath. Electrolytic recovery techniques have been used to recover copper, nickel, zinc, silver, cadmium, gold, and other heavy metals.

These processes have been demonstrated for the recovery of gold, silver, cadmium, nickel, nickel-iron alloy, copper, zinc, and other metals. One of the most common applications of the electrolytic system is the recovery of copper from sulfuric acid solutions. Electrolytic techniques can also be used to recover metal from crystals and sludges and to regenerate process solutions. However, electrolytic techniques are not effective at recovering nickel because of the low standard reduction potential and the high stability constants its cyanide complexes.

The advantages of the electrolytic recovery techniques are these:

- Equipment maintenance is minimal.
- Valuable metals are recovered, reused, or sold.
- Discharge of toxic heavy metals is reduced or eliminated.
- Production of toxic sludges is eliminated.

13.5 ADSORPTION

Adsorption can be used to remove a wide variety of contaminants from liquid or gaseous streams. It is typically employed for organic compounds, although some inorganic species are also efficiently adsorbed. The process is relatively nonspecific and thus is frequently utilized as a broad-spectrum treatment operation when the chemical composition of a stream is not fully understood. Common applications include groundwater treatment, chemical spill response, industrial wastewater treatment, and air-pollution control systems.

Most carbon adsorption systems utilize granular activated carbon (GAC) in flow-through column reactors. These systems are efficient and relatively simple to operate if properly designed. Powdered activated carbon (PAC) can be added to some existing treatment processes or used by itself in slurry reactors.

The adsorption process is normally reversible. It is therefore common to remove the adsorbed contaminants after the adsorption capacity of the carbon has been exhausted. This regenerates the carbon, allowing it to be reused.

13.5.1 Mechanisms of Adsorption

A variety of attractive forces exists between fluid-phase (gas or liquid) molecules and the molecules of a solid adsorbent, all having their origin in the electromagnetic interactions of nuclei and electrons. Traditionally, three loosely defined categories have been distinguished: physical, chemical, and electrostatic interactions.

- *Physical adsorption* results from the action ofVan der Waals forces, relatively weak interactions produced by the motion of electrons in their orbitals.
- *Chemical adsorption*, or *chemisorption*, involves electronic interactions between specific surface sites and adsorbate molecules, resulting in the formation of a bond that can have all the characteristics of a true chemical bond. Chemisorption is typified by a much stronger.

Chemical and physical adsorption result from electrostatic interactions, and the term *electrostatic adsorption* is generally reserved for Coulombic attractive forces between ions and charged functional groups and is synonymous with the term *ion exchange*.

In liquid solutions, consideration must also be given to the nature of the solution state, that is, the extent to which the solvent is capable of accommodating the solute. Solvophobic forces result when there is a substantial chemical incompatibility between the solute and solvent. These forces may be associated with a significant thermodynamic gradient that drives the solute out of solution. This can result in adsorption energies that are considerably higher than those resulting from the surface reaction alone. In aqueous solutions, for example, most nonpolar organic molecules will readily adsorb to any available solid surface in what is often labeled *hydrophobic bonding*.

Adsorption capacity is simply a measurement of the state of thermodynamic equilibrium for a given set of system conditions and, thus, should be related to one or more fundamental thermodynamic properties of the adsorbate. For liquid solutions, adsorption generally increases as the solubility of the compound in the solution decreases. It is also frequently observed that the adsorption of organic compounds usually increases as one moves up a homologous series or as molecular weight increases.

Adsorption equilibrium relationships, commonly called *isotherms*, relate the concentrations of the adsorbed compound in each of the two phases. Isotherms are measured experimentally by equilibrating known quantities of the adsorbent with the compound of interest and plotting the resultant concentrations.This graphical result can be expressed mathematically by fitting the experimental data to one or more mathematical relationships or isotherm models. Fig. 13.3 shows both the mathematical expressions and graphical representations for the three most common isotherm models and their respective linear forms, which are sometimes used for curve fitting.

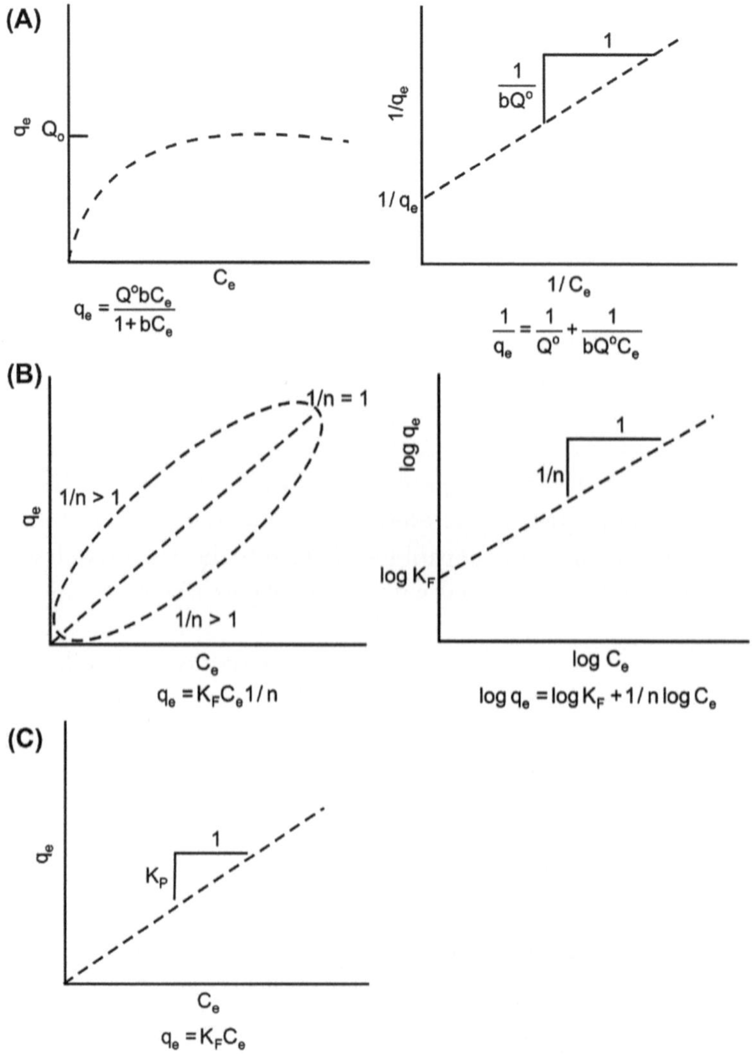

Figure 13.3 Adsorption isotherm models: (A) Langmuir, (B) Freundlich, and (C) linear.

13.5.2 Completely Mixed Reactors

PAC is generally used in either completely mixed batch or completely mixed flow (CMF) reactors. In both cases, mixing is applied to the system to assure that the PAC is kept in suspension, that the slurry of PAC and contaminated water is distributed uniformly throughout the tank, and that no significant spatial concentration gradients are established. In a CMF reactor, wastewater and PAC are continuously added to a tank, and the mixed contents of the

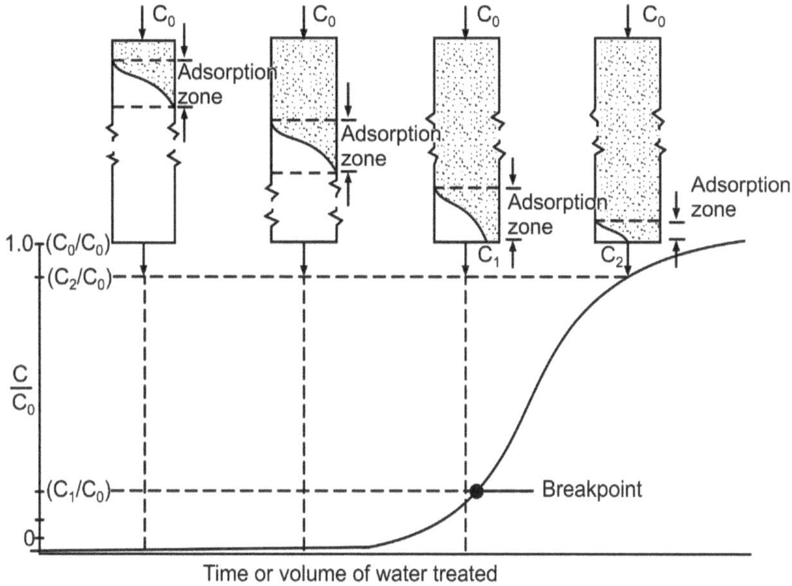

Figure 13.4 Breakthrough curve.

tank are withdrawn at the same rate. Since the system is completely mixed, the tank contents are maintained at a steady-state concentration equal to that desired in the effluent. While this approach does not assure that a desired effluent concentration will always be reached, as changes in the influent will translate to changes in the effluent, it does lend itself to unattended operation. For both types of completely mixed systems, it is necessary to separate the PAC from solution once an acceptable concentration is reached. For example, a PAC slurry can be added at various points in a conventional water treatment plant and removed by coagulation and settling or filtration for control of tastes and odors or removal of specific organics such as the trihalomethanes.

Column or plug flow reactors are not hydraulically feasible for PAC but have several distinct advantages over CMF reactors for GAC adsorption systems. Adsorption process in bed column reactor is shown in Fig. 13.4, where the state of the carbon bed is shown at several points along a breakthrough curve, the plot of effluent concentration as a function of time or volume treated. It can be observed from this figure that when the bed is first brought online, only the very top layer of the bed is exposed to the influent concentration of the adsorbate since the adsorption reaction depletes the solution reaching the layers below. This depletion process establishes a finite depth in the bed, over which the concentration changes from the influent value to essentially zero. The column depth over which this occurs is called the *adsorption zone* or the *mass-transfer zone*. As the top layer continues to adsorb, it eventually reaches saturation, where the concentration in the carbon is in equilibrium with the influent, and the adsorption zone moves downward. The concentration of the effluent exiting the column will remain essentially zero until the adsorption zone reaches the bottom of the column. When the effluent reaches some predetermined concentration, termed *breakthrough*, the adsorber is generally removed from service. The major advantage of the column reactor is its very efficient utilization of the adsorption capacity, since most of the bed is in equilibrium with the influent concentrate in when the carbon is removed from service. Secondarily, any given volume of carbon is exposed to an increasing concentration of solute as the reaction progresses, resulting in a much higher overall reaction rate than can be achieved in completely mixed systems. Finally, since there is always fresh (unexposed) carbon ahead of the adsorption zone, changes in influent concentration generally do not produce changes in the effluent: they simply alter the rate at which the mass-transfer zone moves.

Fixed-bed adsorbers for liquid-phase applications can be operated in either an upflow or downflow mode, with the pressurized downflow approach being the most common option for industrial wastewater, process water, and groundwater treatment applications. Downflow operation also allows the adsorbers to function as filters, although both the design and operation must address the need for frequent and efficient backwashing. A widely used implementation of this approach is the placement of GAC as the granular media in a conventional rapid sand filter for drinking water treatment.

Adsorption is normally a reversible process; that is, under suitable conditions the materials that have accumulated in the carbon can be driven off and the carbon reused.

13.6 ION EXCHANGE

Although ion exchange has become widely accepted as a standard method of purifying water for such applications as boiler feed, pharmaceutical makeup, and semiconductor manufacture, its potential and, indeed, actual use in treating hazardous wastes has not been nearly so widely recognized. Unlike many other separation processes such as evaporation and RO that remove the water from the polluting species, ion exchange usually removes the pollutant from the water. Since the offending pollutant is often present in low concentrations, ion exchange is frequently more efficient in treating large flows of dilute hazardous waste streams than many other processes. Another significant feature of the ion-exchange process is that it has the ability to separate, or *purify*, as well as to concentrate pollutants. Some exchangers are selective for certain metals and can remove low concentrations of toxic metal from a wastewater containing a high background level of a nontoxic metal such as sodium or calcium. Ion exchange can also be utilized in some applications to purity a spent chemical concentrate by removal of low-level contamination.

Ion-exchange process has certain general limitations and disadvantages, including these:

- Chemical wastes are produced if excess regenerant is required.
- There are limitations on the concentration of what can be treated and produced.
- Ion-exchange resins are prone to fouling by some organic substances.

There are a number of possible applications of ion exchange to the treatment of organic wastes, e.g., phenol removal and decolorization of Kraft pulp mill effluents, but usually the mechanism is one of adsorption by the resin as opposed to true ion exchange.

It is the functional groups that actually participate in the ion-exchange reactions.

Cation exchangers: Strong-acid cation resins have sulfonic acid groups. They exchange cations over the entire pH range. Analogous to sulfuric acid, they do not retain hydrogen tightly and consequently are somewhat difficult to regenerate with acid, usually requiring several times the theoretical or stoichiometric dosage of acid to convert them back to the hydrogen form after exhaustion. For example, strong-acid cation-exchange resins can remove a toxic metal, such as copper, from a wastewater by exchange of hydrogen for the copper according to reaction:

$$Cu^{++} + 2HR \rightarrow R_2Cu + 2H^{+}$$

Ion exchange is reversible, so that the copper can be removed from the resin by regeneration with a moderately concentrated to yield a more concentrated solution of the copper. In some cases, it is possible to recycle the metal concentrate produced by regeneration.

$$R_2Cu + 2H^{+} \rightarrow Cu^{++} + 2HR$$

Table 13.4 shows the relative preferences of strong-acid cation resins for various actions.

Table 13.4 Affinities of Strong-Acid Cation Resins for Various Cations

Monovalent	**Ag>Cu>K>NH_4>Na>H**
Divalent	Pb>Hg>Ca>Ni>Cd>Cu>Zn>Fe>Mg>Mn
Trivalent	Fe>Al

Weak-acid resins have carboxylic acid groups and do not function below a pH of about 4. *Chelating resins* are very similar to weak-acid resins. They exhibit a high degree of selectivity for many toxic metals such as copper, mercury, nickel, and lead. Chelating resins utilizing a picolylamine functionality are highly selective for copper and operate at pHs as low as 1 and below. Resins with this functionality have extremely high affinity for mercury and some other metals such as lead, silver, copper, and cadmium.

Anion exchangers: *Strong-base anion resins* have quaternary ammonium groups, which, being analogous to sodium hydroxide, do not readily associate with hydroxyl ions. As a result, they function over the entire pH range but require an excess of strong base (usually sodium hydroxide) to regenerate.

$$ROH + HCl \rightarrow RCl + H_2O$$

$$ROH + NaCl \rightarrow RCl + NaOH$$

Acid sorption is reversible, and the acid can be eluted from the resin simply with water. *Weak-base anion resins*, having a tertiary amine functionality that behaves similarly to ammonium hydroxide, readily associate with hydroxyl, resulting in high regeneration efficiencies.

One of the most important applications of ion exchange is to the electroplating industry. When plated parts are removed from the chrome-plating bath, the adhering film of chromic acid plating electrolyte must be rinsed off. Chromic acid recovery is a good example of the use of ion exchange to purify and concentrate a waste, converting it back into a valuable product. Rinse water, typically containing a few hundred mg/L of chromic acid, is pumped through a strong-acid cation exchanger in the hydrogen form to remove metallic impurities such as trivalent chrome and iron. This is necessary to avoid precipitation of metallic hydroxides in the precipitation of metallic hydroxides in the subsequent anion-exchange bed. Upon exhaustion, the cation resin is regenerated with sulfuric acid.

$$3RH + M^{+++} \rightarrow R_3M + 3H^+$$

The *decationized* chromic acid rinse water is next directed through an anion-exchange resin in the hydroxyl form to remove the chromate. The effluent leaving the anion exchange will be deionized water, which is recycled to the final rinse tank

$$2ROH + H_2CrO_4 \rightarrow R_2CrO_4 + 2H_2O$$

Upon exhaustion, the anion exchanger is regenerated with dilute sodium hydroxide, yielding sodium dichromate according to reaction.

$$R_2CrO_4 + 2NaOH \rightarrow ROH + Na_2CrO_4$$

The cation resin must be subsequently regenerated with sulfuric acid.

$$2RH + Na_2CrO_4 \rightarrow 2RNa + H_2CrO_4$$

$$RH + NaOH \rightarrow RNa + H_2O$$

Recovery of metals from acid copper- and nickel-plating rinse water. The problem with the original ion-exchange process for nickel recovery is that an excess of sulfuric acid is required to regenerate the nickel from the resin, so the recovered nickel sulfate product has a pH of approximately 1. This cannot be recycled to a nickel-planting bath operating at a pH of 4 without adversely affecting the bath chemistry. This can be overcome by using the ion exchange process.

Recovery of metals from mixed rinse waters: Theoretically, it is possible to combine a group of rinse waters, deionize them, and recycle them. Even if low-cost water is in good supply, the concept is appealing, since the amount requiring final treatment is substantially reduced.

Various ion exchangers used in the treatment of industrial wastes include these:

Inclusion polymers: Inclusion polymers are polymeric adsorbent resins that have been impregnated with liquid extractants such as di-2-ethylhexyl phosphate (DEHPA). These polymers are used to remove toxic metals from wastewater streams.

Zeolites: While the first ion exchangers commercially employed for water treatment, *zeolites*, were inorganic in nature, organic ion exchangers quickly came to dominate the field of water and wastewater treatment.

Immobilized biomass: Under specific conditions, it has been established that organisms, such as bacteria, yeast, fungi, peat moss, chitosan, and algae, have the ability to remove heavy metals from solution, even in the presence of high concentrations of alkali and alkaline earth metals.

13.7 AIR AND STEAM STRIPPING

Stripping is a physical unit operation in which dissolved molecules are transferred from a liquid into a flowing gas or vapor stream. The driving force for mass transfer is provided by the concentration gradient between the liquid to the gas phases, with solute molecules moving from the liquid to the gas until equilibrium is reached. *In air stripping*, the moving gas is air, usually at ambient temperature and pressure, and the governing equilibrium relationship is Henry's law.

Steam stripping uses live steam as the gas phase. In this case, the vapor–liquid equilibrium between water and the organic compound is the key equilibrium relationship. Like air stripping, steam stripping has been successfully applied to the removal of hazardous organics from aqueous waste. Steam stripping is more widely applicable in that it can effectively remove less volatile or more soluble compounds not easily removed by air stripping. Air stripping may also be used for the removal of volatiles from industrial aqueous wastes containing traces of dissolved solvents.

Air stripping is of various types and includes the following:

1. Cascade air stripping
2. Membrane air stripping
3. Inclined cascade aeration
4. In situ air stripping

Cascade air stripping: Conventional air stripping (PTA) is the method of choice when stripping highly volatile contaminants from water. However, stripping semi-volatile and low-volatile contaminants is difficult because of the high loadings and substantial packing depths and air flow rates that are required to achieve the necessary mass transfer required to achieve the higher removal efficiencies.

Membrane air stripping: Another technology that seeks to overcome the limitations of conventional air stripping is membrane air stripping. The difference between membrane air stripping and packed tower aeration is the fact that the surface area is supplied by a hollow-fiber membrane. Contaminated water is passed through the inside of the hollow fibers, and air flows counter currently along the outside of the fibers. Membrane air stripping possesses some potential advantages to packed-tower aeration. Lower air flow rates can be used for contaminant removal.

Inclined cascade aeration: Inclined cascade aeration is a process that was originally designed for the oxygenation of sugar refinery wastewater. In this process, water is pumped to the top of an inclined corrugated surface. The water is released and allowed to flow over the surface by gravity. The liquid turbulence of the contaminated water causes mass transfer of the low solubility organics from the water to the air.

In situ air stripping: Air stripping is also used in remediation applications for removal of VOCs from saturated soils.

Like air stripping, steam stripping can be used to remove VOCs from water or aqueous waste streams. However, steam stripping is more broadly applicable, in that it can treat the following:

- Aqueous wastes contaminated with more soluble, less volatile compounds, not readily air-strippable, including acetone, methanol, and pentachlorophenol
- Higher concentrations, up to several percent by weight, of VOCs in an aqueous waste
- Nonaqueous wastes such as spent solvents contaminated with nonvolatile impurities

13.8 PERVAPORATION

In addition to the traditional membrane separation technologies, a new innovative membrane separation technique that has been used in Europe and the United States is pervaporation. This technology, called the *cross-flow pervaporation system*, removes VOCs from water. The technology, which is commercially available, can be used to remediate ground water, leachate, and wastewater that contains solvents, degreasers, and gasoline. In the pervaporation process, organic-permeable membranes made of synthetic polymers such as silicon rubber or polyethylene are used to adsorb VOCs preferentially from contaminated water. Depending on the organic contaminant, concentration factors of 5- to 200-fold are found achievable.

13.9 SOLVENT EXTRACTION

Solvent extraction can occur under three processing approaches. The most common approach employs two phases in contact at ambient conditions of temperature and pressure, in which the contaminants are exchanged between the solid matrix and a liquid solvent. In another approach, liquefied gases, such as propane, are used as the solvent. Finally, critical solution temperature solvent extraction systems use solvents, such as aliphatic amines, in which solubility can be varied over the process operating temperature range. These processes use liquid–liquid extraction at two different temperatures. At the lower operating temperatures, the solvents are miscible, while at the upper temperatures, the two solvents are completely immiscible.

Solvent extraction processes operate in either a batch or continuous mode, and all employ similar unit operations. Solvent extraction consists of the unit operations shown in Fig. 13.5.

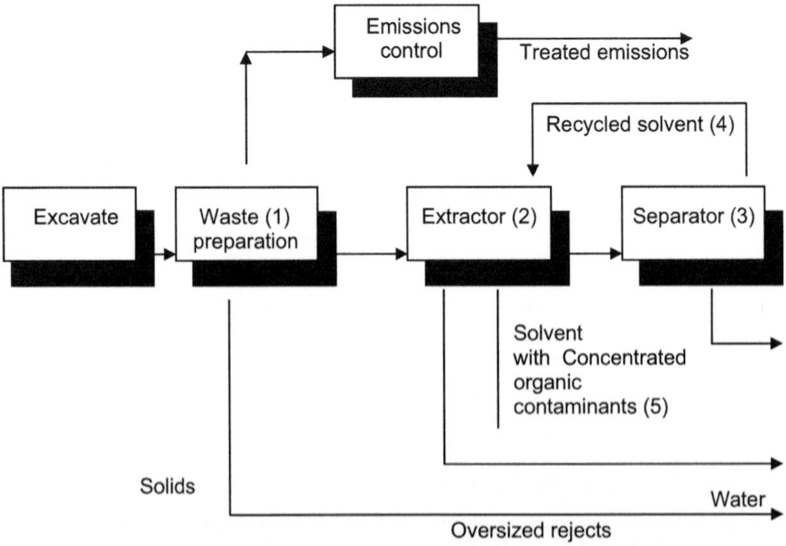

Figure 13.5 Solvent extraction process, simplified process flow diagram.

Solvent extraction systems have been shown to be effective in treating sediments, sludges, and soils containing primarily organic contaminants such as polychlorinated biphenyl, polynuclear aromatic hydrocarbons, VOCs, halogenated solvents, pesticides, and petroleum-refining oily wastes. Solvent extraction is generally not used to treat in organics (acids, bases, salts, heavy metals). Solvent extraction processes are not designed to treat particular compounds, and extraction efficiencies and processing rates are lower when there are high concentrations of indigenous organic compounds (humic and tannic acids in soil). Similarly, extraction efficiencies and processing rates are lower when emulsifiers and water-soluble detergents are in the feed.

13.10 GASEOUS EMISSIONS FROM INDUSTRIAL WASTE WATERS WITH SPECIFIC REFERENCE TO GHGS

Only industrial wastewater with significant carbon loading that is treated under intended or unintended anaerobic conditions will produce CH_4. Assessment of CH_4 production potential from industrial wastewater streams is based on the concentration of degradable organic matter in the wastewater, the volume of wastewater, and the propensity of the industrial sector to treat their wastewater in anaerobic systems. Using these criteria, major industrial wastewater sources with high CH_4 gas production potential can be identified as follows:

- pulp and paper manufacture
- meat and poultry processing (slaughter houses)
- alcohol, beer, starch production
- organic chemicals production
- other food and drink processing (dairy products, vegetable oil, fruits and vegetables, canneries, juice making, etc.)

Both the pulp and paper industry and the meat and poultry processing industries produce large volumes of wastewater that contain high levels of degradable organics. The meat and poultry processing facilities typically employ anaerobic lagoons to treat their wastewater, while the paper and pulp industry also use lagoons and anaerobic reactors. The nonanimal food and beverage industries produce considerable amounts of wastewater with significant organic carbon levels and are also known to use anaerobic processes such as lagoons and anaerobic reactors. Anaerobic reactors treating industrial effluents with biogas facilities are usually linked with recovery of the generated CH_4 for energy. The development of emission factors and activity data is more complex because there are many types of wastewater, and many different industries to track. The most accurate estimates of emissions for this source category would be based on measured data from point sources. Due to the high costs of measurements and the potentially large number of point sources, collecting comprehensive measurement data is very difficult.

Total CH_4 emissions from industrial wastewater are calculated as follows:

$$CH_4 \text{ emissions} = \sum_i [(TOW_i - S_i)\, EF_i - R_i]$$

where CH_4 emissions = CH_4 emissions in inventory year, kg CH_4 per year, TOW_i = total organically degradable material in wastewater from industry I in inventory year, kg COD per year, i = industrial sector, S_i = organic component removed as sludge in inventory year, kg COD per year, and EF_i = emission factor for industry i, kg CH_4/kg COD for treatment/discharge pathway or system(s) used in inventory year. If more than one treatment practice is used in an industry, this factor would need to be a weighted average. R_i = amount of CH_4 recovered in inventory year, kg CH_4 per year.

There are significant differences in the CH_4-emitting potential of different types of industrial wastewater. To the extent possible, data should be collected to determine the maximum CH_4-producing capacity (Bo) in

Table 13.5 Default MCF Values for Industrial Wastewater

Type of Treatment and Discharge Pathway or System	Comments	MCF^{-1}	Range
Untreated			
Sea, river, and lake discharge	Rivers with high organics loadings may turn anaerobic; however, this is not considered here	0.1	0–0.2
Treated			
Aerobic treatment plant	Must be well managed. Some CH_4 can be emitted from settling basins and other pockets	0	0–0.1
Aerobic treatment plant	Not well managed, overloaded	0.3	0.2–0.4
Anaerobic digester for sludge	CH_4 recovery not considered here	0.8	0.8–1.0
Anaerobic reactor (e.g., UASB, fixed film reactor)	CH_4 recovery not considered here	0.8	0.8–1.0
Anaerobic shallow lagoon	Depth less than 2 m, use expert judgment	0.2	0–0.3
Anaerobic deep lagoon	Depth more than 2 m	0.8	0.8–1.0

2006 IPCC Guidelines for National Greenhouse Gas Inventories.

each industry. As mentioned before, the methane correction factor (MCF) indicates the extent to which the CH_4-producing potential (Bo) is realized in each type of treatment method.

CH_4 emission factor for industrial wastewater is calculated as follows:

$$EF_j = B_o \cdot MCF_j$$

where EF_j = emission factor for each treatment/discharge pathway or system, kg CH_4/kg COD, j = each treatment/discharge pathway or system, B_o = maximum CH_4 producing capacity, kg CH_4/kg COD, and MCF_j = methane correction factor (fraction).

Table 13.5 includes default MCF values, which are based on expert judgment as given in IPCC report, 2006.

Industrial production data and wastewater outflows may be obtained from national statistics, regulatory agencies, wastewater treatment associations, or industry associations (Table 13.6).

Table 13.6 Examples of Industrial Wastewater Data

Industry Type	Wastewater Generation W (m^3/ton)	Range for W (m^3/ton)	COD (kg/m^3)	COD Range (kg/m^3)
Alcohol refining	24	16–32	11	5–22
Beer and malt	6.3	5.0–9.0	2.9	2–7
Coffee	NA	NA	9	3–15
Dairy products	7	3–10	2.7	1.5–5.2
Fish processing	NA	8–18	2.5	
Meat and poultry	13	8–18	4.1	2–7
Organic chemicals	67	0–400	3	0.8–5
Petroleum refineries	0.6	0.3–1.2	1.0	0.4–1.6
Plastics and resins	0.6	0.3–1.2	3.7	0.8–5
Pulp and paper (combined)	162	85–240	9	1–15
Soap and detergents	NA	1.0–5.0	NA	0.5–1.2
Starch production	9	4–18	10	1.5–42
Sugar refining	NA	4–18	3.2	1–6
Vegetable oils	3.1	1.0–5.0	NA	0.5–1.2
Vegetables, fruits, and juices	20	7–35	5.0	2–10
Wine and vinegar	23	11–46	1.5	0.7–3.0

NA = not available.
M.R.J. Doorn, R. Strait, W. Barnard, B. Eklund, Estimate of Global Greenhouse Gas Emissions from Industrial and Domestic Wastewater Treatment, Final Report, EPA-600/R-97–091, Prepared for United States Environmental Protection Agency, Research Triangle Park, NC, USA, 1997.

Uncertainty estimates for Bo, MCF, P, W, and COD are provided in Table 13.7.

Similar estimates have to be brought out for oxides of nitrogen. Nitrous oxide (N_2O) emissions can occur as direct emissions from treatment plants or from indirect emissions from wastewater after disposal of effluent into waterways, lakes, or the sea. Direct emissions from nitrification and denitrification at wastewater treatment plants may be considered as a minor source. The simplified general equation is as follows:

N_2O emissions from wastewater effluent

N_2O emissions $= N_{efficient} \cdot EF_{effluent} \cdot 44/28$

where N_2O emissions $= N_2O$ emissions in inventory year, kg N_2O per year, $N_{effluent}$ = nitrogen in the effluent discharged to aquatic environments, kg N per year, $EF_{effluent}$ = emission factor for N_2O emissions from discharged to

Table 13.7 Default Uncertainty Ranges for Industrial Wastewater

Parameter	Uncertainty Range
Emission factor	
Maximum CH_4-producing capacity (B_o)	±30%
Methane correction factor (MCF)	The uncertainty range should be determined by expert judgment, bearing in mind that this is a fraction and uncertainties cannot take it outside the range of 0–1
Activity data	
Industrial production (P)	±25% use expert judgment regarding the quality of data source to assign more accurate uncertainty range
Wastewater/unit production (W) COD/unit wastewater (COD)	These data can be very uncertain as the same sector might use different waste handling procedures at different plants and in different countries. The product of the parameters (W·COD) is expected to have less uncertainty. An uncertainty value can be attributed directly to kg COD/ton of product. −50%, +100% is suggested (i.e., a factor of 2)

Judgment by expert group (co-chairs, editors, and authors of this sector).

wastewater, kg N_2O-N/kg N, and the factor 44/28 is the conversion of kg N_2O-N into kg N_2O.

Total nitrogen in the effluent is calculated as follows:

$$NE_{EFFLUENT} = (P \cdot Protein \cdot F_{NPR} \cdot F_{NON-CON} \cdot F_{IND-COM}) - N_{SLUDGE}$$

where $NE_{EFFLUENT}$ = total annual amount of nitrogen in the wastewater effluent, kg N per year, P = human population, Protein = annual per capita protein consumption, kg/person per year, F_{NPR} = fraction of nitrogen in protein, default = 0.16, kg N/kg protein, $F_{NON-CON}$ = factor for nonconsumed protein added to the wastewater, $F_{IND-COM}$ = factor for industrial and commercial co-discharged protein into the sewer system, and N_{SLUDGE} = nitrogen removed with sludge (default = zero), kg N per year.

N_2O emission from centralized wastewater treatment processes is as follows:

$$N_2O_{PLANTS} = P \cdot T_{PLANT} \cdot F_{IND-COM} \cdot EF_{PLANT}$$

where N_2O_{PLANTS} = total N_2O emissions from plants in inventory year, kg N_2O per year, P = human population, T_{PLANT} = degree of utilization of modern, centralized WWT plants, %, $F_{IND-COM}$ = fraction of industrial and commercial co-discharged protein (default = 1.25, based on data in [1] and expert judgment), and EF_{PLANT} = emission factor, 3.2 g N_2O/person per year.

13.11 WASTE MINIMIZATION AND CLEAN TECHNOLOGIES

The internal process evaluation should typically start with specifying the emission or discharge sources and the discharge points within the factory system. The discharge restroom water and low contaminated water can normally be sewered directly to the receiving waters. Recycling of chemicals and process water leads to less effluent flow and less discharge of pollutants. A process alteration by change of raw materials and chemicals can reduce the amount of pollutants achieving increased yield that corresponds to less discharges. Improved operational routines like cleaning routines (process equipment) and prevention of accidental discharges (process and raw water) and use of an environmental control systems approach like use of simple control parameters (pH, conductivity, turbidity, and temperature), alarm at overflow positions, spill collection systems, and rapid feedback to operators can be used for waste minimization. This can also be achieved by imparting environmental education of effluent treatment operators and other mill operators who have to be aware of environment and of the important of spills. Industry has to realize that in any process, any material being discharged in a waste stream constitutes a raw material of negative cost. It is a sound business practice to recover these materials at economic costs, a situation that can be brought about by the input of right technologies.

13.11.1 In-Plant Survey

No two industries are alike as processes and raw materials used are different. In the same type of industry, two different factories at two different places can discharge waste water of different composition. There will be diurnal, daily, and seasonal variations in the quality and quantity of wastewater discharged from an industry. It is therefore necessary to make flow measurements to determine peak, minimum, and average flows and to find out the composition of the waste water from individual section and the combined final discharge. Before carrying out flow measurement, information on the following have to be obtained:

1. Water using operations, water and material balance
2. Sources and quantities of waste water generated and nature of pollutants in them
3. Water and waste water quality
4. Efficiency of existing treatment plant, if any
5. Ultimate disposal of waste water on land, into sewer, into a stream, or into a marine environment
6. Raw materials/products
7. Site plan, plant layout, drainage map, process flow sheet, and building plan

13.11.2 Flow Measurement

1. *Flow rates measurement based on water level changes*: There are two modes of waste water transport and two corresponding ways to measure the quantity of flow:
   a. Open channels, with the liquid surface exposed to the atmosphere; here the quantity of flow (Q) is a function of the water depth Q = f(n)
   b. Closed system, where the liquid is in a pipe and the pipe is full; here the quantity of flow depends on the flow velocity Q = f(v).

   The most commonly used equipment are wiers or Parshall flumes that are placed in a channel as a constriction of dam causing changed level and velocity of the water. Wires are made of steel plates, planed wood, or plastic. A wier is a constriction placed in a channel over which the water has a free fall. The flow is calculated from the geometry of the wier and the water level upstream. There are wiers without end contraction, wiers with end contraction, and the Thompson wier (V-notched). The edge of the wier plates in contact with the flowing medium is sharp and cut to form an angle of 45 degree in the direction of flow. Wiers are mounted accurately at right angles to direction of flow and with the upper edge horizontal (spirit level). The choice of suitable wier is made on the basis of channel width, the accessible dam height, and the range within which the flow is expected to vary. Designs resulting in submerged flow situations should be avoided. A constriction in a channel can cause an increase in the water level. For a suitably shaped constriction, the flow rate can be obtained with sufficient accuracy by measuring the level upstream the constriction.

Parshall flume is a convenient device for measuring flow in sewers. It is a modified Venturi flume of standard dimensions. It consists of three parts: a converging section, a throat section, and a diverging section. Future increase in flow should be given attention while designing a flume. The level of the floor of the converging section is higher than the floor in the throat and diverging sections. The head of the water surface in the converging section measures the flow through the flumes. The elevation of the water surface should be measured back from the crest of the flume at a distance equal to two-thirds or the length of the converging section. The crest is located at the junction of the throat and converging section. The head should be measured in a stilling well instead in the flume itself as sudden changes in flow are dampened in stilling well. The size of the Parshall flume should be determined during the preliminary survey. The general formula for computing the free discharge from a Parshall flume is this:

$$Q = 4WH^{N}$$

where Q = discharge, CIS, W = width of throat in ft, H = head of water above the level floor in ft in the covering section, and $N = 1.522W^{0.026}$.

The flume may be built of wood, fiber glass, concrete, plastic, or metal and can be installed at convenient locations such as a manhole. The Parshall flume is used for sewer lines where continuous flow measurements are desirable. The main advantage of the Parshall flume over a wier is the self-cleaning property of the flume. It is better to avoid submergence situations as far as possible by proper design.

Formula for Parshall flume flow in m^3/s is this:

$$Q = 0.1132 \times b \times \left(3.28 \times h^1\right) m\ m^3/s$$

where $m = 1.522 \times b0.026$, b = throat width (m), and h_1 = water level at H before throat (m).

Several types of gauges can be used to register water level based on pressure, distance, time, and capacitance. Generally, gauges based on pressure are recommended for permanent installation. Venturi meters installed in a pipeline consists of a throat of known diameter, converging from throat to pipe.

Flow rate measured based on velocity: If the area is known, the flow rate can be circulated from the velocity of flow, as measured with a velocity indicating instrument. Since the velocity varies over a cross-section of the flow, it is necessary to know the approximate velocity distribution across the section. From the velocity distribution the mean velocity is calculated. Velocity can be measured with a pitot tube, which measures the difference between the total and the static pressure in the system. The pitot tube is connected to a differential manometer.

2. *Flow rate measurement based on dilution*: When a salt solution or tracer dye is injected, the time for it to reach a given point or to pass between two given points is measured, the flow can be calculated from the result. A concentrated solution of a substance (inorganic salt or dye tracer) that can easily be determined in low ion concentrations is injected into the medium at a constant known rate. The concentration of tracer is then measured for enough downstream for complete mixing to have taken place. The flow is calculated from the formula:

$$Q = q\,(C_1)\,/\,(C_2 - C_3)$$

where Q = flow in check point, L/mm, q = injected flow, L/min, C_1 = concentration of tracer element in injected flow, mg/L, C_2 = concentration of tracer element in check point, mg/L, and C_3 = concentration of tracer element in zero sample in check point in mg/L.

Good tracer elements are lithium and K-salts. $LiCl_3$ is commonly used for this type of investigations with good results. In this case the resulting concentration in check point should preferably be above I mg/L during a sampling time of about 10 min.

13.11.3 Flow Rate Measurement in Closed Systems

Several methods have been developed for flow rate measurements with a good degree of accuracy in closed systems. The methods may be divided into three main categories:

1. Measurement of a pressure difference
2. Determination of the velocity and area of flow of the liquid
3. Methods based on dilution.

The most common types of measuring devices used for continuous flow measurement in closed system are orifice plates, flow nozzles, and entire tubes. Installation of an orifice plate results in an increase in the seed of the flowing medium and a corresponding drop in its static pressure. The pressure difference is measured by manometers. A standardized orifice plate consists of thin plate with a central, sharp-edged hole that is mounted perpendicular to the direction of flow. The flow rate is highest and the concentration greatest just below the orifice. Orifice plates give a high level of accuracy but at the expense of a large pressure drop, which increases the pumping costs. They do not function satisfactorily in the presence of suspended particulate and are therefore unsuitable for permanent installation in systems with fibers. The liquid flow in a pipe can also be measured by a magnetic flow meter, which consists of a straight pipe section fitted with standard flanges. An electric coil around the tube imposes a magnetic field that changes with the velocity of the liquid. Waste water blowing through a magnetic field produces a field voltage in proportion to velocity that is converted by electrical and mechanical means to indicate and record the flow. The supersonic meter is based on the measurement of velocity according to Doppler principle, i.e., the frequency of sound wave is changed by reflection on air bubbles or particles in the fluid. The dilution methods (salt solutions, chemical, and dye tracer) can be applied to flow rate measurements in closed systems as described in an earlier section.

13.11.4 Waste Volume and Strength Reduction

Volume of waste generated by an industry can be reduced by the following:

1. Segregation
2. Conservation
3. Reuse, recycle
4. In-plant control measures
5. Housekeeping

13.11.4.1 Segregation

It is cheaper to treat low volume of concentrated wastewater than large volume of dilute wastewaters. Cooling water is generally free from pollution. Wastewater from process, cooling boiler blow down, sanitary waste, wastewater from canteen, and storm water should be segregated and treated separately. In the process, wastewaters from some sections are stronger than those from the other sections. All these wastewater have to be sewered separately, if necessary.

13.11.4.2 Conservation of Water

There is lot of scope for conservation of watering many industries like tanneries, textiles, paper mills, etc. In many industries in summer the consumption of water is less. In water-scarce areas the amount of water used is less per unit of product than in industries in other areas. It is therefore possible to reduce water consumption in many industries without affecting the quality of the product. Important steps include these:

1. Prevention of running taps, leaks, spills
2. Alarm at overflow positions
3. Spill collection systems
4. Preventive maintenance
5. No overloading

Modification of equipment and process automation has in many cases minimized operational errors, reduced spills, and reduced waste generation.

13.11.4.3 Reuse and Recycle of Water

The first preference is reuse of wastewater without treatment like reuse of textile mill wash waters. The second preference is reuse of wastewater after partial treatment like reuse of paper machine wastewater. Third preference is reuse of wastewater after complete treatment. Improved operation routines like cleaning routines, analysis, and prevention of spills (accidental discharges) and internal treatment of some separated effluent streams that can result process water closure (recycling) and raw material recycling are important. Reclaiming water from sewage is being practiced using tertiary treatment methods in many countries. In many countries, water reclaimed from sewage is being used in industries for various purposes like cooling, washing, etc.

13.11.4.4 In-Plant Control Measures

In a sugar mill, for example, cooling oil used in roller mills for tandem cooling can be collected in a sump filled with bagasse, which is solid waste from the same industry. Bagasse absorbs large quantities of cooling oil that can later be used in boilers. The overflow from evaporators can be collected in a sump and recycled to clarification section. Press mud can be used as a soil conditioner and should not be allowed into drains. Proper storage of molasses is very important. Molasses spills have high BOD and COD. In a pulp and paper mill, leakages and spillage of black liquor can be collected in a sump and pumped to the soda recovery section. Chemical and process water recycling will result in less effluent flow and less discharge of pollutants.

13.11.4.5 Housekeeping

Preventive maintenance, prevention of leaks and spills, cleaning schedule, and clean environment are important housekeeping measures required to reduce the volume of wastewater generated. Good housekeeping practices are very important and involve alteration of an existing system to limit unnecessary generation of wastes attributable to human intervention.

13.11.4.6 Waste Strength Reduction

Strength of wastewater generated from an industry can be reduced by these:

1. Equipment change
2. Process change
3. Equalization and proportioning
4. By-product recovery

Recycling technologies include (1) distillation of solvent wastes, (2) dechlorination of halogenated, nonsolvent wastes, and (3) metal concentration techniques such as in exchange, evaporation, solvent extraction. Some of the factors encouraging wastewater minimization by the generator are shown in Table 13.8.

Table 13.8 Some Factors Encouraging Waste Minimization by the Generator

Technical	**Legislation/Policy**
New processes available	Bans on specified wastes or raw materials
New chemicals available	Limits on waste production
New plant installed	Compulsory waste audits
Improved product design	Waste minimization is criterion in plan permits
New raw materials	
Operational	**Management**
Regular maintenance	Waste policy adopted
Trained operators	Staff incentives
Printed company directives	Operational directives
Area set aside for collection recovery	Waste audit procedures
Avoid over-ordering	Positive publicity
Storage areas kept safe	Public scrutiny
	Regular monitoring
Disposal	**Economic**
Lack of disposal sites	High disposal fees
Pretreatment by generator	High dumping fines
Required by authorities	High chemical costs
Supplier obliged to accept	Incentives for new plant
Return of surplus stock	

13.12 PERFORMING A WASTE/EFFLUENT MINIMIZATION ASSESSMENT

How does a company go about assessing the status of its operations and developing a waste minimization program? The best way is to conduct a waste minimization assessment.

Step 1: The first and most crucial aspect is to obtain the commitment and support of senior management of the company to undertake a comprehensive and ongoing program:

- To reduce wastes at the source
- To recycle wastes
- To adopt efficient methods for waste treatment and disposal

Step 2: Assessment should be performed by a team led by general manager or manager with good organizational and communication abilities and the knowledge, authority, skill, and credibility to get the required information. Typically the team could include the following.

General Manager
Technical Manager
Administrative Manager

↓

External Expert (Consulting)

↓

Supervisors
Technical Officers
Purchasing

↓

Shop Floor
Union Representative

- whole operation should be taken as benefiting everyone
- hence rewards for good ideas
- no recriminations for revealing errors for identifying poor operating practices

Step 3: Once a team is formed, define goals and identify project for assessment.

- Settling special and achievable goals is important so that programs can be monitored and their success quantified.

Typical goals could include these:

- Reduce the quantity of toxic wastes that are shipped off the site to reduce disposal costs and long-term liability.
- Achieve compliance with discharge or emissions regulations.
- Reduce water usage in a process.
- Find replacement solvent for CFCs because of their planned phase out of production.
- Reduce solid waste generation because of decreasing landfill space.

Step 4: Waste audit

- Conduct site investigation
- Construct process flow charts
- Define-process inputs and outputs
- Construct material balances
- Identify waste sources
- Prioritize wastes in terms of economic value, volume, toxicity, etc.

Step 5: Identification and prioritization of waste minimization alternatives:

Brain storming, technology search, raw material changers, use of by-products, technique changes, housekeeping, use of life cycle analysis, eco design, etc.

- The assessment team is responsible for evaluating and screening the waste minimization alternatives and ranking them in relation to the assessment objectives, cost, and case of implementation.

Priority of options

- Options for source reduction have a high priority, since it minimizes the generation of waste.
- Recycling allows materials to be put to beneficial use but quite often involves significant effort or cost; hence medium priority.
- Treatment options are the lowest priority.

Step 6: Implementation/feasibility analysis:

- Number of alternative choices to the feasibility analysis will depend on the time and resources available for the project.
- Alternatives are evaluated in terms of technical and economic feasibility.
- Economic analysis involves cost/benefit analysis involving payback estimation including purchase cost, running, maintenance, and labor costs.

Technical criteria considered at this stage include:

- worker health and safety
- product quality
- space requirements
- compatibility with existing production processes
- labor requirements
- environmental effects

Implementation involves the following:

- purchase, installation, and commissioning, etc.

Step 7: Assessment report.

At the end of waste minimization assessment, documenting the information and conclusions of the study in a formal report of assessment is extremely important.

The report should include these points:

- a summary of background information
- results of waste minimization audit
- result of assessment
- technical and economic analysis of waste minimization
- alternatives, recommendation, and priorities for implementing waste minimization measures

Step 8:

- The project/program does not end with the issue of a report.
- Monitoring the success of the measures taken, giving feedback to the assessment team and employees, and generally maintaining awareness and enthusiasm for the program are necessary.

The waste minimization program should be dynamic. It needs to be reviewed and evaluated in response to changes in production, new regulations, and advances in technology.

13.13 CLEAN TECHNOLOGIES

These involve a conceptual or procedural approach to production that demands that all phases of the life cycle of a product or of a process should be addressed with the objective of prevention or minimization of short- and long-term risks to human health and environment.

13.13.1 Cleaner Technology

Clean technology refers to avoiding environmental damage at the source through use of materials, processes, or practices to eliminate or reduce the creation of pollutants or wastes.

13.13.2 Remediation

Remediation is repairing damage caused by past human activity or natural disaster's clean-up technology. End of pipe refers to reducing environmental damage by retrofitting, modifying, or adding pollution abatement measures to an established plant of process. The asymptotic behavior of the curve at low environmental load indicates that any human activity involves some environmental impact or resource utilization, so environmental load cannot be completely eliminated. Similarly, the asymptotic behavior at low environmental cost indicates that even the most environmentally inefficient activity has a financial cost (Fig. 13.6).

Curve 1 represents an established technology to which clean-up can be applied.

Curve 2 represents a different technology that is cleaner.

As a typical example, an organization operating at point "C" has three options.

1. It can adopt cleaner technology of curve 2 and can reduce environmental load without increase in cost (D).
2. It can adopt curve 2 and reduce cost while retaining the environmental load (E).
3. It can adopt curve 2 and reduce both environmental load and cost.

The adoptability of curve 2 by a particular organization depends on:

1. Whether the cleaner technology is available
2. The organizational constraints in terms of the existing investment in technology 1 and the investment needed to exploit the technology 2

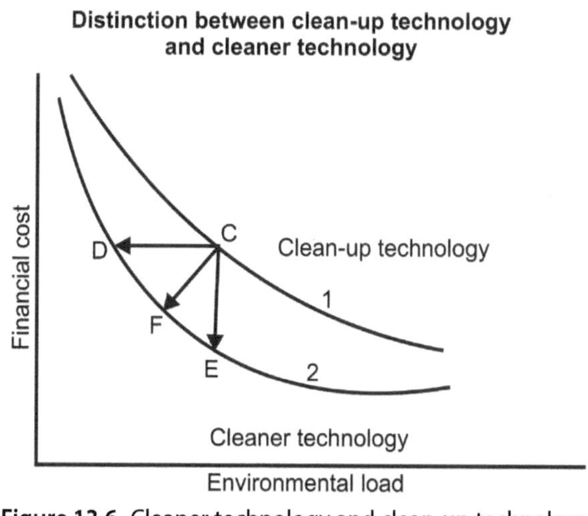

Figure 13.6 Cleaner technology and clean-up technology.

A number of existing organizations can achieve improved environmental performance in a short time by resorting to improvements to existing products and processes than adopting cleaner technology.

From the environmental point of view, the concept of clean technologies means use of minimum resources with maximum efficiency to achieve the twin benefits of resource conservation and environmental protection. From the economic point of view, it means cost effectiveness and increased productivity within available resources.

There are three broad categories of clean technologies:

- Low and non-waste technologies (LNWT) of production aimed at waste minimization at all points in the cycle of production through process changes, good housekeeping, recycle and reuse, equipment design, and product formulations,
- Recycle technologies designed to recover raw materials, energy, water and by-products in the course of end-of-pipe treatment,
- Waste utilization technologies for reclamation and utilization of waste to manufacture products with various end uses. Waste minimization has to be achieved through more selective, environmentally benign methods.

Selection and application of clean technologies require a comparative analysis and evaluation of various competing technologies based on economic, social, technological, and environmental conditions. Already significant modifications have been made in well-established processes, such as in the manufacture of cyclohexanone, sulfuric acid, etc., permitting them to be made through newer feed stocks, conditions, and catalysis. In the manufacture of caustic soda, membrane technology without mercury pollution completely replaced accepted mercury cell technology. Selected examples of LNWT of production are presented in Table 13.9.

Table 13.9 Some Examples of Low and Non-Waste Technologies

S. No	Industry	Process Modification	Recycle/Reuse
1.	Pulp and paper	Substitution of chlorine with chlorine dioxide, oxygen delignification	White water recycling for washing pulp
2.	Textile	1. Counter current washing 2. Thermal printing for cotton cloth	Caustic soda recovery in mercerizing
3.	Iron and steel	1. Mechanical cleaning to replace acid pickling	Blast furnace slag as construction material
4.	Fertilizer	Nitrophosphate process for NPK complete production	Nitric acid recovery in ammonia fertilizer plant

Table 13.10 Raw Material Recovery Options for Selected Industries

Industry	Product	Recovery Scheme
Pulp and paper	Fibers and fillers	Filtration/sedimentation/flotation
Photo processing	Silver	Electrolysis
Dyes and dye intermediate	Dyes	Membrane technology
Cement	Cement dust	ESP/filters
Chrome tannery	Chromium	Physicochemical process
Fertilizers	Ammonia	Wet scrubbing

Source: C.D. Bader, Sorting Through the Best Equipment for Recyclables. MSW Management Elements, 9 (7) 2000.

Recycling/reuse of raw materials is the preferred option, followed by process modification.

The main apprehensions of the industry regarding water reuse relate to the quality of recycled water for use in process and cost of such treatment. The raw material recovery technologies are based on unit operations, namely, screening, filtration, sedimentation, membrane technologies, etc. The recovery options for selected industries are shown in Table 13.10.

Many solid wastes are generated by the industry.

1. Fly ash from the thermal power stations
2. Blast furnace slag from steel manufacture
3. Lime sludge from pulp and paper production
4. Bagasse from sugar industry
5. Phosphogypsum from fertilizer industry
6. Red mud from aluminum industries

The industrial and mineral solid wastes possess immense potential as raw material for manufacture of building materials such as slag cement, Portland pozzolan cement, and lightweight bricks.

Thus, development of clean technologies warrants an interdisciplinary subsystem and environmental subsystem. As against end-of-pipe treatment, cleaner technologies conserve resources, generate less pollution, and provide direct benefit to the industry.

REFERENCE

[1] Metcalf & Eddy, Inc., Wastewater Engineering: Treatment, Disposal, Reuse, McGraw-Hill, New York, 2003, ISBN: 0-07-041878-0.

FURTHER READING

[1] American Public Health Association and American Water Works Association, Standard Methods for the Examination of Water and Wastewater, twentieth ed., Water Environment Federation, 1998, ISBN: 0-87553-235-7.

[2] P. Czepiel, P. Crill, R. Harriss, Nitrous oxide emissions from domestic wastewater treatment, Environmental Science and Technology 29 (9) (1995) 2352–2356.
[3] Destatis, Öffentliche Wasserversorgung und Abwasserbeseitigung 2001, Tabelle 1 "Übersichtstabelle Anschlussgrade, Statistical Office Germany, 2001. http://www.destatis.de/.
[4] M.R.J. Doorn, R. Strait, W. Barnard, B. Eklund, Estimate of Global Greenhouse Gas Emissions from Industrial and Domestic Wastewater Treatment, Final Report, EPA-600/R-97–091, Prepared for United States Environmental Protection Agency, Research Triangle Park, NC, USA, 1997.
[5] M.R.J. Doorn, D. Liles, Global Methane, Quantification of Methane Emissions and Discussion of Nitrous Oxide, and Ammonia Emissions from Septic Tanks, Latrines, and Stagnant Open Sewers in the World. EPA-600/R-99–089, Prepared for U.S. EPA, Research Triangle Park, NC, USA, 1999.
[6] FAO, FAOSTAT Statistical Database, United Nations Food and Agriculture Organization, 2004. Available on the Internet at: http://faostat.fao.org/.
[7] R.G. Feachem, D.J. Bradley, H. Gareleck, D.D. Mara, Sanitation and Disease – Health Aspects of Excreta and Wastewater Management, World Bank, John Wiley & Sons, USA, 1983.
[8] IPCC, in: J.T. Houghton, L.G. Meira Filho, B. Lim, K. Tréanton, I. Mamaty, Y. Bonduki, D.J. Griggs, B.A. Callander (Eds.), Revised 1996 IPCC Guidelines for National Greenhouse Inventories. Intergovernmental Panel on Climate Change (IPCC), IPCC/OECD/IEA, Paris, France, 1997.

CHAPTER FOURTEEN

Air Pollution Control Technologies

14.1 INTRODUCTION

The atmosphere is understood by its composition, temperature structure, and pressure. Air is a fluid mixture, which is constantly changing in its motion (wind), pressure distribution, temperature, and composition or cloud cover. The composition of the air is primarily of permanent gases of clean, dry air, variable gases, green house gases, ozone, and suspended particles (aerosol droplets). The concentration of these gases vary widely; nitrogen (N_2, 78%) and oxygen (O_2, 21%), which are most plentiful and have little or no importance in affecting weather, argon (Ar, 1%); a noble gas with no effect, and green house gases which have a major role in determining the weather. Table 14.1 shows the permanent gases in the atmosphere.

The composition of the atmosphere varies with the vertical increases in height. Typically two layers are identified; homosphere and heterosphere. Homosphere is 0–80 km and the permanent components are generally uniform. Heterosphere is >80 km and the heavier gases deplete with height and the lighter gas components occur as we go higher. These include molecular N_2, atomic oxygen (O), helium atoms (He), and hydrogen atoms (H). The vertical structure of the atmosphere is also identified by the variations in temperatures. The features of the layers in the temperature structure are identified and given as follows:

Troposphere (greek: "overturning"):

- 0–10 km
- Temperature decrease with height:
  ~6.5°C/km (due to adiabatic cooling)
- Strong vertical mixing (cumulonimbus clouds)
- Contains 80% of the atmospheric mass
- Contains almost all atmospheric H_2O
- Called the "weather layer"

Tropopause: Very cold (first cold trap), boundary between troposphere and stratosphere; start of temperature inversion.

Environmental Management
ISBN 978-0-12-811989-1
http://dx.doi.org/10.1016/B978-0-12-811989-1.00014-2

Table 14.1 Permanent Gases of the Atmosphere

Constituent	Formula	Percent by Volume	Molecular Weight
Nitrogen	N_2	78.08	28.01
Oxygen	O_2	20.95	32.00
Argon	Ar	0.93	39.95
Neon	Ne	0.002	20.18
Helium	He	0.0005	4.00
Krypton	Kr	0.0001	83.8
Xenon	Xe	0.00009	131.3
Hydrogen	H_2	0.00005	2.02

Stratosphere (greek: "lying flat"):

- 10–50 km
- Temperature increase with height: temperature inversion, due to absorption of UV-radiation by Ozone: the "ozone layer"
- Temperature inversion: stable layering, reduced vertical mixing

Stratopause: Boundary between stratosphere and mesosphere; upper end of temperature inversion.

Mesosphere (greek: "middle layer"):

- 50–90 km
- Temperature decrease with height (almost adiabatically)

Upper part: coldest part of the atmosphere.

Mesopause: extremely cold (second cold trap), boundary between mesosphere and thermosphere; start of temperature inversion.

Thermosphere (greek: "hot layer"):

- Above ~90 km
- Strong temperature increase with height (temperature inversion), due to absorption of UV-radiation by O_2 and N_2
- Extremely "thin" atmosphere (temperature high, but almost no mass: energy content is low)
- No defined upper end

14.2 CLASSIFICATION OF AIR POLLUTANTS, THEIR SOURCES OF EMISSION, AND AIR QUALITY STANDARDS

14.2.1 Classification

As clean air in the troposphere moves across the Earth's surface, it collects the products of both natural events, dust storms and volcanic eruptions, and human activities (emissions from sources like transportation, fuel combustion, industrial operations, solid waste disposal, and various other activities). These potential pollutants, called, **primary pollutants**, which are emitted directly from the source, mix with the churning air in the troposphere. Some primary pollutants may react with one another or with the basic components of the atmosphere in chemical and photochemical reactions to form new pollutants, which are called **secondary pollutants**.

14.2.2 Sources of Emission of Air Pollutants

Natural sources produce considerable pollutants but most of them are essential components of a balanced ecosystem. None of these natural pollutants normally accumulate to a level that is dangerous for life. However, many industrial activities produce air pollutants in levels that exceed the normal natural assimilation processes. Some of the typical sources of air pollutants are given in Table 14.2.

Table 14.2 Typical Sources of Air Pollutants

Sulfur dioxide	Colorless gas produced by combustion of oil and coal and certain industrial sources
Nitrogen dioxide	Brownish orange gas produced by combustion and major industrial locations and motor vehicles
Hydrogen sulfide	Refineries, chemical industries, and bituminous fuels
Carbon monoxide	Burning of coal, gasoline, and automobile exhausts
Hydrogen cyanide	Blast furnance, fumigation, chemical manufacturing, metal plating, etc
Ammonia	Explosives, dye making, fertilizer plants, and lacquers
Lead	Very small particles emitted by motor vehicles and smelters
Ozone	A colorless gas formed from the reaction of emissions from motor vehicles and sunlight. It forms a major portion in smog
Phosgene or carbonyl chloride	Chemical and dye making
Aldehydes	Thermal decomposition of oils, fats, and gylcerols
Arsines	Process involving metal or acids containing arsenic soldering
Suspended particles (ash, soot, smoke, etc)	Solid or liquid particles produced by combustion and other processes at major industrial sources (steel mills, power plants, chemical plants, incinerators, etc.)

14.3 AIR POLLUTANTS AND THEIR HARMFUL EFFECTS

1. **Particulate pollutants**: Airborne, small, solid particles and liquid droplets are commonly known as particulates. When present in air in excess, they pose a serious pollution threat. The life period of particulates varies from a few seconds to several months; it depends on the setting rate, size, and density of particles and turbulence. Particulates can be inert or extremely reactive materials ranging in size from 100 μm down to 0.1 μm and less. The inert materials do not react readily with the environment nor do they exhibit any morphological changes as a result of combustion or any other process, whereas reaction materials could be further oxidized or may react chemically with the environment.
   Particulates can be further classified as:
   Dust: Particles of size 1–200 μm belong to this category and are formed by the natural disintegration of rocks and soil or by mechanical processes like grinding and spraying. They are removed from the air by gravity and other inertial processes by large settling velocities and also act as centers of catalysis for many of the chemical reactions taking place in the atmosphere.
   Smoke: Particles of size 0.01–1 μm constitute smoke which can be either in the liquid or solid form, and is formed by combustion or other chemical processes. Smoke may have different colors depending on the nature of materials burnt.
   Fumes: Solid particles of size 0.1–1 μm which are normally released from chemical or metallurgical processes belong to this category.
   Mists: Liquid droplets generally smaller than 10 μm, which are formed by condensation in the atmosphere or released from industrial operations represent mist.
   Fog: It is the mist in which the liquid is water and is sufficiently dense to obscure vision.
   Aerosols: All airborne suspensions, either solid or liquid, belong to this category and these are generally smaller than 1 μm.

   Particles of size 1–10 μm have measurable settling velocities but are readily stirred by air movements, whereas particles of size 0.1–1 μm have small settling velocities. Those below 0.1 μm, a sub-microscopic size found in urban air, undergo random Brownian motion resulting from collision among individual molecules.

2. **Sulfur oxides**: Sulfur dioxide (SO_2) is one of the major air pollutants discharged by various pollutant sources. Further, it reacts photochemically or catalytically with other pollutants or normal atmospheric constituents to form sulfur trioxide (SO_3), sulfuric acid, and salts of sulfuric acid. It is estimated that SO_2 remains in the air for an average of 2–4 days; during this time it may be transported a distance of 1000 km before it is deposited on the ground. Thus, the harmful effects of SO_2 and its deposition as sulfuric acid may be felt far away from the source, the pollution problem becomes an international one affecting countries which may not have pollutant sources discharging SO_2.

   Many industries, especially those dealing with petroleum, metallurgy, paper, and pulp contributes substantially to SO_2 pollution. It is perhaps the most damaging among the various gaseous air pollutants. Along with SO_2, SO_3 is discharged at about 1–5% of the SO_2 concentration, and it combines rapidly with moisture in the atmosphere to form sulfuric acid, which has a low dew point. Both of these oxides are rapidly removed from the atmosphere by rain, or settle out as aerosol due to their concentration being less compared to their emissions from human activities. More than 90% of the anthropogenic emissions of SO_2 are over Europe, North America, India, and the Far East. Emissions of SO_2 were highest (77 Mt S a^{-1}) during the late 1970s, but have fallen over the last two decades as a result of emission controls, changes in the patterns of fuel consumption, and economic recession. The taste threshold limit is 0.3 ppm, and SO_2 produces an unpleasant smell at 0.5 ppm concentration. In fact, sulfur oxides in general have been considered as prime candidates for an air pollution index.
3. **Oxides of nitrogen**: Natural stratospheric oxides of N_2 are produced by the action of cosmic rays in the upper atmosphere. Emissions of oxides of N_2 from man-made sources vary in different areas. Nitrogen oxides are 10–100 times greater in the urban atmosphere as compared to rural areas. Oxides of N_2, which include N_2O, NO, NO_2, N_2O_3, and N_2O_5 are usually represented by the symbol NO_x. The two major pollutants among them are nitric oxide, NO, and nitrogen dioxide, NO_2. Any excess artificial fertilizer remaining after application, which has not drained away, is usually removed by denitrification. Increasing global use of artificial fertilizers therefore contributes to higher NO_x levels. Emissions from stationary sources are estimated to be 16 million tons of NO_x per year (computed as NO_2). Stationary sources include power plants, industrial boilers, certain internal combustion engines

used in gas transmission, and small combustion sources. Mobile sources of NO_x pollution are automobiles emitting an "estimated" average of 10.7 million tons per year.

Major man-made activities like combustion of coal, oil, natural gas, and gasoline produce up to 50 ppm of oxides of N_2. Atmospheric nuclear explosions can be another potential serious source of man-made stratospheric oxides of N_2. Anthropogenic impacts upon the global exchange of N are considerable, but the bulk of the global N cycling remains microbial. Soil conditions where O_2 levels are low favor denitrification. Consequently, stagnant, waterlogged, or compacted soils are major sources of N_2O, and one of the main purposes of plowing and draining is to discourage this anaerobic process.

4. **Carbon monoxide**: Carbon monoxide is the most abundant gaseous pollutant emitted through anthropogenic sources into the troposphere; 0.075 Gt per annum which is a very high rate when compared to other gaseous pollutants, and this rate is still rising. Carbon monoxide, a product of combustion processes, is produced from cigarette smoking, household heating, and, more seriously, from the internal combustion engines of automobiles. Incomplete combustion, yielding CO instead of CO_2, results when any of the following four variables are not kept sufficiently high:
   a. Oxygen supply,
   b. Flame temperature,
   c. Gas residence time at high temperature, and
   d. Combustion chamber turbulence.
5. **Hydrocarbons**: As their name indicates, the components of hydrocarbons are hydrogen and carbon. Hydrocarbons constitute the major chemicals in petrol, gasoline, and other petroleum products. In the presence of sunlight, NO_2 reacts with hydrocarbons to give a series of extremely complex reactions; chief among the products of these reactions are peroxyacetyl nitrate (PAN), and ozone:

$$\text{Hydrocarbons} + NO_2 + O_2 + \text{Sunlight} \rightarrow \text{PAN} + O_3 + \text{etc}$$

Natural sources, particularly trees, emit huge quantities of hydrocarbons into the air. Plants mostly emitting terpenes belong to the family *Coniferae* and *Mystaceace* and the genus *Uterus*. Automobile exhausts emit the maximum amount of hydrocarbons into the atmosphere. Some industrial sources, especially refineries, emit hydrocarbons, but the major source of this type of pollution is automotive emissions.

The emissions from incomplete combustion in car engines, along with evaporative emissions from fuel tanks, carburetors, and crank cases, amount to approximately 12 million tons per year. Human activities contribute nearly 20% of the hydrocarbons emitted to the atmosphere every year; animals contribute about 80–85 million tons of methane to the atmosphere every year.

6. **Ammonia**: Ammonia (NH_3), which is a pungent gas, is used as a raw material in large quantities by industries for the synthesis of ammonium nitrate, plastics, explosives, dyes, and drugs. Emission of NH_3 from the biological degradation of proteins on soil surfaces into the atmosphere occurs on a very large scale. This is known as "NH_3 volatilization,t and compared to this, the industrial contribution is negligible. Atmospheric concentrations of NH_3 in temperate rural regions range from 5 to 10 ppm, but are much higher near the equator. In urban regions, higher levels of NH_3 up to 280 ppm are recorded and it may be found in increasing levels close to industrial and intensive agricultural sources.
7. **Organic lead**: Lead is discharged into the atmosphere in the organic form as tetraethyl lead ($(CH_3CH_2)_4Pb$) or trimethyl lead ($(CH_3)_3Pb$) in un-burnt or partially combusted fuel vapors. The amounts involved in developed countries were huge. In the late 1960s, for example, the amounts were 181 kt of lead in the whole of the United States, and 5 kt of lead in Los Angeles alone. In most developed countries, legislation to limit the amounts of tetraethyl and trimethyl lead, used as anti-knocking agents in vehicle fuels, has been progressively introduced and newer vehicles have been redesigned to use low lead or unleaded fuels. This has significantly reduced that fraction of lead entering the environment by airborne emissions in developed countries to less than one-tenth of the highest overall total in the past.
8. **Hydrogen sulfide**: Hydrogen sulfide is emitted into the atmosphere by the degradation of industrial wastes in stagnant waters, swamps, and other areas where bacterial action reduces sulfur compounds to hydrogen sulfide, which is highly insoluble in water. It is estimated that decaying organic matter in the world emits 70 million tons of H_2S. However, industrially, this gas comes from sewage treatment plants and the petroleum industry.
9. **Fluorides**: Hydrogen fluoride, HF, is a highly corrosive and irritant gas. A typical fluoride concentration in the atmosphere is 0.05 mg/m^3. Because of its extreme toxicity, HF is a problem wherever processes involving fluorides take place, such as in the production of phosphate

fertilizers, smelting of certain iron ores, and manufacturing of aluminum. Many fluoride containing minerals such as fluorspar, cyrolite, and certain appetites are used by industry. Some industries also produce HF either as a by-product or to form various useful fluoro-derivatives. Industrial emissions are superimposed upon significant natural background sources. Consequently, levels in both air and water supplies vary widely. The majority of rural and urban air monitoring sites record very low levels of atmospheric fluoride measured as total dissolved fluoride. Near phosphate fertilizer plants, aluminum smelters, or volcanoes, however, levels may rise above 200 ppm.

10. **Radon**: Radon, which occurs naturally, is the heaviest known gas. Radon gas comes from a decay process of underground uranium ore. It is by far the most important source of ionizing radiation to affect humans. In most developed countries, radon accounts for 40–50% of the total ionizing radiation received by the population. There are 27 isotopes of radon ($^{200-226}Rn$), but only three have half-lives longer than an hour (^{210}Rn, 2.4 h; ^{211}Rn, 14.6 h and ^{222}Rn, 3.82 days). Of these, ^{222}Rn is the most important, and arises from the decay of ^{238}U. ^{222}Rn also decays into a series of radionuclides known as radon daughters or progeny. Principal among these are ^{214}Pb (half-life 26.8 min), ^{210}Pb (22.3 years), ^{210}Bi (5 days), and ^{210}Pb (138.4 days). The final product is ^{206}Pb, which is non-radioactive. Radon emerges into the atmosphere by a variety of routes, including from the ground below and around the water supplies, and from natural gas or building materials; being denser than other gases, it tends to concentrate at low points. Consequently, lower stories of buildings in radon-prone areas have higher levels of radon than upper floors.

14.3.1 Acid Rain

"Acid rain" is a broad term used to describe several ways that acids fall out of the atmosphere. A more precise term is acid deposition, which has two parts; wet and dry. Wet deposition refers to acidic rain, fog, and snow. As this acidic water flows over and through the ground, it affects a variety of plants and animals. The strength of the effects depend on many factors, including how acidic the water is, the chemistry and buffering capacity of the soils involved, and the types of fish, trees, and other living things that rely on the water. Dry deposition refers to acidic gases and particles. About half of the acidity in the atmosphere falls back to earth through dry deposition. The wind blows these acidic particles and gases onto buildings, cars, homes, and trees. Dry deposited gases and particles can also be washed from trees and other surfaces by rainstorms. When that happens, the runoff water adds those acids to the acid rain, making the combination more acidic than the falling rain alone. Prevailing winds blow the compounds that cause both wet and dry acid deposition across state and national borders, and sometimes over hundreds of miles. Scientists discovered, and have confirmed, that SO_2 and NO_x are the primary causes of acid rain. In the United States, about two-thirds of all SO_2 and one quarter of all NO_x comes from electric power generation that relies on burning fossil fuels like coal.

Acid rain occurs when these gases react in the atmosphere with water, oxygen, and other chemicals to form various acidic compounds. Sunlight increases the rate of most of these reactions. Rain is slightly acidic because it contains dissolved carbon dioxide (CO_2), SO_2, and NO_x, which are normally present in the air. Acid rain contains more acidity than the normal value because of presence of acid ions due to the dissolution of these gases in higher concentration. Acid rain, therefore, is the direct consequence of air pollution caused by gaseous emissions from industrial sources, burning of fuels (thermal plants, chimneys of brick-kilns or sugar mills) and vehicular emissions. Acid rain will not necessarily occur locally near the sources of air pollution. Due to the movement of air, acid rain may occur far away from the source. For instance, the United Kingdom contributes 26% of the acidic sulfur deposited in the Netherlands, 23% in Norway, and 12% in Sweden. Smokestacks of power plants and a number of industries around the world spew out the basic ingredients of acid rain, namely, SO_2 and NO_x. With time, these oxides combine with water vapor in the atmosphere and return to the Earth's surface in the form of acid rain.

Acid emissions arise naturally from volcanoes, forest fires and biological decomposition, especially in the oceans. But their contribution to acid rain are SO_2, NO_x, and to a lesser extent, CO_2 and hydrocarbon (HC) gas. SO_2 pollution is mostly contributed by thermal power plants, refineries, and industry, and NO_x from road transport, power stations, and industry. The acid gas concentrations in the air will vary according to location, time, and weather conditions. Acid rain is measured using a scale "pH." The lower a substance's pH, the more acidic it is. Pure water has a pH of 7.0. Normal rain is slightly acidic because carbon dioxide dissolves into it, so it has a pH of about 5.5. When the pH is below 5.0, it is expected to lead to acid rain.

Like China, in India to the main threat of an acid rain disaster springs from our heavy dependence on coal as a major energy source. Even though Indian coal is relatively low in sulfur content compared to the nature of coal reserves of other countries like China, what threatens to cause acid rain in India is the concentrated quantity of consumption that is expected to reach very high levels in some parts of the country by 2020. As energy requirements in India are growing rapidly in tune with the growing economy, coal dependence in the country is expected to grow threefold over the current level of consumption, making the clouds of acid rain heavier over many highly sensitive areas in the country like the northeast region, parts of Bihar, Orissa, West Bengal, and coastal areas in the south. Already the soils of these areas have a low pH value, which acid rain will aggravate further, making them infertile and unsuitable for agriculture. As experience stands in Europe and North America, the threat of acid rain was severely dealt with in these regions through heavy spending on SO_2 abatement technologies and cutting down the dependence on coal by shifting to natural gas and nuclear energy. But, action in these regions came only after a considerable amount of ecological damage. In the 1960s, fish populations in the Scandinavian countries were showing a rapid decline as a result of acid rain. The infamous forest dieback in some parts of central Europe was also from acid rain. Thus, experience from elsewhere bears out clearly enough that the whole problem as it confronts India needs proactive handling.

The issue of rapidly growing SO_2 emissions, the resultant sulfur deposition and the threat of acid rain in many areas of Asia is a transboundary problem involving many countries, and therefore, its solution calls for regional initiatives. In Europe, the situation of acid deposition from many countries in the 1970s and the related concerns about the pollution being carried over long distances, led to the signing of an international agreement in 1979 called "The Convention on Long-Range Transboundary Air Pollution." The signing of subsequent protocols led to binding commitments from European countries to limit and reduce their transboundary emissions of air pollutants.

14.3.1.1 Effects of Acid Rain

Acid rain causes acidification of lakes and streams, and contributes to damage of trees at high elevations (for example, red spruce trees above 2000 ft), and many sensitive forest soils. In addition, acid rain accelerates the decay of building materials and paints, including irreplaceable buildings, statues, and sculptures that are part of our nation's cultural heritage. Prior to falling to the

Earth, SO_2 and NO_x gases and their particulate matter derivatives, sulfates and nitrates, contribute to visibility degradation and harm public health.

The most important effects are damage to freshwater aquatic life, damage of vegetation, and damage to buildings and material.

1. **Damage to aquatic life**: The main impact of fresh water acidification is a reduction in diversity and populations of fresh water species. The effect on soil and rock will depend upon the in situ capacity called "buffering capacity" to neutralize the acids. The soil organisms are killed in acid rain where soils have limited buffering capacity. The acidic leaf litter in forest areas adds to the nutrient leaching effects of acid rain. This scavenging from cloud increases the amount of pollution deposited. Trees are quite effective in intercepting the air borne pollutants than other types of upland vegetation.

   In the areas of high acid deposition and poor buffering in the lakes, a pH less than 5 has become common. At pH 5, fish life and frogs begin to disappear. By pH 4.5, virtually all aquatic life has gone. Acid rain releases metals, particularly aluminum, from the soil, which can build up in lake water to levels that are toxic to fish and other organisms. A decline in fish and amphibian population will affect the food chain of birds and mammals that depend on them for food.
2. **Damage to trees and plants**: For a number of years, there has been concern about the apparent deterioration of trees and other vegetation. It is not easy to establish the cause of damage; pollution, drought, frost, pests, and frost management methods can all affect tree health. SO_2 has a direct toxic effect on trees, and in parts of central Europe, for example, where SO_2 levels are very high, extensive areas of forest have been damaged or destroyed.

   Acid deposition may combine with other factors to affect tree health; for instance by making trees more susceptible to attack by pests, or by acidifying soils which may cause loss of essential nutrients such as magnesium, thus impairing tree growth. Secondary pollutants like ozone are also known to exacerbate the effects of acid deposition.
3. **Damage to buildings and materials**: All historic buildings suffer damage and decay with time. Natural weathering causes some of this, but there is no doubt that air pollution, particularly SO_2, also plays an important part. SO_2 penetrates porous stones such as limestone, and is converted to calcium sulfate, which causes gradual crumbling. Most building damage happens in urban areas where there are many SO_2 emitters (domestic chimneys, factories, and heating plants).

14.3.2 Green House Gases: Global Warming

Air pollutants are derived from both natural sources and human activities. Natural sources include forest fires, which add particulates; volcanoes, which add acid gases and particulates; biological processes in soil, which add NO_x; lightning; and dust due to soil erosion. However, a large proportion of air pollutants are caused by human activities, primarily the combustion of fossil fuels. When fossil fuel is burned, primary pollutants are created. These include: CO_2; carbon monoxide (CO); NO_x; SO_2; HCs [also known as volatile organic compounds (VOCs)]; and airborne particulates. Fossil fuels may also contain contaminants or additives in the form of heavy or toxic materials, which are emitted as suspended particles.

The greenhouse effect is primarily a function of the concentration of water vapor, CO_2, and other trace gases in the atmosphere that absorb the terrestrial radiation leaving the surface of the Earth, and act like a blanket over the earth's surface, keeping it warmer than it would otherwise be. Changes in the atmospheric concentration of these gases can alter the balance of energy transfers between the atmosphere, space, land, and the oceans. A gauge of these changes is called the radiative forcing, which is a measure of changes in the energy available to the Earth–atmosphere system. Holding everything constant, increases in greenhouse gas concentrations in the atmosphere will produce positive radiative forcing (i.e., a net increase in the absorption of energy by the Earth).

The global carbon cycle consists of the various stocks of carbon in the earth system and the flow of carbon between these stocks. Carbon in the form of inorganic and organic compounds, notably CO_2, is cycled between the atmosphere, oceans, and terrestrial biosphere. The largest natural exchanges occur between the atmosphere and terrestrial biota and between the atmosphere and ocean surface waters. The carbon cycle has been linked to the changes in climate that have recently been observed on earth, especially the increases in temperature. The Intergovernmental Panel on Climate Change (IPCC), in its report (2002), states that "there is new and stronger evidence that most of the warming observed over the last 50 years is attributable to human activities, and it is likely to have been due to the increase in greenhouse gas concentrations."

Global warming potential (GWP) is defined as the cumulative radiative forcing, both direct and indirect effects, over a specified time horizon resulting from the emission of a unit mass of gas related to some reference gas [CO_2: (IPCC 1996)]. A GWP is intended as a quantified measure of the relative radiative forcing impacts of a particular greenhouse gas. Direct effects occur when the gas itself is a greenhouse gas. Indirect radiative forcing occurs when chemical transformations involving the original gas produce a gas or gases that are greenhouse gases, or when a gas influences the atmospheric lifetimes of other gases.

14.3.3 Stratospheric Ozone Depletion

Ozone (O_3), is poisonous gas, which forms a blanket to our planet, guarding life on it from the harmful effects of ultraviolet radiation from outer space. It is this protective layer which has undergoing fast depletion which, if unchecked, would turn Earth an abode of living beings with deformation of vital organs and eventual peril. It is scattered so thinly through the 35-km stratosphere, that if it were all collected together, it would form a girdle around the Earth no thicker than a shoe sole. Yet, this thinnest of filters efficiently screens out almost all the harmful ultraviolet rays of the sun. The ozone layer stops the UV-B and UV-C rays, which have the shortest wavelength and are dangerous to life, from entering the atmosphere. Meanwhile, UV-A, which is relatively harmless, is let in.

Non-melanoma skin cancers, besides optic disorders including cataract and deformation of eye lens, are caused by UV radiation. Exposure to increased UV-B radiation could also affect the immune system and trigger changes in the chemical composition of several species of plants resulting in decreased crop yield and destruction of forests. Depletion of stratospheric ozone would aggregate photochemical pollution resulting in increase of ozone near the Earth's surface, which is dangerous. The CFCs are being replaced by other materials which are less harmful to ozone.

14.3.4 Photochemical Smog

With the increase in vehicle use in cities, there has been a major change in the character of air pollution. Combustion of fossil fuels contributes to photochemical smogs, so-called due to the role of sunlight in forming a cocktail of harmful chemicals from other gases. They form when NO_x and HCs react in the atmosphere, fueled by solar energy in the form of ultraviolet radiation. The smogs form in a complex series of reactions, forming nitric acid and other secondary oxidizing pollutants such as O_3. The main source of NO_x is vehicle exhausts, and therefore occurs in cities where vehicle density is very high. As such, they are a major problem in many cities. Other vehicle pollutants include particulates and lead, adding to the noxious character of urban smogs.

14.3.4.1 Adverse Effects of Photochemical Smog

The major constituents of smog, with the exception of CO_2, are powerful poisons. Many aspects of their toxic effects, both individually and in combination with one another, are not completely known. The danger, however, to people with susceptibility to respiratory and cardiac difficulties is widely recognized. Those with incipient or advanced bronchial asthma, chronic bronchitis, and pulmonary emphysema are apt to be hard hit. The various components of smog may affect people with different susceptibilities in different ways.

Photochemical smog causes irritation of the eyes, nose, throat, and chest. Eye irritation is not caused by O_3, but by PANs and trace free radical HCs. O_3 is a powerful oxidant that may injure the bronchiolar and alveolar walls of the lungs. Surface epithelial cells of the airways are damaged by O_3 and afterward, these are replaced by thick cuboidal cells with few or small cilia (cell hairs). Apart from loss of cilia, additional changes to the epithelial (surface lining) cells include cytoplasmic vacuolation (formation of internal cellular spaces), and condensation of abnormal mitochondria. Respiratory irritation and difficulty in breathing have been common, and the correlation of the oxidizing power of smog with its irritating action is having implications beyond the simple annoyance factor. The air passageways of the lungs do more than carry the air and waste gases to and from the air sacs; the passageways have been lined with mucus-secreting epithelium, and serve to condition the air before it reaches the alveoli. The air is changed in three important ways; it is warmed to body temperature, it is saturated with moisture, and it is filtered free from most of the dust particles and other foreign substances. The cells that form the epithelial lining are ciliated, that is, they have hair-like protoplasmic appendages that project into the passageways and, by continuous, coordinated, rhythmic action, work the mucous and entrapped foreign material upward to the throat where it is either swallowed or expelled.

14.4 AIR POLLUTION DISPERSIONS: TEMPERATURE LAPSE RATES AND STABILITY

The two most important factors determining the climate of an area are temperature and precipitation. The temperature and precipitation patterns that lead to different climates are caused mostly by the way air circulates over the Earth's surface. The formation of clouds is largely related to the concept of *atmospheric stability*. The types of pollution source (point, line, area sources), and the way in which such pollution is subsequently transported and mixed in the atmosphere is important. The atmosphere is a fluid, and so many aspects of the course cover generic dispersion processes.

If the concentration of some pollutant at a fixed point three main processes which will affect the concentration:

- *Advection*: The transport of the pollution by the mean flow in the atmosphere.
- *Diffusion*: The spreading of the pollution as a result of the random motion of gas molecules in the air.
- *Turbulence*: Random, unresolved, and unpredictable motion of the air. We usually separate the motion of a fluid into the *mean* and *turbulent* parts.

14.4.1 Stability and Instability

Stability is a concept that describes what happens when a system is disturbed. Stable environment is defined as one in which the atmosphere is stable and can be explained as a stone in the bowl kind of environment. Consider a simple situation of a stone in the bottom of a bowl if you push the stone up the side of the bowl, it will fall back down to the bottom, to its original position, that is, stable position. When a ship is tilted sideways by waves, it returns to its upright position quickly, because it is stable. A sheet of cloud at 1500 m level is usually stable. It neither rises when disturbed by updrafts from below nor sinks due to downdrafts from above.

In *stable air (parcel)*; vertical motion is inhibited and if clouds form, they will be shallow, layered clouds like stratus. *Neutral equilibrium* is the special case when a change is neither resisted nor encouraged.

Unstable equilibrium: The slightest disturbance that causes the movement from its position to a position further from its equilibrium position is unstable. A packet of air in unstable equilibrium will keep rising when given a slight push from below.

Unstable environment: If the stone is on the top of the bowl and you give it a little push, it rolls off the bowl, does not come back to its original position (Fig. 14.1). This is an unstable situation. *Unstable air (parcel)*; vertical motion occurs commonly produces cumulus, cumulonimbus clouds.

Factors that play a role include:

- *Pressure* excess inside a parcel of air causes it to expand against the inward pressure of the surrounding air.
- *Density*, which is mass per unit volume, changes inversely as the volume of an air parcel (whose mass is constant).

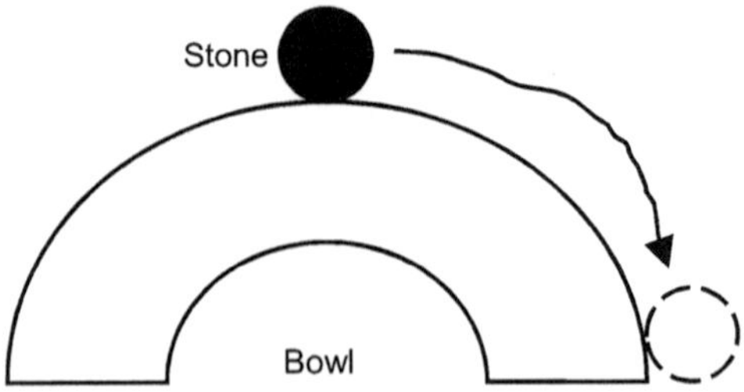

Figure 14.1 Movement of stone in unstable environment.

- *Temperature*, measured in Kelvin for equations, or Celsius for everyday use.
- *Moisture content*, commonly measured by relative humidity or dew point. The absolute value of the water content is not usually the most important issue but its value compared with saturation vapor pressure. That is why *relative humidity* is more relevant than specific humidity.

Changes in parcels of air are governed by the *gas law*, which relates changes in pressure (P), volume (V), and temperature (T). The *ideal gas* is the simplest physical model of a gas. The "ideal gas law" links volume, pressure, and temperature:

$$PV = nRT$$

R is a fundamental physical constant (called the "gas constant" with a value of 8.31 J/mol K), and n is also a constant for a given parcel of air, effectively measuring the amount of air in the parcel. n is the number of "moles," with 1 mol occupying 22.4 L at standard temperature and pressure.

In other words, warmer gas has less density at a given pressure. This relationship gives the key connection between density and temperature for parcels of air at the same pressure, i.e., height.

14.4.1.1 Rising and Falling

The key quantity that determines whether a parcel of air will rise or fall is its *density* relative to that of its surroundings. If a parcel of air is less dense than its surroundings, it will experience an *up-thrust* on it, which is a net force upwards, exerted by its surroundings. This is a direct consequence of *Archimedes Principle*. If something is less dense than its surroundings, it tends to rise. As an air parcel rises, it moves into less dense air *but* it also expands because the pressure exerted on it is less, so the parcel, too, becomes less dense (Fig. 14.2).

If the density of the packet is less than that of the new air it has risen into, the packet keeps rising and it is said to be unstable. If the reverse happens, it stops rising, and may even sink.

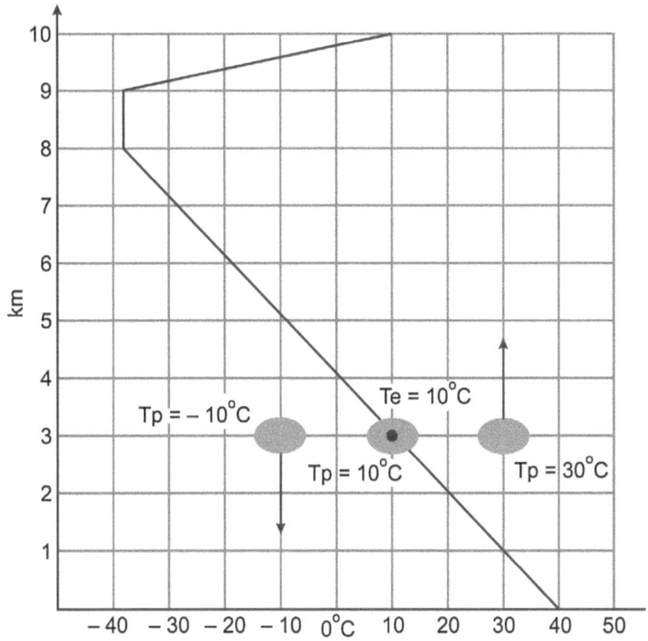

Figure 14.2 Fate of the air parcel.

To determine whether or not a parcel will rise or sink in the atmosphere, one must compare the parcels temperature (T_p) with that of the environment (T_e) at some altitude. Three situations arise:

1. If $T_p > T_e$ the air parcel rises.
2. If $T_p = T_e$ parcel does not move up or down.
3. If $T_p < T_e$ parcel sinks.

Fig. 14.2 explains the state of the air parcel in different environments.

14.4.2 Adiabatic Changes

How much expansion (which is a change in volume) of air occurs depends on how the expansion takes place. Various conditions are possible, which might be volume change at constant temperature, volume change at constant pressure, etc. The usual assumption in meteorology is that the parcels of air are sufficiently big that no significant heat gets in to a parcel as it expands, or no heat leaks out when it is compressed. Such a change is called *adiabatic*. Adiabatic changes are governed by $PV\tilde{a} = \text{constant}$, with $\tilde{a} \sim 1.4$ for air.

Combining the above with the original relationship of $PV = nRT$, we can find out how temperatures change with adiabatic expansion. This explains what happens to the temperatures of air packets as they rise and fall, and that in turn is related to density changes and hence their stability. The adiabatic assumption is reasonable in these circumstances because heat flow is a very slow whereas air subject to changes of pressure adjusts its volume more quickly.

14.4.2.1 Adiabatic Lapse Rates

The temperature changes will result in heat lost from the parcel of air. The parcel of air has to push against its surroundings as it expands, moving them out of the way to allow for the expansion and hence it loses some of its internal energy in doing this. The result is a cooling. Meteorologists concentrate on the temperature changes as a parcel of air rises or sinks. The change in temperature with height when adiabatic changes are occurring is called the *adiabatic lapse rate*. It is a bit different for dry and moist air, because when moisture condenses as the temperature drops, latent heat comes out to reduce the temperature drop.

Supposing the parcel of air moves up from the ground to where the pressure has dropped by 15%; if in doing so it expands by 15%, then PV will remain constant and the temperature will not change. If the volume V expands by less than 15%, the left-hand side of the equation will become smaller, and so the corresponding right-hand side should also be smaller so that temperature will drop. In an adiabatic change, PV1.4 remains constant, and hence V1.4 changes by 15%, which means that the volume increases by 8.1%. The product PV drops and so the temperature also drops. To convert this into a lapse rate, the height of the pressure drop of 15% has to be calculated. The pressure at height h, denoted $P(h) = P(0)e - h/8$, where h is in km and the lapse rate calculated.

14.4.3 Determining Air Parcel Temperature: Rising Air Parcels and Adiabatic Cooling

Consider a rising parcel of air; as the parcel rises, it will *adiabatically expand and cool* (Fig. 14.3). *Adiabatic* is a process where the parcel temperature changes due to an expansion or compression; no heat is added or taken away from the parcel. The parcel expands since the lower pressure outside allows the air molecules to push out on the parcel walls since it takes energy for the parcel molecules to "push out" on the parcel walls, they use up some of their internal energy in the process. Therefore, the parcel also cools since temperature is proportional to molecular internal energy.

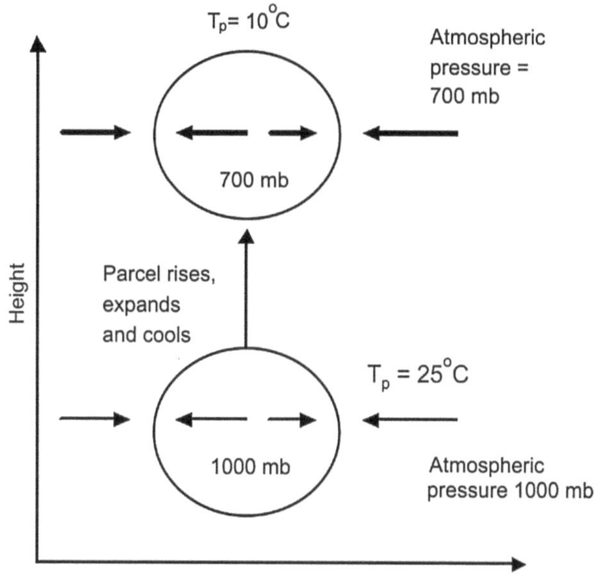

Figure 14.3 Air parcel as it expands.

As the parcel sinks, it will *adiabatically compress and warm adiabatic*; a process where the parcel temperature changes due to an expansion or compression, and no heat is added or taken away from the parcel. The parcel compresses since it is moving into a region of higher pressure due to the parcel compression, the air molecules gain internal energy, hence the mean temperature of the parcel increases.

Suppose the air pressure outside a conventional jet airliner flying at an altitude of 10 km is 250 mb. Further, suppose the air inside the aircraft is pressurized to 1000 mb. If the outside air temperature is −50°C, the temperature of this air if brought inside the aircraft and compressed at the dry adiabatic rate to an altitude of 0 m, if is also assumed that the parcel's temperature at 2 km then its dry lapse rate is approximately 20°C (Fig. 14.4).

14.4.3.1 Atmospheric Temperature Profile

The actual temperature profile depends on the history of the air present. It can be measured by balloon-born thermometer. The result is called the *environmental lapse rate*. This lapse rate is one of the most important properties of the atmosphere when predicting the weather. The lapse rate is the slope of the profile, measured from the vertical on the standard graph.

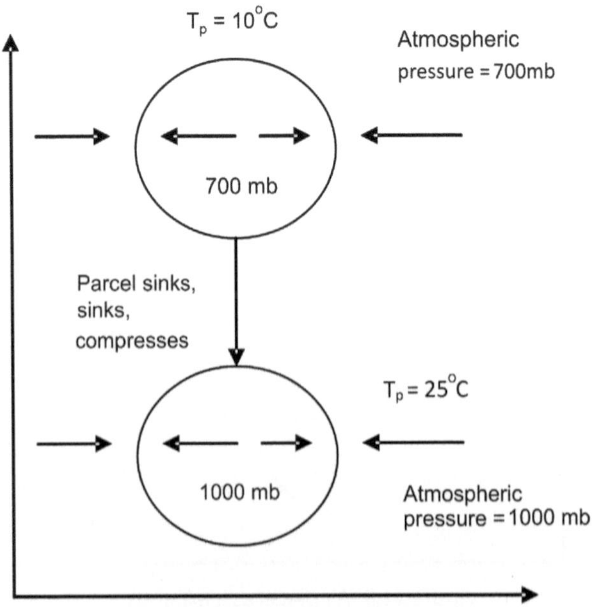

Figure 14.4 Air parcel as it compresses.

14.4.3.2 Moist Adiabatic Lapse Rate

For a saturated parcel of air, i.e., when its $T = T_d$, then it cools at the *moist adiabatic lapse rate* = 6°C/km. If the parcel is at 3 km, then its temperature will be 14°C.

Dry versus Moist-Adiabatic Process.

The moist adiabatic lapse rate is less than the dry adiabatic lapse rate because as vapor condenses into water (or water freezes into ice) for a saturated parcel, latent heat is released into the parcel, mitigating the adiabatic cooling.

The condition for absolute stability is:

$$\Gamma_d > \Gamma_m > \Gamma_e$$

Γ_d is the *dry adiabatic lapse rate* (10°C/km), Γ_m is the *moist adiabatic lapse rate* (6°C/km), and Γ_e is the *environmental lapse rate* (variable – 0°C/km in this case).

14.4.4 Conditions of Various Atmospheric Stability and Air Pollution

Lapse rate: Change of temperature with altitude = $\Delta T/\Delta H$ (Fig. 14.5).

Adiabatic lapse rate: Change of temperature with a change in altitude of an air parcel without gaining or losing any heat to the environment surrounding the parcel.

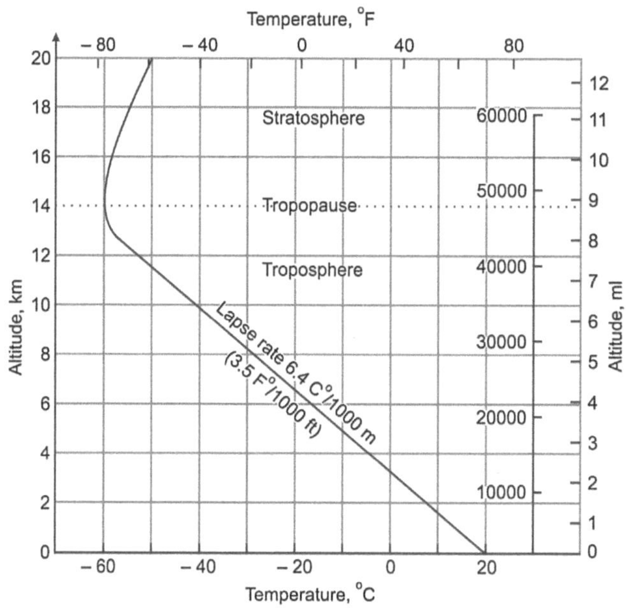

Figure 14.5 Lapse rate.

Dry adiabatic lapse rate: Assumes a dry parcel of air. Air cools 3°C/100 m rise in altitude (5.4°F/1000 ft).

Wet adiabatic lapse rate: As parcel rises, H_2O condenses and gives off heat, and warms air around it. Parcel cools more slowly as it rises in altitude, ≈6°C/1000 m (≈3°F/1000 ft).

Ambient or prevailing lapse rate: The actual atmospheric temperature change with altitude; not only does water content modify lapse rates, but wind, sunlight on the Earth's surface, and geographical features change actual lapse rates.

A comparison of dry or wet adiabatic lapse rates to prevailing lapse rates gives a sense of the stability and mixing conditions of the atmosphere (Fig. 14.6).

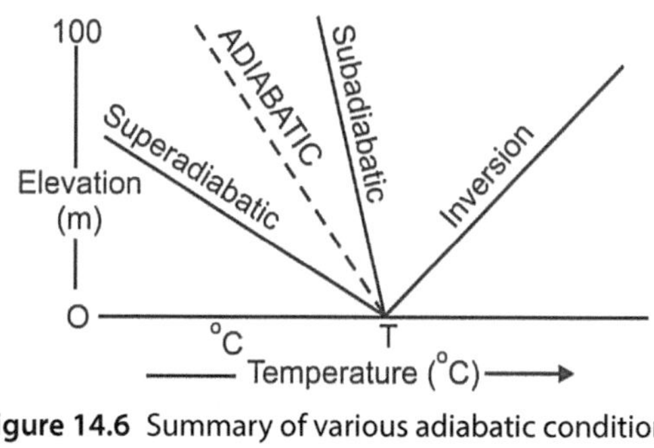

Figure 14.6 Summary of various adiabatic conditions.

Superadiabatic: Ambient lapse rate > adiabatic indicates unstable atmosphere. Vertical motion and mixing processes are enhanced. Dispersion of pollution plume is enhanced.

Subadiabatic: Ambient lapse rate < adiabatic. It indicates stable atmosphere, vertical motion, and mixing are suppressed. Dispersion is suppressed, and contamination is trapped.

Inversion: An extreme case of subadiabatic, where temperature actually increases with altitude near the ground before it begins to decrease with altitude. This results in warm, low-density air riding on top of cool high density air; a very stable air column that traps pollution near the ground, like that which occurs during the winter in the Platte Valley of Denver.

14.5 TEMPERATURE LAPSE RATES

The decrease in the observed temperature with height is called the environmental or observed lapse rate. Traditionally, lapse rates are defined as the negative of the temperature gradient with respect to height, i.e.,

$$\Gamma = -\,dT/dz$$

The average lapse rate in the middle latitude troposphere is about 0.6°C/100 m. It is also known that the pressure decreases with height, so that a parcel of air that changes its altitude will undergo a pressure change, and therefore a temperature change as well. If one assumes that vertical displacements occur adiabatically, one can calculate the change in temperature associated with these displacements. This temperature change is described in terms of a process lapse rate. If the parcel moves vertically with no water vapor phase change, the process is called dry adiabatic, and the rate of cooling per unit rise is called the dry adiabatic lapse rate. Since we assume the process to be reversible, we can derive this process lapse rate from the reversible form of the first law of thermodynamics:

$$\Delta H = c_p dT - vdp$$

If adiabatic, $\Delta H = 0$ and:

$$c_p dT = vdp \text{ or } dT = vdp/c_p$$

If hydrostatic:

$$dp = -\,\rho g dz\,,\ (\text{note } \rho = 1/v) \text{ so that}$$

$$dT = -\,g/c_p dz$$

The dry adiabatic processes lapse rate is then:

$$\Gamma = -dT/dz = g/c_p = 5.4\,°F/1000\ ft = 0.98\,°C/100\ m$$

This then describes how the temperature changes during an adiabatic lifting or sinking of an unsaturated air parcel. If the environmental lapse rate is greater than the dry adiabatic rate, that means that as a parcel rises from a given point in the environment, the parcel will cool at a slower rate than the environment cools. Therefore once it begins to rise, it will be warmer than the air around it and it will buoyantly accelerate upwards. If the parcel is displaced downwards, it will not warm as quickly as the environment, will be cooler than its surroundings, and accelerate downward. This situation in which a vertical displacement in any direction results in an acceleration away from the original position is termed "unstable." The environmental temperature profile depicted by curve $\Gamma_{env,1}$ is unstable. The temperature distribution labeled $\Gamma_{env,2}$ in Fig. 14.7 depicts a "stable" condition in that for this case any vertical displacement results in an acceleration back toward the original position. If the environmental and process lapse rates $\Gamma_{env} = \Gamma_{DALR}$ are equal, vertical displacement of a parcel produces no buoyancy forces and the condition is called "neutral." If the temperature increases with height, $\Gamma_{env,3}$ the lapse rate is very stable. (Note that it is not necessary for there to be an inversion for the atmosphere to be stable.)

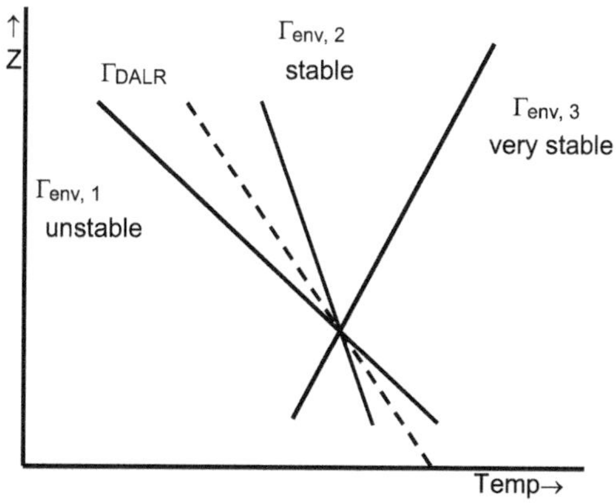

Figure 14.7 Temperature-height plot illustrating the temperature profile describing a dry adiabatic process (*dashed line*) and three observed environmental lapse rates (*solid lines*). Curve 1 is "unstable," curve 2 is "stable," and curve 3 is very "strongly stable." The dry adiabatic lapse rate Γ_{DALR} corresponds to "neutral" stability.

14.6 TEMPERATURE INVERSION

Inversion occurs when temperature increases with altitude, and the lapse rate in such a condition is known as the negative lapse rate. This is the extreme case of stable atmosphere. Practically no mixing of the pollutants takes place under stable atmospheric conditions. Temperature inversions play a significant role in air pollution meteorology. Within an inversion, the air is stable against buoyant vertical motion. That stability also lessens the exchange of wind energy between the air layer near the ground and high altitude winds, so that both horizontal and vertical dispersion of pollutants are hindered.

Inside the inversion, the situation is extremely stable; vertical disturbances are strongly damped out. Above the inversion, in the region with the standard lapse rate, the situation is mildly stable; vertical disturbances are damped, but not nearly as strongly as inside the inversion (Fig. 14.8). Inversion layers based on their nature of formation can be termed as *subsidence inversions*, radiation inversions, and advection inversions. Subsidence inversion is usually associated with subtropical anticyclones, where the air is warmed by compression as it descends in a high-pressure system and achieves temperature higher than that of the air underneath. If the temperature increase is sufficient, an inversion will result. Anticyclones are semi-permanent in nature and can cause subsidence inversions to last for months at a time (Fig. 14.9). These semi-permanent highs are the result of the general atmospheric circulation patterns.

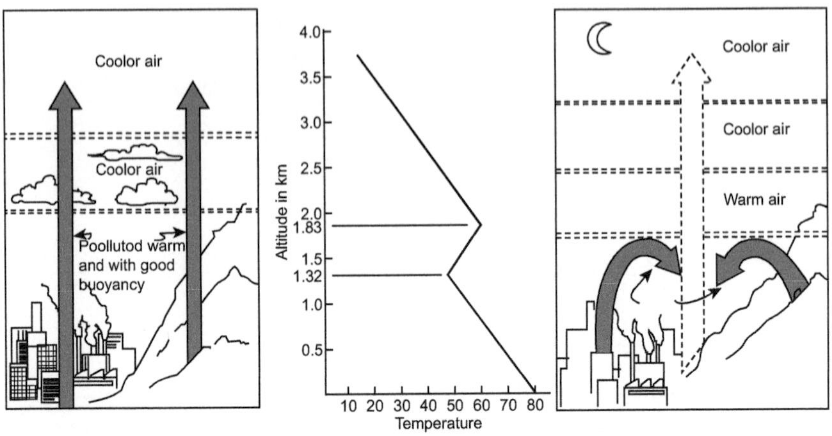

Figure 14.8 Temperature inversion upsets the natural tendency for upward air movement.

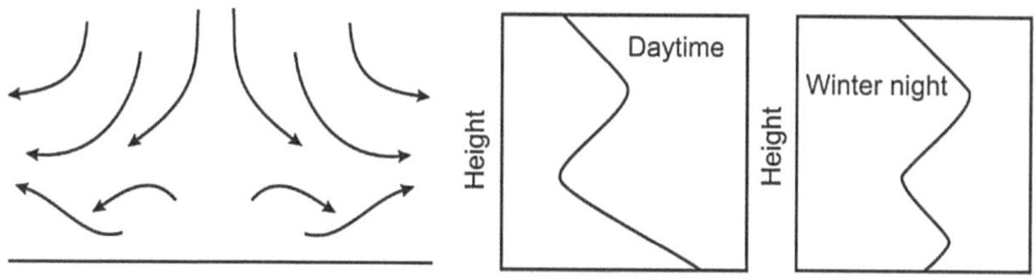

Figure 14.9 Subsidence inversions (A) As air descends it is warmed adiabatically creating clear skies and an upper level inversion; (B) during the day, air under the inversion may be unstable due to solar warming of the surface; and (C) radiation inversions may form at night under subsidence inversion, especially in winter.

Air in the middle of a high-pressure zone and descending air on the edges, will be rising. Air near the ground moves outward from the center while air aloft moves toward the center from the edges. The result is a massive vertical circulation system as air in the center of the system falls, it experiences greater pressure and is compressed and heated. If its temperature at elevation $z2$ will be heated adiabatically to $T2 = T1 + \Delta(z1 - z2)$ as is often the case, this compressive heating warms the descending air to a higher temperature than the air below, whose temperature is dictated primarily by conditions on the ground. The result is an inversion, located anywhere from several hundred meters above the surface to several thousand meters, that lasts as long as the high-pressure weather system persists. Since subsiding air is getting warmer, it is more and more able to hold water vapor as it descends. Without sources of new moisture, its relative humidity drops; thus there is little chance for clouds to form. The result is that high-pressure zones create clear, dry weather with lots of sunshine during the day and clear skies at night. Clear skies allow solar warming of the Earth's surface. This helps create super-adiabatic conditions under the inversion during the daytime and hence, good mixing. At night, the surface can cool quickly by radiation, which may result in a radiation inversion located under the subsidence inversion.

Radiation inversion occurs due to the normal diurnal cooling cycle (Fig. 14.10). The ground cools quickly by radiational heat transfer after sunset due to the lowest layer of air being in contact with the surface losing sensible heat through conduction and small scale mixing; this will result in temperature inversion between the cool, low level air and the warmer air above, in the first few hundred meters above the land surface on a clear or slightly cloudy day with low or average winds.

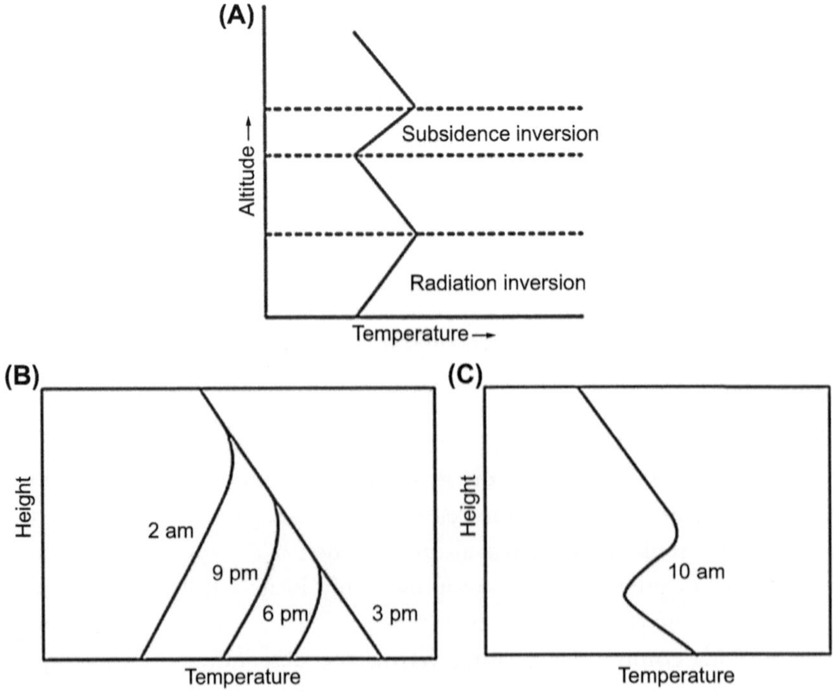

Figure 14.10 (A) Temperature–altitude profiles for radiation and subsidence inversion; (B) development of a radiation inversion; and (C) the subsequent erosion of the inversion.

The subsidence inversion is potentially more serious than radiation inversion because the latter usually dissipates quite rapidly after sunrise.

14.7 PLUME CHARACTERISTICS AND PLUME BEHAVIOR

A plume is a distribution of pollutant from a continuous source. Here we will predominantly discuss 1-D sources, e.g., smoke from a chimney stack. The dispersion of a plume is influenced by various plume properties including:

1. Rate of release
2. Temperature of release (buoyancy)
3. Height of release

And also by environmental properties including:

1. Wind speed
2. Turbulence
3. Atmospheric stability

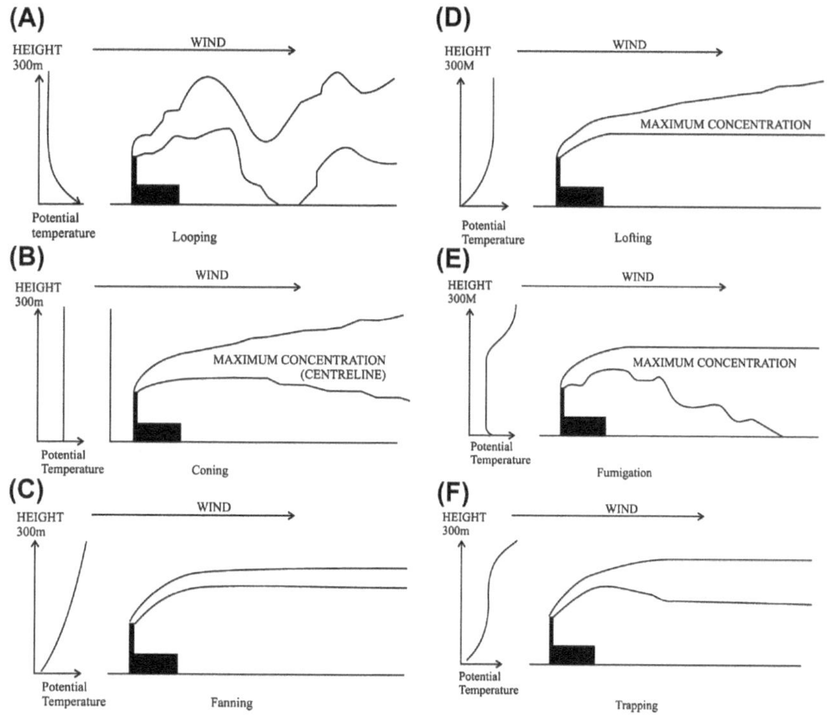

Figure 14.11 Plume characteristics.

All industrial emissions are released through stacks (chimneys) of varying heights. These represent the stationary sources of air pollution with the exhaust gases dispersed over a region. Normally the plume or parcel of gaseous effluents released from a stack will be at a higher temperature and have drifted to the surrounding environment due to which it will be moving upward with buoyancy action. Atmospheric stability and the wind turbulence conditions existing will determine the behavior of the plume emitted from a stack of any industry.

There are a well-known set of characteristic behaviors observed for plumes (Fig. 14.11). These are:

Looping: In a well-mixed turbulent boundary layer on a hot day (forced by buoyancy), the turbulent eddies may be large and intense enough to advect the whole plume down to the ground. This can result in extremely high plume concentrations in the vicinity of the source.

Coning: This is the kind of form assumed for a Gaussian plume, when the boundary layer is well mixed, and turbulent eddies are smaller than the plume scale. The plume forms a cone downstream.

Fanning: In a stable boundary layer, the plume spreads out horizontally at its level of neutral buoyancy. Vertical motion is weak, so there is little upward spread, but the plume forms a "fan" when viewed from above. The plume is not well mixed in the vertical, which implies relatively slow dilution, but there are not likely to be high plume concentrations at the ground. Unfortunately, this kind of plume may be the precursor to a "fumigation" event (see Fig. 14.11E below) if the inversion is subsequently mixed to ground level.

Lofting: At early evening, if a surface inversion is developing, vertical motion may be inhibited below the plume while remaining active above; the plume is diluted but does not reach the ground. This is a favorable situation.

Fumigation: There is a strong inversion restricting mixing above, and the plume is mixed throughout the boundary layer. This can occur quite rapidly. For example, after sunrise when the nocturnal inversion is being eroded from below by buoyant eddies, plume-level air of high concentration may be brought down to the surface over a wide area.

Trapping: There is a relatively well-mixed layer with inversions above and below it, which trap the plume at a particular height. This might occur at night when there is a low-level inversion above the Nocturnal Boundary Layer (NBL) and a higher level inversion left over from the previous days Convective Boundary Layer (CBL). In between is the relatively well-mixed residual layer.

14.8 GAUSSIAN PLUME MODEL

For the steady-state concentration downwind from a continuous point source, the Gaussian Plume Model was presented by Sutton and further developed by Pasquill and Gifford. In this model, the concentration distribution perpendicular to the plume axis is assumed to be Gaussian distribution in both horizontal and vertical planes, and the extent of plume growth in these planes is measured by the standard deviation.

In the Gaussian plume approach the expanding plume has a Gaussian, or normal, distribution of concentration in the vertical (z) and lateral (y) directions. Fig. 14.12 depicts the nature of the plume coming out. The coordinate system has been set up to show a cross-section of the plume, with z representing the vertical direction and x being the distance directly downwind from the source.

As stack emissions will be at elevated temperatures, they will have some initial upward velocity and buoyancy, and the plume, before it takes up a symmetrical path about the center line, will travel some distance downwind,

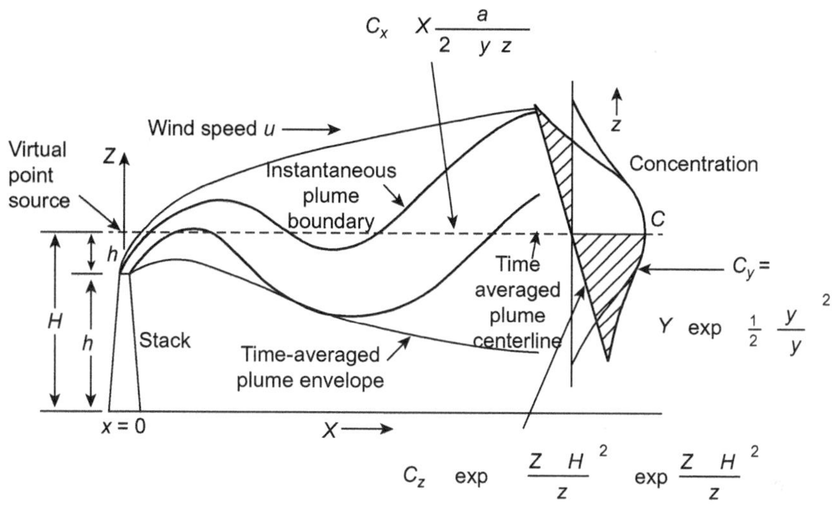

Figure 14.12 The instantaneous plume boundary and the plume envelope.

and the center line will be above the actual stack height. The highest concentration of pollutants will be spread along this center line, which will decrease in the concentration values as we move further and further away. This model assumes that the pollutant concentration follows a normal distribution in both the vertical plane, as is shown in the figure, and in the horizontal plane and treats the emissions as if they came from a virtual point source along the plume center line, at an effective stack height H.

The average steady state pollutant concentrations are related to the source strength, wind speed, effective stack height, and atmospheric conditions in Gassuian plume source earlier. These can be derived from the nature of various factors involving gaseous diffusion in three-dimensional space. The Gaussian plume equation for the concentration C (in units of gm-3 for example) at any point (x, y, z) in the three-dimensional coordinate system of the plume is then given by Turner:

$$C(x, y, z) = \underset{\text{central line}}{\frac{Q}{2\pi u \sigma y \sigma z}} \exp\left[\begin{matrix}(-y^2/2\sigma y)\\ \text{Horizontal}\end{matrix}\right] \left(\begin{matrix}\exp - (z-H)^2/2\sigma z + \exp - (z+H)^2/2\sigma z\\ \text{Vertical}\end{matrix}\right)$$

where $C(x,y,z)$ = center line concentration of pollutant A at any point (x, y, z) in space g/m^3; Q = Source strength i.e., pollutant emission rate, μg/s;

u = average wind speed at the source level (at stack height) m/s; H = Effective stack height, m

$$H = h + \Delta h,$$

where h = physical height of stack, m and Δh = Plume rise, m.

Δy and Δz are the standard deviations of the concentration of the pollutant at A in the horizontal crosswind and vertical directions respectively. These dispersion coefficients vary with the time of sampling, the basic time of sampling being 1 h Δy and Δz increase with downwind distance, x signifying that the dilution increases with distance.

14.8.1 Advantages Gaussian Model

This model is easy to understand, efficient in computer running time, appealing conceptually, and results agree well with experimental data.

The Gaussian plume model is based on the following assumptions:

1. The pollutant plume coming from a stack will have Gaussian distribution in both horizontal and vertical planes.
2. y and z represent the standard deviations of the concentrations of the plume in horizontal crosswind and vertical directions respectively.
3. The plume will be affected by the wind speed, which is existing at the source level.
4. The rate of emission of the pollutant from the source, i.e., Qg/s will be continuous and uniform.
5. Steady–state conditions prevail through the dispersion, and factors affecting the dispersion of the pollutant do not change with time and space.
6. The pollutant emitted from the stack are inert and do not undergo any chemical reaction in the atmosphere.
7. Perfect reflection of the plume at the underlying surface, i.e., no ground adsorption.
8. The turbulent diffusion in the x–direction is neglected relative to advection in the transport (x) direction, which implies that the model should be applied for average wind speeds of more than 1 m/s ($u > 1$ m/s).
9. The coordinate system is directed with its x-axis into the direction of the flow, and the v (lateral) and w (vertical) components of the time averaged wind vector are set to zero.
10. The Gaussian plume equation is derived based on the assumption that terrain in which plume is dispersed is flat.

11. All variables are averaged over a period of about 10 min, which implies that for different averaging times, corrections to equation have to be made.

14.8.2 Limitations

1. Gaussian plume model do not take into consideration the formation of different stability layers at different heights in the atmosphere and different time of the day.
2. For varying terrain characteristics such as terrain roughness, mountains, valleys, land increases, and bodies of water, corrections have to be incorporated in the Gaussian plume model.
3. Free convection regions and strong wind shears like change of wind directions and change of wind speed with heights are not taken into consideration in Gaussian plume model.
4. Downwind concentrations of the pollutant can be calculated for shorter distances and of shorter travel times.

Though this model gives good values for downwind distances between approximately 100 and 2000 m for distances below 100 m, the value of has to be corrected by considering the details of the wind flow pattern around the source structure. Similarly, for distances greater than 1000 m, the local terrain features and meteorological variability have to be taken into consideration.

14.9 AIR POLLUTION CONTROL TECHNOLOGIES

14.9.1 Particulate Control Equipment

The various technologies for particulate control include gravitational settling chambers, cyclone separators, fabric filters, electrostatic precipitators, and wet collectors (scrubbers).

14.9.1.1 Selection of Particulate Control Equipment

The selection of a specific particulate control system for cleaning industrial effluent gases depends on the nature of effluent gaseous discharge, the extent of purification required and gaseous discharge characteristics which affect the performance of the purification system. Particle sizes range generally from 100 mm down to 0.1 mm and even less. To remove particulate matter from gas streams, various types of control equipment are available which can be selected based on the following information:

1. Nature of the gas phase and particulate matter
2. The physical and chemical characteristics of particulates

3. The particulate size and concentration in the gas
4. Volume of particulates to be handled
5. Temperature and humidity of gaseous medium
6. Factors like toxicity and inflammability must be taken into consideration when evaluating column efficiency
7. Quality of the treated effluent, i.e., efficiency of the removal of particulate matter

14.9.1.2 Gravitational Settling Chambers

A gravity settler is simply a long chamber through which the contaminated gas passes slowly, allowing time for the particles to settle by gravity to the bottom. Solid or liquid particles suspended in the gas reach a terminal free falling velocity which is given by Stokes' Law for small particles, and is proportional to the product of the square of the particle diameter and the density difference between the particle and the carrier gas, and inversely proportional to the viscosity of the carrier gas as discussed in detail earlier. In practice, this implies that the method can only by applied for removing particles greater than 100 μm in diameter. As most of the troublesome particles have sizes smaller than 50 μm, these devices are normally used as pre-cleaners prior to passing the gas stream through high efficiency collection devices. The cross-sectional area of (WH) of the settling chamber is much larger than that of the duct through which the pollutant gas stream enters this settling chamber. The gas stream, containing the particulate matter is allowed to flow at a low velocity in the settling chamber allowing sufficient time for the particles to settle down.

The gravity settling chamber is one of the simplest and oldest methods of dust collection. The large cross-sectional area of the chamber (A_c) relative to the cross-sectional area of the entrance duct (A_d) serves to reduce the gas and particle velocity within the chamber:

$$\text{Velocity of gas, in chamber} = \left[\frac{\text{Cross-section area, duct}}{\text{Cross-sectional area, chamber}}\right]$$

$$[\text{Velocity of gas, induct}]$$

$$\upsilon_{gc} = \begin{bmatrix} A_d \\ \upsilon_{gc} \\ A_c \end{bmatrix}$$

For best results, the gas flow through the chamber should be uniform and less than 60 fpm.

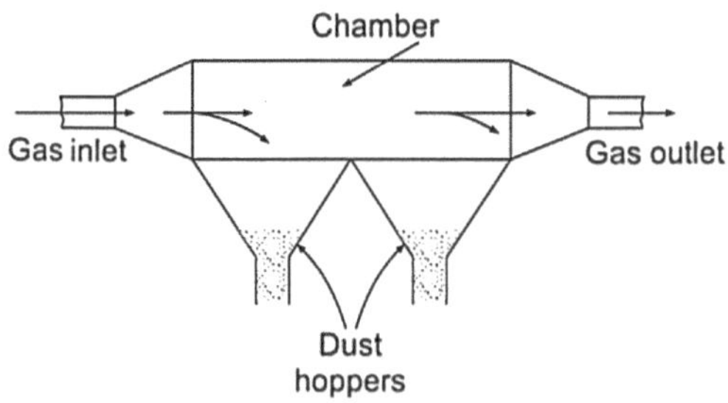

Figure 14.13 Horizontal flow settling chamber.

A simple horizontal flow gravity settler (Fig. 14.13), which has a cross-sectional area much larger than that of the duct, brings the dust laden gas stream into the settling chamber. The gravitational setting chambers usually operate with velocity between 0.5 and 2.5 m/s, although for best operating results, the gas flow should be uniformly maintained at less than 0.3 m/s. Some settling chambers have simply enlarged conduits and some have horizontal shelves and baffles, about 2.5 cm apart. The collection efficiency of the chamber depends upon the type of flow also.

In Howard type of settling chambers (Fig. 14.14), horizontal trays will be fixed in the chambers at about 1–3 cm height intervals. The increase in the efficiency of the Howard type gravitational chambers is dependent on the number of trays added. A maximum of five to seven trays can be added to increase efficiency.

14.9.1.3 Cyclone Separators

In a cyclone, the particulate laden gas is accelerated through a spiral motion which imparts centrifugal force to the particles. The centrifugal force forces the particles out from the spinning gas, and makes them fall on the wall

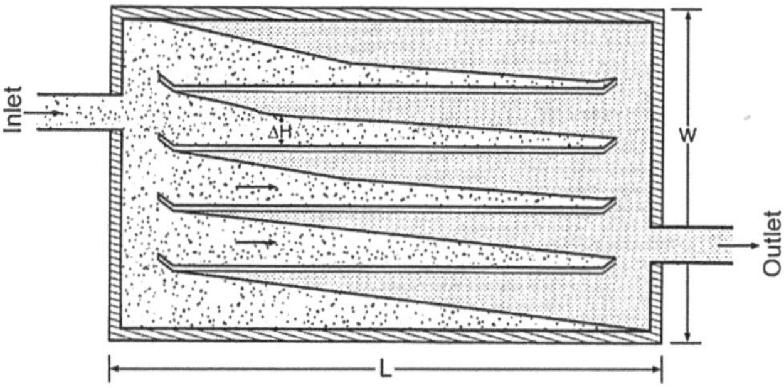

Figure 14.14 Howard settling chamber with five trays.

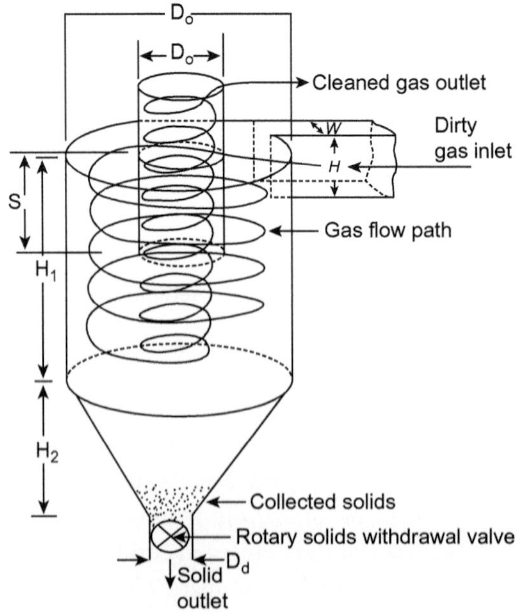

Figure 14.15 Cyclone separator.

of the cyclone. The particles will finally slide down. The cyclone separator shown in Fig. 14.15 is merely a gravity settler made in the form of two concentric helices and utilizes the centrifugal force generated by the spinning gas to separate the particulate matter from the polluted gas stream. A simple cyclone collector consists of a cylinder with a tangential inlet and an inverted cone attached to the base. Gas enters the cyclone through the tangential inlet, which imparts a whirling motion to the gas. Suspended particles will be forced toward the wall on which they collect and slide down into the conical collector. Near the bottom of the cone, the gas turns abruptly upward and forms an inner spiral, which leaves through the pipe or duct extending into the center of the cyclone body.

There are three types of cyclones which are under industrial use. They are: (1) high-throughput cyclone, (2) conventional cyclone, and (3) high-efficiency cyclone. High-throughput cyclones process lush volumes of waste air as input, but are opening at low efficiencies. The conventional cyclones are in between high throughput and high-efficiency cyclones. Cyclones can also be arranged to operate in multiples to produce higher efficiencies. The efficiency of removal is not only a function of the size of the particle, but also of the other variables such as the airflow.

The centrifugal force (F_c), which separates a particle of mass (m) from the gas stream, is given by the equation:

$$\text{Centrifugal force} = \frac{[\text{particle mass}]\ [\text{particle tangential velocity}]^2}{\text{Cyclone radius}}$$

$$F_c = \frac{(m_p)\,(\upsilon_{pt})^2}{\gamma_c}$$

According to this equation, the magnitude of the centrifugal force which can be exerted on a given particle, in a given size cyclone, is a function of the particle mass and tangential velocity. The latter is dependent on the gas velocities in the cyclone. As velocities increase, energy losses due to friction and turbulence also increase, and the pressure drop across the cyclone rises. In general, dust collection efficiency becomes greater as pressure drop increases.

High Efficiency Cyclones

Small diameter cyclones (9 in. or less) are more efficient than conventional. Theoretically, the efficiency of particulate removal of a given cyclone design increases as the ratio of centrifugal force to drag force increases:

$$\text{Ratio} = \frac{\text{Centrifugal force}}{\text{Drag force}}$$

After substituting in and simplifying the ratio reduces to:

$$\text{Ratio} = \frac{[\text{particle diameter}]^2\ [\text{particle density}]\ [\text{particle tangential velocity}]}{[\text{gas viscosity}]\ [\text{cyclone diameter}]}$$

$$= \frac{(D_p)^2\,(\rho_p)\,(\upsilon_{pt})}{(\mu_g)\ (D_c)}$$

Thus, it can be seen that for a given set of conditions the operating efficiency will increase with a decrease in cyclone diameter. However, this does not mean that a small diameter cyclone is inherently more efficient than one of large diameter, although it does mean that a smaller diameter cyclone of particular proportions is inherently more efficient than a large diameter cyclone of the same proportions.

	Efficiency (%)	
Particle Size Microns	**Conventional Cyclone**	**High Efficiency Cyclone**
5	–	50–80
5–20	50–80	80–95
15–50	80–95	95–99
40	95–90	95–99

In a high efficiency cyclone, gas enters tangentially through the inlet cone, which provides the swirling motion necessary to throw the dust particles to the outer perimeter of the tube. Acceleration takes place in the cone, and then it remains constant from there to the bleed-off for the secondary circuit. There is no increase in velocity and dust concentration. The clean air travels out through the end of the tube without changing direction, and 10% of the primary air circuit, plus the collected dust, is bled off and thrown into the primary dust hopper. Any number of small diameter cyclones can be operated in parallel to achieve the desired capacity.

14.9.1.4 Electrostatic Precipitator

The basic principle of an electrostatic precipitator (ESP) is to give particles an electrostatic charge and then put them in an electrostatic field that drives them to a collecting wall. A gas stream containing suspended particulates is allowed to pass between two electrodes electrically insulated from each other, and between which there is a considerable difference in electric potential. The high-voltage electrode usually has a small cross-section and some curvature, e.g., a wire. The other electrode could be a plate or a surface of only slight curvature. The high voltage on the electrode of small cross-section ionizes the gas and aerosol particles. A corona is formed of ions, which attach to the particulates making them ionic; they are then attracted to the large collecting electrode. There they are allowed to flow down or are drained by gravity and collected at the bottom. The ESP is like a gravity settler or centrifugal separator, but as the electrostatic force drives the particles more powerfully to the collecting plate, it is more effective on smaller particles than the gravity settlers and cyclone separators.

An ESP uses electric forces to separate suspended particles from gases. This is accomplished in two basic steps:

1. A corona-charging field provides the particles with an electrical charge.
2. A high-voltage collecting field attracts the charged particles to the collecting electrodes.

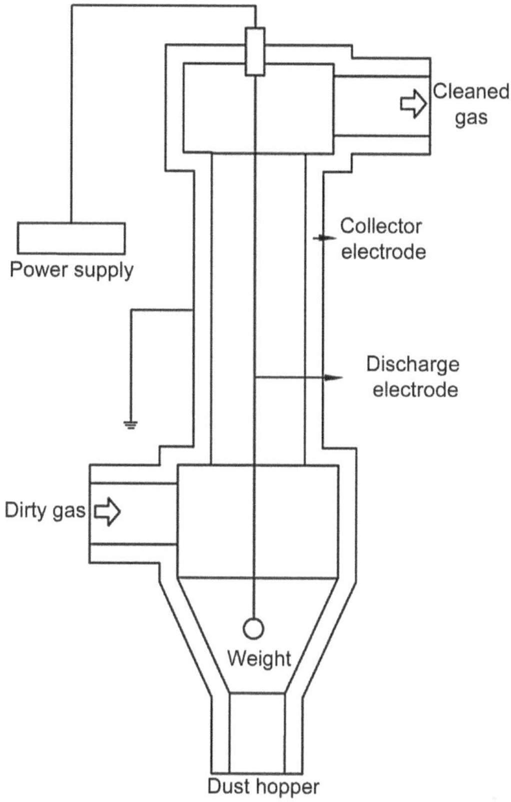

Figure 14.16 Electrostatic precipitator.

Fig. 14.16 shows one "gas passage" of a horizontal electro-filter. Many of such lanes are arranged in parallel in one casing so that the actual gas volume finds a sufficient section to flow through. In order to obtain an effective corona discharge, the discharge electrode is supplied with 15,000 to 80,000V negative DC current. The corona-discharge charges the particles in the gas stream.

The particle charge is given by:

$$\text{Particle charge} = 3\ [\text{charging field}]\ [\text{particle radius}]^2$$

$$q = 3\,(E_o)\,(r_p)^2$$

Under the influence of the electric field built up between the discharge electrodes and the grounded collector electrodes, the ionized particles accelerate toward the collector electrodes.

In general, in all these particulate collecting systems—gravitation settling chambers, cyclone separators, or ESPs—the driving force acting on the particle to push it to the collecting wall against the resistance offered by the viscous medium which is proportional to the particle diameter. For gravity settlers and cyclone separators, this force is proportional to the cube of the diameter of particles of constant density. In ESPs, though the resistance force offered by the viscous medium is Stokes' drag, the force driving the particle to the wall is electrostatic which is proportional to the square of the particle diameter.

The parallel plate two-stage ESP is shown in Fig. 14.17. The gas passes between the plates, which are electrically grounded. Between the plates are rows of wires, held at a voltage of typically −40,000V. The power is obtained by transforming ordinary alternating current to high-voltage, and then rectifying it through some kind of solid-state rectifier. This combination of charged wires and grounded plates produces both the free electrons to charge the particles and the field to drive them against the plates. On the plates the particles lose their charge and adhere to each other and the plate, forming a "cake." The cleaned gas then passes out of the far side of the precipitator.

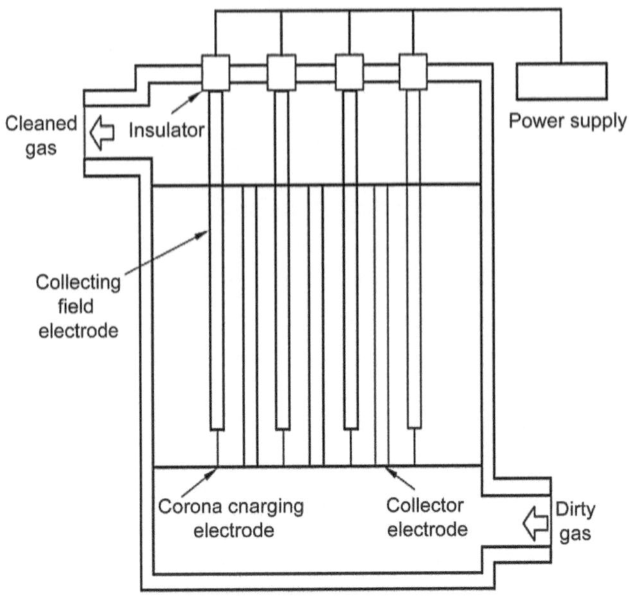

Figure 14.17 Two stage electrostatic precipitator.

Dust collection efficiency of ESP varies greatly according to the electrical resistivity of the collected dust. In the case of high resistivity dust, an abnormal phenomenon known as the back corona problem may occur, and this seriously reduces the dust collection efficiency of ESP.

Advantages of Electrostatic Precipitators

1. Pressure drop and hence power requirement is small compared to the other devices; economical and simple to operate.
2. 99% efficiency obtainable.Very small particles can be collected, wet or dry.
3. Can handle both gases and mists for high volume flow.
4. Can be operated at high temperatures and pressures.

14.9.1.5 Gas Filtration for Particulate Removal

Two main types of gas filters are in use for the removal of particulate matter from industrial gases:

1. Packed bed or depth filters.
2. Fabric elastic filter.

Fabric filters are used for cleaning industrial gases with a particulate loading gravity of 1 g/m^3 while depth filters are used only for gases containing particles of the order of 1 ng/m^3. The performance efficiency of any filter is determined from the following criteria: (1) pressure loss, (2) collection efficiency, and (3) lifetime or dust holding capacity.

1. Pressure loss is expressed in mm of water and gives the required fan power. Air is a major factor that determines the operational cost.
2. The collection efficiency (mass percentage of dust collected) determines the performance of a fiber, and is measured instantaneously or cumulatively since the efficiency varies with dust adhered to the filter which varies with filtering time.
3. The lifetime or dust holding capacity of any filter is determined by actual experimentation under operational conditions and this determines the major portion of the initial costs. It is the maximum capacity of the filter to hold the dust per unit area of the filter medium for a certain range of pressure drop. Normally filters will have 0.3–2 kg of dust per square meter of filter paper.

14.9.1.6 Classification of Fabric Filters

Classified according to the methods of cleaning, there are three types of baghouses; shaker, reverse-air, and pulse-jet baghouses. The filtering element is usually shaped in tubular form as shown in Fig. 14.18, called filter bags.

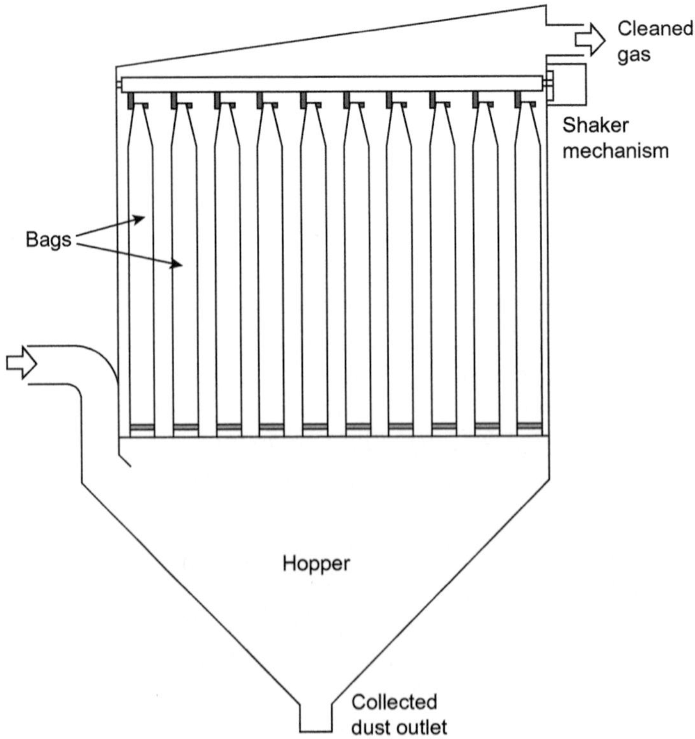

Figure 14.18 Cross-section of baghouse.

Filter bags may range from 1.8 to 9 m in length, and may have diameters of around 20 cm. Several of these bags are grouped together and put in to a compartment, and several of these compartments are put together and assembled inside a structure called a baghouse. The capacity of a compartment is determined by the area of the fabric filter.

Bag Filter

The bag filter is designed and engineered for filtering dust in gas emissions from all industry areas through a bag-shaped fiber filter so that the dust can be separated from the gases.

Furthermore, by pre-coating, such as injecting calcium hydroxide in just ahead of the bag filter, oil mist or noxious gases such as hydrogen chloride, too, is possible to remove simultaneously. The bag filter is grouped into the pulse jet type, reverse air type, dual reverse, and shaking blow-off type according to the method of shaking off the dust, which are selected in consideration of gas temperature, properties of substances, moisture, and so on.

Dust removal efficiency can be reached up to 99.9% or larger, and if the bag is pre-coated, gases, too, can be adsorbed simultaneously with up to 90% or larger efficiency.

- A wide variety of gases are possible to collect.
- The $0.05\,g/m^3N$ dust concentration is possible to attain regardless of concentration at inlet.
- Pressure loss in 80–200 mm H_2O.
- Fine particles down to 0.1 μm are possible to collect.
- Required power except for fan power is small.
- Continuous operation is possible to perform and maintenance is easy to control.
- Both facility and maintenance cost are low because of the simple construction.
- Recovered dust is possible to re-use because of dry type dust collecting design.

The bag filter is ideal in the following fields of industry:

- Battery production
- Electrical and electronic production
- Electrical equipment production
- Metal machining
- Automobile component production
- Other industrial processes requiring the collection of heavy metal dust

The engineering performance of the bag filter can also be put to effective use in the treatment of combustible and explosive dust such as wood dust, coal dust, or hydrogen storage alloy dust, and in the design of hoods at dust generating sources.

Absorption

Sorption comprises the general phenomenon of the assimilation of a gas by a solid or a liquid. When the gas is taken on only at the surface or in the capillaries of the solid to form a surface compound or condensate, the phenomenon is designated as adsorption. When the sorbed gas forms a homogeneous solution with the liquid or is taken deep into the internal structure of a solid, the transformation is called absorption.

Spray Tower

In spray towers the water is introduced through nozzles, which direct the spray toward the bottom of a circular or rectangular chamber (Fig. 14.19). The polluted gas enters at the bottom, and as it flows upward the dust

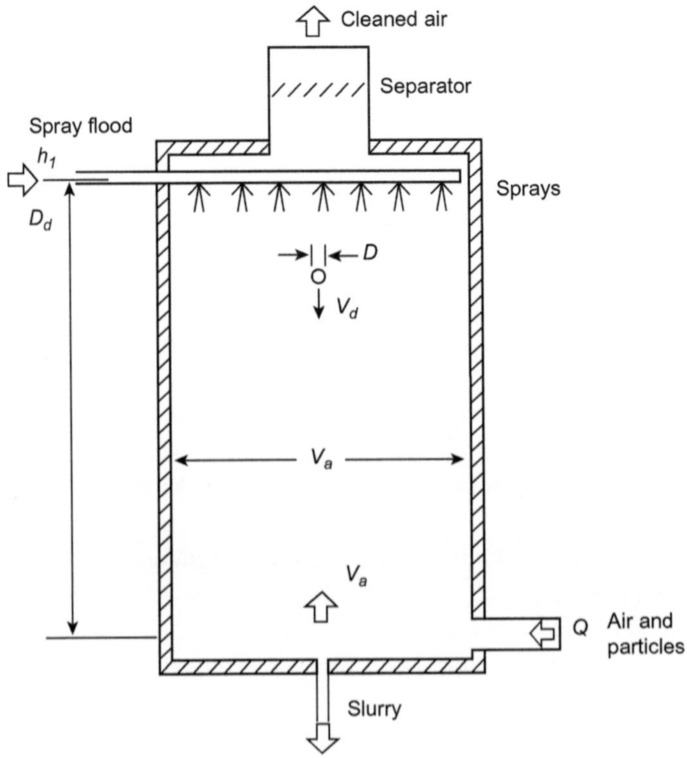

Figure 14.19 Spray tower.

particles collide with the liquid droplets generated by the nozzle spray. The liquid droplets and their captured dust particles impinge on the water eliminator and are removed from the gas stream. This is the simplest type of counter flow gas scrubber with a moderate contact between phases and so used only for remaining coarse dust when high efficiency is not necessary.

The pressure drops in a spray tower are typically low, on the order of 1–2 inches of water. The liquid:gas ratio is generally around 20–100 gallons per 1000 cubic feet of polluted air. The liquid used in the scrubbing process is normally recirculated in order to minimize the need for excessive amounts of water. Low energy is normally required, and these simple spray towers are easy to operate and maintain.

One of the major disadvantages is the relatively large amount of water used and the low efficiency in removing mists smaller than 5 μm in diameter. The towers generate a polluted water stream, which is the by-product of the operation. This wastewater carries in the contaminants which are in the air. It must be further treated before discharge. Installation and operation

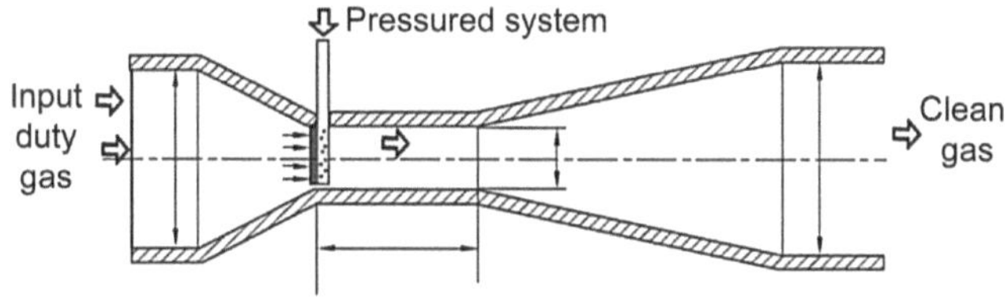

Figure 14.20 Venturi scrubber.

of spray towers is an inexpensive matter when compared to other processes. The problem is in the high cost of treating the wastewater which is generated during their use. In addition, operating expenses will include the pump and fan power requirements.

Venturi Scrubbers

Venturi scrubbers offer high-performance collection of fine particles, usually smaller than 2–3 mm in diameter. They are particularly suitable when the particulate matter is sticky, flammable, or highly corrosive. The high performance of the venturi scrubbers is achieved by accelerating the gas stream velocities, of the order of 60–120 m/s. Due to the high speed action, the feed liquid is atomized with a uniform fashion across the throat through several low pressure spray nozzles directed radially inward as shown in Fig. 14.20. The droplets accelerate in the throat section, and due to the velocity difference between the particles and the droplets, the particles are impacted against the slow moving droplets. The acceleration continues to some extent into the diverging section of the venturi. The gas–liquid mixture is then directed to a separation device such as a cyclone separator where particulate matter is separated from the gas stream. The affecting mechanisms for collection of particulates in the scrubber are inertial impaction, diffusion, electrostatic phenomenon, and condensation and agglomeration, the principal mechanism is inertial impaction. The application of venturi scrubbers is more often in Kraft mill furnaces, metallurgical furnaces, sulfuric acid concentrators, etc., for removing mists and dusts from gases.

Centrifugal Scrubbers

The wet centrifugal collector utilizes centrifugal force and the wet impingement principle in collecting light to heavy loading of all size granular dust (Fig. 14.21). Dry dust particles impinge against the wetted peripheral surfaces, thus reducing re-entertainment of the particle into the gas stream. As the incoming gas is spun down the inlet cones, the water and collected solids are separated from the gas stream.

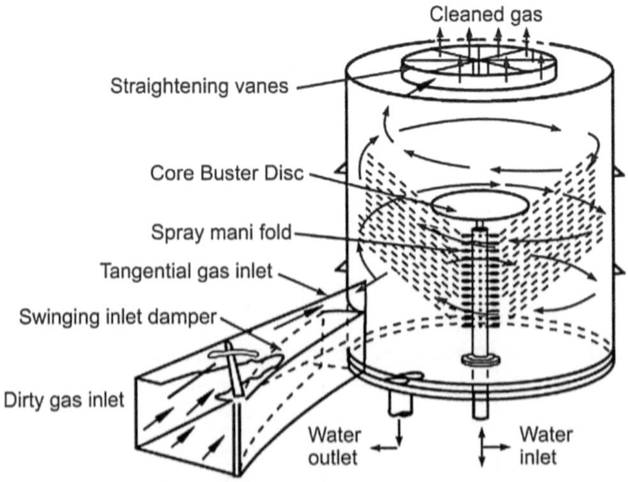

Figure 14.21 Centrifugal scrubbers.

Packed Beds and Plate Column Scrubbers

Packed towers are very efficient absorption devices involving a continuous contact of two phases. These use a variety of packing materials ranging from specially designed ceramic packing to crushed rock. The liquid is distributed over the packing, which provides high interfacial surface area, and flow down the packing surface in the form of thin films or sub-divided streams.

Packed Scrubber

In a packed scrubber, fiberglass, saddles, coke, or broken stone, etc., are used as the collection material (Fig. 14.22). The polluted gas stream moves upward in a countercurrent flow packed scrubber, and comes in contact with the scrubbing liquid stream which is moving downward over the packing in a film. The gas stream passes through the packing pore spaces and captures the particles by the inertial impaction. Because of the good mass transfer characteristics of the packing, efficient collection of fine particles by diffusion is also possible. Smaller packing increases the efficiency of collection but its shape does not appear to affect the collection efficiency. Sometimes packing towers encounter plugging problems, which can be reduced by employing sprays to wash the packing or by using low density spheres, etc., agitated by gas flow chemical foam packing is being employed recently which drains slowly from the scrubber with captured particles and is replaced with fresh material.

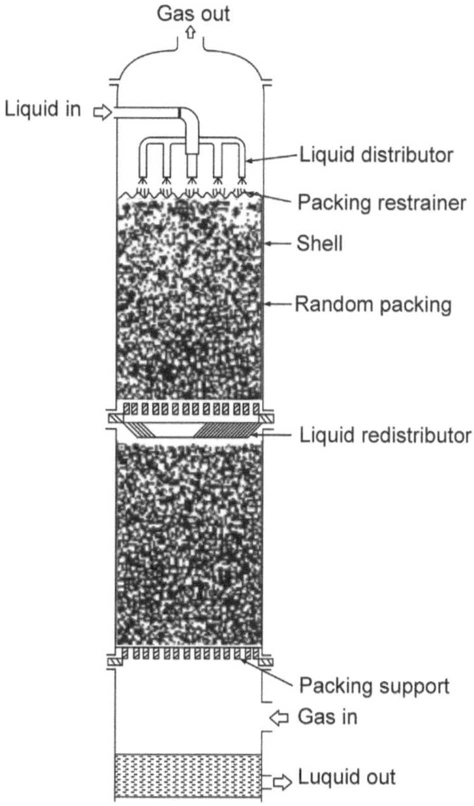

Figure 14.22 Packed tower.

In an ideal situation, the liquid is distributed uniformly and wets the surfaces of the packing. The tower packing provides a large area of contact between the liquid and the gas, encouraging intimate contact between the two phases. The inlet liquid is called fresh or "weak" liquid. The outlet liquid is called rich, or "strong liquor," because it has picked up some of the pollutant gas and particulate matter. The entering gas at the bottom is called the *rich gas*, since it is rich with pollutants, and the cleaner air exiting at the top is called *lean gas*, since it has lost most of the pollutants. The countercurrent operation is most beneficial for separation, because the fresh liquid contacts the lean or cleaner gas at the top and the strong liquor contacts the rich gas at the bottom. In this manner, the gas with a weak concentration of pollutants is scrubbed with fresh water, and the strong liquor, which has already absorbed some pollutants, comes into contact with a dirtier gas at the bottom. From a practical point of view, mass transfer is good when concentration gradients are appreciable. The amount of liquid present at any time in the packed section of the tower is called the total holdup. This increases with increasing flow rate of liquid, but it is independent of the gas flow rate up to the loading point. For a given liquid flow rate, the loading point is the gas velocity above which the total holdup increases substantially, with further increase in gas velocity, the liquid fills up all the voids and the packing tower becomes a bubbling unit, and it is said to be flooded. It is difficult to keep a stable operation under these conditions, and the liquid will become entrained and carried by the exit stream of air.

Pressure drops in packed towers range between 1 and 8 inches of water. The typical liquid to gas ratio are 10–20 gallons per 1000 cubic feet. Higher removal efficiencies are obtained here, and a lower water consumption rate is the norm; it is about 0.75–3.7 gallons per 1000 cubic feet per minute of polluted air treated. However, one of the major disadvantages of packed towers is the high-pressure drop in the system (more than 3 times higher than the pressure drop in a spray tower). There are also greater clogging and fouling possibilities and more maintenance costs, along with wastewater treatment and disposal. These packed scrubbers are more expensive to install and operate than spray towers. Maintenance of the packing adds another cost to the operation.

14.9.1.7 Tower Packings

There are several types of tower packings. Some are made from cheap, inert, and light materials such as porcelain, graphite, or clay. Some are made from fibrous materials such as glass or steel wool. However, plastic packings are becoming widely used. They are light and essentially unbreakable. With plastic packings, towers of light construction may be used, and breakage is not likely as is the case with ceramic packings.

Polyethylene packings called tellerettes, of a helical-torpid shape, are especially effective. A good tower packing should be:

1. Inexpensive.
2. Chemically inert to the gas and liquid in the tower.
3. Strong enough and light.
4. Made to provide adequate passages for the two phases without excessive pressure drop.
5. Durable.
6. Made in a shape providing a large contact area per unit volume.

Packings may be dumped at random when they are in the size range 0.25–2 inches. Packings are usually stacked when their size is in the range 2–8 inches, of course, these large packings are normally used in large towers.

In general, packed towers vary in diameter from 1.1 to 20 inches. The depth of the packing may vary from an inch to several feet depending on the application and type of packing used. Coarse beds are used to remove particles larger than 5 μ at superficial gas velocities varying between 1 and 15 ft/s. Fine beds are used to remove particles smaller than 5 μ at low superficial gas velocities on the order of 1–50 ft/min.

Packed towers are limited to handling low concentrations of particulate matter, since excessive concentrations will cause accumulation on the packing, and eventually plug the bed. Packed towers are more efficient in transfer operations than spray towers, but for very short heights, spray towers approach the performance of a packed tower, and thus, they are preferred due to lower pressure drops.

Tray or Plate Column

The tray or plate column is cylindrical tower housing with perforated plates. The liquid enters from the side of the column at the top and cascades down in a zigzag flow. The polluted air enters at the bottom. The clean air exists at the top, and the liquid leaves at the bottom of the column. The horizontal perforated plates are called sieve trays with typical three-sixteenth inch holes on half inch square centers. The trays are spaced 1–3 ft apart. The liquid, discharged onto one side of the top tray, flows across it and over a weir, then to a downcomer, which directs the liquid to the next tray down. This process is repeated until the liquid reaches the bottom. The polluted air entering at one side of the bottom rises up through the openings in each tray and through the liquid preventing it from draining through the holes. This is repeated until the clean air emerges at the top of the column. The function of the trays or "special packing" is to facilitate the contact between the liquid and gas streams. This column may be used to separate dust and gaseous pollutants from air.

Adsorption

Certain solids have different affinities for different gases. If a gaseous mixture comes in contact with an adsorbent solid, some of the components of the gas stream will concentrate on the surface of the solid and separate from the main stream. Thus, adsorption is a process dealing with the contacting of a solid with a fluid mixture to remove one or more components of the mixture. There are basically three types of adsorption. The first type is physical adsorption or condensation of toxic or obnoxious gases and vapors on solids at temperatures above the dew point. The second type is chemical adsorption or chemisorption, where a chemical bonding occurs between the adsorbed pollutants and the solid adsorbent. The last type is ion exchange solids, where an exchange of ions takes place. The solids give up ions for the ions they adsorb from the fluid stream. This is mostly applied in water softening where the solids give up sodium ions to the water in exchange for magnesium and calcium ions, and this exchange decreases or removes water hardness. Of the three types of adsorption, physical adsorption is the most important in air pollution control.

Some of the important adsorbents and their respective applications are listed in Table 14.3 to aid in the proper selection of these solids in removing pollutants.

Table 14.4 shows the various devices and the efficiency with which the particular particle size can be removed.

Table 14.3 Important Adsorbents

Adsorbent	Application
Activated alumina	Drying of air, gases and liquids
Activated carbon	Gas purification, solvent recovery, elimination of odors
Anhydrous calcium sulfate ($CaSO_4$ desiccant)	Drying and purification of gases
Bauxite	Drying of gases and liquids
Lithium chloride solution	Drying and purification of gases, reducing odor, dust, and smoke
Silica gel	Dehydration and purification of gases; adsorption of organic solvent (acetone, benzene); separation of hydrocarbons.

Table 14.4 Devices That Can Be Used for Particulate Matter Removal

Device	Minimum Particle Size (m)	Efficiency (%)
Venturi	>0.5	<99
Electrostatic precipitator	>1	<99
Bag filter	<1	>99
Spray chamber	>10	<80
Cyclonic spray chamber	>3	<80
Impingement scrubber	>3	<80
Centrifugal settler	>5	<80
Gravitational settler	>50	<50

14.10 METHODS OF CONTROL OF EMISSIONS FROM POINT SOURCES FOR OXIDES OF SULFUR, NITROGEN, AND CARBON

The control equipment to be discussed would be oxides of sulfur, oxides of nitrogen, oxides of carbon and hydrocarbons.

The principal gaseous pollutants of concern in air pollution are SO_x, NO_x, CO, the hydrocarbons, and other organic and inorganic gases. In general, these emissions may be controlled by absorption, adsorption, and incineration. The general methods by which gaseous pollutants can be removed from the industrial gaseous effluent stream are by sorption of gaseous pollutants by absorption in a liquid, by adsorption on a solid, and by decomposition or conversion of the pollutant, from toxic form to non-toxic stable form, chemically either by combustion or by catalytic incineration.

Liquid absorption is one of the most versatile techniques used for reducing the pollutant levels from various industrial gaseous effluents. In this process due to a concentration gradient the pollutant from the gaseous phase is transferred across the gas–liquid interface and the concentration decreases in the direction of mass transfer.

Removal of gases by adsorption is another technique that is used. This is a mass transfer process in which the gas is bonded to a solid and a surface phenomenon. The gas (adsorbate) penetrates into the pores of the solid (adsorbent) but not into the lattice itself. The bond may be physical binding, where electrostatic forces hold the pollutant gas or chemical bonding by reaction with the surface. The adsorption process is endothermic, while the desorption process is exothermic, and both are reversible. Gases with higher molecular weights and lower boiling points are preferentially adsorbed and these displace gases which are less preferentially adsorbed effecting separation. For example, organic vapor in air can be preferentially adsorbed by activated carbon than air.

For destruction of gaseous pollutants by combustion, one has to consider the safety aspects of combustion and improper combustion products. Many times the industrial gaseous effluents are not directly combustible and require addition of a fuel. However, certain gaseous effluents from petroleum refinery and blast furnaces are flammable directly without the addition of any fuel. Thermal incineration is one of the best ways to destroy them. This incineration process burns organic compounds and renders them harmless by giving off carbon dioxide and water. The proper length of residence time must be given so that complete combustion takes place. The temperature should be high enough so that total destruction of the toxic wastes occurs. Incinerators require residence times ranging from 0.5 to 1.0 s, with temperatures ranging from 1200 to 1600°F. In general, destruction efficiencies are above 95%.

14.10.1 Removal of Oxides of Sulfur

Fossil fuel burning in power plants and industrial furnaces are the major sources of SO_2 production in addition to the mobile sources. The following are the control technologies adopted for reducing SO_x emissions:

1. Dilution of SO_x emissions by increased stack height
2. Use of alternate fuels
3. Fuel desulfurization
4. Reduction of sulfur in the combustion process
5. Flue gas desulfurization

There are a number of leading processes for removing SO_2 from flue gases. These include limestone injection (wet absorption), wet flue gas desulfurization, the Lurgi adsorption process, and the catalytic oxidation (catox) process.

SO_x reduction technology wet type desulfurization system—flue gas desulfurization system (limestone–gypsum process).

Limestone-water slurry is mainly used as absorbent in the absorber installed in the flue. In the absorber, flue gas is brought into contact with the absorbent to remove SO_x in the gas. Reaction products are collected in the form of gypsum. This is normally used in fossil power stations (coal, heavy oil), etc. The volume of SO_x emissions per unit of electricity generated by fossil power stations in Japan in 1988 is 0.38 g/kWh, which is far below the average of major countries in Europe and America (United States, France, Canada, United Kingdom, former West Germany) of 6.70 g/kWh.

Limestone injection is one of the most popular processes under development. SO_x produced from burning oil and coal react with the calcined products of limestone and dolomite to give some stable removable salts of calcium and sulfur. In this process, limestone and/or dolomite is pulverized along with coal in a mill; it is then fed to the high temperature zone of a furnace, where it is calcined to give CaO and MgO. The air polluted with SO_x enters a preheater and then goes to the furnace where the temperature is above 1200°F. The alkaline earth additives react with SO_2 and O_2 to give calcium sulfate, $CaSO_4$, called gypsum, some calcium sulfite, $CaSO_3$, is also produced. At this stage 20–30% of the sulfur oxides are removed. The sulfates, fly ash, and the unreacted lime pass through the air preheater on their way to the scrubber; here, water spray create an intimate contact with the unreacted lime and the remaining SO_2. Sulfates and sulfites are formed and removed along with particulate matter. The spent liquor from the scrubber and the sulfates, sulfites, and ash are allowed to a settling tank. The scrubber liquor is recycled in order to reduce water requirements and minimize water pollution. The sulfates, sulfites, and ash are, most likely, used as landfill.

14.10.2 Fuel Gas Desulfurization

Wet flue gas desulfurization processes include the use of lime or dolomite and/or sodium alkali. When lime and alkali are used, the process is called dual-alkali, simply because two alkali feeds are used, namely, pebble lime and soda ash.

When using lime or limestone, a slurry is prepared of calcium hydroxide $Ca(OH)_2$ or calcium carbonate ($CaCO_3$). It is used to remove the sulfur oxide from the flue gas stream. SO_x reacts with the lime to form $CaSO_3$ and calcium sulfate ($CaSO_4$). The two products precipitate from the solution. The lime slurry moves on to the scrubber (see Fig. 14.23), where the polluted gas makes intimate contact with it and reacts with the calcium. Then the slurry and products of the reaction, along with the gas, move on to a settling tank or a particulate removal unit, where residence time will be essential for the particulate matter to settle down. The solids are removed into a container, while some of the slurry is recycled to go

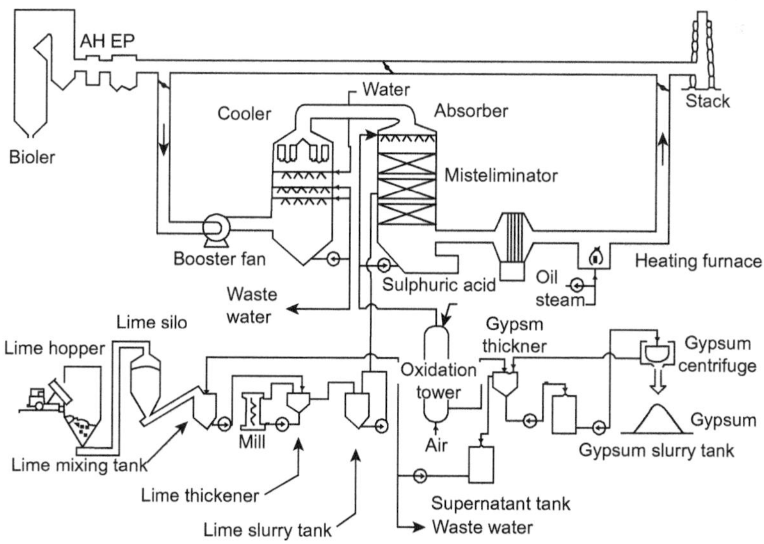

Figure 14.23 Wet type desulfurization system.

through the process again. The clean gas moves up the stack. Fresh quantities of lime are added as needed. The solids removed are disposed of in a sanitary landfill. The clarified liquid is recycled to be used in the scrubber unit again.

14.10.3 CATOX Process

CATOX process is an **oxidation/reduction process**. This process basically consists of passing the flue gas through a fixed catalyst bed where SO_2, in the presence of O_2, is converted to SO_3 (Fig. 14.24). The SO_2 is then absorbed in recirculated sulfuric acid in an absorption tower. Since sulfuric acid is a salable product, there is no solid waste product, assuming that a market can be found. In the "CATOX" process, fly ash is first removed from the flue gas by a high-temperature ESP, and SO_2 is then catalytically oxidized to SO_3 which is then recovered as sulfuric acid. For achieving good combustion efficiency, V_2O_5 at 400–500°C is used as a catalyst. In another improved efficiency process, SO_2 and O present in the stack gas are passed through the surface of an active carbon catalyst where they are absorbed and the catalyst catalyzes the oxidation of SO_2 to SO_3. The moisture present in the pores of active carbon

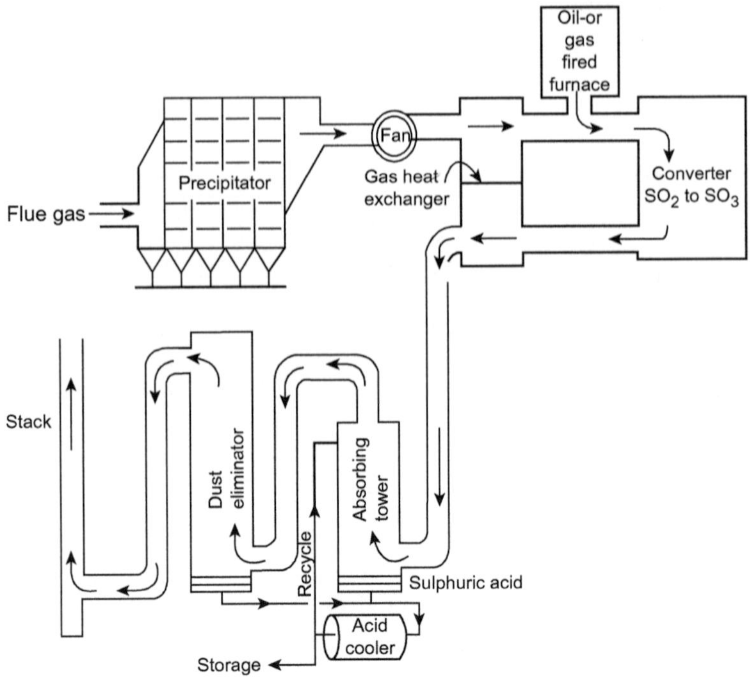

Figure 14.24 CATOX process.

reacts with SO_3 to form H_2SO_4. Complete combustion of SO_2 to SO_3 can be achieved by combined effect of adsorption and catalysis of active carbon.

14.10.4 Alkalized Alumina Process

In the alkalized alumina process, developed by U.S. Bureau of Mines, dust-free flue gas is fed to a reactor in which the porous form of sodium aluminate ($Na_2O{\cdot}Al_2O_3$), adsorbent sorbs SO_2 at a temperature of 315°C (Fig. 14.25). In this process, the SO_2 and O_2 in the flue gas react with the adsorbent. The spent material is then contracted with a reducing gas such as hydrogen in a regenerator at about 680°C to produce hydrogen sulfide.

14.10.5 Simplified Limestone/Lime Gypsum Process

The simplified wet-type desulfurization system is made cost-effective by simplifying the conventional limestone/lime gypsum process. The open spray tower is adopted for the absorber with high performance spray nozzles and simple internal structures. Gypsum treatment section is also simplified by using continuous gypsum centrifuge.

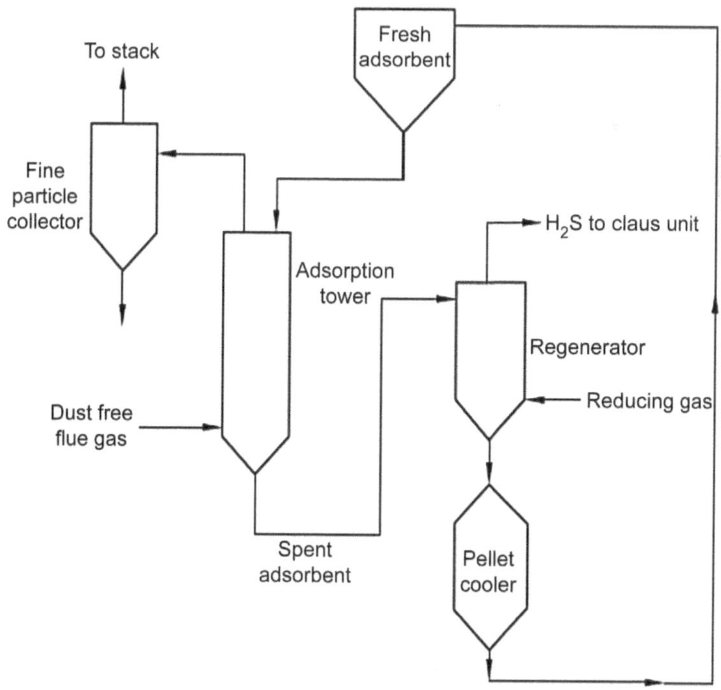

Figure 14.25 Alkalized alumina process.

1. Absorber Tower Section

$$SO_2 + H_2O + (1/2)\,O_2 \rightarrow H_2SO_4$$

2. Absorber Sump Section

$$CaCO_3 + H_2SO_4 + H_2O \rightarrow CaSO_4 \cdot 2H_2O + CO_2\,(g)$$

or

$$Ca(OH)_2 + H_2SO_4 \rightarrow CaSO_4 \cdot 2H_2O$$

Simplified configuration of equipment and using inexpensive $CaCO_3$ or $Ca(OH)_2$ as absorbents are to be used as effective absorption of oxides of sulphur. Process by-products, mixtures of gypsum and ash, are normally utilized as retardants for cement manufacturing.

14.10.6 Wet Limestone–Gypsum Flue Gas Desulfurization System

Wet limestone–gypsum flue gas desulfurization (FGD) system removes SO_2 contained in the flue gas in contact with limestone slurry droplets as an absorbent when the flue gas containing SO_2 passes through the absorber. Limestone slurry absorbs SO_2, then it is oxidized by air at the lower part of absorber to produce calcium sulfate which is extracted from the absorber as gypsum slurry, and finally dewatered and reused in the form of gypsum powder.

Chemical reaction: $CaCO_3 + SO_2 + 2H_2O + (1/2)O_2 \rightarrow CaSO_4{\cdot}2H_2O + CO_2$

1. Regardless of SO_2 concentration, SO_2 removal efficiency of more than 90% can be achieved.
2. By-product gypsum can be reused as material of cement or wallboard.
3. Limestone, which is supplied stably, is used as an absorbent.
4. Single absorber tower, which has integrated function of pre-scrubbing, absorption, and oxidation, is adopted. Moreover, it also removes particulate in the flue gas with high removal efficiency.

14.10.7 Removal of Oxides of Nitrogen

It is relatively easy to remove SO_2 from combustion gases by dissolving SO_2 in water and reacting it with alkali. Aqueous SO_2 quickly forms sulfurous acid, which reacts with alkali and then is oxidized to sulfates. Collecting NO_x is not nearly as easy this way because NO, the principal NO_x present in combustion gas streams, has a very low solubility in water.

When gas, coal, or fuel oils are burned with air, the nitrogen in the air combines with some of the oxygen according to the reversible reaction:

$$N_2 + O_2 \rightleftharpoons 2NO$$

Whenever NO is formed, the rate of its decomposition becomes very slow under reaction conditions. It is usually formed at high temperatures, and an equilibrium concentration of approximately 2% in air is obtained at 3800°F. Unlike sulfur oxides, which quickly react with water to form acids, NO must undergo a two-step process to form an acid.

There are two possible approaches for controlling NO_x in combustion gases:

1. Modification of the combustion processes to prevent the formation of NO_x.
2. Treatment of the effluent gases to convert the NO_x to N_2.

14.10.8 Scrubbing Methods for Effluent Gas Treatment for Reduction of NO_x

The scrubbing techniques for NO_x control can be further sub-divided into (1) absorption by liquids and (2) adsorption by solids.

14.10.8.1 Absorption by Liquids

From power plants with 2000–15,000 ppm levels of NO_x in the effluent gas, the following process can be adopted for NO_x control. Further these methods can also remove SO_2:

1. Absorption by lime slurry when gypsum and nitric acid can be recovered as by-products.
2. Magnesium hydroxide scrubbing recovering concentrated NO as by-product.
3. Absorption by sulfuric acid, which produces nitric acid and sulfuric acid.
4. Absorption by lime/sodium hydroxide.

From nitric acid plants, tail gas emissions vary from 0.1 to 0.69 vol% with an average of 0.37%. Using sodium and calcium hydroxide solutions, a number of scrubbing technologies have been developed.

14.10.8.2 Catalytic Reduction

This process is usually used to treat the tail gas from nitric acid plants, which has been a nuisance for many years, especially because of the offensive reddish brown plume that is usually associated with nitric acid plants. The catalytic treatment usually uses platinum, palladium, or rhodium as catalysts

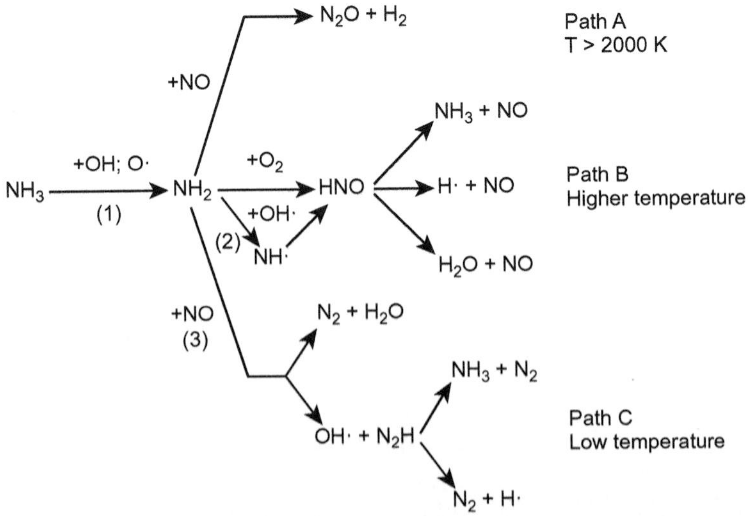

Figure 14.26 Simplified reaction mechanism of NO reduction or NO formulation with NH_3.

for speeding the reaction of the NO_x with the reducing agent, which is normally methane or natural gas. The reactions involved are:

$$CH_4 + 4NO_2 \rightarrow 4NO + CO_2 + 2H_2O \text{ (Decolorization)}$$

$$CH_4 + 4NO \rightarrow CO_2 + 2H_2O + 2N_2 \text{ (Abatement)}$$

$$CH_4 + 2O_2 \rightarrow CO_2 + 2H_2O \text{ (Combustion)}$$

Selective Catalytic Reduction for Removal of NO_x

Selective catalytic reduction (SCR) is one of the most efficient methods for removal of NO_x pollutants. Here, anhydrous aqueous ammonia, which is a selective reductant, is injected into the polluted flue gas through a bed containing the catalyst. Both ammonia and NO_x combine to form an ammonium salt intermediate by the action of the catalyst. The intermediate compound then decomposes to given elemental nitrogen and H_2O. The NO_x reactions take place at a small range of temperatures between 500 and 950°F, and the details are shown in Fig. 14.26. The type of catalyst determines this range. Operating below the optimum range will not give the energy necessary to initiate the desired reaction; while operating at a temperature above the maximum, oxidation of ammonia to either NO_x or ammonium nitrate and ammonium nitrite would take place. It is very

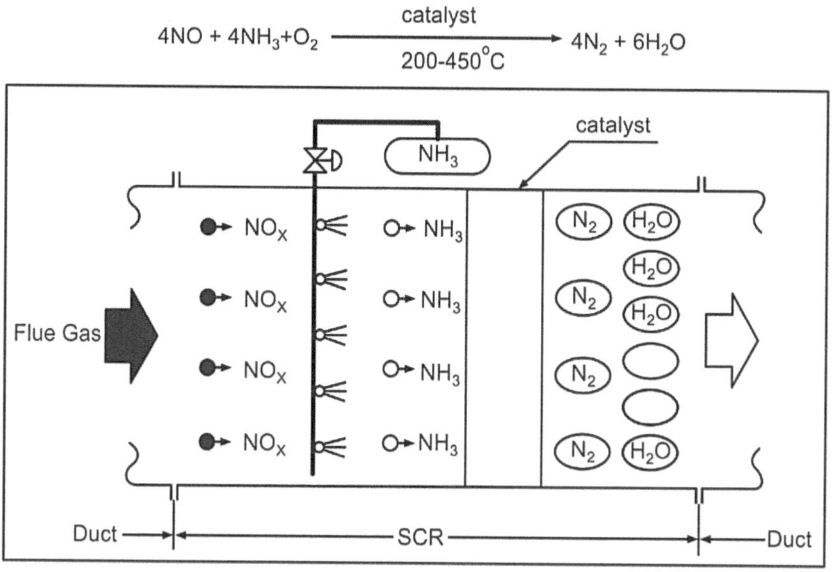

Figure 14.27 Typical configuration of selective catalytic reduction system.

important to operate in the proper temperature range. This is usually gained from pilot plant tests and experience. The flue gas from industrial boilers is usually the input feed in this type of catalytic reduction process.

Outline of Selective Catalytic Reduction System

The flue gas NO_x removal process developed by Kawasaki is a selective catalytic NO_x reduction system, in which NO_x in the flue gas is decomposed to N_2 and H_2O by NH_3, while flue gas passing through the catalyst layer as shown in Fig. 14.27.

14.10.8.3 Three-Stage Low NO_x Combustion System

1. In case of three-stage combustion system, reducing combustion is limited only to the secondary combustion zone, so that reducing atmosphere space is very narrow (Fig. 14.28). Furthermore, since measures are taken so that combustion gas with a strong reducing capability does not directly strike the water wall, adverse effects from such things as reducing corrosion and slagging on the heating surfaces inside the furnace due to reducing atmosphere, even in the case of super low NO_x combustion, are extremely small compared to other NO_x reduction systems such as Overfire Air (OFA) combustion system, which expose a large part of the furnace inside to reducing atmosphere.

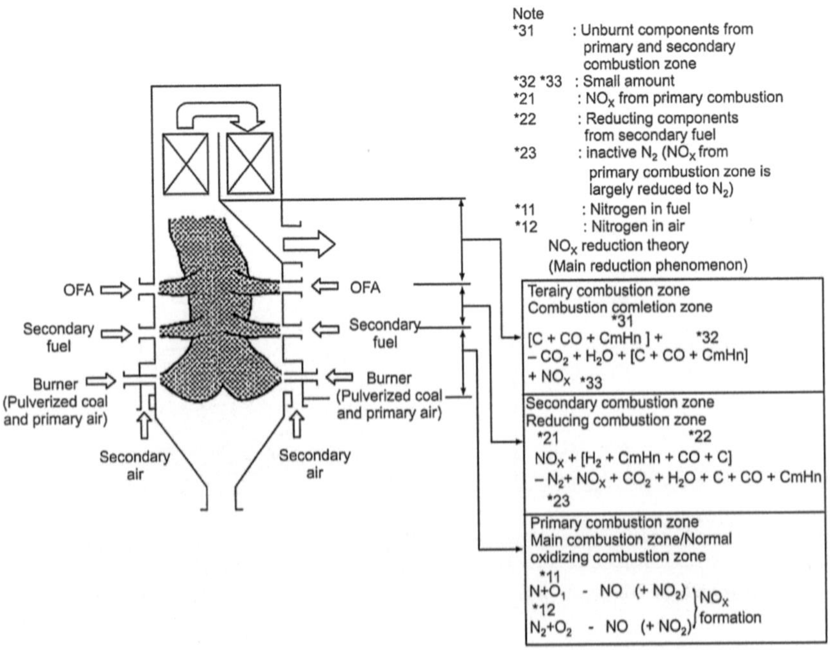

Figure 14.28 Three-stage combustion type low NO_x system.

2. Normal oxidizing combustion is performed in the primary combustion zone (main combustion zone), where most of the fuel burns and combustion efficiency is very high. The unburnt parts from the secondary fuel in the secondary combustion zone (reducing combustion zone) is burnt in the tertiary combustion zone (combustion completion zone). Consequently, combustion efficiency in a three-stage type low NO_x boiler is as high as in ordinary boilers without NO_x reduction.
3. Due to the features mentioned above, the NO_x value in a three-stage combustion type low NO_x boiler can reach to 100 ppm or less with ordinary bituminous coal. The NO_x value decreases if the combustion gas remains longer in the secondary combustion zone.

Primary Combustion Zone

In the "normal combustion type" non-low x burner, normal oxidizing combustion with an air ratio of 1.0 or over is performed. The amount of fuel here is 65–75% of total boiler fuel and the flame is raised to a very high temperature and a high combustion efficiency is obtained because the minimum combustion air needed for complete combustion is applied, and also flame length is adjusted to be as short as possible. Consequently, the flame in this zone is a bright "golden color" and good combustion can be confirmed at a glance. Generation of NO_x in the primary combustion area is not controlled at all and the aim is only to reach 100% combustion efficiency.

Secondary Combustion Zone

25–35% of the fuel for the boiler is blown in from the secondary fuel port. This secondary fuel includes air for transport but any other air is strictly excluded to improve its performance as a denitration agent. The denitration reaction is as follows:

1. The secondary fuel. Consequently, the longer the gas remains in the secondary combustion zone, the lower NO_x becomes.

Tertiary Combustion Zone

30–40% of all the air used for combustion is injected from the OFA port and the unburnt components from the secondary combustion zone are burnt. The OFA port is positioned so that it is in the ideal gas temperature zone from the viewpoint of improved combustion of the unburnt components, and for control of new generation of thermal NO_x. OFA is injected in stripes on the horizontal cross-section of the furnace, and there are a number of OFA ports so the OFA is completely mixed with the combustion gas from the secondary combustion zone. In the tertiary combustion zone, combustion efficiency increases in accordance with the increase of the space between the OFA ports and the furnace exit. However, this increases the size and weight of the boiler, so the size of this space is determined from the total cost including facility costs and running cost of the boiler.

14.10.8.4 Incineration

Incineration is widely used in industry when the fumes from a stack contain toxic pollutants, especially those with offensive odors. A wide range of odorous air pollutants are destroyed by the incineration process. Among these are mercaptans, hydrogen sulfide, cyanide gases, and hydrocarbons. Three types of incineration processes are encountered, depending on the range of temperature used to effect the desired reaction.

Flame Incineration

If the contaminated gas stream contains pollutants which approach lower limit of flammability, this stream is passed through a combustion chamber, a certain quantity of natural gas and fresh air to support combustion are fed to the chamber; the operating temperature is of the order of 2500°F or higher. Hydrocarbons are burnt to CO_2 and H_2O. The exit gas stream is odorless, and the polluted air is preheated by the countercurrent exchange of heat with the odorless stream.

Thermal Incineration

When the combustible contaminants exist in very small concentrations, and thus cannot support combustion, thermal incineration becomes more economical to use than flame incineration, since the temperature, since the temperature needed here is of the order of 1250°F. A gas burner raises the temperature of the polluted stream to 1000–1500°F, causing thermal degradation of the fumes.

VOCs are very serious air pollutants. Thermal incineration is one of the best ways to destroy them. This incineration process burns organic compounds and renders them harmless by giving off CO_2 and water. The proper length of residence time must be given so that complete combustion takes place. The temperature should be high enough so that total destruction of the toxic wastes occurs. Incinerators require residence times ranging from 0.5 to 1.0 s, with temperatures ranging from 1200 to 1600°F. In general, destruction efficiencies are above 95%.

One of the major advantages of thermal incinerators is the near total destruction of the hazardous waste without any residual waste being generated. Maintenance of the equipment and the initial cost are relatively reasonable. If the toxic stream contains halogenated compounds such as chlorine and fluorine, materials of construction must be carefully chosen because of the prevailing high corrosivity feed, materials such as stainless steel or fiber-reinforced plastics are used. Many times, a scrubber issued prior to the release of the effluent gas, especially since any halogenated acid gases must not be released into the atmosphere.

Catalytic Incineration

A catalyst is a compound which changes the rate of a chemical reaction without being changed chemically itself. The desired reactions take place at the surface of the catalyst at temperatures ranging between 600 and 900°F. The catalytic oxidation of ammonia gives NO_x, which are in turn absorbed by water to give nitric acid. SO_2 is oxidized catalytically to give SO_3, which is absorbed in weak acid to give stronger sulfuric acid. A large number of hydrocarbons are catalytically decomposed to give innocuous compounds. Sulfur and metal-bearing compounds poison the catalyst rapidly and render its use less attractive economically. Again, noxious odors are practically eliminated with this process.

Catalytic incineration destroys organic fumes and renders them harmless by giving CO_2 and water as products of combustion. Here, the catalyst plays its role in speeding up the reaction by allowing the combustion to take place at a lower temperature, around 600°F. The inlet stream of toxic wastes is heated in the recovery unit; it then moves on to a catalyst where it is converted into the harmless substances. A plate-and-frame arrangement of the catalyst is the normal way of using it. This way the toxic waste will come into good contact with the catalytic substance, where it is burned and rendered harmless at a relatively low temperature. Catalysts used are normally platinum alloys, copper chromite, or platinum itself along with some others like copper oxide, chromium, manganese, or nickel. They are usually placed on chemically inert materials like ceramics.

FURTHER READING

[1] B. Bastam, U.S. CPI start to feel effects of Bhopal tragedy, Chemical Engineering (March 18, 1985) 27–33.
[2] K.P. Checherov, Development of ideas about reasons and processes of emergency on the 4th unit of Chernobyl NPP 26.04.1986 Slavutich, Ukraine: International Conference "Shelter-98", November 25–27, 1998.
[3] T.R. Detwyler, Man's Impact on Environment, McGraw-Hill Book Co., New York, 1971.
[4] S. Draggon, J.J. Cohressen, R.E. Morrison (Eds.), Environmental Impacts on Human Health, the Agenda for Long-term Research and Development, Praeger, New York, 1987.
[5] D.M. Elsom, Atmospheric Pollution: Causes, Effects, and Control Policies, Basil Blackwell, Inc., New York, 1989.
[6] Air pollution: plant killer, Environmental Science and Technology 4 (8) (1970) 635–643.
[7] E.H. Haskell, The Politics of Clean Air, EPA Standards for Coal-Burning Power Plants, Praeger Publishers, New York, 1982.
[8] M. Heylin, India's chemical tragedy: death toll at Bhopal still rising, Chemical Engineering (December 10, 1984) 5–7.

CHAPTER FIFTEEN

Noise Pollution and Its Control

15.1 INTRODUCTION

Sources of noise are numerous but may be broadly classified into two classes such as industrial and nonindustrial. The industrial may include noises from various industries operating in cities like transportation, vehicular movement such as cars, motors, trucks, trains, tempo, motorcycles, aircraft, rockets, defense equipment, and explosions. Among the nonindustrial sources, important ones are loudspeakers, automobiles, aircraft, trains, construction work, radios, microphones, etc. The maximum permissible sound level at a worker's ears and the time of exposure are not related directly to the noise produced by any one machine but depend on the total noise in the area, where the workers are located with respect to the machine and other factors. For this reason, noise emission standards or their intent must be confined with product-oriented noise emission regulations.

15.2 SOURCES OF NOISE

The following are some of the sources of noise pollution with which we are quite familiar:

- Appliances in the home such as mixers, grinders, vacuum cleaners, washing machines, etc., together cause a cumulative sound of about 87 dB. This itself is above the sound limits in most areas. On top of that, if loudspeakers, television sets, and music systems are used with high volumes, then we can well imagine how much noise pollution is being created.
- Factories using single- or multiple-unit machines would cause a sound of about 98 dB and above. The sound will definitely go higher as the number of machines increase.
- Airplanes cause the highest sound among all: 150 dB. But road vehicles are also great contributors of noise pollution. These vehicles include the trucks, buses, tractors, SUVs, and even motorcycles and most cars.

Environmental Management
ISBN 978-0-12-811989-1
http://dx.doi.org/10.1016/B978-0-12-811989-1.00015-4

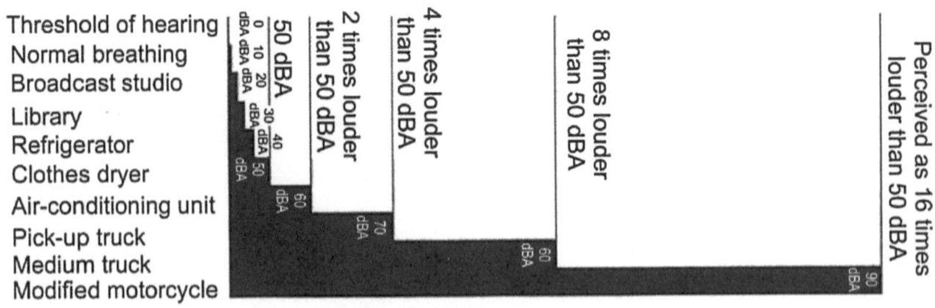

Figure 15.1 Noise levels.

- Then there are lots of environmental sources of noise pollution that cannot be ignored. Continuous noises are the most distressing. Noise coming from sources such as dripping taps and ticking of clocks can contribute to environmental noise pollution.

It has been reported that high intensities, high frequencies, and intermittent nature of noise are the factors of annoyance for workers. Such situations not only bring about physical and psychological damages but also impair workers' efficiency, giving rise to their low production and causing dissatisfaction. Community response to industrial noise and hence the setting of acceptable limits for community areas is difficult to establish precisely because of the variety and complexity of the different factors involved.

The A-weighted decibel scale begins at zero. This represents the faintest sound that can be heard by humans with very good hearing. The loudness of sounds (that is, how loud they seem to humans) varies from person to person, so there is no precise definition of loudness. However, based on many tests of large numbers of people, a sound level of 70 is twice as loud to the listener as a level of 60 (Fig. 15.1).

15.3 EFFECTS OF NOISE POLLUTION

15.3.1 The Physiology of Hearing

The physiology of our hearing mechanism can conveniently be divided into three topics:

1. The outer ear (auricle or pinna) and ear canal
2. The middle ear
3. The inner ear

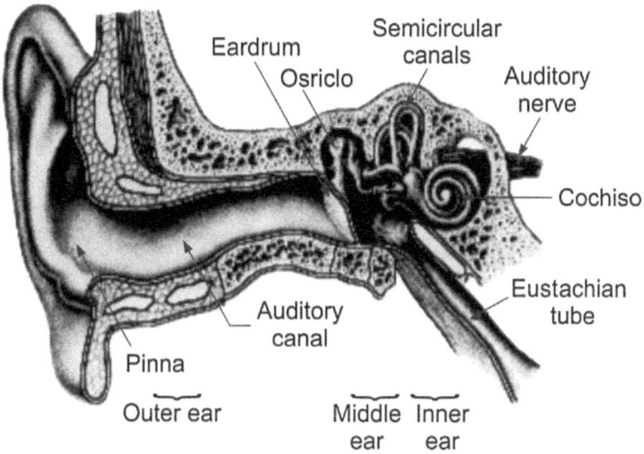

Figure 15.2 The ear canal.

15.3.1.1 The Auricle and Ear Canal

Each hole in the side of the skull leads into an ear canal. The ear canal is an irregular cylinder with an average diameter of less than 0.8 mm and is about 2.5 cm long. The ear canal (Fig. 15.2) is open at the outer end, which is surrounded by the pinna (or auricle). The pinna plays an important spatial-focusing role in hearing. The canal then narrows slightly and widens toward its inner end, which is sealed off by the eardrum. Thus the canal is a shaped tube enclosing a resonating column of air with the combination of open and closed ends. This makes it rather like an organ pipe.

The ear canal supports (resonates or enhances) sound vibrations best at the frequencies that the human ears hear most sharply. This resonance amplifies the variations of air pressure that make up sound waves, placing a peak pressure directly at the eardrum. For frequencies between approximately 2 and 5.5 kHz, the sound pressure level at the eardrum is approximately 10 times the pressure of the sound at the auricle.

15.3.1.2 The Eardrum: Interface Between Outer and Middle Ear

Airborne sound waves reach only as far as the eardrum. Here they are converted into mechanical vibrations in the solid materials of the middle ear. Sounds (air pressure waves) first set up sympathetic vibrations in the taunt membrane of the eardrum, just as they do in the diaphragm of some types of microphones. The eardrum passes these vibrations on to the middle ear structure.

Although we recognize noise pollution as a major environmental problem, it is difficult to quantify the effects it has on human health. Exposure to excessive noise has been shown to cause hearing problems, stress, poor concentration, productivity losses in the workplace, communication difficulties, fatigue from lack of sleep, and a loss of psychological well-being.

15.4 EFFECTS ON HEALTH

Noise pollution can take a severe toll on human health in the long run. These effects will not become apparent immediately, but there could be repercussions later on.

The following is a list of the kinds of effects noise pollution will have on human health after continuous exposure for months, and even years:

- The most immediate effect is a deterioration of mental health. As an example, people who are living too close to airports will probably be quite jumpy. Continuous noise can create panic episodes in a person and can even increase frustration levels. Also, noise pollution is a big deterrent in focusing the mind to a particular task. Over time, the mind may just lose its capacity to concentrate on things.
- Another immediate effect of noise pollution is a deterioration of the ability to hear things clearly. Even on a short-term basis, noise pollution can cause temporary deafness. But if the noise pollution continues for a long period of time, there is a danger that the person might go permanently deaf.
- Noise pollution also takes a toll on the heart. It is observed that the rate at which heart pumps blood increases when there is a constant stimulus of noise pollution. This could lead to side-effects like elevated heartbeat frequencies, palpitations, breathlessness, and the like, which may even culminate into seizures.
- Noise pollution can cause dilation in the pupils of the eye, which could interfere in ocular health in the later stages of life.
- Noise pollution is known to increase digestive spasms. This could be the precursor of chronic gastrointestinal problems.
- Noise can awaken people from sleep, and it can keep them awake, frequently awakening, or awakening for long periods, which can be very disruptive. Even if not awakened by noise, a person's sleep pattern can be significantly disturbed, and a reduced feeling of well-being can result the next day. Frequent and prolonged sleep disturbances can result in physical, mental, or emotional illness.

- External sounds are able to interfere with conversations and use of the telephone as well as the enjoyment of radios, television programs, and like pastimes. It can thus effect the efficiency of offices, schools, and other place where communication has been of vital importance. The maximum acceptable level of noise under such conditions has been 55 dB. 70 dB is considered very noisy, and serious interference with verbal communication is inevitable.

Table 15.1 lists the effects of high intensity noise on human beings.

Noise hazards are classified into several stages based on the quantum of impact they cause. Table 15.2 lists some of the health issues and the quantum of impact they cause.

Table 15.1 Effects of High Intensity Noise on Human Beings

Noise (dB)	Effects Observed
0	Threshold of audibility
150	Significance change in pulse rate
110	Stimulation of reception in skin
120	Pain threshold
130–135	Nausea, vomiting, dizziness, interference with touch and muscle sense
140	Pain in ear, extreme limit of human noise tolerance
150	Prolonged exposure causing burning of skin
160	Minor permanent damage if prolonged
190	Major permanent damage in a short time

Table 15.2 Health Issues Related to Noise Pollution

A. Noise Hazards	
Stage: I **Threat to survival** 1. Communication interference 2. Permanent hearing loss	Stage: II **Causing injury** 1. Neural-humoral stress response 2. Temporary hearing loss
B. Noise Nuisances	
Stage: III **Curbing efficient performance** 1. Mental stress 2. Task interference 3. Sleep interference	Stage: IV **Diluting comfort and enjoyment** 1. Invasion of privacy 2. Disruption of social interaction 3. Hearing loss

Basic noise levels for an industrial zone should not exceed 55 dB at night and 65 dB during the daytime. Noise contributes to development of cardiovascular problems like heart diseases and high blood pressure. Workers exposed to high noise levels are having more circulatory problems, cardiac disturbances, neuro-sensory, motor impairment, and even more social conflicts at home and at work.

15.4.1 Physiological Responses

Physiological responses accompanying a response and other noise exposures include:

1. a vascular response characteristic by peripheral vaso-constriction, changes in heart rate and blood pressure,
2. various glandular changes such as increased output of adrenaline evidenced as chemical changes in blood during circulation,
3. slow, deep breathing,
4. a change in the electrical resistance of skin with changes in activity of the sweat glands,
5. brief changes in skeletal muscle tension.

According to environmentalist Thomsa G. Ayles Worth, "constant noise may cause our blood-vessels to contract, our skin to become pale, our muscles to contract and adrenaline to be shot into our blood stream." This adrenaline is responsible for both excretory and inhibitory responses in living beings. This is the reason that factory workers develop abnormal heartbeat rates and suffer from insomnia, nervousness, and impaired motor coordination. The US Government has kept 90 dB as a health hazard for an 8-h-day working environment. It has been proved that high noise is bad particularly for those suffering from hypertension and diabetics. Noise also produces startling effects on babies, and they may even develop a fear psychosis as a result of sharp and sudden noise.

15.4.2 Effects on Communication

External sounds are able to interfere with conversations and use of the telephone as well as the enjoyment of radios, television programs, and like pastimes. It can thus effect the efficiency of offices, schools, and other place where communication has been of vital importance. The maximum acceptable level of noise under such conditions has been 55 dB. 70 dB is considered very noisy and serious interference with verbal communication is inevitable.

15.4.3 Epidemiological Studies

Several researchers have conducted field studies testing industrial workers and/or collating their health records in an attempt or to overcome the limitations of duration and realism in laboratory studies. The difference between very noisy industries and less noisy industries is the two groups indicate a higher incidence of problems among the high noise group than by the low noise group. The high noise group came from the light industries such as textile industries. There are other numerous differences that could have effects on health, such as heat, physical work load, anxiety, and the type of people.

Noise exposure can cause two kinds of health effects: **non-auditory effects** and **auditory effects**. Non-auditory effects include stress, related physiological and behavioral effects, and safety concerns. Auditory effects include hearing impairment resulting from excessive noise exposure. Noise-induced permanent hearing loss is the main concern related to occupational noise exposure.

15.4.4 Auditory Health Effects

The main auditory effects include these:

Acoustic trauma: sudden hearing damage caused by short burst of extremely loud noise such as a gunshot,

Tinnitus: ringing or buzzing in the ear,

Temporary hearing loss: also known as temporary threshold shift, which occurs immediately after exposure to a high level of noise; there is gradual recovery when the affected person spends time in a quiet place, and complete recovery may take several hours,

Permanent hearing loss: Permanent hearing loss, also known as permanent threshold shift (PTS), progresses constantly as noise exposure continues month after month and year after year. The hearing impairment is noticeable only when it is substantial enough to interfere with routine activities. At this stage, a permanent and irreversible hearing damage has occurred. Noise-induced hearing damage cannot be cured by medical treatment and worsens as noise exposure continues.

When noise exposure stops, the person does not regain the lost hearing sensitivity. As the employee ages, hearing may worsen as "age-related hearing loss" adds to the existing noise-induced hearing loss.

15.4.5 Characteristics of Noise-Induced Permanent Hearing Loss

The main characteristics of noise-induced hearing loss are these:

- Noise-induced hearing loss is a cumulative process: both level of noise and exposure time over a worker's work history are important factors.
- At a given level, low-frequency noise (below 100 Hz) is less damaging compared to noise in the mid-frequencies (1000–3000 Hz).
- Noise-induced hearing loss occurs randomly in exposed persons.
- Some individuals are more susceptible to noise-induced hearing loss than others.
- In the initial stages, noise-induced hearing loss is most pronounced at 4000 Hz, but it spreads over other frequencies as noise level and/or exposure time increases.

Hearing sensitivity declines as people become older. This medical condition is called presbycusis. Age-related hearing loss adds to noise-induced hearing loss. Hearing ability may continue to worsen even after a person stops working in a noisy environment. Noise affects the hearing organs (cochlea) in the inner ear. That is why noise-induced hearing loss is a sensory-neural type of hearing loss. Certain medications and diseases may also cause damage to the inner ear resulting in hearing loss. Generally, it is not possible to distinguish sensory-neural hearing loss caused by exposure to noise from sensory-neural hearing loss due to other causes. Medical judgment, in such cases, is based on the noise exposure history. Workers in noisy environments who are also exposed to vibration (e.g., from a jack hammer) may experience greater hearing loss than those exposed to the same level of noise but not to vibration. Some chemicals are ototoxic; that is, they are toxic to the organs of hearing and balance or the nerves that go to these organs. This means that noise-exposed workers who are also exposed to ototoxic chemicals (e.g., toluene and carbon disulfide) may suffer from more hearing impairment than those who have the same amount of noise exposure without any exposure to ototoxic chemicals.

15.4.6 Measurement of Hearing Loss

Hearing loss is measured as a threshold shift in dB units using an audiometer. The 0 dB threshold shift reading of the audiometer represents the average hearing threshold level of an average young adult with disease-free ears. The PTS, as measured by audiometry, is the decibel-level of sounds of different frequencies that are just barely audible to that individual. A positive threshold shift represents hearing loss, and a negative threshold shift means better than average hearing when compared with the standard. Several methods of calculating the percentage of hearing disability are in use. The American Medical Association (AMA)/American Academy of Otolaryngology (AAO) formula is widely accepted in North America. The current method recommended by AMA/AAO is as follows:

1. The average hearing threshold level at 500, 1000, 2000, and 3000 Hz should be calculated for each ear.
2. ultiplying, one should calculate the percentage of impairment for each ear (the monaural loss) as 1.5 times the amount by which the aforementioned average exceeds 25 dB (low fence). Hearing impairment is 100% for the 92-dB average hearing threshold level.
3. The hearing disability (binaural assessment) is calculated by multiplying the smaller percentage (better ear) by 5, adding it to the larger percentage (poorer ear), and dividing the total by 6.

15.4.7 Relationship Between Noise Exposure and Hearing Loss

From the scientific data accumulated to date, it is possible to determine the risk of hearing loss among a group of noise-exposed persons. To do this we need the following data:

- a measure of daily noise exposure level,
- duration of noise exposure (months, years),
- age of person,
- Hearing loss is defined as average threshold shift at 500, 1000, 2000, and 3000 Hz (Fig. 15.3).

15.5 INDUSTRIAL NOISE

No environmental factor has caused so much confusion regarding its effect on workers efficiency and workers health as industrial noise. Noise in industry originates from processes causing impact, vibration or reciprocation movements, friction, and turbulence in air or gas streams. Noise emission standards have only an indirect control over the noise radiated by a machine. They state maximum permissible sound levels in

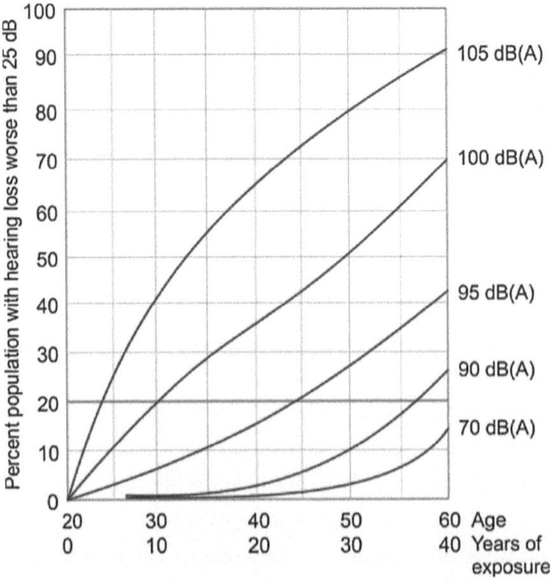

Figure 15.3 Relation between exposure and hearing loss.

work places; acceptable daytime and nighttime levels in residential, commercial, and industrial areas; and maximum permissible noise crossing industrial and construction site boundaries. The maximum permissible sound level at a worker's ears and the time of exposure are not related directly to the noise produced by any one machine but depend on the total noise in the area, where the workers are located with respect to the machine, and other factors. For this reason, noise emission standards or their intent must be confined with product-oriented noise emission regulations. It has been reported that high intensities, high frequencies, and the intermittent nature of noise are the factors of annoyance for the workers. Such a situation not only brings about physical and psychological damages but also impairs workers' efficiency, giving rise to their low production and causing dissatisfaction. Community response to industrial noise and hence the setting of acceptable limits for community areas is difficult to establish precisely because of the variety and complexity of the different factors involved.

The measured noise in such cases may be produced by a single machine or by a combination of many kinds of machinery. Fig. 15.4(A) shows the interrelationship between the elements of noise. Fig. 15.4(B) shows the various paths for the management of noise.

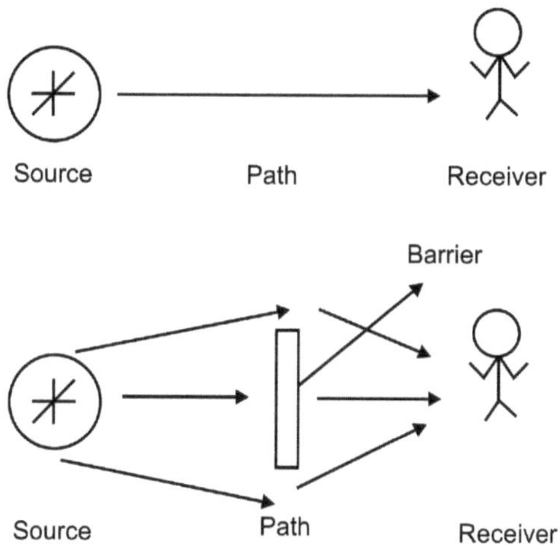

Figure 15.4A Interrelation between elements of noise.

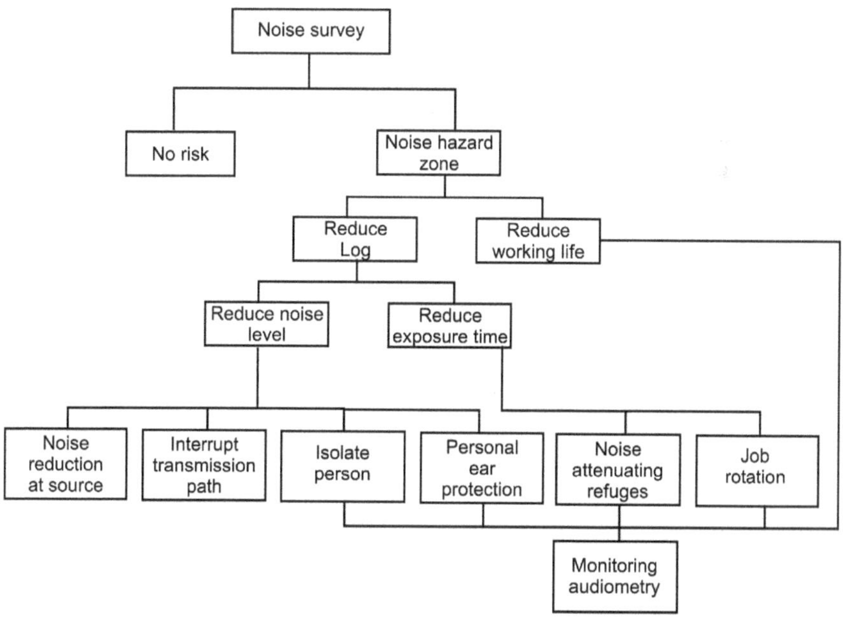

Figure 15.4B Management of noise.

15.6 NOISE SOURCE FROM TRANSPORTATION SECTOR

1. **Aircraft noise**: The noise spectra of a wide-body fan jet reveal that sound pressure levels are higher on takeoff than during the approach to land. This is typical of all aircraft. The annoyance criteria for aircraft operations are based on extensive field measurements.
2. **Highway vehicle noise**: The level of highway traffic noise depends on three things: (1) the volume of the traffic, (2) the speed of the traffic, and (3) the number of trucks often in the flow. Generally, the loudness of traffic noise is increased by heavier traffic volumes, higher speeds, and greater numbers of trucks. Vehicle noise is a combination of the noises produced by the engine, exhaust, and tires: the loudness of traffic noise can also be increased by defective mufflers or other faulty equipment on vehicles. Any condition (such as a steep incline) that causes heavy laboring of motor vehicle engines will also increase traffic noise levels. In addition, there are other more complicated factors that affect the loudness of traffic noise. For example, as a person moves away from a highway, traffic noise levels are reduced by distance, terrain, vegetation, and natural and manmade obstacles. Traffic noise is not usually a serious problem for people who live more than 500 ft from heavily traveled freeways or more than 100–200 ft from lightly traveled roads.

For most automobiles, exhaust noise constitutes the predominant source for normal operation below about 55 km/h. Although tire noise is much less of a problem in automobiles than trucks, it is the dominant noise source at speeds above 80 km/h. While not as noisy as trucks, the total contribution of automobiles to the noise environment is significant because of the large number of vehicles in operation. Diesel trucks are 8–10 dB noisier than gasoline-powered ones. At speeds above 80 km/h, tire noise is the source on trucks. The "crossbar" tread is the noisiest. Motorcycle noise is highly dependent on the speed of the vehicle. The primary source of noise is the exhaust. The noise spectra of two-cycle and four-cycle engines are of somewhat different character. The two-cycle engines exhibit more frequency spectra energy content:

1. based on estimates of the total number of units in operation per day
2. equivalent level for evaluation of relative hearing damage risk
3. during engine trimming operation.

Table 15.3 Summary of Noise Characteristics of Internal Combustion Engines

Source	A-Weighted Noise Energy (kWh/day)[a]	Typical A-Weighted Noise Level at 15.2 m [dB(A)]	8-h Exposure Level [dB(A)][b] Average	8-h Exposure Level [dB(A)][b] Maximum	Typical Exposure Time (h)
Lawn mowers	63	74	74	82	1.5
Garden tractors	63	78	N/A	N/A	N/A
Chain saws	40	82	85	95	1
Snow blowers	40	84	61	75	1
Lawn edgers	16	78	67	75	½
Model aircraft	12	78	70[c]	79[c]	¼
Leaf blowers	3.2	76	67	75	¼
Generators	0.8	71	—	—	—
Tillers	0.4	70	72	80	1

[a]Based on estimates of the total number of units in operation per day.
[b]Equivalent level for evaluation of relative hearing damage risk.
[c]During engine trimming operation.

15.6.1 Sources of Traffic Noise

At low speeds, vehicle engines, transmissions, exhausts, and brakes cause most traffic noise. The stop–start braking and acceleration during peak-hour congestion also increases noise levels. On freeways where speeds are high and relatively constant, most noise is caused by a combination of tire contact with the road and aerodynamic drag over the vehicle.

Trucks and motorcycles combine to make up 7% of vehicles on our roads, but they are largely responsible for the peak noises that stand out from the steady background rumble. It is these sharp and intermittent noises that are more likely to cause sleep disturbances and to contribute to other physical and psychological problems. The noise levels of internal combustion engines of various sources are given in Table 15.3.

15.7 THE NOISE POLLUTION (REGULATION AND CONTROL) RULES, 2000, IN INDIA

In exercise of the powers conferred by clause (ii) of subsection (2) of section 3, subsection (1) and clause (b) of subsection (2) of section 6 and section 25 of the Environment (Protection) Act, 1986 (29 of 1986) read with rule 5 of the Environment (Protection) Rules, 1986, the central government hereby makes the rules for the regulation and control of noise producing and generating sources, namely:

1. Short-Title and Commencement:

(1) These rules may be called the Noise Pollution (Regulation and Control) Rules, 2000.
(2) They shall come into force on the date of their publication in the Official Gazette.

2. Definitions: In these rules, unless the context otherwise requires,

(a) "Act" means the environment (protection) Act, 1986 (29 of 1986);
(b) "Area/zone" means all areas which fall in either of the four categories given in the schedule annexed to these rules;
(c) "Authority" means and includes any authority or officer authorized by the central government, or as the case may be, the state government in accordance with the laws in force and includes a district magistrate, police commissioner, or any other officer not below the rank of the deputy superintendent of police designated for the maintenance of the ambient air quality standards in respect to noise under any law for the time being in force;
(d) "Court" means a governmental body consisting of one or more judges who sit to adjudicate disputes and administer justice and includes any court of law prescribed over by a judge, judges, or a magistrate and acting as a tribunal in civil, taxation, and criminal cases;
(e) "Educational institution" means a school, seminary, college, university, professional academies, training institutes, or other educational establishment, not necessarily a chartered institution and includes not only buildings, but also all grounds necessary for the accomplishment of the full scope of educational institution, including those things essential to mental, moral, and physical development;
(f) "Hospital" means an institution for the reception and care of sick, wounded, infirm, or aged persons, and it includes government or private hospitals, nursing homes, and clinics.
(g) "Person" shall include any company or association or body of individuals, whether incorporated or not.
(h) "State government" in relation to a union territory means the administrator thereof appointed under article 239 of the constitution.

3. Ambient air quality standards in respect to noise for different areas/zones:

(1) The ambient air quality standards in respect to noise for different areas/zones shall be such as specified in the schedule annexed to these rules.
(2) The state government shall categorize the areas into industrial, commercial, residential, or silence areas/zones for the purpose of implementation of noise standards for different areas.
(3) The state government shall take measures for abatement of noise including noise emanating from vehicular movements and ensure that the existing noise levels do not exceed the ambient air quality standards specified under the rules.
(4) All development authorities, local bodies, and other concerned authorities while planning developmental activity or carrying out functions relating to town and country planning shall take into consideration all aspects of noise pollution as a parameter of quality of life to avoid noise menace and to achieve the objective of maintaining the ambient air quality standards in respect of noise.
(5) An area comprising not less than 100 meters around hospitals, educational institutions, and courts may be declared as a silence area/zone for the purpose of these rules.

4. Responsibility as to enforcement of noise pollution control measures:

(1) The noise levels in any area/zone shall not exceed the ambient air quality standards in respect to noise as specified in the schedule.
(2) The authority shall be responsible for the enforcement of noise pollution control measures and the due compliance of the ambient air quality standards in respect to noise.

5. Restriction on the use of loudspeakers/public address system:

(1) A loudspeaker or a public address system shall not be used except after obtaining written permission from the authority.
(2) A loudspeaker or a public address system shall not be used at night (between 10:00 p.m. and 6:00 a.m.) except in closed premises for communication within, e.g., auditoria, conference rooms, community halls, and banquet halls.
(3) Notwithstanding anything contained in sub-rule (2), the state government may subject to such terms and conditions as are necessary to reduce noise pollution a permit for the use of loud speakers or public address systems during night hours (between 10:00 p.m. and 12:00 midnight) on or during any cultural or religious festive occasion of a limited duration not exceeding 15 days in all during a calendar year (*added vide S.O. No. 1088(E) dated 11th October, 2002*).

6. Consequences of any violation in silence zone/area:

Whoever, in any place covered under the silence zone/area, commits any of the following offenses shall be liable for penalty under the provision of the Act:

(i) whoever plays any music or uses any sound amplifiers,

(ii) whoever beats a drum or tom–tom or blows a horn either musical or pressure, or trumpet or beats, or sounds any instrument,

(iii) whoever exhibits any mimetic, musical, or other performances of a nature to attract crowds.

7. Complaints to be made to the authority:

(1) A person may, if the noise level exceeds the ambient noise standards by 10 dB(A) or more given in the corresponding columns against any area/zone, make a complaint to the authority.

(2) The authority shall act on the complaint and take action against the violator in accordance with the provisions of these rules and any other law in force.

8. Power to prohibit, etc., continuance of music sound or noise:

(1) If the authority is satisfied from the report of an officer in charge of a police station or other information received by him/her that it is necessary to do so to prevent annoyance, disturbance, discomfort, or injury or risk of annoyance, disturbance, discomfort, or injury to the public or to any person who dwells or occupies property on the vicinity, he/she may by a written order issue such directions as he/she may consider necessary to any person for preventing, prohibiting, controlling, or regulating the following:

   (a) the incidence or continuance in or upon any premises of

      (i) any vocal or instrumental music,

      (ii) sounds caused by playing, beating, clashing, blowing or use in any manner whatsoever of any instrument including loudspeakers, public address systems, appliance, or apparatus or contrivance that is capable of producing or re-producing sound

   (b) the carrying on in or upon any premises of any trade, avocation, or operation or process resulting in or attended with noise.

(2) The authority empowered under sub-rule (1) may, either on its own motion or on the application of any person aggrieved by an order made under sub rule (1), either rescind, modify, or alter any such order: Provided that before any such application is disposed of, the said authority shall afford to the applicant an opportunity of appearing before it either in person or by a person representing him/her and showing cause against the order and shall, if it rejects any such application either wholly or in part, record its reasons for such rejection.

15.8 GENERAL NOISE CONTROL

15.8.1 Source Path Receiver Concept

It you have a noise problem and want to solve it, you have to find out something about what the noise is doing, where it comes from, how it travels, and what can be done about it. A straightforward approach is to examine the problem in terms of its three basic elements:

1. Sound arises from a source.
2. Travels over a path.
3. Affects a receiver or listener.

The source may be one or any number of mechanical devices that radiate noise or vibratory energy. Such a situation occurs when several appliances or machines are in operation at a given time in a home or office.

The most obvious transmission path by which noise travels is simply a direct line-of-sight air path between the source and the listener. Noise can travel from one point to another via any one path or a combination of several paths. The receiver may be, for example, a single person or a group of people.

Solution of a given problem might require alternation or modification of any or all these three basic elements.

1. Modifying the source to reduce its noise output,
2. Altering or controlling the transmission path and the environment to reduce noise level reaching the listener,
3. Providing the receiver with personal equipment.

15.8.2 Control of Noise Source by Design

15.8.2.1 Reduce Impact Factors

Many machines and items of equipment are designed with parts that strike forcefully against other parts, producing noise. Often, this striking action or impact is essential to the machine function. Several steps can be taken to reduce noise from impact forces. The particular remedy to be applied will be determined by the nature of the machine in question.

Some of the obvious design modifications are as follows:

1. Reduce the weight, size, or height of fall of the impacting mass.
2. Cushion the impact by inserting a layer of shock-absorbing material between the impacting surfaces.
3. Whenever practical, one of the impact heads or surfaces should be made of nonmetallic material to reduce resonance.
4. Substitute the application of a small impact force over a long time period for a large force over a short period to achieve the same result.

5. Smooth out acceleration of moving parts by applying accelerating forces gradually. Avoid high, jerky acceleration or jerky motion.
6. Minimize overshoot, backlash, and loose play in cams, followers, gears, and other parts.

15.8.2.2 Reduce Speeds and Pressures

Reducing the speed of rotating moving parts in machines and mechanical systems results in smoother operation and lower noise output. Likewise, reducing pressure and flow velocities in air, gas, and liquid circulation systems lessens turbulence, resulting in decreased noise radiation. The following suggestions can be implemented:

1. Fans, impellers, rotors, turbines, and blowers should be operated at the lowest blade tip speeds that will still meet job needs.
2. All other factors being equal, centrifugal, squirrel-cage type fans are less noisy than vane, axial, or propeller type fans.
3. In air ventilation systems, a 50% reduction in the speed of the air flow may lower the noise output by 10–20 dB, or roughly one-quarter to one-half of the original loudness.

15.8.2.3 Reduce Frictional Resistance

Reducing friction between rotating, sliding, or moving parts in mechanical systems frequently results in smother operation and lower noise output. The other points to be checked include these:

1. *Alignment*: Proper alignment of all rotating, moving, or contacting parts results in less noise output.
2. *Polish*: Highly polished and smooth surfaces between sliding, meshing, or contacting parts are required for quiet operation, particularly where bearings, gears, cams, rails, and guides are concerned.
3. *Balance*: Static and dynamic balancing of rotating parts reduces frictional resistance and vibration, resulting in lower noise output.
4. *Eccentricity*: Off-centering of rotating parts such as pulleys, gears, rotors, and shaft/bearing alignment causes vibration and noise.

15.8.2.4 Reduce Radiation Area

Generally speaking, the larger the vibrating part or surface is, the greater the noise output will be. The rule of thumb for quiet machine design is to minimize the effective radiating surface areas of the parts without impeding their operation or structural strength. This can be done by making parts smaller, removing excess material, or cutting openings, slots, or perforations in the parts.

15.8.2.5 Reduce Noise Leakage

In many cases, machine cabinets can be made into rather effective soundproof enclosures through simple design changes and the application of some sound-absorbing treatment.

15.8.2.6 Isolate and Damper Vibrating Elements

Generally, vibration problems can be considered in two parts. First, we must prevent energy transmission between the source and surfaces that radiate the energy. Second, we must dissipate or attenuate the energy somewhere in the structure. The first part of the problem is solved by *isolation*. The second part is solved by *damping*.

15.8.2.7 Provide Mufflers/Silencers

There is no real distinction between mufflers and silencers. They are often used interchangeably. They are, in effect, acoustical filters and are used when fluid flow noise is to be reduced. The devices can be classified into two fundamental groups: *adsorptive mufflers* and *reactive mufflers*.

15.8.3 Noise Control in the Transmission Path

The next method is to set up devices in the transmission path to block or reduce the flow of sound energy before it reaches your ears. This can be done in several ways:

1. absorb the sound along the path,
2. deflect the sound in some other direction by placing a reflecting barrier in its path,
3. contain the sound by placing the source inside a sound-insulating box or enclosure.

15.8.3.1 Separation

The use of the absorptive capacity of the atmosphere can be made use of as well as divergence, as a simple, economical method of reducing the noise level. Air absorbs high-frequency sounds more effectively than it absorbs low-frequency sounds.

If we can double the distance from the point source, we will succeed in lowering the sound pressure level by 6 dB.

15.8.3.2 Absorbing Materials

Noise, like light, will bounce from one hard surface to another. In noise control work, this is called *reverberation*. Sound absorbing materials are rated either by their *Sabin absorption coefficients* (α_{SAB}) at 125, 500, 1000, 2000, and 4000 Hz or by a single number rating called the *noise reduction coefficient*. Sound absorbing materials such as acoustic tile, carpets, and drapes placed on ceilings, floors, or wall surfaces can reduce the noise level in most rooms by about 5–10 dB for high-frequency sounds, but only by 3 or 3 dB for low-frequency sounds.

15.8.3.3 Acoustic Lining

Noise transmitted through ducts, pipes chases, or electrical channels can be reduced effectively by lining the inside surfaces of such passageways with sound-absorbing materials. A comparable degree of noise reduction for the lower frequency sounds is considerably more difficult to achieve because it usually requires at least a doubling of the thickness and/or length of acoustic treatment.

15.8.3.4 Barriers and Panels

Placing barriers, screens, or deflectors in the noise path can be an effective way of reducing noise transmission, provided that the barriers are large enough in size, and depending upon whether the noise is high frequency or low frequency. High-frequency noise is reduced more effectively than low-frequency noise. The effectiveness of a barrier depends on its location, its height, and its length. Fig. 15.5 shows the frequencies and the center frequencies of the octave band. The barrier may be either close to the source or receiver, subject to the condition of $R \ll D$, or in other words, to increase the traverse length for the sound wave. It should also be noted that the presence of the barrier itself can reflect sound back toward the source. At very large distances, the barrier becomes less effective because of the possibility of refractive atmospheric effects.

15.8.3.5 Transmission Loss

When the position of the noise source is very close to the barrier, the diffracted noise is less important than the transmitted noise. If the barrier is in fact a wall panel that is sealed at the edges, the transmitted noise is the only one of concern. The ratio of the sound energy incident on one surface of a panel to the energy radiated from the opposite surface is called the *sound transmission loss* (TL). The actual energy loss is partially reflected and partially absorbed.

15.8.3.6 Enclosures

Sometimes, it is much more practical and economical to enclose a noisy machine in a separate room or box than to quiet it by altering its design,

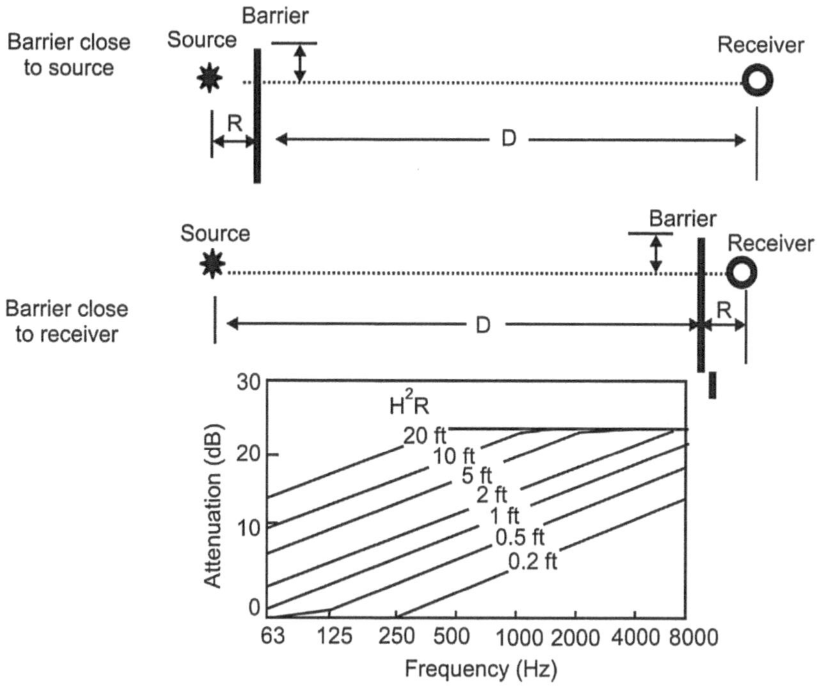

Figure 15.5 Attenuation of noise levels using barriers.

operation, or parts. The walls of the enclosure should be massive and airtight to contain the sound. Absorbent lining on the interior surfaces of the enclosure will reduce the reverberant build-up of noise within it. Structural contact between the noise source and the enclosure must be avoided, or else the source vibration will be transmitted to the enclosure walls and, thus, short-circuit the isolation.

15.8.3.7 Control of Noise Source by Redress

The best way to solve noise problems is to design them out of the source. Hence, we are frequently faced with an existing source that, either because of age or poor design, is a noise problem. The result is that we must redress, or correct the problem, as it currently exists.

Balance Rotating Parts

One of the main source of machinery noise is structural vibration caused by the rotation of poorly balanced parts, such as fans, fly wheels, pulleys, cams, and shafts. Measures used to correct this condition involve the addition of counterweights to the rotating unit or the removal of some weight from the unit.

Reduce Frictional Resistance

A well-designed machine that has been poorly maintained can become a serious source of noise. General cleaning and lubrication of all rotating, sliding, or meshing parts at contact points should go a long way toward fixing the problem.

Apply Damping Materials

Since a vibrating body or surface radiates noise, the application of any material that reduces or restrains the vibrational motion of that body will decrease its noise output. The basic types of redress vibration damping materials are available:

1. liquid mastics, which are applied with a spray gun and harden into relatively solid materials, the most common being automobile "undercoating,"
2. pads of rubber, felt, plastic foam, leaded vinyls, adhesive tapes, or fibrous blankets, which are glued to the vibrating surface,
3. sheet metal viscoelastic laminates or composites, which are bonded to the vibrating surface.

Seal Noise Leaks

Small holes in an otherwise noise tight structure can reduce the effectiveness of the noise control measures. If the designed TL of an acoustical enclosure is 40 dB, an opening that comprises only 0.1% of the surface area will reduce the effectiveness of the enclosure by 10 dB.

Perform Routine Maintenance

Regular maintenance is required. For example, studies of automobile tire noise in relation to pavement roughness show that maintenance of the pavement surface is essential to keep noise at minimum levels. Normal road wear can yield noise increases on the order of 6 dB.

Protect the Receiver

When all else fails: When exposure to intense noise fields is required and none of the measures are practical, then the following two techniques are commonly employed to limit noise.

Alter work schedule: Limit the amount of continuous exposure to high noise levels. In terms of hearing protection, it is preferable to schedule an intensely noisy operation for a short interval of time each day over a period of several days rather than a continuous 8-h run for a day or two. Inherently noisy operations such as street repair, municipal trash collection, factory operation, and aircraft traffic should be curtailed at night and early morning hours to avoid disturbing the sleep of the community.

Ear protection: Molded and pliable earplugs, cup type protectors, and helmets are commercially available as hearing protectors. Such devices may provide noise reductions ranging from 15 to 35 dB. Earplugs are effective only if they are properly fitted by medical personnel. These devices should be used as a last resort after all other methods have failed to lower the noise level to acceptable limits.

15.9 CONTROL OF NOISE FROM INDUSTRY

Acoustic design criteria: One of the first decisions that should be made before starting a noise reduction program on a machine has been to establish an acoustic design goal this should be slightly lower, because the cost of noise reduction rapidly as design goals are lowered.

They must be set somewhat below what is actually needed, however, to allow for errors in measurement, unexpected acoustic leaks, differences in quality of acoustic materials, and if a group of machines is involved, to allow for difference between machines. Even though the objective may be to meet a certain noise omission standard or to comply with the same community noise ordinance, the design criteria must be either the sound power level of the machine or the sound pressure level at a fixed distance from the machine.

The main feature of effective noise control has been the provision of adequate sound insulation between the noise source and the desired environment. Sound-insulating materials must be impervious and of high density as the insulation value increases by about 5 dB for each doubling of the weight according to a relationship known as mass law. The maximum attenuation is expected from a barrier that has been about 15 dB, although in most industrial applications, reductions would be much lower than this. The attenuation requirement for a majority of industrial noise problems will usually be in the range of 15–25 dB. This order of attenuation on machines can only be attained by complete enclosure if only insulation principles have been used. Attention to detail in the design, construction, and erection of an enclosure is extremely important as any gaps or leaks severely limit the insulation potential.

1. **Screen the receiver**: In most of the industries, noise is an essential part of their machines used, and if a worker is continuously exposed to sound hazards for a long period, he may suffer from annoyance, loss of efficiency, and damage to hearing. It has become necessary that employees be given ear defenders or earplugs to protect them from such losses.

2. **Internal design changes**: Centrifugal compressor noise can be controlled to a limited extent by internal design changes, but large, high-speed high-horsepower machines require additional external sound control when they must meet low noise criteria. Noise reduction can be accomplished by selecting the proper combination of blades and by controlling clearance between rotating and stationary vanes.
3. **External design techniques**: The EPA has designated profitable air compressors as a major noise source. Their noise can be reduced by external design techniques. It would be a waste of time and money to attempt reduction by internal design changes in compressor because the driving internal combustion engine is the major noise source. The situation will probably remain this way until quieter engines can be obtained. Small compressors can be reduced to acceptable sound levels without much trouble, but the large units require complete enclosures including sound absorption and sound isolation.
4. **Silencers**: Generally speaking, two types of silencers are known
   a. reactive
   b. absorptive

   The reactive type has been analogous to an electrical jetter. It is frequently selective, and it does not absorb energy but either transmits or reflects it back to the source depending on the frequency of sound. Most reactive silencers for industrial applications have been high-pass filters that alternate low-frequency noise only; the main reason for this has been that the dimensions of the reactive elements must be small compared with the wavelength of sound to be alternated, and high-frequency attenuation is more readily achieved by absorptive means.
5. **Vibration isolation**: A vibration machine could be isolated from the surrounding structure by supporting it on resistant mountings. This then may be associated as a simple mass spring.
6. **Miscellaneous methods**: Impact and vibration noises are considerably reduced by the mass, careful design of shape, and arrangement of parts and machines so that resonance is avoided. Nevertheless certain machines will remain inherently noise, and demand to be surrounded with absorbent or insulating screens. Noise caused by gas streams can be alternated or even eliminated by the use of suitable ducts and by correct design and positioning of inlets and outlets.

The Bureau of Indian Standards (BIS) has published several code books for sampling and analysis of noise pollution and guidelines for control of noise pollution from domestic and industrial sources. The reader is advised to refer to the BIS code books (Table 15.4) for a better understanding of methods of noise sampling. For sampling of noise levels from industrial sources, noise levels in the different octave bands are measured by a sound level meter in conjunction with octave band filters at the worker's ear level or at about a distance of 1 m from the source of noise.

15.10 CONTROL OF NOISE FROM TRANSPORTATION

Control of airport noise: Noise complaints from people living beneath flight paths have reached record levels across the world. Most of the time a "three-legged stool" approach to the problem has been adopted. The first is to ensure that aircraft emit the lowest possible noise levels, compatible with airline safety. Aircraft such as the Boeing 727 and 757 are as much as 20 dB quieter than the Boeing. The second is to impose controls on airport operations, such as restricting the number of arrivals and departures, imposing night curfews, and minimizing flight paths over populated areas. The third is to control urban development near existing airports and the site of future airports. Homes, schools, hospitals, and commercial and public buildings all need protection from excessive aircraft noise.

Traffic noise has become a serious problem now because of inadequate urban planning in the past. Homes, schools, hospitals, churches, libraries, and other community buildings were routinely built on main roads without buffer zones or adequate soundproofing. The problem has been compounded by increases in traffic volumes far beyond the expectations of our early urban planners.

Table 15.4 Selected BIS Code Books on Noise Pollution

BIS Code	Description
IS—4954-1968	Noise abatement in town planning recommendations
IS—3098-1990	Noise emitted by moving road vehicles, measurement
IS—10399-1982	Noise emitted by stationary road vehicles, methods of measurement of
IS—6098-1971	Airborne noise emitted by rotating electrical machinery, method of measurement of
IS—4758-1968	Noise emitted by machines, methods of measurements of
IS—3483-1965	Noise reduction in industrial buildings, code of practice for
IS—1950-1962	Sound insulation of nonindustrial buildings, code of practice
IS—9167-1979	Ear protectors

15.10.1 Structure-Borne and Airborne Noise

Noise can pass from one room to another either through the building structure itself (structure-borne noise) or through the surrounding air (airborne noise). Airborne noise is the more common and occurs, for example, when loud music in a living area interferes with people sleeping in bedrooms. Airborne noise can pass from one room to another along a variety of paths such as open doors and windows, openings in walls separating the rooms, stairwells, or heating and air conditioning ducts. Structure-borne noise occurs when the building structure itself is made to vibrate, for example, a washing machine in contact with a wooden floor, a saucepan falling to the kitchen floor, and the impact of footsteps on hard floors.

There is excessive noise from trucks that use engine braking as a backup to conventional wheel braking. While engine braking is an indispensable safety device, fitting trucks with improved and relatively inexpensive mufflers can significantly reduce the noise levels.

The installation of noise barriers along major roads and freeways is another way to combat traffic noise. The barriers deflect noise from adjoining urban areas and can be made from relatively light and inexpensive materials such as timber, fibro-cement sheet, or Perspex. An effective barrier needs to be long enough and high enough to deflect noise from the area that is to be protected. Usually, this means that the barrier blocks the line of sight to the noise source. Any gaps in the barrier (e.g., driveways) decrease its effectiveness.

15.10.2 Control of Highway Traffic Noise

The level of highway traffic noise depends on three things:

1. the volume of the traffic
2. the speed of the traffic
3. the number of trucks in the flow of the traffic

Generally, the loudness of traffic noise is increased by heavier traffic volumes, higher speeds, and greater numbers of trucks. Vehicle noise is a combination of the noises produced by the engine, exhaust, and tires. The loudness of traffic noise can also be increased by defective mufflers or other faulty equipment on vehicles. Any condition (such as a steep incline) that causes heavy laboring of motor vehicle engines will also increase traffic noise levels. In addition, there are other more complicated factors that affect the loudness of traffic noise. For example, as a person moves away from a highway, traffic noise levels are reduced by distance, terrain, vegetation, and natural and manmade obstacles. Traffic noise is not usually a serious problem for people who live more than 500 ft from heavily traveled freeways or more than 100–200 ft from lightly traveled roads.

1. Two-thousand vehicles per hour sound twice as loud as 200 vehicles per hour.
2. Traffic at 65 miles per hour sounds twice as loud as traffic at 30 miles per hour.
3. One truck at 55 miles per hour sounds as loud as 10 cars at 55 miles per hour.

15.10.3 Determining Noise Impact

Highway traffic noise is never constant. The noise level is always changing with the number, type, and speed of the vehicles that produce the noise. Traffic noise variations can be plotted on a graph, as shown in Fig. 15.6. However, it is usually inconvenient and cumbersome to represent traffic noise in this manner. A more practical method is to convert the noise data to a single representative number. Statistical descriptors are almost always used as a single number to describe varying traffic noise levels. The two most common statistical descriptors used for traffic noise are L_{10} and L_{eq} L_{10} is the sound level that is exceeded 10% of the time.

In Fig. 15.6, the shaded areas represent the amount of time that the L_{10} value is exceeded. Adding each interval during which this occurred shows that during the 60-min measuring period the L_{10} was exceeded 6 min $(1/2+2+2+11/2=6)$ or 10% of the time. The calculation of L_{eq} is more complex. L_{eq} is the constant, average sound level, which over a period of time contains the same amount of sound energy as the varying levels of the traffic noise. L_{eq} for typical traffic conditions is usually about 3 dB less than the L_{10} for the same conditions.

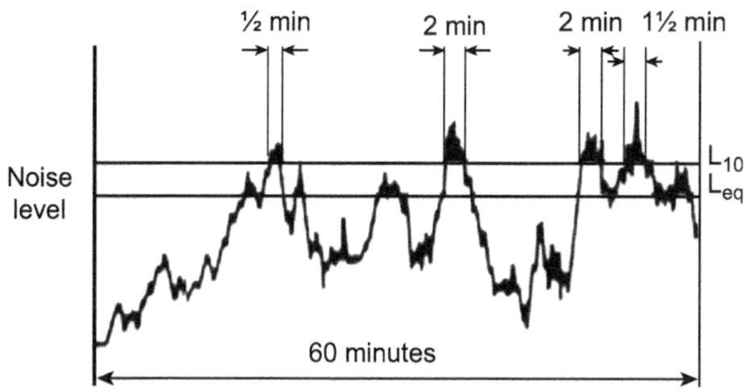

Figure 15.6 L_{10} and L_{eq} values of noise levels.

If a project causes a significant increase in the future noise level over the existing noise level, it is also considered to have an impact. Highway noise is being attacked with a three-part strategy: motor vehicle control, land use control, and highway planning and design. The responsibilities for implementing these strategies are to be shared by all levels of government: national, state, and local. The following two sections briefly describe how traffic noise impacts can be reduced or prevented through efforts to obtain quieter vehicles and efforts to control future development near highways.

The following methods can be employed:

1. **Highway planning and design**: In the planning stages of most highway improvements, highway agencies do a noise study. The purpose of this study is to determine if the project will create any noise problems. First, the existing noise levels of a highway are measured or computed by models and suitable alternatives sought.
2. **Noise reduction on existing roads**: Some noise reduction measures that are possible on existing roads or on roads that are being rebuilt include creating buffer zones, constructing barriers, planting vegetation, installing noise insulation in buildings, and managing traffic. Buffer zones are undeveloped open spaces that border a highway. Buffer zones are created when a highway agency purchases land or development rights, in addition to the normal right of way, so that future dwellings cannot be constructed close to the highway. This precludes the possibility of constructing dwellings that would otherwise experience an excessive noise level from nearby highway traffic. An additional benefit of buffer zones is that they often improve the roadside appearance. However, because of the tremendous amount of land that must be purchased and because in many cases dwellings already border existing roads, creating buffer zones is often not possible.
3. **Open space can be left as a buffer zone between residences and a highway**: Noise barriers are solid obstructions built between the highway and the homes along the highway. Effective noise barriers can reduce noise levels by 10–15 dB, cutting the loudness of traffic noise in half. Barriers can be formed from earth mounds along the road (usually called earth berms) or from high, vertical walls. Earth berms have a natural appearance and are usually attractive. However, an earth berm can require quite a lot of land if it is very high. Walls take less space. They are usually limited to 25 ft in height for structural and aesthetic reasons. Noise walls can be built of wood, stucco, concrete, masonry, metal, and

other materials. Many attempts are being made to construct noise barriers that are visually pleasing and that blend in with their surroundings.

However, barriers do have limitations. For a noise barrier to work, it must be high enough and long enough to block the view of a road. Noise barriers do very little good for homes on a hillside overlooking a road or for buildings that rise above the barrier. Openings in noise walls for driveway connections or intersecting streets destroy the effectiveness of barriers. In some areas, homes are scattered too far apart to permit noise barriers to be built at a reasonable cost.

- earth berm noise barrier
- wooden noise barrier
- concrete noise barrier with woodgrain texture

Vegetation, if high enough, wide enough, and dense enough (cannot be seen through), can decrease highway traffic noise. A 200-foot width of dense vegetation can reduce noise by 10 dB, which cuts the loudness of traffic noise in half. It is often impractical to plant enough vegetation along a road to achieve such reductions; however, if dense vegetation already exists, it could be saved. If it does not exist, roadside vegetation can be planted to create psychological relief, if not an actual lessening of traffic noise levels.

15.10.4 Vegetation and Noise Reduction

Insulating buildings can greatly reduce highway traffic noise, especially when windows are sealed and cracks and other openings are filled. Sometimes, noise-absorbing material can be placed in the walls of new buildings during construction. However, insulation can be costly because air conditioning is usually necessary once the windows are sealed.

Managing traffic can sometimes reduce noise problems. For example, trucks can be prohibited from certain streets and roads, or they can be permitted to use certain streets and roads only during daylight hours. Traffic lights can be changed to smooth out the flow of traffic and to eliminate the need for frequent stops and starts. Speed limits can be reduced; however, about a 20 mile-per-hour reduction in speed is necessary for a noticeable decrease in noise levels.

Pavement is sometimes mentioned as a factor in traffic noise. While it is true that noise levels do vary with changes in pavements and tires, it is not clear that these variations are significant when compared to the noise from exhausts and engines, especially when there are a large number of trucks on the highway. More research is needed to determine to what extent different types of pavements and tires contribute to traffic noise. Until this research is completed, the use of different types of pavement cannot be depended upon to reduce traffic noise.

15.10.5 Noise Reduction on New Roads

All of the measures described earlier can be employed on both existing roads and new roads. There are, however, some additional measures that can usually be used only on new roads.

First, a new road can be located away from noise-sensitive areas, such as schools or hospitals, and placed near non-sensitive areas, such as businesses or industrial plants. New roads can also be located in undeveloped areas.

Second, a new road can be constructed below ground level. Much of the noise from vehicles traveling on this type of road is deflected into the air by embankments on the side of the road. Thus, these embankments function in much the same way as noise barriers.

15.11 COMPARISON OF AIR AND NOISE POLLUTION

There are a number of difference between air pollution and noise pollution.

1. The source of air pollution persists for some time, while the source of noise pollution can be immediately controlled or removed.
2. Air pollution is more serious than noise pollution because air pollution adversely affects human health and can even be very fatal. The effects caused by noise pollution are not as fatal.
3. Noise pollution is not as serious as air pollution is because air pollution may be extremely serious, leading to death.
4. Noise pollution is local, while air pollution is international.
5. Noise pollution does not contain harmful substances, while air pollution is due to various toxic substances such as CO_2, CO, SO_x, NO_x, particulate matter, organic matter, fly ash, etc.
6. Noise pollution cannot be fatal, even if not treated properly, but air pollution may cause havoc, and so control measures are extremely necessary.

FURTHER READING

[1] V. Agarwal, M.K. Jagetia, Introduction to Geological and Mining Environment and Related Issues, Aravalli Research Development Society, Udaipur, 2001. 161 pp.

[2] N.K. Garg, V.K. Gupta, R.K. Vyas, Noise pollution and its impact on urban life, Journal of Environmental Research and Development 1 (3) (2007) 290–296.

[3] D. Singh, B.D. Joshi, Study of the noise pollution for three consecutive years during Deepawali festival in Meerut city, Uttar Pradesh, New York Science Journal 3 (6) (2010) 40 ISSN:1554-0200.

[4] M. Sørensen, M. Hvidberg, B. Hoffmann, Z.J. Andersen, R.B. Nordsborg, K.G. Lillelund, et al., Exposure to road traffic and railway noise and associations with blood pressure and self-reported hypertension: a cohort study, Environmental Health 10 (2011) 92.
[5] S. Agarwal, B.L. Swami, Road traffic noise, annoyance and community health survey—a case study for an Indian city, Noise Health 13 (2011) 272–276.
[6] Ministry of Environment and Forests, Government of India. No. 14 of 1981, [29/03/1981]—The Air (Prevention and Control of Pollution) Act, 1981, amended 1987. Available from: http://www.moef.nic.in/legis/air/air1.html.
[7] D. Banerjee, S.K. Chakraborty, S. Bhattacharyya, A. Gangopadhyay, Evaluation and analysis of road traffic noise in Asansol: an industrial town of eastern India, International Journal of Environmental Research and Public Health 5 (2008) 165–171.
[8] V. Pathak, B.D. Tripathi, V.K. Mishra, Dynamics of traffic noise in a tropical city Varanasi and its abatement through vegetation, Environmental Monitoring and Assessment 146 (2008) 67–75.

CHAPTER SIXTEEN

Solid Waste Management

16.1 INTRODUCTION

Solid waste management is a major challenge in urban areas throughout the world. Without an effective and efficient solid waste management program, the waste generated from various human activities, both industrial and domestic, can result in health hazards and have a negative impact on the environment. Understanding the waste generated, the availability of resources, and the environmental conditions of a particular society are important to developing an appropriate waste management system. Solid waste is defined as material that no longer has any value to its original owner and is discarded. The main constituents of solid waste in urban areas are organic waste (including kitchen waste and garden trimmings), paper, glass, metals, and plastics. Ash, dust, and street sweepings can also form a significant portion of the waste. Solid waste management may be defined as the discipline associated with controlling the generation, storage, collection, transfer and transport, processing, and disposal of solid waste in a manner that is in accordance with the best principles of health, economics, engineering, conservation, aesthetics, and other environmental considerations, and that is also responsive to public attitudes.

The amount of solid waste generated in the cities is much higher than in rural areas. The generation rate in rural areas can be as low as 0.15 kg/cap/day, while in the urban areas the rate can be above 1.0 kg/cap/day. The generation rates of major cities reported by the participating member countries are listed in Table 16.1.

In India the average solid wastes (MSW) generation is approximately 100,000 MT/day. Out of that, only 60% (60,000 MT/day) is collected by municipal corporations and councils. The rest is disposed of in an unscientific manner.

Environmental Management
ISBN 978-0-12-811989-1
http://dx.doi.org/10.1016/B978-0-12-811989-1.00016-6

Table 16.1 Solid Waste Generation Rates of Major Asian Cities

City	Country	Generation Rate (kg/cap/day)
Delhi	India	0.47
Dhaka	Bangladesh	0.50
Urban	Islamic Republic of Iran	0.80
Penang	Malaysia	0.98
Katmandu	Nepal	0.30
Manila	Philipines	0.66
Singapore	Singapore	0.94
Colombo	Sri Lanka	0.62
Taipei	Republic of China	0.95
Bangkok	Thailand	0.88
Hanoi	Vietnam	0.63

Table 16.2 Contrast of Typical Waste Characteristics in Low- and High-Income Countries

	Low-Income Country	High-Income Country
Generation per household	0.5 kg	2 kg
Density	500 kg per cubic meter	100 kg per cubic meter
Composition:		
Organic	Up to 80%	30%
Paper	5%	40%
Metals	Less than 1%	10%
Plastic	Less than 1%	2%
Glass	Less than 1%	10%
Moisture content	High	Low

16.2 SOLID WASTE AND ITS COMPOSITION

Waste generation and composition may vary seasonally and also between various countries. Table 16.2 shows the contrasts between typical waste characteristics in low- and high-income countries. Table 16.3 shows the typical sources and types of wastes generated in India.

Solid waste constitutes essentially the following:

- *garbage*: waste from cooking, handling, storage, serving food
- *rubbish*: cartons, barrels, boxes, paper and furniture, which are all combustible, and the noncombustible rubbish, which is metal, tin cans, glass, etc.
- *trash from streets*: leaves, dirt, sweepings
- *ashes*: residue from fires for cooking and incineration
- dead animals

Table 16.3 Sources and Types of Waste

Source of Waste	Type of Waste
Residential areas	Food waste, paper, cardboard, plastic, textiles, glass, metal and nonhazardous waste, batteries, construction debris, and demolition waste
Commercial area (general store, restaurant/hotel)	Paper, cardboard, plastic waste, glass, metal
Institutional area (school, hospital, government offices)	Paper, cardboard, plastic waste, glass, metal, hazardous waste, processing waste, ashes, infectious and toxic waste
Industrial areas (light, medium, and major plants)	Paper, cardboard, plastic, metal, e-waste, hazardous waste, and nonhazardous waste
Municipal services (street cleaning, parks, water, wastewater treatments)	Green trash, silt/ashes, construction, and demolition waste, sludge

Table 16.4 General Characteristics

Population of Cities (in Million)	Number of Cities	Moisture	Organic Matter	N%	P_2O_5%	K_2O%
Above 10	3	38.7	39.07	0.56	0.52	0.52
2–10	10	21.03	25.06	0.56	0.69	0.78
1–2	22	26.98	26.89	0.64	0.82	0.72

- abandoned vehicles
- *construction wastes*: gravels, concrete, scrap lumber
- *demolition waste*: brick, lumber, masonry, destroyed buildings

The major sources of solid waste are divided into five categories. Leading among them is animal waste, constituting approximately 42.7% of the total solid waste on a weight basis; mining waste, constituting 31.3%; agricultural, 15.7%; commercial, municipal, and household 7.2%; and industrial waste 3.1%. On a tonnage basis, the animal waste is estimated to be approximately 1500 million tons per year. The composition of solid waste varies significantly in the different cities in the region. Even within a city the composition varies with location and time. In general, the solid waste contains more organic components than other materials. The average percentages of organic matter in the solid waste in major cities in Asian countries ranged from 50% to 70%.

The general characteristics of urban waste (based on population) are described in Table 16.4, and Table 16.5 describes the comparison of biodegradable and nonbiodegradable waste generation per day and the calorific value of waste across 35 Indian cities.

Table 16.5 Comparison of Biodegradable and Nonbiodegradable Characteristics

Population of Cities (in Million)	Number of Cities	Waste Generation (MT)	Biodegradable (%)	Nonbiodegradable (%)	Calorific Value (kcal/kg)
Above 10	3	15,150	35–39	65–61	500–700
2–10	10	14,175	10–60	90–40	800–1050
1–2	22	8,952	10–60	90–40	500–1500

16.3 TYPICAL ISSUES IN SOLID WASTE MANAGEMENT

The main disposal methods for municipal solid waste in Asian developing countries are open dumping and sanitary landfill. Overall the environmental condition of the uncontrolled dumpsites is extremely vulnerable, with severe environmental pollution. On open dumping grounds, foul odors and air pollution are dangerously affecting the surroundings. Rodents are spreading pathogens in the surrounding areas, and the workers are highly exposed to disease and hazardous waste. Some cities dispose of their waste in sanitary landfills. The landfills are generally well operated and maintained.

16.3.1 MSW in India

The growth in the amount of solid waste generation in India poses many threats to the environment and to occupational health. The improper and manual handling of solid waste and the transfer of waste in open vehicles create unhygienic conditions. Disposal of waste in low-lying areas without proper liners, leachate collection, and treatment systems creates groundwater pollution, and the disposal of solid waste into streams and rivers creates water pollution. Air pollution is created by odor nuisances and the generation of greenhouse gases from most of the landfill sites.

16.3.1.1 Current Practices

The existing practices in solid waste management can be classified at three levels, depending upon the quantity of solid waste and the physical area covered.

- *Rural level*: Rural people generally do not use plastic or metal containers to keep waste segregated into biodegradable and nonbiodegradable. Instead, they throw it in the open fields. Sometimes it is naturally composted at the local level.

- *Town level*: In most towns in India, the practices for the collection and transportation of waste are not defined. No specific mode of collection, transportation, and disposal exists. The garbage is generally dumped in low-lying areas and burned openly.
- *Big city level*: A more defined system of collection, transportation, and disposal/composting exists. People send their waste through locally hired waste collectors and organizations to the community bin. From the community bin, it is transported by various methods to sanitary landfill sites. Rag pickers can be seen at waste collection and disposal points.

16.3.2 Institutional Framework in India

The institution framework for handling municipal solid waste in India is shown in Fig. 16.1. Both central and state departments have a role to play in the handling of these works.

16.3.2.1 Selection of Candidate Sites

This phase is very crucial in the siting process and can be carried out through a multilevel screening process.

Level I, constraint mapping: Constraint mapping eliminates environmentally unsuitable sites and narrows down the number of sites for further consideration. Certain features termed as "exclusionary factors" identified for constraint mapping are given in Fig. 16.2.

Level II, potential site selection: The level II factors include land use and infrastructure facilities. Land use includes target land area

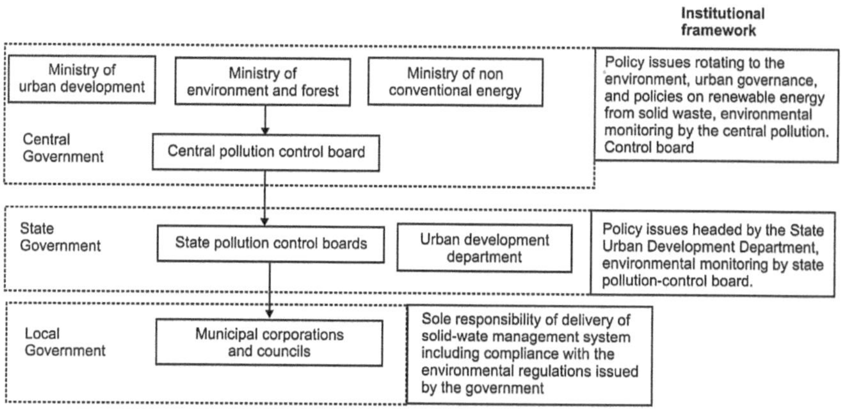

Figure 16.1 Institutional framework in India.

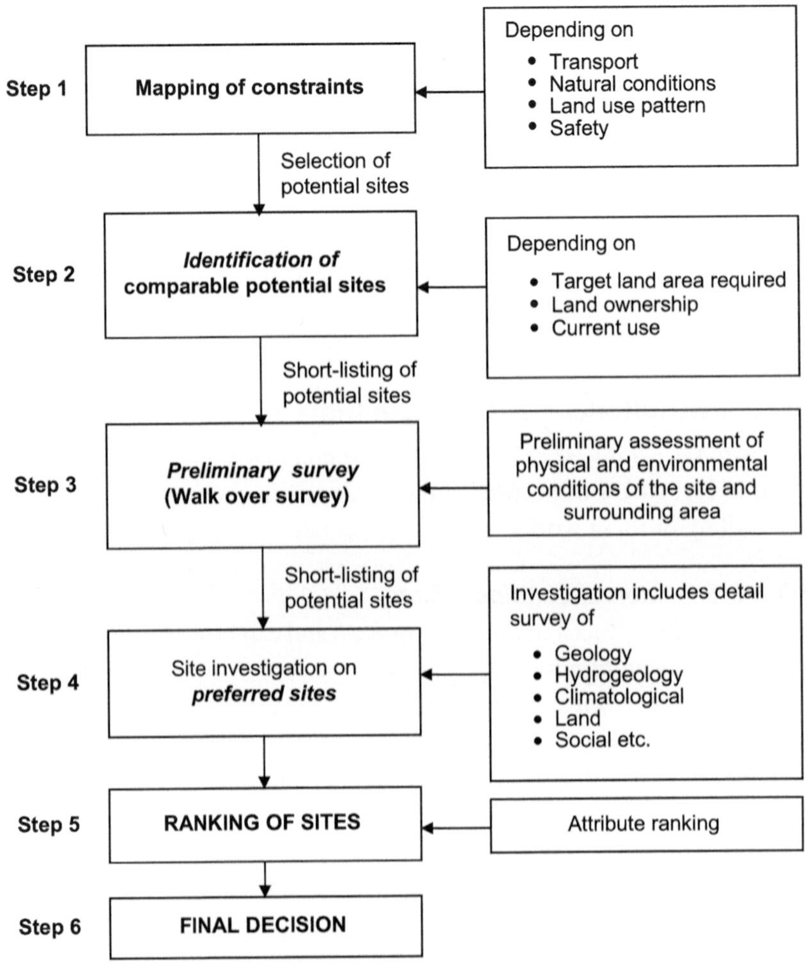

Figure 16.2 Site selection criteria.

required, land ownership, and its current use. Infrastructural facility includes major highway access, sites of existing/former waste disposal facilities, and land designated for industrial use. These provide the basis for highlighting promising sites within the candidate areas remaining after level I analysis.

Level III, preliminary survey: The sites selected in level II are further scrutinized to eliminate areas that fail to meet additional socio-economic and environmental concerns at the site and surrounding areas. The objectives of the walkover survey (preliminary survey) are to identify sufficient constraint to reduce the number of possible sites. This may be carried out by surveying the areas and collecting data regarding:

- existing zones of development
- agricultural land preserves
- areas of mineral development
- freshwater wetlands
- visual corridors of scenic rivers
- riverine and dam-related flood hazard areas

16.3.3 Site Sensitivity Index

The development of site sensitivity index is given in Table 16.6. The most important parameters include environmental, geological, and climatological parameters.

16.4 STEPS IN SOLID WASTE MANAGEMENT

Solid waste management (SWM) involves the collection, storage, transportation, processing, treatment, recycling, and final disposal of waste. Systems need to be simple, affordable, sustainable (financially, environmentally, and socially) and should be equitable, providing collection services to all.

Many developed countries have formal door-to-door collection systems. However, in low-income countries, waste generators (e.g., householders), domestic helpers, or private waste collectors carry waste to transfer points. This stage is referred to as "primary collection." A transfer point is an intermediate place at which waste is deposited and stored before being transported to the final disposal site. Local authorities then collect waste from the transfer point and convey it to the final disposal site, and this is referred to as "secondary collection." Finally, waste is disposed of to a variety of standards according to available resources and knowledge. This stage is called final disposal. Recyclable materials may be extracted from the waste stream from the points of generation, transfer, or disposal. Table 16.7 shows the stakeholders in SWM.

16.4.1 Primary Collection

Primary collection is what helps ensure waste enters the waste management process without ending up on streets or blocking drains. Primary collection may be undertaken by waste generators themselves, domestic helpers, or paid waste collectors.

Table 16.6 Development of Site Sensitivity Index

S. No.	Attribute	0.0–0.25	0.25–0.5	0.5–0.75	0.75–1.0
Accessibility Related					
1.	Type of road	National highway	State highway	Local road	No road
2.	Distance from collection area	<10 km	10–20 km	20–25 km	>25 km
Receptor Related					
3.	Population within 500 m	0–100	100–250	250–1000	>1000
4.	Distance to nearest drinking water source	>5000 m	2500–5000 m	1000–2500 m	<1000 m
5.	Use of site by nearby residents	Not used	Occasional	Moderate	Regular
6.	Distance to nearest building	>3000 m	1500–3000 m	500–1500 m	<500 m
7.	Land use/zoning	Completely remote (zoning not applicable)	Agricultural	Commercial or industrial	Residential
8.	Decrease in property value with respect to distance	>5000 m	2500–5000 m	1000–2500 m	<1000 m
9.	Public utility facility within 2 km	Commercial and industrial area	National heritage	Hospital	Airport
10.	Public acceptability	Fully accepted	Acceptance with suggestions	Acceptance with major changes	Non-acceptance

Environmental Related					
11.	Critical environment	Not a critical environment	Pristine natural areas	Wetlands, flood plains, and preserved areas	Major habitat of endangered or threatened species
12.	Distance to nearest surface water	>8000 m	1500–8000 m	500–1500 m	<500 m
13.	Depth to ground water	>30 m	15–30 m	5–15 m	<5 m
14.	Contamination	Air, water, or food contamination	Biota contamination	Soil contamination only	No contamination
15.	Water quality	Highly polluted	Polluted	Potable	Conforming to standard
16.	Air quality	Highly polluted	Polluted	Conforming to industrial standards	Conforming to residential standards
17.	Soil quality	Highly contaminated	Contaminated	Average	No contamination
Socioeconomic Related					
18.	Health	No problem	Moderate	High	Severe
19.	Job opportunities	High	Moderate	Low	Very low
20.	Odor	No odor	Moderate odor	High odor	Intensive foul odor
21.	Vision	Site not seen	Site partly seen (25%)	Site partly seen (75%)	Site fully seen
Waste Management Practice Related					
22.	Waste quantity/day	<250 tons	250–1000 tons	1000–2000 tons	>2000 tons
23.	Life of site	>20 years	10–20 years	2–10 years	<2 years

Continued

Table 16.6 Development of Site Sensitivity Index—cont'd

S. No.	Attribute	0.0–0.25	0.25–0.5	0.5–0.75	0.75–1.0
Climatological Related					
24.	Precipitation effectiveness index[a]	<31	31–63	63–127	>127
25.	Climatic features contributing to air pollution	No problem	Moderate	High	Severe
26.	Soil permeability	$>1 \times 10^{-7}$ cm/s	1×10^{-5} to 1×10^{-7} cm/s	1×10^{-3} to 1×10^{-5} cm/s	$<1 \times 10^{-3}$ cm/s
27.	Depth to bedrock	>20 m	10–20 m	3–10 m	<3 m
28.	Susceptibility to erosion and runoff	Not susceptible	Potential	Moderate	Severe
Geological Related					
29.	Physical characteristics of rock	Massive	Weathered		Highly weathered
30.	Depth of soil layer	>5 m	2–5 m	1–2 m	<1 m
31.	Slope pattern	<1%	1–2%	2–5%	>10%
32.	Seismicity	Zone I	Zone II	Zone III	Zone IV and V

[a]Precipitation effectiveness index is the ratio of annual precipitation to annual evaporation.

Table 16.7 Stakeholders in Solid Waste Management

	Generation	Primary Collection	Secondary Transportation	Recycling	Sweeping	Disposal	Awareness Raising	Monitoring	Policy/ Planning
Householders and other generators									
Informal sector									
Private sector									
NGOs									
Municipalities									
Government bodies									
		Common roles			Occasional roles				

Rouse and Ali (2002).

Often primary collection is undertaken by informal sector entrepreneurs who charge a fee for periodic removal of waste. The vehicles used for primary collection are often small and low-cost, such as wheelbarrows, handcarts, or tricycles carts.

16.4.2 Transfer Points

One of the most visible aspects of SWM is the transfer point, providing an interface between primary and secondary collection. These are often poorly designed, involving double handling of waste (once to unload tricycles, again to reload trucks) and unsanitary conditions where transfer points are not properly cleaned.

Where waste generators carry their own waste, transfer points need to be located within easy walking distance (a good guide is ~50 m) to discourage indiscriminate dumping. All transfer points and areas should be cleared daily and cleaned as necessary to prevent odors and keep rats and other disease vectors under control. They should also be designed to ensure minimum double handling. There are a number of approaches for achieving this, including "ramp transfer points," which raise primary collection vehicles up to the loading level of demountable containers. Sometimes, however, space does not allow this. Another innovative solution is to use carts that carry a series of small containers that can be easily and safely lifted and emptied into containers.

16.4.3 Secondary Collection

Secondary collection entails the removal and transportation of waste from transfer points to processing and disposal facilities. This is often one of the most costly elements of SWM systems. Waste characterization study data will play an important part in planning secondary collection as it informs us how much waste requires collection, its weight and volume (affecting payloads), where it is located (affecting collection routes), and so on. Key planning questions for secondary collection include these:

- What resources (including staff and hardware) exist at present and how efficiently/inefficiently are they being used?
- What percentage of total waste generated is being collected?
- Are the collection routes as efficient as possible?

16.4.4 Landfill and Incineration

Disposal is one of the most problematic aspects of SWM in low-income countries. Waste characterization data will prove useful in planning disposal options; the quantities of waste generated will help decide what volume of landfill site is required and, according to trends in waste production, enable to project its lifespan. Waste composition data will also help guide your decisions about suitable options: for example the presence of toxic waste will indicate the need for particular care in disposal, and a high organic/moisture content could make incineration very difficult.

Key planning issues include these:

- identifying and understanding practices at present, including issues (e.g., corruption) that may impact performance now and in the future;
- for landfilling, where could a disposal site be situated, what are the pervading geological conditions, where does the water table lie, and what regulations must be adhered to;
- for incineration, is the waste composition suitable is safe technology affordable and would it be feasible to maintain and operate it to sufficiently high standards?

Fully engineered sanitary landfill is the safest disposal option in terms of human health and the environment, and methane can be recovered for electricity generation. However, engineered landfill sites are usually very expensive to construct and operate so may be a more suitable long-term objective. Incineration, involving high-temperature complete combustion of organic material, can be used to reduce volume, and in some situations, energy can be recovered. Open burning of waste is not incineration and is not recommended because it releases toxic smoke. Controlled incinerators can be developed relatively cheaply, but the high moisture content of waste in many low-income countries can cause problems. Most waste incinerators are costly to develop and operate.

In many cases a carefully managed basic landfill site will be more achievable in the short run, involving the following practices:

- preventing all burning;
- locating the site at least 500 m downwind from housing and water sources on a geologically suitable site;
- daily compacting and covering with soil, to increase stability and discourage vermin;
- basic monitoring of dumping, ideally using a weigh-station if available;
- fencing to prevent waste blowing outside the site;
- basic leachate control.

Leachate is liquid runoff, and it is a particular problem in rainy conditions. It has a high "biological oxygen demand" and can damage ecosystems in water bodies. It may also contain toxins such as heavy metals, which can pollute groundwater sources.

16.5 METHODS OF WASTE DISPOSAL

The methods for solid waste disposal fall under two segments.

- waste management: a throwaway or high-waste approach encouraging waste production
- pollution prevention: a low-waste approach using waste as resources

Solid waste disposal methods fall directly under SWM, and they include the following:

1. open dumping, sea dumping,
2. compaction and baling,
3. sanitary landfill,
4. composting,
5. incineration,
6. reclamation and recycling.

16.5.1 Open Dumping

Open dumping is not only practiced in urban areas, but it is also practiced in many parts of the countryside.

16.5.2 Compaction and Baling

With the compaction and baling method, solid waste is compressed, without any prior treatment, for the purpose of reducing the volume of the waste into manageable size, with the specific advantage in mind being the long-haul transportation of the refuse. The original volume of the waste is reduced approximately 8% by using hydraulic presses, and it may be reduced as much as 20% when mechanical compactors are used. It must be mentioned that large items such as old refrigerators, stoves, washing machines, and heaters are removed and treated separately.

16.5.3 Sanitary Landfill

This is an operation where basic principles of engineering are applied to render a successful approach to the disposal of solid waste, much of which is biodegradable. A sanitary landfill consists of placing the solid waste in a selected site, spreading, and compacting it into layers. The refuse is then covered with soil, typically 6 in. deep. This coverage by earth poses no threats to the environment, or to the health of the community. The soil coverage varies, but it is on the order of 6 in. to 1 ft of soil cover for each 4-ft layer of compacted solids; but ratios as high as 1:8 have been used. The final cover of landfill should be at least 2 ft of compacted soil, to prevent the problems that are usually associated with open dumps. The temperature in the layers usually rises to as much as 150°F due to microbial action. The gas emitted is methane, which is essentially odorless and colorless. It is emitted in very small quantities, and thus sometimes it is not beneficial to harness it.

One major problem that must be taken into consideration in a sanitary landfill is the formation of leachate. These are the extracts by water from solid waste seeping through the ground. When water falls and seeps through a landfill, it will go through the layer of dirt and the solid waste, extracting from the solid waste some undesirable substances, which under improper conditions may reach underground water and pollute it. Thus precautions must be taken to guard against this possibility. The bottom of the sanitary landfill cell or unit should be covered with a layer of clay or plastic sheet for preventing gas flow and the seepage of the leachate into ground water. The top of the sanitary landfill cell may be covered with approximately 2 ft of clay; since gas may seep through the layers, some vent is provided so that the gas escapes to the atmosphere, and the explosion hazard is eliminated.

There are three major methods of landfill. These are area, trench, and ramp (or slope) landfills. All of these have in common the same basic procedure, namely, spreading the solid waste and covering it with dirt. The area sanitary landfill type is best suited for flat or gently sloping land; it may be used in ravines, quarries, valleys, or where other suitable depressions exist. The earth material for covering is obtained from adjacent areas, or it may be hauled from other spots.

In the *trench type sanitary landfill*, a hole is dug, and the solid waste is dumped into it. This method is best used for flat land where the water table is not very close to the surface. Usually, the dirt near the trench is used to cover the solid waste.

The *ramp or slope method* makes use of dumping the solid waste on an already existing slope. The ramp or slope is essentially a variation of the area and trench landfills.

As mentioned earlier, ground water may become polluted due to the washing action and extraction by water seeping through solid waste; a potential for chemical and bacteriological pollution may become a reality. The contract of these pollutants with groundwater makes it unfit for either human consumption or for irrigation. However, the proper selection of a site, combined with ingenuity and good engineering judgment, will prevent such a problem, and the design of sanitary landfills takes into consideration the following protective measures:

- Sites are usually located at a safe distance from streams, wells, lakes, and other water bodies.
- Avoid the location of a site above subsurface stratification that will lead the leachate from the landfill to water sources, i.e., avoid a fractured landfill area.
- Use an earth cover that is essentially impervious, such as clay.
- Provide proper drainage trenches so the surface water is carried away from the site.

Landfilling (80% of municipal waste)

- sanitary landfills, typically clay-lined depressions,
- garbage is covered with thin soil layer daily,
- modern site selection based on groundwater geology, soil type,
- since 1993, EPA requirements are an impermeable bottom layer (several layers of clay, thick plastic, sand),
- liner collects leachate or rainwater contaminated as it percolates down,
- leachate is monitored and/or collected by wells, pumped, and stored in tanks or sent to a treatment plant,
- filled landfills are completely covered by clay, sand, gravel, and soil then monitored by wells to detect any leakage into groundwater,
- methane gas produced collected or burned to produce steam/electricity.

Landfilling benefits

- reduced odor, rodents, pests
- low operating costs, handles large amounts of waste
- put into operation quickly
- landscaped aesthetically for recreation after use

Landfilling disadvantages

- traffic, noise, dust
- emit toxic gases (H_2S, *methane*, *organic gases*)
- contamination of groundwater

Landfill design

As mentioned before, proper site selection and design alone are insufficient to result in a landfill that provides for the protection of public health and the environment. To achieve such protection, operation of a landfill should be based upon proper guidelines or other equivalent practices. The following are the steps involved in the site preparation and construction:

1. Clear site.
2. Remove and stockpile topsoil.

3. Construct berms.
4. Install drainage improvements.
5. Excavate fill areas.
6. Stockpile daily cover materials.
7. Install environmental protection facilities (as needed):
   a. landfill liner with leachate collection system,
   b. groundwater monitoring system, gas control equipment, and gas monitoring equipment.
8. Prepare access roads.
9. Construct support facilities:
   a. service building
   b. employee facilities
   c. weigh scale
   d. fueling facilities
10. Install utilities:
    a. electricity
    b. water
    c. sewage
    d. telephone
11. Construct fencing:
    a. perimeter
    b. entrance
    c. gate and entrance
    d. litter control
12. Prepare construction documentation (continuously during construction)

Development of the complete landfill may be divided into stages, some of which are completed many years after the opening of the site.

Acceptable wastes: In general, only wastes for which the facility has been specifically designed should be accepted for disposal.

Cover material: Cover material should be applied, as necessary, to minimize fire hazards, odors, blowing litter, vector food, and harborage; control gas venting and infiltration of precipitation; discourage scavenging; and provide an aesthetic appearance.

- A minimum of 6 in. of soil cover material should be applied daily.
- Cells that will not have additional wastes placed on them for 3 months or more should be covered with 12 in. of cover material.
- Most soil materials can satisfy the purpose of cover soil. However, if minimization of infiltration is necessary, relatively low-permeability cover material should be utilized and placed at the steepest allowable

grade to encourage runoff. Low-permeability soils will remain effective only if the soil has a low shrink/swell potential or if the soil moisture can be maintained to prevent cracks from shrinkage and swelling.

- The completed landfill should be covered with 6 in. of clay or other suitable material with permeability equal to or less than 1×10^{-7} cm/s or equivalent, followed by a minimum cover of 18 in. of additional soil to complete the final cover and support vegetation. Deeply rooted vegetation may require an even greater depth of suitable soil.

  Compaction: To conserve landfill disposal site capacity and preserve land resources, solid wastes should be incorporated into the landfill in the smallest practicable volume.
  - For most solid waste materials, landfill compaction equipment is necessary for volume reduction.
  - Compaction or other volume reduction may take place at or before delivery to the landfill, by utilizing balers, shredders, or stationary compactors.
  - Compaction of solid waste and cover soil reduces the attraction of rodents and vectors and the potential for fires.
  - Open burning of solid waste for volume reduction should not be practiced at landfill disposal facilities.

Compaction requirements: Degree of compaction is a critical parameter for extending the useful life of a landfill. For achieving high in-place waste densities, a compactor may be necessary. A minimum in-place compaction density of 1000 pounds per cubic yard is recommended. The number of passes that the machine should make over the wastes to achieve optimum compaction depends upon machine wheel pressure, waste compressibility, land and fuel requirements, labor costs, and work load. Generally, three to five passes are recommended to achieve optimum in-place waste densities.

Although additional passes will compact the waste to a greater extent, the return on the effort diminishes beyond six passes. An experienced operator will know if additional passes will result in greater compaction.

Fig. 16.3 shows the relationship between waste layer thickness, number of passes, and the compacted waste density found in a field test for a particular type of machine and operating procedure. Each landfill will have different results, but the shape of the curves will be similar. The most efficient solid waste compaction should be in a number of thin layers up to the total cell thickness and not in layers greater than 2 ft thick. The working face slope will also affect the degree of compaction achieved. As the

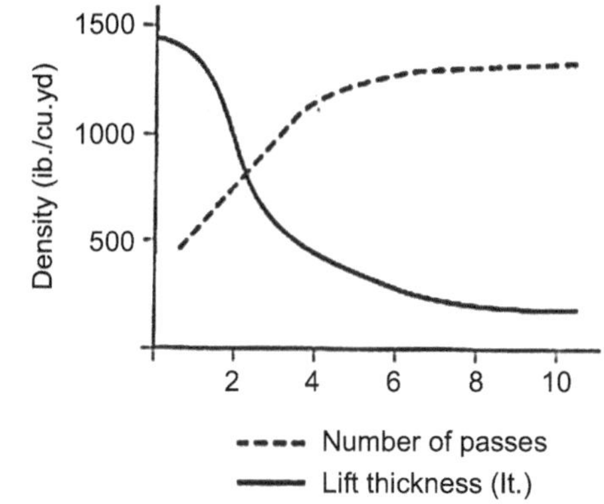

Figure 16.3 Effect of lift thickness and compactor passes on density.

slope increases, vertical compaction pressure decreases. The highest degree of compaction is achieved at the grade with the least slope. However, the feasibility of flat working face grades has to be weighed against the larger area over which the solid wastes must be spread.

16.5.3.1 Sanitary Landfilling Methods

The designer of a sanitary landfill should prescribe the method of construction and the procedures to be followed in the disposing of the solid waste, because there is no "best method" for all sites. The method selected depends on the physical conditions involved and the amount and types of solid waste to be handled. The two basic landfilling methods are **trench and area**; other approaches are only modifications. In general, the trench method is used when the ground water is low and the soil is more than 6 ft deep. It is best employed on flat or gently rolling land. The area method can be followed on most topography and is often used if large quantities of solid waste must be disposed of. At many sites, a combination of the two methods is used:

a. *Cell construction*: The building block common to both methods is the cell. All the solid waste received is spread and compacted in layers within a confined area. At the end of each working day, or more frequently, it is covered completely with a thin, continuous layer of soil, which is then also compacted. The compacted waste and soil cover constitute a cell. A series of adjoining cells all of the same height makes up a lift (Fig. 16.4). The completed fill consists of one or more lifts.

The dimensions of the cell are determined by the volume of the compacted waste, and this, in turn, depends on the density of the in-place solid

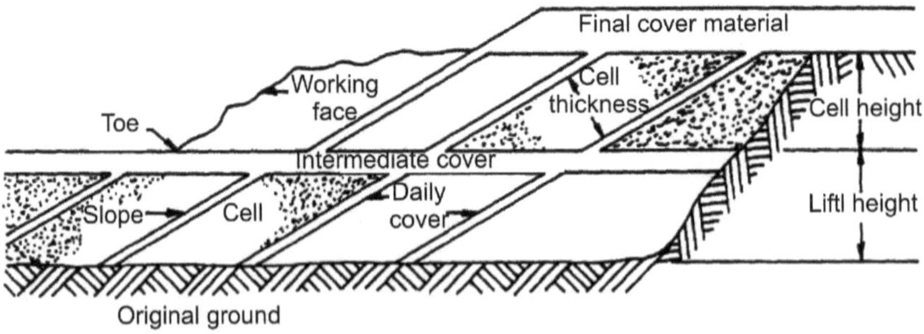

Figure 16.4 Landfill construction.

waste. The field density of most compacted solid waste within the cell for mobilization work should be at least 600 pounds per cubic yard. Higher figures may be difficult to achieve if trimmings from bushes and trees, plastic turnings, or synthetic fibers predominate. Because these materials normally tend to rebound when the compacting load is released, they should be spread in layers up to 2 ft thick, then covered with 6 in. of soil. Over this, mixed solid waste should be spread and compacted. The overlying weight keeps the fluffy or elastic materials reasonably compressed. An orderly operation should be achieved by maintaining a narrow working face (that portion of the uncompleted cell on which additional waste is spread and compacted). It should be wide enough to prevent a backlog of trucks waiting to dump, but not be so wide that it becomes impractical to manage properly, never over 150 ft. The height of a cell is not restricted. However, operations must be such that the required cover material is placed and compacted on a daily basis.

Cover material: Cover material volume requirements are dependent on the surface area of waste to be covered and the thickness of soil needed to perform particular functions. Cell configuration can greatly affect the volume of cover material needed. The surface area to be covered should therefore be kept minimal. In general, the cell should be about square, and its sides should be sloped as steeply as practical operation will permit. Side slopes of 20–30 degrees will not only keep the surface area, and hence the cover material volume, at a minimum but will also aid in shredding and obtaining good compaction of solid waste, particularly if it is spread in layers not greater than 2 ft thick and worked from the bottom of the slope to the top.

Trench method: Waste is spread and compacted in an excavated trench. Cover material, which is taken from the spoil of the excavation, is spread and compacted over the waste to form the basic cell structure. In this method, cover material is readily available as a result of the excavation. Spoil material not needed for daily cover may be stockpiled and later used as a cover for an area fill operation designed for the top of the completed trench fill operation. Cohesive soils, such as glacial till or clayey silt, are desirable for use in a trench operation because the walls between the trenches can be thin and nearly vertical. The trenches can, therefore, be spaced very closely. Weather and the length of time the trench is to remain open also affect soil stability and must, therefore, be considered when the slope of the trench walls is being designed. If the trenches are aligned perpendicularly to the prevailing wind, this can greatly reduce the amount of blowing litter. The trench can be as deep as soil and ground water conditions safely allow, and it should be at least twice as wide as any compacting equipment that will work in it. The equipment at the site may excavate the trench continuously at a rate geared to land filling requirements.

Area method: In this method, the waste is spread and compacted on the natural surface of the ground, and cover material is spread and compacted over it. The area method is used on flat or gently sloping land and also in quarries, strip mines, ravines, valleys, or other land depressions.

Combination methods: A sanitary landfill does not need to be operated by using only the area or trench method. Combinations of the two are possible, and flexibility is, therefore, one of sanitary landfilling's greatest assets. The methods used can be varied according to the constraints of a particular site.

1. One common variation is the progressive slope or ramp method, in which the solid waste is spread and, compacted on a slope. Cover material is obtained directly in front of the working face and compacted on the waste. In this way, a small excavation is made for a portion of the next day's waste. This technique allows for more efficient use of the disposal site when a single lift is constructed than the area method does because cover does not have to be imported, and a portion of the waste is deposited below the original surface.
2. Both methods might have to be used at the same site if an extremely large amount of solid waste must be disposed of. For example, at a site with a thick soil zone over much of it but with only a shallow soil over the remainder, the designer would use the trench method in the thick soil zone and use the extra spoil material obtained to carry out the area method over the rest of the site. When a site has been developed by either method, additional lifts can be constructed using the area method by having cover material hauled in. The final surface of the completed

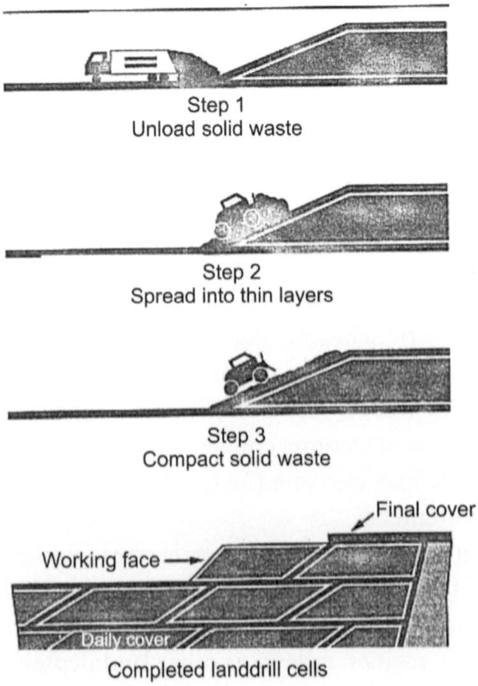

Figure 16.5 Solid waste placement and compaction.

landfill should be so designed that ponding of precipitation does not occur. Settlement must, therefore, be considered. Grading of the final surface should induce drainage but not be so extreme that the cover material is eroded. Side slopes of the completed surface should be three to one or flatter to minimize maintenance (Fig. 16.5).

Landfill Equipment Selection

Selection of type, size, quantity, and combination of machines required to spread, compact, and cover waste depends on the following factors:

- amount and type of waste to be handled;
- amount and type of soil cover to be handled;
- distance cover material is to be transported;
- weather conditions;
- compaction requirements;
- landfill method utilized;
- site and soil conditions such as topography, soil moisture, and difficulty in excavation;

- supplemental tasks such as maintaining roads, assisting in vehicle unloading, and moving other materials and equipment around the site.

Types: Equipment at sanitary landfills falls into five functional categories: site construction, waste movement and compaction, cover transport, placement and compaction, and support functions. Landfill site construction is often done by contractors employed by the site developer. Whether the work is done by contractors or site personnel, good construction management and coordination of equipment is essential. The most common equipment used on sanitary landfills is the crawler or rubber-tired tractor. The tractor can be used with a dozer blade, trash blade, or a front-end loader. A tractor is versatile and can normally perform all the operations: spreading, compacting, covering, trenching, and even hauling the cover material. The decision on whether to select a rubber-tired or a crawler-type tractor and a dozer blade, trash blade, or front-end loader must be based on the conditions at each individual site and the equipment's availability. Other equipment used at sanitary landfills are scrapers, compactors, draglines, and graders. This type of equipment is normally found only at large sanitary landfills where specialized equipment increases the overall efficiency. For example, at a site receiving a high proportion of hard-to-compact, heavy industrial waste (bricks and concrete), a compactor might not achieve normal compaction densities and the pushing and gripping ability of a track-type tractor may be needed. However, a small track-type tractor has more difficulty compacting bulky wastes than a landfill compactor. Landfills accepting only shredded wastes are operated much like landfills handling unprocessed wastes, although there may be less need for daily soil cover, and there will usually be less trouble with waste compaction. Landfills handling baled wastes have substantially different operating procedures and requirements. Not only are soil cover requirements often less stringent, but the bales can be handled with forklifts or similar types of equipment, without the need for compaction equipment.

Size: The size of the equipment is dependent primarily on the size of the operation. Small sanitary landfills for camps of 15,000 or fewer, or sanitary landfills handling 46 tons of solid wastes per day or less, can operate successfully with one tractor of the 5- to 15-ton range. Heavier equipment in the 15- to 30-ton range or larger can handle more waste and achieve better compaction. Heavy equipment is recommended for sanitary landfill sites serving more than 15,000 people or handling more than 46 tons/day.

Amount: Sanitary landfills servicing 50,000 people or fewer, or handling about 155 tons of solid wastes per day or less, normally can manage well with one piece of equipment, but provisions must be made for standby equipment. It is preferable that a second piece of equipment be used for replacement during breakdown and routine maintenance periods of the regular equipment. At large sanitary landfills serving more than 100,000 persons, or handling more than 310 tons of solid wastes per day, more than one piece of equipment will be required. At these sites, specialized equipment can be utilized to increase efficiency and minimize costs.

Effect of climate on sanitary landfill: Adverse climate can severely limit the capability of the sanitary landfill, but this can be partially overcome by pre-planning and operational techniques.

During site construction, a quality control program should be followed to assure the landfill is built in accordance with the design plans. An inspector should be on-site to approve construction work as each structure is completed. Compliance with specifications should be checked by soil tests before waste is placed over the liner. **Grades** and elevations can be measured with surveying equipment to document the as-built features of the landfill.

Some operational records that should be maintained include waste quantity by tons or, preferably, by volume, since landfill capacity is by volume; cover material used and available; equipment operation and maintenance statistics; landfill costs; labor requirements; safety statistics; and environmental monitoring data. Data on waste loading allow the site operator to predict the useful remaining site life of special equipment or personnel requirements.

16.5.4 Composting

Composting is bioconversion of organic matter by heterotrophic microorganisms (bacteria, fungi, actinomycetes and protozoa) into humus-like material called compost. The process occurs naturally, provided the right organisms, moisture, aerobic conditions, feed material and nutrients are available for microbial growth. By controlling these factors the composting process can occur at a much faster rate.

In composting, discarded waste is decomposed into a mix for conditioning of soil. From this, the amount entering landfills will be reduced by nearly 18%. Composting is a key in developing this approach and reducing solid waste. It may take about a year for the composting to take place in a compost form. Composting structures could be of two general types; the holding unit and the returning units. Holding units do not require much labor or much effort, while returning units will require turning a series of three or more bins every so often to speed up the process of composting.

Urban wastes as a source of compost: A major constituent of domestic solid wastes is dead animal and vegetable matter, mainly in the form of kitchen wastes and garden wastes. Urban areas generate domestic and shop wastes on a large scale, between 300 and 800 g/person/day. At the mean of this range, a city with a population of one million would generate greater than 500 tons/day. At least 25% and up to 75% of this weight comprises vegetables and putrescible matter.

Character and use of compost: The aim of composting is to convert a major proportion of solid wastes into a marketable product. It is necessary to begin, therefore, with some understanding of the properties, and the limitations, of compost.

Qualities: Compost is a brown, peaty material, the main constituent of which is humus. It has the following physical properties when applied to the soil:

- the lightening of heavy soil
- improvement of the texture of light sandy soil
- increased water retention
- enlarging root systems of plants

Compost also makes available additional plant nutrients in three ways:

- it contains N, P, and K, typical percentages being N, 1.2%; P, 0.7%; K, 1.2%; but with fairly wide variations;
- when used in conjunction with artificial fertilizers, it makes the phosphorus more readily available and prolongs the period over which the nitrogen is available, thus improving nutrient take-up by plants;
- all trace elements (micro-nutrients) required by plants are available in compost.

Composting systems: The minimum requirements that have been listed form the basis for all composting systems, although additional processes may be added, and the order of some of the activities may be changed. Whatever the level of mechanization and sophistication, almost all current composting systems rely on windrowing for all or part of the decomposition process. They fall into four main categories.

- windrowing of untreated wastes, followed by separation of contraries;
- windrowing of size-reduced wastes, preceded by separation processes;
- windrowing of size-reduced, separated wastes, which have been partially decomposed within an enclosed vessel;
- total, or almost total, decomposition of pretreated wastes within a digestor.

1. *Windrowing of crude wastes*: This comprises only the minimum requirements already described. The wastes are delivered in the collection vehicle direct to the windrow area, which should be paved. The windrows are turned by hand or machine, and watered by a hose. The decomposed wastes are passed through a posttreatment process for removal of contraries before being stacked in a storage area to await sale. The system is practicable only when the initial particle size is relatively small, say 60% under 50 min, but it has the following advantages:
   - **a.** The incorporation of coarse materials in the windrows increases the amount of interstitial air.
   - **b.** Both capital cost and energy consumption are greatly reduced by avoiding the size reduction process.
   - **c.** The removal of contraries after decomposition is cheaper because decomposition reduces the volume of material to be handled by about one-third, and the size of plant required is proportionately smaller.
   - **d.** Separation of contraries when the compost has a low moisture content after decomposition is much more efficient than with raw wastes with a high moisture content.
   - **e.** The contraries, and any salvage that may be recovered at this stage, are free of pathogens and present no problems in disposal.
2. *Windrowing of size-reduced wastes*: This is the most common type of plant in use today and has the following main features:
   - **a.** reception and storage; this can take the form of a deep bunker with a grab crane, or a hopper with slat conveyor,
   - **b.** elevation to attain sufficient height to enable subsequent processes to be fed by gravity; this usually takes the form of an inclined elevator belt, but sometimes the hopper slat conveyor is inclined and extended outside the hopper,
   - **c.** picking belt for the removal of salvage and contraries,
   - **d.** over-band magnet, for the extraction of ferrous metal,
   - **e.** size reduction, normally a hammer mill,
   - **f.** transport of the shredded wastes to the windrows; this can be by overhead conveyor belts or tractors and trailers,
   - **g.** windrow turning system; usually front-end loaders are used, but other machines are available,
   - **h.** storage area.

   Shredding is certainly an essential treatment for the wastes of temperate, industrialized countries, for such wastes contain many objects that would be highly resistant to decomposition, but it may be a questionable requirement in countries where the proportion of vegetable/putrescible matter is very high, and where lower living standards and intensive private salvaging tend to minimize the proportions of salvageable materials and contraries in the wastes.

3. *Windrowing of partly fermented wastes*: This type of system has all the features of the preceding type, except that instead of a hammer mill a very large rotating drum is employed; this has a capacity of several hundred tons, and the wastes are retained within it for up to 7 days, during which the following functions are performed:
   - **a.** Size reduction of most of the constituents is achieved by slow attrition as materials of differing hardness rub together in the rotating drum.
   - **b.** The wastes are mixed and achieve homogeneity.
   - **c.** The initial moisture content is adjusted, and subsequently the drum is supplied with air under pressure to maintain aerobic conditions.
   - **d.** The gaseous products of decomposition are withdrawn from the drum and passed through a filter to minimize odors.
   - **e.** Provided that the retention time is adequate, the wastes enter the thermophilic stage so that pathogens are eliminated before the contents of the drum are discharged and delivered to windrows.

     It is the custom to windrow drum-treated wastes for at least 2 weeks, during which it is turned at intervals.
4. *Digesters*: The aim of this type of system is to complete the decomposition process within a totally enclosed vessel after passing the wastes through all the separation and size reduction treatments. The vessel can take the form of a vertical silo with multiple floors, the wastes being fed in at the top and being kept in motion by revolving arms. Apertures in the floors allow the wastes to descend through the silo over a retention period of about a week, during which period they are supplied with forced air and the required moisture level is maintained. It is usual to provide posttreatment to break up clumps and to remove dense particles. Finally the product is partially dried to prevent further decomposition, and it may be supplemented by artificial fertilizers to provide a guaranteed nutrient content before being bagged by machine.

   For most cities the capital and operating cost of plants of this kind would be prohibitive. If operated on a commercial basis the selling price of the compost would be so high as to exclude its use for agriculture, and the market would therefore be restricted to domestic horticulture. There is an impressive list of failures of such enterprises.

16.5.4.1 Vermicomposting

Vermicomposting is the process by which worms are used to convert organic materials (usually wastes) into a humus-like material known as vermin-compost. The goal is to process the material as quickly and efficiently as possible.

Vermicompost appears to be generally superior to conventionally produced compost in a number of important ways:

- Vermicompost is superior to most composts as an inoculant in the production of compost.
- Worms have a number of other possible uses on farms, including value as a high quality animal feed.
- Vermicomposting and vermiculture offer potential to organic farmers as sources of supplemental income.

Vermicompost has the following advantages over chemical fertilizers:

- it restores microbial population, which includes nitrogen fixers, phosphate solubilizers, etc.,
- provides major and micro-nutrients to the plants,
- improves soil texture and water holding capacity of the soil,
- provides good aeration to soil, thereby improving root growth and proliferation of beneficial soil microorganisms,
- decreases the use of pesticides for controlling plant pathogens,
- improves structural stability of the soil, thereby preventing soil erosion,
- enhances the quality of grains/fruits due to increased sugar content.

At the same time, the beginning of vermicomposting process is a more complicated process than traditional composting:

- It can be quicker, but to make it so generally requires more labor.
- It requires more space because worms are surface feeders and will not operate in material more than a meter in depth.
- It is more vulnerable to environmental pressures, such as temperature, freezing conditions, and drought.
- Perhaps most importantly, it requires more start-up resources, either in cash (to buy the worms) or in time and labor (to grow them).

Vermicomposting of wastes in field pits

- It is preferable to go for optimum-sized ground pits of 20 ft length, 3 ft width, and 2 ft deep for effective vermin-composting beds.
- A series of such beds are to be prepare at one place.

Vermicomposting of wastes on ground heaps

- Instead of open pits, vermicomposting can be taken up in ground heaps.
- Dome shaped beds (with organic wastes) are prepared, and vermicomposting is taken up.
- The optimum size of ground heaps may be 10 ft length × 3 ft width × 2 ft high.

Materials required for vermicomposting

- farm wastes (straw from wheat, soybean, chickpea, mustard, etc.) were used for vermicomposting
- fresh dung
- wastes: dung ratio (1:1 on dry weight basis)
- earthworm: 1000–1200 adult worms (about 1 kg per quintal of waste material)
- water: 3–5 L in every week per heap or pit

Vermi-Compost Preparation Under Tree Shade by Pit and Heap Methods

Open permanent pits of 10 ft length, 3 ft width, and 2 ft deep were constructed under the tree shade, which was about 2 ft above ground to avoid entry of rainwater into the pits. Brick walls were constructed above the pit floor and perforated into 10-cm-diameter five to six holes in the pit wall for aeration. The holes in the wall were blocked with nylon screen (100 mesh) so that earthworms may not escape from the pits. Partially decomposed dung (dung about 2 months old) was spread on the bottom of the pits to a thickness of about 3–4 cm. This was followed by addition of layer of litter/residue and dung in the ratio of 1:1 (w/w). A second layer of dung was then applied followed by another layer of litter/crop residue in the same ratio up to a height of 2 ft. Moisture content is maintained at 60–70% throughout the decomposition period. Jute bags (gunny bags) were spread uniformly on the surface of the materials to facilitate maintenance of suitable moisture regime and temperature conditions. Watering by sprinkler was often done. The materials were allowed to decompose for 15–20 days to stabilize the temperature because to reach the mesophilic stage, the process has to pass the thermophilic stage, which comes in about 3 weeks. Earthworms were inoculated in the pit or heap with 10 adult earthworms (1.160.3 g each) per kg of waste material, and a total of 500 worms were added to each pit or heap. The materials were allowed to decompose for 110 days. The forest litter was decomposed much earlier (75–85 days) than farm residue (110–115 days). In the heap method the waste materials and partially decomposed dung (1:1 w/w) are made in heaps of dimension of 10 ft length × 3 ft width × 2 ft high, and during inoculation, channels are made by hand and earthworm at 1 kg per quintal of waste and inoculated, and then watering is done by sprinkler method. Jute cloth pieces are used as a covering material.

16.5.4.2 Vermiculture

Vermiculture is the culture of earthworms. The goal is to continually increase the number of worms to obtain a sustainable harvest. The worms are either used to expand a vermicomposting operation or sold to customers who use them for the same or other purposes.

16.5.5 Incineration (17% is Incinerated)

- trash-to-energy incineration; trash burned as fuel for steam/electricity
- most incinerators use unprocessed solid waste, not efficient
- some use refuse-derived fuel; processed into pellets before combustion
- only 90% removal of potential air pollutants

Incineration benefits

- Incineration kills germs and reduces amount of solid waste to landfills.
- Energy is produced.
- Electrostatic precipitators, scrubbers, and filters reduce air pollution.

Incineration disadvantages

- Pollutants are released to atmosphere.
- Ash produced can contain small amounts of toxic heavy metals.
- Cost ($30–200 m) and siting are of major concern to communities.

16.5.6 Reclamation and Recycling

Pollution prevention and source reduction

- 17% MSW recycled or composted
- use less materials in making products
- conversion from heavy packaging to lighter ones
- making consumer products in concentrated forms
- making products that last longer
- using reusable containers: refillable glass bottles are 11% of market

Recycling (10% of waste)

- conserves resources; goal is for 25% recycling by 2000
- primary recycling (*closed loop*) recycles products to produce new products of the same type
- secondary recycling (*open loop*) with waste producing different products
- high-technology materials recovery facilities shred and automatically separate mixed urban waste

- recover glass, aluminum, iron, paper for sale
- recycle or burn remaining waste for heat, electricity
- expensive to build and maintain
- low technology recovery uses homes, businesses to separate waste
- collection by municipalities or contractors for resale

Recycling concerns

- technology for recycling differs with different plastics
- milk containers may be high density polyethylene; egg containers may be polystyrene; soft drink containers are polyethylene terephthalate
- economics also problematical; supply/demand

Waste-to-energy management option for municipal solid waste

- Waste-to-energy facilities are incinerators that burn waste to create heat or electricity for nearby consumers.

16.6 GREEN PRODUCTIVITY OF SOLID WASTE

Organic waste should not be seen as a source of environmental pollution that has to be gotten rid of by putting it in landfills or burned in incinerators, as this could cause other pollution problems. It should be seen as a valuable resource that can be transformed into marketable products providing employment and profits. Given any situation we could prioritize our strategy as follows:

Priority 1: Processing organic waste

Priority 2: Possible activities for farming areas and communities

Organic waste can constitute as much as 70% of the total waste stream. Selected organic waste can either be reduced or transformed into organically beneficial products through the application of new and innovative approaches and technology for the reuse of these resources for energy, organic fertilizers, and animal feed. This will ultimately create new methods for improving the quality of life of the people. In addition, such an approach is in line with the principles of sustainable development where the efficient utilization of resources is closely linked with poverty alleviation goals (Fig. 16.6).

These come from the following dominant sources:

- municipal waste, e.g., restaurant and kitchen waste, domestic organic waste and sewerage, and waste from the food processing industry,
- agricultural and crop processing, e.g., crop and garden waste, sawdust and fruit waste, chicken and other animal manure, and waste from abattoirs.

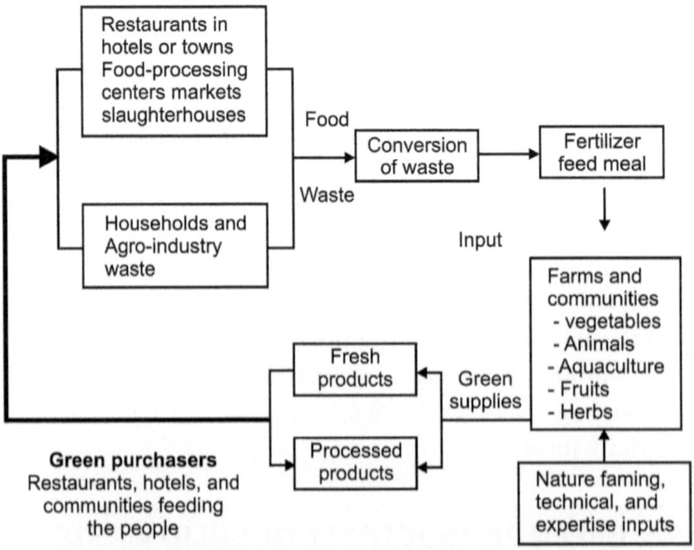

Figure 16.6 Flowchart of a green and productive community.

FURTHER READING

[1] S. Kumar, A.N. Mondal, S.A. Gaikwad, S. Devotta, R.N. Singh, Qualitative assessment of methane emission inventory from municipal solid waste disposal sites: a case study, Atmospheric Environment 38 (29) (2004) 4921–4929.

[2] IPE, Study of Management of Solid Waste in Indian Cities, Twelfth Finance Commission, Government of India, 2004. http://www.infrastructureindia.com/monitoreval_cs2.htm.

[3] A. Agarwal, A. Singhmar, M. Kulshreshtha, A.K. Mittak, Municipal solid waste recycling and associated markets in Delhi, India, Resources, Conservation and Recycling 44 (1) (2005) 73–90.

[4] A. Haque, I.M. Mujtaba, J.N.B. Bell, A simple model for complex waste recycling scenarios in developing economies, Waste Management 20 (8) (2000) 625–663.

[5] C.A. Boyle, Solid waste management in New Zealand, Waste Management 20 (7) (2000) 517–526.

[6] F. Eiland, M. Klarner, M. Lind, E. Baath, Influence of initial C/N ratio on chemical and microbial composition during long term composting of straw, Microbial Ecology 41 (2001) 272–280.

[7] IPE, J.R. and Ali, Vehicles for People or People for Vehicles? Issues for waste collection in low-income countries, WEDC, UK. ISBN: 1 84380 012 8, 2002. www.lboro.ac.uk/WEDC.

CHAPTER SEVENTEEN

Hazardous Waste Management

17.1 INTRODUCTION

Hazardous waste is material, disposal of which poses a threat to the environment, which is explosive, flammable, oxidising, poisonous/infectious, radioactive, corrosive and/or toxic. For managing hazardous waste issues such as sustainability, cleaner production, health, safety, and environmental protection needs to be considered. India has witnessed a fivefold increase in industrial production in the last 3 decades. Nearly 4000 medium and large chemical and allied industries are presently estimated to be in operation. The alkali industry alone has grown tenfold since 1947, whereas total dyestuff production has increased by 4.5 times. This industry involves the use of as many as 139 organic chemicals and heavy metals such as zinc, lead, chromium, copper, mercury, molybdenum, and their compounds and also various acids and alkalies. Most of the hazardous wastes generated in India are disposed in waste dumpsites. As the constituents of these wastes are transient and migrate through soil, they have the potential to contaminate soil and groundwater. The widespread use of hazardous organic chemicals like pesticides, aromatic hydrocarbons, bulk drugs, and dyes has led to their inadvertent introduction into different segments of the environment, leading to ecological degradation.

17.2 DEFINITION OF HAZARDOUS WASTE

The Environmental Protection Agency of the United States has defined, under the authority of the Resource Conservation and Recovery Act, hazardous substances in terms of characteristics of ignitability, corrosivity, reactivity, and toxicity. Anything that because of its quality, concentration or physical, chemical, or infectious characteristics may cause, or significantly contribute to, an increase in mortality, or pose a substantial hazard to human health and the environment when improperly treated, stored, transported, or disposed of or otherwise managed is termed a hazardous chemical.

- *Toxicity* is an adverse effect upon a living species by a constituent administered at a toxic concentration in a biologically available form by a particular route and during a fixed exposure of time.

Environmental Management
ISBN 978-0-12-811989-1
http://dx.doi.org/10.1016/B978-0-12-811989-1.00017-8

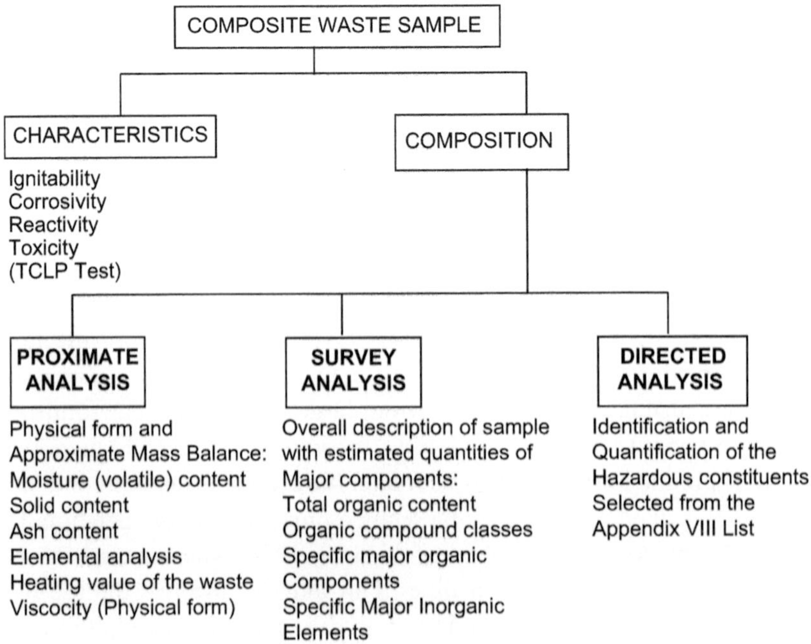

Figure 17.1 Analytical approach for hazardous waste characterization.

- *Hazard* is a possibility of a toxic effect being manifested upon exposure.
- *Risk* is a probability of hazard occurring.

A chemical is considered as hazardous if it possesses any of the following four characteristic attributes (Fig. 17.1).

1. *Ignitability*: *Ignitable* substances are easily ignited and burn vigorously and persistently.
2. *Corrosivity*: *Corrosive* substances include liquids with pH less than 2 or greater than 12.5 and those that are capable of corroding metal containers.
3. *Reactivity*: *Reactive* substances are unstable under normal conditions. They can cause explosions and/or liberate toxic fumes, gases, and vapors when mixed with water.
4. *Toxicity*: *Toxic* substances are harmful or fatal when ingested or absorbed.

17.2.1 Commonly Used Terms in Hazardous Waste Management

MPC (*maximum permissible concentration*): Levels of radioisotopes in waste streams, which if continuously maintained would result in maximum permissible doses to occupationally exposed workers and which may be regarded as indices of the radio toxicity of different radionuclides.

NFPA (National Fire Protection Association): Materials that include highly flammable gases, very volatile flammable liquids, and materials that in the form of dusts or mists readily form explosive mixtures when dispersed in air.

LD50 (lethal dose 50): A calculated dose of a chemical substance that is expected to kill 50% of a population of experimental animals exposed through a route other than respiration. LD50 concentration is expressed in mg/kg of body weight.

LC50 (lethal concentration 50): A calculated concentration that when administered by the respiratory route is expected to kill 50% of the population of experimental animals during an exposure of 4 h. The ambient concentration of LC50 is expressed in mg/L.

Grade 8 dermal irritation: An indication of necrosis resulting from skin irritation caused by the application of a 1% chemical solution.

96-h TLM (median threshold limit): That concentration of a material at which it is lethal to 50% of the test population over a 96-h exposure period. The ambient concentration is expressed in mg/L.

Phytotoxicity: It is the ability to cause poisonous or toxic reactions in plants.

ILM (median inhibitory limit): That concentration at which a 50% reduction in the biomass, cell count, or photosynthetic activity of the test culture occurs compared to a control culture over a 14-day period. The ambient concentration is expressed in mg/L.

Genetic changes: They are molecular alternations of the DNA or RNA of mitotic or meiotic cells resulting from chemicals or electromagnetic or particulate radiation.

EPA designates substances assigned as hazardous with an EPA hazardous waste number in the format of a letter followed by three numerals, where a different letter (F, K, P, or U) is assigned to substances from each of the four following lists:

- wastes from nonspecific sources (F-type wastes)
- wastes from specific sources (K-type wastes)
- acute hazardous wastes (P-type wastes)
- generally hazardous wastes (U-type wastes)

The major industries that generate hazardous chemicals are petroleum refining, plastic materials, synthetic fibers, paper and pulp, leather tanning and finishing, and iron and steel. The methods of disposal for hazardous wastes generally practiced by the Indian industry are given in Table 17.1.

Table 17.1 Hazardous Chemical Disposal Practices in Indian Chemical Industry

Product	Quantity and Nature of Waste		Mode of Disposal
	Solid	Liquid	
Acrylonitrile	---	1.3 m^3/ton product with HCN, aceto-and acrylo-nitriles, ammonia, etc.	Liquid waste incinerator using auxiliary fuel with 20% excess air, capacity 10 m^3/h, and 0.16 ton of fuel oil per ton of waste
Caprolactam	---	1.1 m^3/ton	Liquid waste incinerators with waste heat recovery boiler and no auxiliary fuel
Acrylates	---	Liquid waste, pH:4.0, HCN: 0.9%, acetone 1.8%, acrylic acid 1.5% others 0.5% water 95.5% off gases containing SO_2 1% HCN 4% N_2 55%	Burnt at the rate of 215 kg/h in a vertical cylindrical incinerators with fuel oil + LPG + LSHS as auxiliary fuels; off gases at 2.27 kg/h are also burnt in the same incinerators
Naphtha cracker	---	24 m^3/day spent caustic with 2.9% NaOH 3.47%, Na_2CO_3 0.5–4% Na_2S, oil phenols 30–50 mg/L	Shell wet air oxidation 5.5 kg/cm^3 pressure and 105°C temperature
Chemicals and pharmaceuticals	---	250 kg/ton product pH 7–7.5 containing combustible organic chemicals	Incinerated with light diesel oil as auxiliary fuel, incinerators rating is 200 kg liquid per day
Organo phosphorus pesticides	---	250 kg/ton product pH 7–7.5 containing combustible organic chemicals	Static, down fired incinerators with wastewater injectors, ejectors for cooling and dilution; thermal capacity 175,000 kcal/h
Pharmaceuticals, rubber chemicals agroche	5 tons/day solid tars, semisolid tars, pesticides, solvents, wastewater sludge, and general waste	---	Processing and feeding systems for solids, semisolids and solvents; rotary kiln incinerator with post-combustion chamber membrane panel boiler; quench/scrubber/adsorber tower, thermal capacity 2.5×10^6 kcal/h; boiler capacity 2–2.5/L

Bulk drugs	---	25 kg/h organic liquids; 50 kg/h wastewater and 30 kg/h off gases	Static, down fired liquid incinerators; dilution chamber centri-scrubber; thermal capacity 140,000 kcal/h
Agro-chemicals, malathion	2.4 tons/day organo phosphophatic solid waste	1.7–1.9 tons/day waste solvents	Conveying and feeding system for solid and liquid waste, rotary kiln incinerator with post-combustion chamber quench/scrubber absorption tower, thermal capacity 650,000 kcal/h
Chlorinated pesticides	---	1.2 tons/day organic liquids and 1.6 tons/day off gases	Static, down fired incinerator, quenched absorption tower thermal capacity 330,000 kcal/h
Agro-chemicals	General solid waste	2.5 tons/day, liquid chlorinated hydrocarbon	Static liquid incinerator, quench and scrubber tower pyrolytic solid waste incinerators with post-combustion chamber; drum detoxification chamber under erection
Organochemicals nitrobenzene nitrotoluene aniline	---	Combined waste 450 m^3/day containing nitroaromatics and phenols and is acidic	countercurrent solvent extraction with solvent recovery by distillation; the distillation column residue 0.6 ton/day along with 5 m^3/day dinitrobenzene wastewater is treated in a liquid waste incinerator with no auxiliary fuel; extractor capital cost is more
Ortho-chloro aniline	300 kg/ton product (resinous mass containing polymeric chloroamine)	---	Open pit burning
Phthalate plasticizers	---	0.05 m^3 sludge/ton product pH:6–10, sodium salt of *p*-toluene sulphonic acid and traces of plasticizers	Open pit burning with other garbage and plant trash

Continued

Table 17.1 Hazardous Chemical Disposal Practices in Indian Chemical Industry—cont'd

	Quantity and Nature of Waste		
Product	**Solid**	**Liquid**	**Mode of Disposal**
Ethanolamine	Polyethanolamine sludge 5 kg/ton and ethanolamine sludge 250 kg/ton product	---	Contained in drums and stored underground/disposed of by incinerators
Paints, varnish by lacquers and synthetic resins	1.5/ton product floor cleanings filter sludge containing phthalic anhydride TiO_2, chrome pigments, etc., 5 TPA of redundant or date expired paint	1.3 m^3/ton product pH 7–7.5 containing combustible organic chemicals	Solid waste is collected at point of generation in barrels and given to contractors for disposal; date expired paints are solid to different parties; waste treatment sludges are disposed of along with other solid waste
Naphthaline-based dye intermediates H-acid, C-acid peracid	9 tons/day consisting of gypsum cake and iron powder contaminated with organics	---	Manually collected and disposal in low lying areas
Petrochemical wastes	Biological sludges, oily sludge, and solid wastes	---	All sludges and solid wastes from plant are collected and sent to a disposal site, which is a ravine land outside the factory

17.3 EFFECT ON HEALTH

The following are some of the effects on health due to constant exposure to hazardous wastes.

1. lead poisoning, including poisoning by any preparation or compound of lead or their sequelae
2. tetraethyl lead poisoning
3. phosphorus poisoning or its sequelae
4. mercury poisoning or its sequelae
5. manganese poisoning or its sequelae
6. arsenic poisoning or its sequelae
7. poisoning by nitrous fumes
8. carbon bisulphide poisoning
9. benzene poisoning, including poisoning by any of its homologues, their nitro or amido derivatives or its sequelae
10. chrome ulceration or its sequelae
11. anthrax
12. silicosis
13. poisoning by halogens or halogen derivatives of the hydrocarbons of the aliphatic series
14. pathological manifestations due to (1) radium or other radioactive substances, (2) X-rays
15. primary epitheliomatous cancer of the skin
16. toxic anemia
17. toxic jaundice due to poisonous substances
18. dermatitis due to mineral oils and compounds containing mineral oil base
19. byssinosis (brown lung)
20. asbestosis
21. primary irritants and allergic sensitizers
22. hearing loss
23. beryllium poisoning
24. carbon monoxide
25. phosgene poisoning
26. occupational cancer
27. isocyanate poisoning
28. toxic nephritis
29. coal miner's pneumoconiosis

17.4 SAMPLING AND ANALYSIS OF HAZARDOUS WASTE

Sampling techniques: Hazardous waste can be solids, sludges, liquids, or gases. A waste sampling plan is needed to determine qualitative and quantitative waste characteristics. Depending on the degree of desired characterization information, *random sampling*, *stratified random sampling*, or *systemic random sampling* can be adopted.

Random sampling is used for heterogeneous waste with no heterogeneous information available on its chemical properties.

Stratified random sampling begins with identification of the locations or strata that are the sources of nonrandom heterogeneity. Simple random samples are then taken from each stratum or location.

Systemic random sampling begins with a random sample from the population at a predetermined location, followed by samples at regular time intervals or fixed space intervals. In sample collections, composite sampling is accomplished by taking a number of grab samples and then combining them into a single sample.

There are many types of sampling devices for hazardous wastes. Liquid wastes or free-flowing slurries may be sampled with a dipper or weighted bottle or from a tap. Solid wastes or sludges may require a scoop or a shovel. After collection the samples must be properly preserved, stored, and prepared for subsequent analysis. For trace analysis of organic hazardous wastes, the most common analytical methods are gas chromatography (GC), liquid chromatography (LC), mass spectrometry (MS), and combinations such as GC/MS and LC/MS.

17.5 TREATMENT, STORAGE, AND DISPOSAL FACILITIES

Hazardous waste management refers to a carefully organized system in which wastes go through appropriate pathways to their ultimate elimination or disposal in ways that protect human health and the environment. It involves generation, treatment, and disposal of hazardous wastes, as illustrated in Fig 17.2.

Treatment means any method, technique, or process, including neutralization, designed to change the physical, chemical, or biological character or composition of any hazardous waste to neutralize such

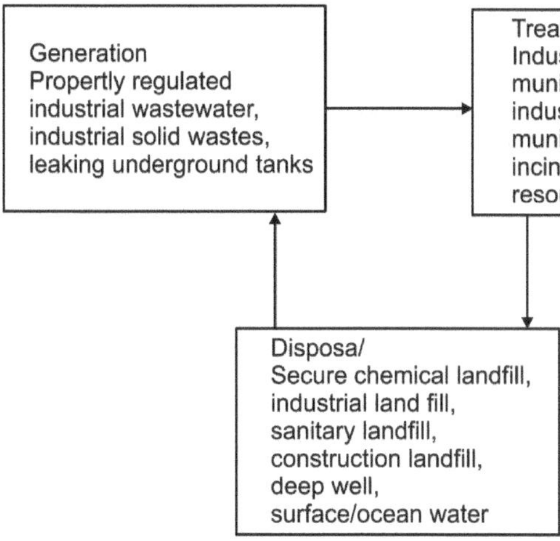

Figure 17.2 Summary of the treatment, storage, and disposal facilities.

waste, or to render such waste nonhazardous, or less hazardous; safer to transport, store, or dispose of; or amenable for recovery, amenable to storage, or reduction in volume. There are many options for the treatment of hazardous wastes, and these include industrial waste water treatment, hazardous waste incinerators, industrial furnaces and boilers, and resource recovery such as solvent reclamation. The ideal treatment process reduces the quality of hazardous waste material to a small fraction of the original amount and converts it to a nonhazardous form. However, most treatment processes yield material, such as sludge from waste water treatment or incinerator ash, which requires disposal and which may be hazardous to some extent. Direct disposal of minimally treated hazardous wastes is becoming more severely limited with new regulations being imposed. Fig 17.3 represents a scheme for the treatment of hazardous wastes.

Treatment technologies: The following are various treatment techniques available for the different types of hazardous wastes that are adoptable to specific wastes and economic feasibility.

- *physical methods*: phase separation (filtration/sedimentation), phase transition (distillation, evaporation and physical precipitation), phase transfer (extraction, adsorption), membrane separation (reverse osmosis, hyper and ultrafiltration)
- *chemical methods*: acid/base neutralization, chemical extraction and leaching, chemical precipitation, coagulation/flotation, oxidation/reduction, ion exchange, electrolysis, hydrolysis, photolysis

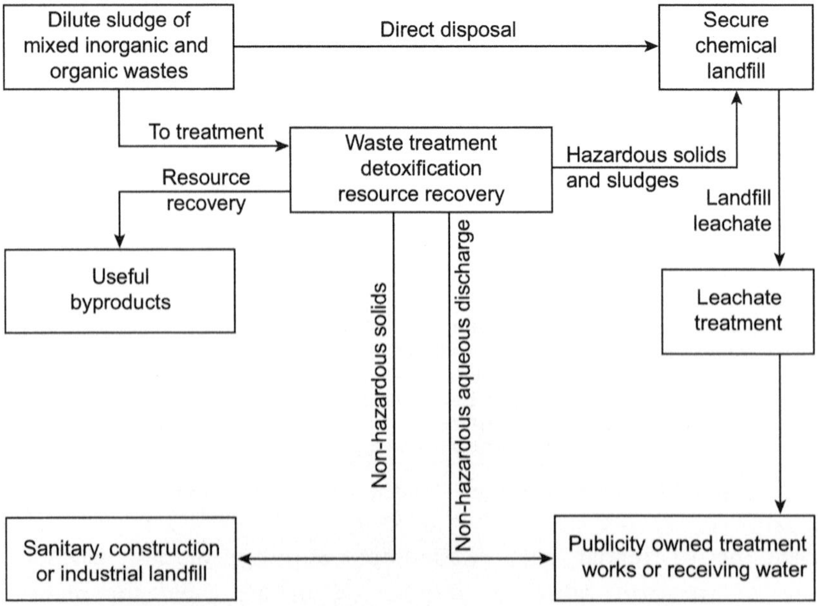

Figure 17.3 Treatment options for mixed hazardous wastes.

- *biological* treatment: aerobic, anaerobic, advanced, composting
- adsorption techniques, activated carbon, zeolites, activated clays
- air stripping
- recycling
- thermal treatment, incineration, wet oxidation, plasma arc, molten salt, superheated water
- bio*technical methods*

Hazardous waste management system consists of three major operations:

- storage upon generation
- collection and transportation
- final treatment to disposal, which depends upon the physical form of the waste

17.5.1 Storage Upon Generation

Storage is done in containers or bulk tanks. Containers are very portable, suitable for any physical state of waste and flexible as to means of filling. Empty containers that contain raw materials may be suitable for storing the wastes, depending on the compatibility of a waste with the container material; for example, a plastic container should not be used to store solvent wastes. The containers may be filled by any available method like pumping, shoveling, or tipping. Some of the disadvantages of containers are that they are easily damaged or toppled, they get accumulated easily and may lead to over-storage, and large groups of stacked ones are difficult to inspect for leaks and spills. The lining material for the tank or the containers can be steel, magnesium, lead, copper, zinc, tin, titanium, fiber glass reinforced plastic, PVC, chlorinated rubbers, epoxy, polyesters, and silicons. A separate area should be earmarked in the facility for the storage of different types of hazardous wastes.

17.5.2 Collection and Transportation

Before transportation, certain measures should be taken to ensure that the waste is properly packaged and information communicated to all the handlers. Reuse of empty containers should be done after proper inspection for deterioration or defects.

17.5.3 Final Treatment and Disposal

The treatment and disposal of the wastes depend on factors like *onsite* or *offsite* discharge/disposal standards, degree of treatment, availability of treatment and disposal technologies, and cost implications.

Storage is "the holding of hazardous waste for a temporary period, at the end of which the hazardous waste is treated, disposed of, or stored elsewhere."

Disposal is "the discharge, deposit, injection, dumping, spilling, leaking or placing of any solid waste, or hazardous waste, or any constituent thereof may enter the environment or be emitted into the air or discharged into any waters, including ground waters."

The following are methods used for the disposal of hazardous wastes.

- land disposal of wastes
- incineration
- dumping in the sea
- underground disposal
- deep well injection

Options for hazardous waste treatment and disposal involve decisions with respect to the following:

- onsite or offsite methods
- degree of treatment before disposal
- selection of treatment/disposal technology

Land treatment is a waste treatment and disposal process whereby a waste is mixed with or incorporated into surface soil and is degraded, transformed, or immobilized through proper management. Hazardous oils sludges from petroleum refineries and wastes from a few chemical industries are treated by land forming. Biodegradable wastes are suitable for land treatment. Highly volatile, reactive, or flammable liquids and inorganic wastes such as heavy metals, acids and bases, cyanide, and ammonia are not considered candidates for land treatment.

The advantages and disadvantages of land treatment are discussed next.

17.5.4 Advantages

- Waste is degraded, transformed, or immobilized; thus, the long term liability is lower than for other land options.
- The treatment area is continually monitored; thus remedial action can be taken immediately where there are signs of waste migration from the treatment zone.
- Cost of land treatment is lower than for landfills and incineration.
- Closed land treatment site can be converted to beneficial uses, such as parks and playgrounds.

17.5.5 Disadvantages

- waste storage will be required periodically during bad weather and equipment failure
- land intensive and management intensive
- air emissions and odor nuisance
- negative environmental impacts, e.g., groundwater pollution
- suitable for only selected wastes
- site selection and permitting is time consuming

17.6 CREATION OF TREATMENT, STORAGE, AND DISPOSAL FACILITIES

For the creation of a TSDF, the principle components are these:

- storage facilities
- stabilization unit
- landfill
- leachate treatment facility
- incinerator

The supporting infrastructure and utilities are these:

- weighbridge

- laboratory
- heavy and light earthmoving machinery
- vehicle wash/container wash unit
- wheel/tire wash unit

17.6.1 Storage Facilities

Covered storage sheds are required for different purposes:

- temporary storage: for storage of wastes till the issue of laboratory advice giving the pathway of disposal
- intractable waste storage: for the long-term storage of wastes, whose disposal path ways are not available
- storage during monsoon period: temporary storage when the landfill is closed during monsoon or breakdown of incinerator or issues like that

While constructing the storage sheds the following provisions should be made:

- adequate spacing with wide entry openings on both sides for the free movement of the machinery
- adequate ventilation and lighting
- suitable flooring to avoid seepage of leachate
- leachate collection system with a sump
- provisions for separation of incompatible wastes
- labels having the details of the wastes

17.6.2 Stabilization Unit

There should be facilities for well mixing of stabilization reagents and waste mechanically such that the toxic ingredients/components should be made immobilized or detoxified. Sometimes the slurries have to be solidified using water-absorbing reagents. At other times, toxic chemical compounds have to be destructed to nonharmful substances based on the mechanisms given by the laboratory. A curing bay has to be created for the wastes, which require curing after mixing. The post-stabilized samples should be tested by the laboratory and approved for disposal to landfill/incineration.

There should be provisions for appropriate flooring, adequate openings, and space for the operation of heavy machinery, storage of stabilization reagents, reagents weighing machine, and leachate collection drains with sump. Standard operating procedures and trained personnel should be deployed. The compatibility of waste categories with waste solidification and stabilization is given in Table 17.2.

Table 17.2 Compatibility of Selected Waste Categories With Waste Solidification and Stabilization Techniques

	SS Treatment Type			
Waste Categories	**Cement Base**	**Lime Base**	**Thermoplastic Solidification**	**Organic Polymer (UF)**
Organics Organic solvents and oils	Many impede setting may escape as vapor	Many impede setting may escape as vapor	Organics may vaporize on heating	Many retard setting of polymers
Solid organics (e.g., plastics, resins, tars)	Good, often increase durability	Good, often increase durability	Possible use as binding agent	Many retard setting of polymers
Inorganics acids	Cement will neutral-ize acids compatible	Compatible	Can be neutralized before incorporation	Compatible
Oxidizers	Compatible	Compatible	May dehydrate and rehydrate, causing splitting	Compatible
Sulfates	Many retard setting and cause spalling unless special cement is used	Compatible	May dehydrate and rehydrate, causing splitting	Compatible
Halides	Easily leached from cement: may retard setting	May retard setting; most are easily leached	May dehydrate	Compatible
Heavy metals	Compatible	Compatible	Compatible	Acid pH solubilizers metal hydroxides
Radioactive	Compatible	Compatible	Compatible	Compatible

A waste encountered at the disposal sites can be a complex mixture of organic and inorganic hazardous chemicals. Such waste can be in the form of solid, sludge, or liquid or a mixture of all these. The factors influencing the movement of chemicals through soil are discussed next.

17.6.2.1 Physical and Chemical Factors

The physical and chemical factors are (1) leaching, (2) adsorption/desorption, (3) volatilization, and (4) bioaccumulation.

These factors can act in accomplices and interrelated series of reactions that may be dependent on the geochemical characteristics of a disposal area. Many hazardous organic compounds display low water solubilities. However, the presence of potentially water-mixable solvents such as chloroform and acetic acid can enhance the leaching of organic compounds from the landfill sites and also the formation of emulsions. For the purpose of predicting sorption of dilute organic contaminants, it is generally recognized that distinct processes are possible (Kariekoff, 1984). These include both the hydrophobic and non-hydrophobic bonding. Hydrophobic bonding is important for neutral organics and some polar compounds. Non hydrophobic bonding may contribute to or dominate sorption when the sorbate is or has a highly polar or ionizable functional group that readily binds to a specific site on the sorbent surface consisting of polar groups or charged sites, or when organic content in a sorbent is low, especially when the clay content is also large [3]. Adsorption of compounds on soil particles or waste material is an important phenomenon that tends to restrict the movement of organic chemicals from a landfill site. The soil adsorption coefficient of chemicals gives their potential for binding to the soil particles. For organic compounds the partitioning between water and organic carbon is the most important factor. A compound with low adsorption coefficient will generally tend to migrate from the landfill site. In the cases when it is appropriate to assume that equilibrium has been achieved, Karickhoff [3] has shown, by examining the contaminants fugacities for dilute concentration, that the relationship between the dissolved and sorbed contaminants can be approximate by a linear isotherm [3]:

$$C_s = Kp \cdot C$$

where C_s is sorbed phase concentration, and C is dissolved phase concentration.

Volatilization is a potential route by which hazardous waste constituents migrate out of a landfill, especially one having a high vapor pressure. The diffusion coefficient also has an important influence on transport of compounds in a landfill. Disposal sites characteristics such as temperature, soil moisture, and soil pH as well as water solubility of the compound influence the extent of volatilization.

For organic compounds, the octanol–water partition coefficient is often used as an index of the bioaccumulation potential in the environment. This coefficient is also proportional to the soil adsorption coefficient, although not linearly, reflecting the importance of the soil organic matter in the adsorption of organic compounds.

17.6.2.2 Degradation of Chemicals

The persistence of hazardous organic chemicals is an important determinant of their environment fate. Certain compounds can undergo either chemical or biological degradation at land disposal sites, while others are resistant to any transformation and can even be toxic to soil microorganisms. Most manufacturing chemicals will pass through anoxic environments, and in some cases, they will reside in these habitats for long periods of time. These anoxic habitats include sediments of all types, anaerobic waste treatment systems, some landfills, and groundwater. To make an environmental risk assessment for a chemical, it is important to determine the chemical's susceptibility to anaerobic biodegradation.

The anaerobic conditions, which predominate at landfill sites, favor the bacterial reduction of sulfates, nitrates, and carbohydrates, which result in production of landfill gas. Generally the major components of landfill gas are CO_2 and CH_3, but traces of H_2S can be a minor constituent. In case of organic contaminants, reaction with an oxidant will result in decreased chemical concentration in the aqueous phase with corresponding changes in the kinetics of the process driven by dissolved concentration. Organic contaminants may be lost from water by reaction with an oxidant, and the reaction can be represented as a second-order reaction. In addition to oxidation, organic reduction has been observed.

Hydrolysis leads to transformation of pollutants by cleavage of a chemical bond and subsequent formation of a new bond. It is mostly promoted by acid or base, although it can occur in a neutral environment. With organic pollutants, hydrolysis proceeds through reaction of an organic compounds, R-X, with water resulting in the formation of new carbon oxygen bond and dissociation of the RX bond in the compound:

$$RX + H_2O = ROH + HX\,(\text{or } H^+, X^{-+})$$

The rate of the hydrolysis reaction for various organic compounds is pH and temperature dependent with a half-life varying from a few seconds to 10^6 years [1].

17.6.2.3 Waste Site Interactions

Soil is not a homogenous substance. The soil matrix contains inorganic solids, organic humic solids, soil gases, and liquids [2]. When a material enters the soil, it can undergo a wide variety of physical, chemical, and biological transformations, both on a micro- and a macroscale. On a microscale, these transformations can result in the partitioning of the hazardous organics within the soil matrix. The hazardous organics may be distributed as a gas in the soil atmosphere, dissolved in pore water, or associated with soil particles as well as in the form of free product. On a macroscale the material can be transformed via biotic and abiotic processes and transported within the soil en1vironment, or it may leave the soil environment, entering either the atmosphere (volatilization/evaporation) or the groundwater (leaching).

Leaching of hazardous organics from the subsurface poses the potential for hazardous organics of aquifers and groundwater supplies. Materials tend to leach downward to the groundwater under the influence of gravity flow of water. Lateral spread of the hazardous organics can result from such processes as advection and flow through channels and pores in the soil matrix. The migration of hazardous organics away from the source is influenced by such factors as the porosity and permeability of the soil, solubility of the material in the water, the specific gravity of the material, and the bulk velocity of the water. Leaching of hazardous organics from the subsurface to the saturated zone is the primary mechanism for the hazardous organics of groundwater supplies.

The study of waste site interactions is the key to assess the fate of land-disposed hazardous waste. These interactions determine the potential for off-site and onsite contamination. Organic contaminants are frequently part of hazardous waste disposed on land. Hence, to assess the movement of organics through soil, understanding of waste soil interaction is required. The hazardous waste once disposed on land invariably results in the formation of leachate. Leachate is generated primarily due to infiltration of water into waste layers in a landfill. Leachate can also be generated due to high original moisture in waste such as sludges and waste materials. Precipitation falling on the surface will either infiltrate the cover soil or leave the site as surface runoff depending on the surface conditions. The infiltrated water that is not lost by evapotranspiration will percolate down through the waste deposit. When the field capacity of the waste is exceeded, leachate will be produced. At waste disposals with no provision for collection, this leachate can contaminate underlying groundwater or nearby surface streams. Leaching, therefore, is a function not only of the downward water flux but also of the concentration of chemical in the liquid phase. Prediction of the extent of leaching for organic compounds depend most critically on knowledge of the partitioning of these compounds between the solid and solution phases of the soil. The soil-water adsorption isotherm provides a direct measure of the partitioning and of the effect of the environmental factors and nature of organic compounds, its concentration, temperature soil-water content, and soil properties that influence the distribution. For predicting the leachability of organic compounds, the linear adsorption isotherm presents the simplest approach. When the isotherm is linear, the slope of isotherm, is called the distribution coefficient K_D. Karickhoff et al. (1979) have shown that for low aqueous concentration of weakly polar compounds, which is the case with leachates, the linear isotherm is an accurate representation of the adsorption isotherm.

Soil organic matter has been shown to be the single most important soil property affecting the adsorption of nonionic organic compounds. The large variability in distribution coefficient K_D of such compounds due to adsorption by different soils can be greatly reduced by normalizing adsorption based on the organic carbon content of the soil (K_{oc}). The octanol–water partition coefficient has had wide application in predicting the partition of weakly polar organic compounds in biological systems.

Rao and Davidson recommended the following regression:

$$\text{Log } K_{ow} = 1.029\text{Log } K_{oc} - 0.21; r^2 0.91$$

The regression utilizes the octanol–water partition coefficient values for nonionic organic compounds to predict their adsorption in soil. Since the adsorption coefficient could not be determined for the target in all the soil-water systems, determination of adsorption coefficient (K_D) was carried out utilizing the octanol–water partition coefficient data for these target analytes for soil-water systems using formula proposed by Rao and Davidson [4] for movement of nonionic organics through soil.

17.6.2.4 Soil Properties Affecting the Leaching

Soil properties that influence the leaching of nonionic organic compounds at the land disposal waste site include texture, structure, and organic matter content.

Soil texture and structure are known to have substantial influence on leachability of the compounds. Leaching can be substantial from sandy soil due to its low exchange capacity, low clay content, low organic matter content, and high number of large pores and resultant high permeability. Clay soils can limit leaching due to their cation exchange capacity, high clay content, high organic matter content, and number of small ultra-aggregate pores and resultant low permeability. Both type and amount of clay present in soil have been found to affect the hazardous organics. Mobility of non-ionic organic compounds was found to be inversely related to clay content.

Presence of sand or loamy layers in the soil profile at waste disposal sites within 3 m of the surface overlying the fine texture clay creates a potential for horizontal flow and may contaminate adjacent areas. Several studies have suggested that movement of organic chemicals in soil is inversely related to the organic matter content of the soil. Helling (1971) found that the retardation of organic chemical movement through soil was highly correlated to the adsorption of these organic chemicals by native soil organic matter.

17.6.2.5 Persistence of Hazardous Organics in Soil

"Persistence" is expressed as the property of hazardous compounds that retain their biological activity for a much longer period than expected. It may be interpreted as the residence time of hazardous organics in the soil environment. The residence time may be considered as the period in which the hazardous organics remain in soil regardless of the method by which it is quantified. It may be expressed in terms of units of time. The persistence of soil-applied chemical is dependent on a host of conditions such as soil type, organic matter content, clay content pH, nature of soil colloids, structure of hazardous molecules, and environmental factors like moisture and temperature.

The concept of half-life is widely used for prediction of persistence of compounds in soil. The half-life has the characteristic of being a constant inversely proportional to the rate constant and independent of concentration. It is usually calculated on the presumption that the disappearance of the compounds occurs through the first-order kinetics. The practical indices such as 50% disappearance time (DT_{50}) have been found to give an idea of persistence of a given chemical.

Soil moisture and temperature play an important role in hazardous organics dissipation in fine loamy soil. Briggs et al. [5] suggested a model for degradation in soil, temperature, and moisture effects on degradation and effects of soil properties on the persistence of insecticide esters. Degradation in clay, loam soil in the laboratory and the constants derived from laboratory results were used in conjunction with meteorological data recorded during field experiments to construct a model simulating the atrazine persistence in the field. They obtained good correlation between persistence from the field as predicted by the model. Walker et al. [6] used energy of activation and moisture levels for predicting persistence of hazardous compounds. Activation energy is determined from the effect of temperature using Arrhenius equation, and moisture effects are characterized using an empirical equation.

17.6.2.6 Characterization of Waste Sludge

A variety of basic organic chemicals are being made in India using both indigenous and foreign technologies. The installed capacity is being expended continuously to meet the increasing domestic needs and overseas demand. The manufacture of organic chemicals produces a variety of solid and semisolid residues, many of which fall under the category of hazardous waste per the Government of India Hazardous Waste (Management and Handling) Rules, 1989, and its amendments. Solid and semisolid wastes generated from the industry are classified as follows:

- back ends/residues from distillation and purification processes
- spent catalysts
- sludges resulting from the treatment of wastewater

The wastes, forementioned, are known to contain nitro, amino, choro, and hydroxy substituted aromatics. Some of these compounds such as chlorinated aromatics are known to persists in the environment and are toxic even at low concentrations.

17.6.3 Landfill

An engineered/secure landfill is one that is capable of containing hazardous wastes with no seepage or percolation of leachate into the ground. Based on the landfill design criteria, the basic requirements are a natural mineral liner followed by synthetic liner and a drainage system to collect leachate. Surveillance monitoring bore wells have to be constructed in the upstream and downstream directions of the landfill to monitor the groundwater quality. The quality of water in these wells has to be assessed prior to the operations also to assess the impact.

17.6.4 Leachate Treatment Facility

The leachate/waste water collected from the landfill, waste storage areas, stabilization unit, laboratory, or another areas has to be treated and disposed.

Different options are these:

- solar evaporation
- forced evaporation
- treatment in common effluent treatment plants

17.6.5 Incineration

Incineration is a process of thermal oxidation. While designing the hazardous waste incinerators, meticulous care has to be taken with respect to the gas cleaning for meeting the emission standards. The prime requirements of a hazardous waste incinerator are that it should be capable of destroying the toxic constituents at high temperatures while maintaining the safety and standards.

17.6.6 Onsite Infrastructure

The individual units are these:

- waste feeding systems
- primary chamber or kiln
- secondary chamber
- flue gas cooling system
- bag filters and scrubbers
- stack

The characteristics of the ash resulting from the process of incineration has to be tested for the proper functioning of the incinerator, and the ash has to be disposed of in the landfill. Stack monitoring on a regular basis has to be conducted.

17.6.6.1 Weigh Bridge

A weigh bridge with suitable capacity has to be installed to weigh the quantity of wastes received at the facility. The bridge has to be maintained properly and calibrated in regular frequencies.

17.6.6.2 Laboratory

The laboratory should be capable of conducting fingerprint analysis of all wastes received at the facility and compared with the comprehensive analysis and ultimately decide the pathway of disposal of the waste based on the waste disposal criteria. The lab should maintain a QA/QC protocol and maintain the records and samples for the desired retention times.

17.6.6.3 Earthmoving Machinery

The compulsory earthmoving equipment include grader, dozer, excavator, roller with sheep foot or pad foot arrangement, tip trucks, etc.

17.6.6.4 Vehicle/Container and Wheel/Tire Wash

A low-volume, high-pressure water spray gun is desired to clean the vehicles and containers to avoid cross-contamination, spillages on roads, unpleasant odors, and to improve esthetic appeal.

17.6.6.5 Green Belt

A green belt with a minimum of four times the built-up area has to be developed and maintained.

17.7 DESIGN OF LANDFILL

One of the major objectives in designing a landfill should be to reduce environmental and health risks to reasonable levels that will mitigate local, regional, and national concerns. A properly designed and constructed landfill should incorporate proven technology to guard against the release or contaminants into air, surface water, and groundwater, reducing environmental concerns over its entire operating and post-closure life. The two main objectives of the design criteria are to achieve the following:

- prevention of leachate percolation into the ground or surface water from the bottom of the landfills,
- prevention of infiltration of rain water into the landfills from the top.

The other major components that must be addressed include the following:

- type of volume of hazardous and nonhazardous wastes to be landfilled
- life expectancy of the landfill during its active operating period
- topography and soil characteristics at the site and in its vicinity
- climatic conditions throughout the year
- surface water and groundwater in the vicinity
- collection and treatment of surface runoff
- soil cover requirements for individual containment cells
- anticipated quality and volume of leachate
- selection of leachate collection and treatment systems
- monitoring of groundwater and surface water during operation and beyond
- selection of venting systems for gaseous products
- selection of flexible membrane liners and other impermeable liners
- closure and post-closure plans
- alternative uses during the post-closure period
- effect on human health and the environment

17.7.1 Landfill Development

The landfills can be developed below the ground or above based on the groundwater table in that area and the earth excavation conveniences. It must be ensured that the groundwater table must be at least 1.5 m below the landfill. The subbase has to be well prepared and graded to uniform level with desired slope. Then, all the liners per the design have to be placed.

The key components are these:

Bottom	Cover
• natural mineral liner system at the base and the side slopes • synthetic liner again at the base and the side slopes • liner protection • drainage and leachate collection system • second synthetic liner (for double liner systems)	Gas media network with retrofitting • mineral layer • synthetic layer • drainage layer • top soil and vegetative soil • green cover

17.7.1.1 Liners

While designing, the following aspects have to be taken into consideration:

- stability of the liners at the base and along the side slopes
- strength to withstand all the loads that come over and above
- permeability and material properties and compatibility
- compatibility with leachate and wastes
- protection to the flexible membrane liners
- transition filter between waste and the media to prevent clogging of the leachate collection system

17.7.1.1.1 Clay Liners

The thickness of the clay liner is in fact based on the national/international standard for composite liner systems such as 1–1.5 m. However, it can be designed based on Darcy's law:

$$T = KH/Q$$

where T = thickness of the liner, K = hydraulic conductivity of the clay, H = head of leachate above the liner base, Q = flow rate of the leachate m^0/annum, and H/T = hydraulic gradient.

With regard to the slope stability, a factor of safety has evolved based on the probable occurrences of failures such as base failure, toe failure, and slope failure.

The clay with all the desired properties has to be selected and conditioned before the placement to ensure desired moisture content and uniformity of the matrix. The preconditioned soil has to be placed on the well-prepared subbase in layers and compacted. The first layer can be a little bit thick, on the order of 200–250 mm compacted to 150–200 mm. Subsequent layers have to be placed in thicknesses of 150 mm and compacted to 100 mm.

17.7.1.1.2 Geomembrane Liner

Geomembranes are relatively thin sheets of flexible thermoplastic or thermoset polymeric materials. The utmost impervious nature of these membranes has resulted in their widespread selection as a synthetic liner material for safe disposal of hazardous waste. High-density polyethylene is the best liner material of this group. Thickness is based on the national/international standards applicable for composite liners. The requirements are these:

- thickness generally 1.5 or 2 mm
- chemically compatible with leachate, landfill gas, and other expected environmental conditions within the landfill
- capable of withstanding all the stresses that come over and above
- minimum possible weld joints (maximum width seems to be selected)

17.7.1.2 Drainage and Leachate Collection Layer

- thickness 300 mm
- hydraulic conductivity >0.001 cm/s
- slope >2%
- spacing between the leachate collection lateral pipes
- spacing = [k(h12-h22)]1/2

where the following are true:

- k = permeability of drainage layer
- h1 = maximum head on the liner
- h2 = head at the collection pipe
- q = leachate infiltration rate

The number of lateral pipes required will be based on the size of the landfill. All the laterals have to be connected to a header pipe, which will open into a sump at the lowest point in the landfill. This leachate collection sump has to be progressively raised as the waste deposition progresses.

A transition filter like geo-textile has to be placed above the media to avoid choking of the drainage media with the solid particles from the waste. Needle-punched, non-oven, 250–300 gsm (grams per square meter) geotextile material will do.

17.7.1.3 Compaction

The soil has to be compacted using a sheep foot or pad foot roller. Depending on the requirement vibration, water spray and adequate passes have to be given. Always, it is recommended to have the moisture content +5% of the optimum moisture content (OMC). The in situ density has to be checked in each grid to achieve 95–100% from that of the $Yd_{max.}$ The compaction is to be done in horizontal layers only, even in the slopes. In the side slopes the layers should be placed in benches and trimmed vertically to the desired thickness.

17.7.1.4 Bonding Between the Lifts

The two layers must be joined properly to avoid gaps between the layers. Kneading or blending a new lift with a previously compacted lift has to be done using the sheep foot roller.

17.7.1.5 Media and Leachate Network

Proper care has to be taken in selecting the size and shape of the pebbles, so they do not damage the synthetic liner under any given possibility and also that they should not pass through the perforations of the leachate pipes. The perforated laterals have to be placed and welded to the header onsite. The leachate sump in the bottom up to the height of the media should have more openings with a cover with a geo-filter to allow the entire media to also collect leachate into the sump.

17.7.1.6 Transition Filter

On the top of the media filter, material has to be placed so the particles in the waste do not clog the drainage media. In case of a double liner system, protection has to be provided to the geo-membrane, which will be placed above the first media.

The synthetic liners and the geo-textile have to be properly anchored in the trench above the bund. Finally, these liners are to be joined to the liners that come in the cover/cap.

17.8 OPERATION

17.8.1 Reception and Weighing

The waste along with the comprehensive analysis report, manifest forms, and the card are to be received at the main gate, and the preliminary details like senders name, address, vehicle, and container numbers are to be recorded.

Then the truck is to move on the weigh bridge, and the gross weight and empty weight after unloading have to be noted, and ultimately the net weight of the waste is to be recorded.

17.8.2 Sampling and Analysis

A representative sample is to be collected from the truck using appropriate sampling equipment. Care has to be taken in avoiding the cross-contamination of the samples. The samples have to be properly registered, and a reference sample has to be preserved for the desired retention time.

Fingerprint analysis of the waste has to be carried out, and the results have to be compared with the comprehensive analysis provided by the generator. The disposal pathway has to be indicated following the disposal criteria. The laboratory is supposed to advise the site managers on the aspects like critical characteristics of the waste, compatibility, and special precautions if any to be taken while handling, etc.

If the waste is not fit for direct disposal, the suitable stabilization mechanism has to be identified by the laboratory, and the same has to be advised. The post-stabilized samples have to be tested and advise accordingly. The spent samples and the retained samples have to be disposed of properly.

17.8.3 In Situ Testing and QA/QC

The core samples have to be collected in each grid and each lift/layer and tested to ensure maximum density and OMC. The hydraulic conductivity is also to be tested in situ or in the laboratory. The entire procedure has to be documented well and preserved for any future verification.

17.8.4 Placement and QA/QC

These sheets have to be placed with an overlap of 15 cm, and the two seems have to be joined through wedge welding. All the weld joints have to be tested by pressure test to ensure no leaks in the liner. In the corners a different mechanism such as extrusion welding has to be adopted. These types of joints have to be tested using a vacuum test. Complete records of seam details, test results, and walk over inspection remarks have to be well documented and preserved.

17.9 POST-MONITORING

The monitoring of the disposal sites for the groundwater, surface water, and ambient air quality is a very important aspect to understand the effectiveness of the containment process. The monitoring is in fact to be done in three stages.

1. before commencement of the operations: baseline monitoring
2. during the course of the operations
3. after closure of the facility: post-monitoring

The baseline monitoring data is essential to compare and study the variances in the levels of selected parameters with respect to the potential chances of contamination due to operations.

The monitoring is to be done during the life of operation of the facility in regular frequencies to create a database and understand the variances.

The monitoring of the facility after complete closure of all operations is called post-monitoring. This should be done on a continuous basis in regular frequencies. The period of monitoring is dependent on several aspects like design of the landfills, type of materials disposed, previous history of the sites, proposed use of the sites after closure, etc. Generally, it will be in the range of 25–30 years.

The parameters to be selected for the water quality monitoring/assessment depend on the constituents that are expected to come in the leachate. For the monitoring of the groundwater, bore wells in the selected locations have to be drilled based on the groundwater flow, contour levels, and other hydrogeological patterns. These surveillance monitoring wells have to be planned so some of them are upstream and some are down-stream with reference to the landfill. These wells have to be well protected, capped all the time, and locked if necessary.

In the same way for the air quality assessment, in addition to the primary parameters like SPM, SO_2, and NO_X, any other specific parameters like hydrocarbons, CO, toxic gases, etc., have to be measured. An automatic weather monitoring station has to be installed to be able to monitor at least temperatures, wind direction, wind velocity, relative humidity, and rainfall. Noise levels in day and night times to be measured in specified frequency. All the environmental monitoring data has to be studied regularly and compared with the base line data and the permissible standards.

17.10 SAFETY AND OCCUPATION HYGIENE

Safety and occupational hygiene is ultimate in the maintenance of hazardous waste management facilities. The following safety equipment has to be provided at all strategic points, regular trainings have to be imparted, and mock drills have to be conducted. Regular health checks of employees have to be made. Trainings on first aid are to be imparted to the selected staff with qualified faculties.

- personal protective equipment
- safety shower with eye wash arrangements
- safety ladders
- relevant fire extinguishes
- preferably a fire hydrant system
- breathing apparatus
- first aid kits

17.10.1 Prevention of Environmental Hazard Risks

Environmental hazard risks in the disposal sites are mainly from the threat due to the following:

- fire
- rain/storm
- air emissions
- ground and surface water contamination
- soil contamination
- explosion and earthquakes

17.10.1.1 Fire

Fire in the landfill may arise to due to disposal noncompatible wastes together or wastes having high ignitability calorific value and reactivity with water, air, acids, and bases. High caloric value (say beyond 2500 cal/kg) wastes may accelerate the process and sustain the fire. Every care has to be taken while deciding the pathway of the disposal as landfill. Fire due to other operations is very remote.

17.10.1.2 Rain/Storm

Heavy rains or storms are not desirable due to the potential of generation of high amounts of leachate. Depending on the rainfall patterns of the area, landfills have to be completely closed or proper temporary cover arrangements have to be made. The storm water drains have to be constructed and maintained to see that they are operational. It is to discourage logging of water in areas where groundwater levels are high.

17.10.1.3 Air Emissions

Dust is the main problem from the point of air emissions. The other emissions due to volatile organics and unpleasant/obnoxious odors may be very well controlled if the organic content in the wastes is restricted. Proper compaction and daily cover combat the entire problem of dust, odors, and volatile gases effectively. The esthetic appeal of the landfill also improves. Spraying of leachate back to landfill in the summer season not only helps in evaporation of leachate but also suppresses the dust.

17.10.1.4 Ground, Surface Waste, and Soil Contamination

The worst scenario is due to failure of a landfill liner system and seepage of leachate into the ground, which ultimately spoils the groundwater. To avoid this problem, experience in design, construction, and operation of landfills is desirable. No activity has to be initiated/executed in the landfill or nearby areas that may cause or is likely to cause damage to the liners. Any spillages on the roads have to immediately be collected and disposed into the landfill time to time. All the storage areas and the operations at stabilization have to be properly monitored for effective functioning in mitigating the contact of waste to the soil and generation of leachate.

17.10.1.5 Explosion and Earthquakes

The possibility of explosion and earthquakes are the ultimate failure cases of the landfills. However, both these possibilities are almost remote. The explosive wastes can be well restricted if proper control on the initial characterization of the wastes and subsequent reactivity tests are conducted for every truck load at the TSD facility. With regard to the earthquakes, the seismologic studies have to be conducted at the stage of site selection itself and landfills have to be developed in non-seismic zones.

17.10.2 List of Potentially Noncompatible Wastes

Many hazardous and toxic wastes, when mixed with other wastes or materials at a hazardous waste storage, treatment, or disposal facility, can produce effects that are harmful to human health and the environment:

1. heat or pressure
2. fire or explosion
3. violent reaction
4. toxic dusts, mists, fumes, or gases
5. flammable fumes or gases

Next are examples of potentially noncompatible wastes, waste components, and materials, along with the harmful consequences that result from mixing materials in one group with materials in another group. The list is intended as a guide to occupiers and owners or operators of storage, treatment, and/or disposal facilities, and to enforcement and permit granting officials, to indicate the need for special precautions when managing these potentially noncompatible waste materials or components.

An occupier and owner or operator must, as the regulations require, adequately analyze the waste so one can avoid creating uncontrolled substances or reactions of the type listed next, whether they are listed or not.

It is possible for potentially incompatible wastes to be mixed in a way that precludes a reaction (e.g., adding acid to water rather than water to acid) or that neutralize them (e.g. a strong acid mixed with strong base), or that control substances produced (e.g., by generating flammable gases in a closed tank equipped so that ignition cannot occur, and burning the gases in an incinerator).

In the lists following, the mixing of a Group A material with a Group B material may have the potential consequences as mentioned.

Group 1A	Group 1B
Acetylene sludge	Acid sludge, acid tar
Alkaline caustic liquids	Acid and water
Alkaline cleaner	Battery acid
Alkaline corrosive liquids	Chemical cleaners
Alkaline corrosive battery	Electrolyte, acid
Fluid caustic water	Etching acid liquid or solvent
Lime sludge and other corrosive	Picking liquor and other corrosive
Alkalies	Acids
Lime wastewater	Spent add
Lime and water	Spent mixed acid
Spent caustic	Spent sulfuric acid

Potential consequences: heat generation, violent reaction

Group 2A	Group 2B
Asbestos waste and other toxic wastes	Cleaning solvents
Beryllium wastes	Data processing liquid
Unrinsed pesticide containers	Obsolete explosives
Waste pesticides	Petroleum waste
	Refinery waste
	Retrograde explosives
	Solvents
	Waste oil and other flammable and explosive wastes

Potential consequences: release of toxic substances in case of fire or explosion

Group 3A	Group 3B
Aluminum Beryllium Calcium Lithium Magnesium Potassium Sodium Zinc powder and other reactive metals and metal hydrides	Any waste in group 1A or 1B

Potential consequences: fire or explosion, generation of flammable gas

Group 4A	Group 4B
Alcohols water	Any concentrated waste in group 1A or 1B Calcium Lithium Metal hybrids Potassium Sodium SO_2Cl_2, $SOCl_2$, PCl_3 CH_3 $SiCl_2$, and other water reactive wastes

Potential consequences: fire explosion or heat generation, generation of flammable or toxic gases

Group 5A	Group 5B
Alcohols Aldehydes Halogenated hydrocarbons Nitrated hydrocarbons and other reactive organic compounds and solvents unsaturated hydrocarbons	Concentrated group 1A or 1B wastes Group 3A wastes

Potential consequences: fire explosion or violent reaction

Group 6A	Group 6B
Spent cyanide solutions	Group 1B wastes

Potential consequences: generation of toxic hydrogen cyanide gas

Group 7A	Group 7B
Chlorates and other strong	Acetic acid and other organic
Oxidizers	Acids
Chlorine	Concentrated mineral acids
Chlorites	Group 2B wastes
Chromic acid	Group 3A wastes
Hypochlorites	Group 5A wastes and other
Nitrates	flammable and combustible waste
Perchlorates	
Permanganates	
Peroxides	

REFERENCES

[1] C. Berkaloff, E. Caadevall, C. Largeau, P. Metzger, S. Peraccca, J. Virlet, The resistant polymer of the walls of the hydrocarbon-rich algae *Botryococcus braunii Phytochemistry* 22 (1983) 389–397.
[2] E.P. Paul, F.E. Clark, (Eds.), Soil Microbiology and Biochemistry. Academic Press New York, 1989, 273 pp.
[3] S.W. Karickhoff, Organic pollutant sorption in aquatic systems, J. Hydraul. Eng. Div. Am. Soc. Civ. Eng. 110 (1984) 707–735.
[4] P.S.C. Rao, J.M. Davidson, Estimation of pesticide retention and transformation parameters required in nonpoint source pollution models, in: M.R. Overcash, J.M. Davidson (Eds.), Environmental Impact of Nonpoint Source Pollution, Ann. Arbor. Science. Pub. Ann. Arbor, MI, 1980, pp. 23–27.
[5] G.G. Briggs, M. Elliott, N.F. Janes, Present status and future prospects for synthetic pyrethroids, in: J. Miyamoto, P.C. Keareny (Eds.), Pesticide Chemistry; Human Welfare and the Environment, Pergam on Press, New York, 1983.
[6] A. Walker, P.A. Brown, A.R. Entwistle, Enhanced degradation of iprodione and vinclozolin in soil. Pesticide Science 17 (1986) 183–193.

FURTHER READING

[1] C.A. Wentz, Hazardous Waste Management, McGraw-Hill International, 1995.
[2] S.E. Manahan, Hazardous Waste Chemistry, Toxicology and Treatment, Lewis Publishers, 1990.
[3] United Nations Environment Programme, Sec.B.C., Geneva, The Basel Convention for the Control of Hazardous Waste and Their Disposal, 1989.
[4] K.L. Shah, Basics of Solid and Hazardous Waste Management Technology, Prentice Hall, NJ, 2000.
[5] G.Tchobanoglous, H.Theisen, S.Vigil, Integrated Solid Waste Management: Engineering Principles and Management Issues, McGraw-Hill, Boston, Mass, 1993.
[6] U.S. Environmental Protection Agency, Code of Federal Regulations, Title 40, Part 264 Subtitle C, Standards for Owners and Operators of Hazardous Waste Treatment, Storage, and Disposal Facilities.
[7] Department for Environment, Food and Rural Affairs (Defra), Methane Emissions from Landfill Sites in the UK. Report Prepared for Defra by Land Quality Management Ltd., 2003. http://www.naei.org.uk/reports.php.
[8] M.R. Allen, A. Braithwaite, C.C. Hills, Trace organic compounds in landfill gas at seven UK waste disposal sites, Environmental Science and Technology 31 (4) (1997) 1054–1061.
[9] Environment Agency, Understanding the Meaning of Operator under IPPC. IPPC Regulatory Guidance Series No. 3, Environment Agency, Bristol, 2001.

CHAPTER EIGHTEEN

Analytical Methods for Monitoring Environmental Pollution

18.1 INTRODUCTION

This chapter deals with the various analytical techniques available for the analysis of the quality of the water, wastewater, and industrial effluents. It also gives the methods of analysis for assessment of soil and air quality.

18.2 STATISTICAL APPROACH

Units: Analytical results are generally expressed either as milligrams per liter (mg/L) or as parts per million (ppm) assuming that 1 L of water, sewage, or industrial waste weighs 1 kg, milligrams per liter is equivalent to parts per million. When dealing with industrial wastes of high specific gravity, results are expressed in milligrams per liter, with the specific gravity given. Only the significant figures are to be recorded. If the concentrations are less than 1 mg/L, it may be more convenient to express the results in micrograms per liter (μg/L) or parts per billion (ppb), where billion is understood to be 10^9. If the concentration is greater than 10,000 mg/L, the results should be expressed as percentage, 1% being equivalent to 10,000 mg/L. For brines or wastes of high specific gravity, results obtained in mg/L can be converted to ppm by the equation:

$$\text{ppm by weight} = \frac{\text{mg/L}}{\text{sp gr}}$$

In reporting analyses of stream pollution or evaluation of plant operation and efficiencies, it is desirable to express the results on a weighted basis, including both the concentration and the volume of flow in cubic feet per second (cfs) or million gallons daily (mgd). These weighted results may be expressed as *quantity units (QU)* according to the practice of the U.S. Public

Environmental Management
ISBN 978-0-12-811989-1
http://dx.doi.org/10.1016/B978-0-12-811989-1.00018-X

Health Service, as *pounds per 24h,* or as *population equivalents* based on biochemical oxygen demand (BOD). Totals of the weighted units may be converted to the weighted average mg/L.

The various units are calculated as follows:

$$QU_1 = (\text{mg/L}) \times 1000\ \text{cfs}$$
$$QU_2 = (\text{mg/L}) \times (\text{mgd})$$
$$\text{lb/24 h} = (\text{mg/L}) \times (\text{mgd}) \times 8.34$$
$$\text{lb/24 h} = (\text{mg/L}) \times (\text{cfs}) \times 5.39$$
$$\text{Population equivalent} = (\text{mg/l 5-day BOD}) \times (\text{mgd}) \times 8.34/0.17$$

Annexure 2 gives some important conversion factors The equivalent weight, in turn, is defined as the weight of the ion (sum of the atomic weights of the atoms making up the ion) divided by the number of charges normally associated with the particular ion. The factors for converting results from mg/L to me/L were computed by dividing the ion charge by the weight of the ion. Conversely, the factors for converting results from me/L to mg/L are calculated by dividing the weight of the ion by the ion charge.

18.2.1 Significant Figures

Any report should include only such figures as are justified by the accuracy of the work. Rounding off is accomplished by dropping the digits which are not significant. If the digit 6, 7, 8, or 9 is dropped, then the preceding digit must be increased by one unit; if the digit 0, 1, 2, 3, or 4 is dropped, the preceding digit is not altered. If the digit 5 is dropped, the preceding digit is rounded off to the nearest even number: thus 2.25 becomes 2.2, and 2.35 becomes 2.4.

Ambiguous zeros: The digit 0 may either record a measured value of zero, or it may serve merely as a spacer to locate the decimal point. If the result of a sulfate determination is reported as 540 mg/L, the recipient of the report may be in doubt whether the zero is significant or not, because the zero cannot be deleted. If an analyst calculates a total-solids content of 1268 mg/L, but realizes that the 4 is somewhat doubtful and that therefore the 7 has no significance, he will round off the answer to 1260 mg/L and so report, but here, too, the recipient of the report will not know whether the zero is significant.

The plus-or-minus (±) notation: If a calculation yields as a result "1555 mg/L" with a standard deviation estimated as ±40 mg/L, it should be reported as 1560 ± 40 mg/L. But if the standard deviation is estimated as ± 100 mg/L, the answer should be rounded off still further and reported as 1600 ± 100 mg/L.

Precision and accuracy: A clear distinction should be made between the terms "precision" and "accuracy" when applied to methods of analysis. *Precision* is a measure of the reproducibility of a method when repeated on a homogeneous sample under controlled conditions, regardless of whether or not the observed values are widely displaced from the true value as a result of systematic or constant errors present throughout the measurements. Precision can be expressed by the standard deviation. *Accuracy* is a measure of the error of a method and may be expressed as a comparison of the amount of element or compound determined or recovered by the test method and the amount actually present. A method may have very high precision but recovery of only a part of the element being determined. An analysis, although precise, may be in error because of poorly standardized solutions, inaccurate dilution techniques, inaccurate balance weights, or improperly calibrated equipment. Thirdly, a method may be accurate but lack precision because of low instrument sensitivity, variable rate of biologic activity, or other factors beyond the control of the analyst.

Standard deviation (σ): If a determination is repeated a large number of times under essentially the same conditions, the observed values, x, will be distributed at random about an average as a result of uncontrollable or experimental errors. If there are an infinite number of observations of causes, a plot of the relative frequency against magnitude will produce a symmetrical bell-shaped curve known as the Gaussian or normal curve. The shape of this curve is completely defined by two statistical parameters: (1) the mean or average, x of n observations; and (2) the standard deviation, σ, which fixes the width or spread of the curve one side of the mean. The formula is:

$$\sigma = \pm \sqrt{\Sigma}(\mathrm{X} - \mathrm{X})^2/\mathrm{n} - 1$$

The proportion of the total observations lying within any given range about the mean is related to the standard deviation.

Range (R): The difference between the smallest and largest of n observations is also closely related to the standard deviation. When the distribution of errors is normal in form, the range, R, of n observations exceeds the standard deviation times a factor d_n only in 5% of the cases. Values for the factor d_n are:

n	d_n
2	2.77
3	3.32
4	3.63
5	3.86
6	4.03

As it is rather general practice to run replicate analyses, use of these limits is very convenient for detecting faulty technique, large sampling errors, or other assignable causes of variation.

Rejection of experimental data: Quite often in a series of observations, one or more of the results deviate greatly from the mean, whereas the other values are in close agreement with the mean value. The problem arises at this point as to rejection of the disagreeing values. Theoretically, no results should be rejected, since the presence of disagreeing results shows faulty techniques and therefore throws doubts on all of the results.

Laboratory apparatus: Volumetric glassware may be calibrated either by the analyst who will use it or by some competent laboratory which can furnish certificates of accuracy. Volumetric glassware is calibrated either "to contain" (TC) or "to deliver" (TD). Glassware which is designed TD will do so with accuracy only when the inner surface is so scrupulously clean that water wets it immediately and forms a uniform film upon emptying. Pyrex glassware should be used whenever possible. For general laboratory use, the most suitable material for containers is resistant borosilicate glass, commonly "pyrex." Special glasses are available with such characteristics as high resistance to alkali attack, low boron content, or exclusion of light. Stoppers, caps, and plugs should be chosen to resist the attack of material contained in the vessel.

It is recommended that samples be collected and stored in bottles made of pyrex, hard rubber, polyethylene, or other inert material. For relatively short storage periods, or for constituents which are not affected by storage in soft glass, such as calcium, magnesium, sulfate, chloride, and perhaps others, the newer type of 2.5-L acid bottle, "bell closure," is satisfactory.

Sample bottles must be carefully cleaned before each use. Glass bottles may be rinsed with a chromic acid cleaning mixture, made by adding 1 L of cone H_2SO_4 slowly, with stirring, to 35 mL saturated sodium dichromate solution, or with an alkaline permanganate solution followed by an oxalic

Table 18.1 Preparation of Common Acid Concentrations Used

Item	HCl	H_2SO_4	HNO_3
Specific gravity (20/4°C) of ACS grade conc acid	1.174–1.189	1.834–1.836	1.409–1.418
Percent of active ingredient in conc reagent	36–37	96–98	69–70
Normality of conc reagent	11–12	36	15–16
Volume (mL) of conc reagent to prepare 1 L of:			
18 N solution	—	500	—
6 N solution	500	167	380
1 N solution	83	28	64
0.1 N solution	8.3	2.8	6.4
Volume (mL) of 6 N reagent to prepare 1 L of 0.1 N solution	17	17	17
Volume (mL) of 1 N reagent to prepare 1 L of 0.02 N solution	20	20	20

acid solution. Rinsing with other concentrated acids may be used to remove inorganic matter. The newer detergents are excellent cleansers for many purposes; either detergents or concentrated HCl can be used for cleaning hard-rubber and polyethylene bottles. After having been cleaned, bottles must be rinsed thoroughly with tap water and then with distilled water.

Distilled Water: Distilled water is supersaturated with carbon dioxide, because of the decomposition of raw water bicarbonates to carbonates in the boiler.

Demineralized water from a mixed bed ion exchanger is satisfactory for many applications. However, as ion exchange fails to remove such non-electrolytes and colloids as plankton, non-ionic organic materials, and dissolved air, it is not suitable for determinations where such constituents interfere. Very-high-purity water, with less than 0.1 μ℧/cm conductance, can be produced by passing ordinary distilled water through a mixed-bed exchanger. Water prepared in this way is often satisfactory for use in the determination of trace cations and anions. Table 18.1 gives the details of preparation of common acid concentrations used.

18.3 INSTRUMENTAL METHODS OF ANALYSIS

Instrumental methods of analysis are finding wide usage for routine monitoring of air quality, groundwater and surface water quality, and soil contamination, as well as during the course of water or waste treatment.

Such methods have allowed analytical measurements to be made immediately at the source, and have permitted the recording of such measurements to be made at some distance from the place of actual measurement. In addition, they have extended considerably the range of inorganic and organic chemicals that can be monitored and the concentrations that can be detected and quantified. A variety of instrumental methods are now routinely used both for investigating the extent of contamination and for routine monitoring of treatment effectiveness.

Ultraviolet spectroscopy: When a molecule absorbs radiant energy in either the ultraviolet or the visible region, valence or bonding electrons in the molecule are raised to higher-energy orbits. Some smaller molecular changes can also take place, but these are usually masked by the above electronic excitations. The result is that fairly broad absorption bands are usually observed in both the ultraviolet and the visible regions. The ultraviolet region is of more limited general usage, although it is particularly suitable for the selective measurement of low concentrations of organic compounds such as benzene-ring-containing compounds or unsaturated straight-chain compounds containing a series of double bonds.

Infrared spectroscopy: Nearly all organic chemical compounds show marked selective absorption in the infrared region. However, infrared spectra are exceedingly complex compared to ultraviolet or visible spectra. Infrared radiation is of low energy, and its absorption by a molecule causes all sorts of subtle changes in the vibrational or rotational energy of the molecule.

Potentiometric titration: Growing in acceptance for titrimetric work are electrical instruments called titrimeters or electrotitrators. If used discreetly, with knowledge of their limitations, these instruments can be applied to many of the titrimetric determinations, including those for acidity and the alkalinities. In addition, titrimetric precipitation reactions such as those for chloride, on complexometric and oxidation–reduction reactions, can be performed with these instruments. To be suitable for these extensive applications, an instrument must be equipped with all of the necessary special electrodes.

Emission methods: Several heavy metals are of environmental concern because of adverse effects they cause to human health and the environment. Thus, analytical methods for their detection and quantification are of great interest. It has long been known that many metallic elements, when subjected to suitable excitation, will emit radiations of characteristic wavelengths. This is the basis for the flame test for sodium (which emits a yellow light) and for other alkali and alkaline-earth metals. When a much more powerful method of excitation is used in place of the flame, most metallic and some nonmetallic elements will emit characteristic radiations. Under properly controlled conditions, the intensity of the emitted radiation at some particular wavelength can be correlated with a quantity of the element present. Thus, both a quantitative and a qualitative determination can be made.

Among the advantages of arc-spark spectrographic analyses are: (1) the minute size of sample required; (2) the elimination of the necessity for bringing solids, such as precipitates and corrosion products, into solution; (3) the detection of all determinable elements present in a sample, whether specifically looked for or not; and (4) the unexcelled sensitivity for some elements. Among the disadvantages of spectrographic analyses are: (1) the high cost of first-class equipment; (2) the necessity for special training and experience; (3) the possible occurrence of severe interferences which must be taken into account if reasonable accuracy is to be achieved; and (4) the inability to distinguish between different valence states of an element, as, for instance, between chromic and chromate or ferric and ferrous.

Atomic emission spectroscopy: This method in its simplest form as a flame photometer is used in water analysis for determining the concentration of alkali and alkaline-earth metals such as sodium, potassium, and calcium. A number of fuel gases, such as acetylene, hydrogen, or the normal natural or manufactured gas used for heating in most laboratories can be used for the flame. The oxidant used is usually tank oxygen rather than air. Flames yielding higher temperatures are capable of exciting more elements, and so are more versatile.

Flame photometry: Flame photometry is used for the determination of sodium, potassium, lithium, and strontium. To some extent, it is also useful for the determination of calcium and other ions.

Atomic absorption spectroscopy: Although this is really an absorption method, it is included under emission spectroscopy because of its similarity to flame photometry. Atomic absorption spectroscopy has gained wide application for environmental analysis because of its versatility for the measurement of trace quantities of most elements in water. Elements such as copper, iron, magnesium, nickel, and zinc can be measured accurately to a small fraction of a milligram per liter.

Electrical methods of analysis make use of the relationships between electrical and chemical phenomena. *Chromatography* is the general term used to describe the set of different procedures used to separate components in a mixture based upon their relative affinity for partitioning between different phases. Two different phases are generally used in modern chromatography to affect separation of a mixture; one is the stationary phase and the other the moving phase. The stationary phase may be either a liquid or a solid, and the moving phase may be either a liquid or a gas. When the moving phase is a gas, the procedure is termed *gas chromatography*, and when it is a liquid, it is called *liquid chromatography*. Other descriptive terms are used depending upon the nature of the stationary phase. For example, when the moving phase is a gas and the stationary phase is a liquid, the procedure is termed *gas–liquid chromatography*. When a liquid is the moving phase, then we may have *paper chromatography* with paper as the stationary phase. Liquid chromatography has advanced considerably in recent years, leading to advanced instruments to which the name *high-performance liquid chromatography* (HPLC) is now applied. Some instruments for "HPLC" use ion exchange resins as the stationary phase and are called *ion chromatographs*.

Chromatographic instruments are now widely used in the environmental engineering and science fields, and permit rapid quantitative measurements to be made of chemicals present in complex mixtures and at concentrations in the ppb or μg/L range and lower.

Pesticides, trihalomethanes, and chlorinated solvents in concentrations of a microgram per liter or less can be measured after extraction from water by use of *electron capture*, *coulometric*, or *electrical conductivity* detectors.

Polarography: Polarography is suggested for scanning industrial wastes for all the various metal ions, especially where the possible interferences in the precise colorimetric procedures are unknown. The older polarographic method for dissolved oxygen remains from the past. A method closely allied to polarography is amperometric titration, which is suitable for the determination of residual chlorine and other iodometric methods by titrimetry.

Gas chromatography: Some comparatively simple gas chromatographic equipment is available for limited procedures such as those described for sludge digester gas. Gas chromatography also shows strong promise for the determination of such organic materials as pesticides in the nanogram range. A number of methods for certain pesticides in water have been published in the literature. The use of a gas chromatograph in conjunction with a mass spectrometer (GC/MS) can allow one to identify hundreds of organic compounds in air or water, and is in common usage to determine the compounds present at sites throughout the country that have been contaminated with hazardous wastes.

High-performance liquid chromatography: HPLC is especially useful for separating non-volatile species and those that are thermally unstable; thus, it extends the range of compounds that can be separated for analysis beyond that possible with gas chromatography. Among the compounds that can be separated by this procedure are amino acids, proteins, nucleic acids, hydrocarbons, fatty acids, carbohydrates, phenols, pesticides, antibiotics, metal-organic species, and a great variety of inorganic substances. Mass spectrometry when used in conjunction with gas chromatography (GC/MS), HPLC (LC/MS), or IPC (IPC/MS) can give positive identification and quantification for a large number of individual organic and inorganic compounds present in water and wastewater, soils, or air.

18.4 WATER QUALITY ANALYSIS

Some of the analyses which are of importance in water quality and treatment are given as follows:

18.4.1 Estimation of Alkalinity

The alkalinity of water is the capacity of that water to accept protons. Alkalinity is usually imparted by the bicarbonate, carbonate, and hydroxide components of a natural or treated water supply. It is determined by titration with a standard solution of a strong mineral acid to the successive bicarbonate and carbonic acid equivalence points, indicated electrometrically or by means of color. Phenolphthalein indicator enables the measurement of that alkalinity fraction contributed by the hydroxide and half of the carbonate. Indicators responding in the pH range of 4 to 5 are used to measure the alkalinity contributed by hydroxide, carbonate, and bicarbonate. The phenolphthalein alkalinity and total-alkalinity titrations are useful for the calculation of chemical dosages required in the treatment of natural water supplies. The stoichiometric relationships between hydroxide, carbonate, and bicarbonate are valid only in the absence of significant concentrations of weak acid radicals other than hydroxyl, carbonate, or bicarbonate.

Interference: Free available residual chlorine markedly affects the indicator color response in some water supplies through bleaching action. The addition of minimal volumes of sodium thiosulfate eliminates this interference without significant loss of accuracy. Where an error is demonstrated, the appropriate correction should be applied. Ultraviolet irradiation also removes residual chlorine. Another interfering factor is the finely divided calcium carbonate and magnesium hydroxide produced during the lime-soda softening process, which may cause a fading endpoint. This suspended material should be removed by filtering the sample through the filter paper prior to titration. Salts of weak inorganic (phosphoric, silicic) and organic acids may contribute to alkalinity.

18.4.2 Reagents

Carbon dioxide-free distilled water: All stock and standard solutions are prepared with distilled water, which has a pH of not less than 6.0.

Phenolphthalein indicator solution: Either the aqueous (1) or alcoholic (2) solution may be used.

1. 5 g phenolphthalein disodium salt is dissolved in distilled water and dilute to 1 L. If necessary, add 0.02 N NaOH dropwise until a faint pink color appears.
2. 5 g phenolphthalein is dissolved in 500 mL 95% ethyl alcohol or isopropyl alcohol, and 500 mL distilled water is added. Then add 0.02 N NaOH dropwise until a faint pink color appears.

Standard sulfuric acid or hydrochloric acid titrant, 0.02 N: Prepare stock solutions approximately 0.1 N by diluting either 9.5 mL concentrated HCl or 3 mL concentrated H_2SO_4 to 1 L. Dilute 200 mL of the 0.1 N stock solution to 1 L with CO_2-free distilled water. Standardize the 0.02 N acid against a 0.02 N sodium carbonate solution which has been prepared by dissolving 1060 g anhydrous Na_2CO_3 (primary standard grade), oven dried at 140°C, and diluting to the mark of 1-L volumetric flask with CO_2 free distilled water. The standardization is performed exactly as the typical alkalinity titration, using the identical volumes of final solution, sodium thiosulfate, phenolphthalein, and total-alkalinity indicator, and the same time interval as for the sample determination. Since the response to endpoint color changes varies among individuals, each analyst should perform the standardization independently and use the normality factor appropriate for his own technique. A standard acid solution, exactly 0.0200 N, is equivalent to 1.00 mg $CaCO_3$ per 1.00 mL.

Mixed bromocresol green-methyl red indicator solution: Either the aqueous (1) or alcoholic (2) solution may be used.

1. 0.02 g methyl red sodium salt and 0.10 g bromocresol green sodium salt is dissolved in 100 mL distilled wsater.
2. 0.02 g methyl red and 0.10 g bromocresol green is dissolved in 100 mL 95% ethyl alcohol or isopropyl alcohol.

Methyl orange indicator solution: 0.5 g methyl orange is dissolved in 1 L distilled water.

Sodium thiosulfate, 0.1 N: 25 g $Na_2S_2O_3 \cdot 5H_2O$ is dissolved and dilute to 1 L with distilled water.

18.4.3 Procedure

Sample volumes requiring less than 25 mL of titrant yield the sharpest color changes at the endpoint and, therefore, are recommended. If indicator methods are used, remove the free residual chlorine by adding 0.05 mL (1 drop) 0.1 N sodium thiosulfate solution or by ultraviolet irradiation.

Phenolphthalein alkalinity: 0.1 mL (2 drops) phenolphthalein indicator is added to a sample of suitable size, in an Erlenmeyer flask. Titration is carried out over a white surface with 0.02 N standard acid to the coloration corresponding to the proper equivalence point of pH 8.3.

Total alkalinity by mixed bromocresol green-methyl red indicator method: 0.15 mL (3 drops) indicator is added to the solution in which the phenolphthalein alkalinity has been determined, or to a sample of suitable size. Titration is performed with 0.02 N standard acid to the proper equivalence point. The indicator yields the following color responses: above pH 5.2, greenish blue; pH 5.0, light blue with lavender gray; pH 4.8, light pink gray with blush cast; pH 4.6, light pink.

Total alkalinity by methyl orange indicator method: 0.1 mL (2 drops) indicator is added to the solution in which the phenolphthalein alkalinity has been determined, or to a sample of suitable size. Titration with 0.02 N standard acid is carried out to the proper equivalence point. The indicator changes to orange at pH 4.6 and pink at 4.0.

18.4.4 Calculation

$$\text{Phenolphthalein alkalinity as mg/L } CaCO_3 = \frac{A \times N \times 50{,}000}{\text{mL sample}}$$

$$\text{Total alkalinity as mg/L } CaCO_3 = \frac{B \times N \times 50{,}000}{\text{mL sample}}$$

where A, mL titration for sample to reach the phenolphthalein endpoint; B, total mL titration for sample to reach second endpoint; and N, normality of acid.

Low alkalinity by potentiometric titration: Low alkalinities of less than 10 mg/L may be determined more accurately by potentiometric titration than by indicator methods. Potentiometric titration eliminates the error due to the sliding endpoint caused by free CO_2 in the sample at the end of the titration.

A sample of suitable size, 100–200 mL, is titrated carefully to pH 4.5 using a microburet. The required volume, X_1, of standard acid is recorded.

Table 18.2 Alkalinity Relationships

Result of Titration	Hydroxide Alkalinity as $CaCO_3$	Carbonate Alkalinity as $CaCO_3$	Bicarbonate Alkalinity as $CaCO_3$
P = 0	0	0	T
P < ½ T	0	2P	T−2P
P = ½ T	0	2P	0
P > ½ T	2P−T	2(T−P)	0
P = T	T	0	0

P, phenolphthalein alkalinity; *T*, total alkalinity.

The titration is then continued to pH 4.2 and the total volume, X_2, of acid is again recorded. The total alkalinity is calculated as shown in the following equation. Precise standardization of the pH meter is not required.

$$\text{Total alkalinity as mg/L CaCO}_3 = \frac{(2X_1\ \text{mL} - X_2\ \text{mL}) \times N \times 50,000}{\text{mL sample}}$$

where N, normality of acid.

The relationship between the alkalinity is mentioned in Table 18.2.

If the pH value of the water has been determined accurately by electrometric means, and from this the mg/L OH^- as $CaCO_3$ is calculated, then the mg/L CO_3^- and HCO_3^- may also be calculated as $CaCO_3$ from the mg/L OH^- and the phsenolphthalein and total alkalinities by the following equations:

$$CO_3^- = 2P - 2\left[OH^-\right]$$

$$HCO_3^- = T - 2P + OH^-$$

Similarly, if difficulty is experienced with the phenolphthalein "end-point," or if it is desired to check the phenolphthalein titration, the phenolphthalein alkalinity may be calculated as $CaCO_3$ from the results of the nomographic determinations of the carbonate $\left(CO_3^-\right)$ and hydroxide (OH^-) ion concentrations, by the use of the relationship:

$$P = \frac{1}{2}\left[CO^-\right] + OH^-$$

18.4.5 Estimation of Calcium

18.4.4.1 Ethylenediaminetetraacetic Acid Titrimetric Method

When EDTA (ethylenediaminetetraacetic acid or its salts) is added to water containing both calcium and magnesium, it combines first with the calcium that is present. Calcium can be determined directly using EDTA when the pH is made sufficiently high so that the magnesium is largely precipitated as the hydroxide and an indicator is used which combines with calcium only. Several indicators are available that will give a color change at the point where all of the calcium has been complexed by the EDTA at pH of 12–13.

Interference: Under conditions of this test, the following concentrations of ions cause no interference with the calcium hardness determination; copper, 2 mg/L; ferrous iron, 20 mg/L; ferric iron 20 mg/L; manganese, 10 mg/L; zinc, 5 mg/L; lead, 5 mg/L; aluminum, 5 mg/L; tin, 5 mg/L. Orthophosphate will precipitate calcium at the pH of the test. Strontium and barium interfere with the calcium determination, and alkalinity in excess of 30 mg/L may cause an indistinct endpoint with hard waters.

Sodium hydroxide, 1N: 40 g of NaOH is dissolved and diluted to 1 L with distilled water.

Indicators: Many indicators are available for the calcium titration. Murexide (ammonium purpurate) was the first indicator available for the detection of the calcium endpoint, and directions are presented for its use in this procedure. Individuals who have difficulty recognizing the murexide endpoint may find the indicator eriochrome blue black R [sodium-1-(2-hydroxy-1-naphthylazo)-2-napthol-4-sulfonic acid] or solochrome dark blue an improvement because of its color change from red to pure blue. Other indicators specifically designed for use as endpoint detectors in the EDTA titration of calcium may be employed.

Murexide (ammonium purpurate) indicator: This indicator changes from pink to purple at the endpoint. An indicator solution is prepared by dissolving 0.15 g of the dye in 100 g of absolute ethylene glycol. Water solutions of the dye are not stable for longer than a day. A ground mixture of the dye powder and sodium chloride provides a stable form of the indicator. It is prepared by mixing 0.20 g of murexide with 100 g of solid NaCl and grinding the mixture to 40–50 mesh. The titration should be performed immediately after the addition of the indicator because it is unstable under alkaline conditions. Endpoint recognition is facilitated by the preparation of a color comparison blank containing 2.0 mL NaOH solution, 0.2 g solid indicator mixture (or 1 to 2 drops if a solution is used), and sufficient standard EDTA titrant (0.05–0.10 mL) to produce an unchanging color.

Eriochrome blue black R indicator: A stable form of the indicator is prepared by grinding together in a mortar 0.20 g powdered dye and 100 g solid NaCl to a 40–50 mesh which is stored in a tightly stoppered bottle. 0.2 g of the ground mixture is used for the titration in the same manner as murexide indicator. During the course of the titration the color changes from red through purple to bluish purple to a pure blue without any trace of reddish or purple tint.

Standard EDTA titrant, 0.01 M: Standard EDTA 0.01 M, is equivalent to 0.4008 mg Ca per 1.00 mL.

18.4.4.1.1 Procedure

Because of the high pH used in this procedure, the titration should be performed immediately after the addition of the alkali. 50.0 mL of sample, or a smaller aliquot diluted to 50 mL is used, so that the calcium content is about 5–10 mg. For hard waters with alkalinity higher than 300 mg/L $CaCO_3$ the endpoint may be improved by taking a smaller aliquot and diluting to 50 mL, or by neutralization of the alkalinity with acid, boiling 1 min, and cooling before continuing the test.

2.0 ml NaOH solution, or a volume sufficient to produce a pH of 12–13 is added with stirring. 0.1–0.2 g of the indicator mixture is added. EDTA titrant is added slowly, with continuous stirring to the proper endpoint. When using murexide, the endpoint may be checked by adding 1 or 2 drops of titrant in excess to make certain that no further color change occurs.

18.4.4.1.1.1 Calculation

$$\text{mg/L Ca} = \frac{A \times B \times 400.8}{\text{mL sample}}$$

$$\text{Calcium hardness as mg/L CaCO}_3 = \frac{A \times B \times 1000}{\text{mL sample}}$$

where A, mL titration for sample; and B, mg $CaCO_3$ equivalent to 1.00 mL EDTA titrant at the calcium indicator endpoint.

18.4.5 Estimation of Chloride

Chloride is one of the major anions in water and sewage. The salty taste produced by chloride concentrations is variable and dependent on the chemical composition of the water. Some waters containing 250 mg/L chloride may evidence a detectable salty taste with sodium ions. On the other hand, the typical salty taste may be absent in waters containing as much chloride as 1000 mg/L when there is a predominance of calcium and magnesium ions. High chloride content also exerts a deleterious effect on metallic pipes and structures, as well as on agricultural plants.

18.4.5.1 Argentometric Method

In a neutral or slightly alkaline solution, potassium chromate can be used to indicate the endpoint of the silver nitrate titration of chloride. Silver chloride is quantitatively precipitated before red silver chromate is formed.

18.4.5.1.1 Regents

Chloride-free water: If necessary, remove any chloride impurity from distilled water by redistillation from an all-pyrex apparatus or passage through a mixed bed of ion-exchange resins.

Potassium chromate indicator solution: 50 g K_2CrO_4 is dissolved in a little distilled water. Silver nitrate solution is added until a definite red precipitate is formed which is allowed to stand 12 h, filter, and dilute filtrate to a 1 L with distilled water.

Standard silver nitrate titrant, 0.0141 N: Dissolve 2.396 g $AgNO_3$ in distilled water and dilute to 1000 mL. Standard silver nitrate solution, exactly 0.041 N, is sequivalent to 0.500 mg Cl per 1.00 mL.

Standard sodium chloride 0.0141 N: Dissolved 0.8241 g NaCl (dried at 140°C) in chloride-free water and dilute to 1000 mL; 1.00 mL = 0.500 mg Cl.

18.4.5.1.2 Removal of Interference

1. **Aluminum hydroxide suspension:** 125 g of potassium or ammonium alum, $K_2Al_2(SO_4)_4 \cdot 24H_2O$ or $(NH_4)_2Al_2(SO_4)_4 \cdot 24H_2O$, is dissolved in 1 L of distilled water and warmed to 60°C and 55 mL of concentrated NH_4OH is added slowly with stirring. After allowing to stand about 1 h, transfer the mixture to a large bottle and wash the precipitate by successive additions, with thorough mixing, and decantations of distilled water, until free from chloride. When freshly prepared, the suspension occupies a volume of approximately 1 L.
2. *Phenolphthalein indicator solution*
3. *Sodium hydroxide*, 1 N.
4. *Sulfuric acid*, 1 N.
5. *Hydrogen peroxide*, 30%.

18.4.5.1.3 Procedure

Use a 100 mL sample or a suitable aliquot diluted to 100 mL. If the sample is highly colored, 3 mL $Al(OH)_3$ is added suspension; mix, allow to settle, filter, wash, and combine filtrate and washing. If sulfide, sulfite, or thiosulfate is present, make the water alkaline to phenolphthalein with sodium hydroxide solution. 1 mL H_2O_2 is added and stirred and neutralize with sulfuric acid.

Samples in the pH range 7–10 may be titrated directly. Samples not in this range are adjusted with sulfuric acid or sodium hydroxide solution. 1.0 mL K_2CrO_4 indicator solution is added and titrated against standard silver nitrate titrant to a pinkish-yellow endpoint. The means of consistent endpoint detection are left to the individual analyst.

Standardize the silver nitrate titrant and establish the reagent blank value by the titration method outlined above. A blank of 0.2–0.3 mL is usual for the method.

18.4.5.1.4 Calculation

$$\text{mg/L Cl} = \frac{(A - B) \times N \times 35,450}{\text{mL sample}}$$

where A, mL titration for sample; B, mL titration for blank; and N, normality of $AgNO_3$.

$$\text{mg/L NaCl} = \text{mg/L Cl} \times 1.65$$

18.4.6 Estimation of Chlorine (Residual)

The chlorination of water supplies accomplishes a number of treatment objectives. The destruction of microorganisms is a primary function. However, overall improvement in the finished water can result from chlorine's reaction with ammonia, iron, manganese, sulfide, and protein substances. With some taste and odor producing compounds such as phenols, chlorine may intensify the problem or under careful control may improve the water quality.

Most methods for the determination of free or combined available chlorine are based on reactions with reducing agents that are not specific for these materials. Chlorine in water may be present as free available chlorine (in the form of hypochlorous acid or hypochlorite ion, or both); or as combined available chlorine (chloramines and other chloro-derivatives). Both free and combined chlorine may be preset simultaneously.

Some oxidizing agents, including free halogens other than chlorine, will appear quantitatively as free chlorine; this is also true of chlorine dioxide. A small proportion of nitrogen trichloride (trichloramine) will titrate as free chlorine. These substances rarely appear in sufficient quantity to introduce a significant error; nevertheless, their action should be familiar to the analyst as they will affect all the chlorine methods described.

Selection of method: The iodometric method is employed as a standard and is the basis for the standardization of chlorine water used in preparing temporary standards. It is also suitable for determining high chlorine residuals, which are more frequently encountered now than heretofore. The iodometric method is more precise than the orthotolidine method when the residual chlorine concentration is greater than 1 mg/L.

The amperometric titration method appears to be one of the most accurate available for the determination of free or combined available chlorine. The method is usually unaffected by the presence of various oxidizing agents, temperature variations, and turbidity and color which interfere with the accuracy of the other methods.

Sampling and storage: Chlorine in aqueous solution is not stable, and the chlorine content of samples or solutions, particularly weak solutions, will rapidly decrease. Exposure to sunlight or other strong light or agitation will accelerate the reduction of chlorine present in such solutions. Therefore, it is recommended that chlorine determinations be started immediately after sampling, avoiding excessive light and agitation. Samples to be analyzed for chlorine cannot be stored.

18.4.7 Estimation of Chlorine Demand

The "chlorine demand" of water is caused by such inorganic reductants as ferrous, manganous, nitrite, sulfide, and sulfite ions. Ammonia and cyanide consume considerable chlorine during the free residual chlorination process. Chlorine substitutes on phenols and other similar aromatic compounds to form chloro-derivative compounds, but may also oxidize the aromatic compounds when larger amounts of chlorine are added. It may also react with ammonia and naturally occurring amino compounds to form chloramines with an active or oxidizing chlorine atom. The destruction of the chloramines compounds can be achieved by the addition of more chlorine and subsequently with the addition of enough chlorine a free available residual (hypochlorous acid) may be attained.

The chlorine demand of water is the difference between the amount of chlorine applied to a treated supply and the amount of free, combined, or total available chlorine remaining at the end of the contact period. The chlorine demand of any given water varies with the amount of chlorine applied, time of contact, pH, and temperature. For comparative purposes it is imperative that *all test conditions be stated*. The smallest amount of residual chlorine considered significant is 0.1 mg/L Cl.

The laboratory method is designed to determine the so-called immediate demand as well as other demands at longer contact periods. Chlorine demand determinations are made to determine the amount of chlorine that must be applied to a water to produce a specific free, combined, or total available chlorine residual after a selected period of contact. If the amount of chlorine applied to waters containing ammonium or organic nitrogen compounds is not sufficient to reach what is termed the "breakpoint," chloramines and certain other chloro-derivatives react as combined available residual chlorine are produced. When sufficient chlorine has been added to reach the breakpoint, which depends on pH, ratio of chlorine to nitrogenous compounds present, and other factors, subsequent additions of chlorine remain in the free available state.

Standard chlorine solution: A suitable solution may be obtained from the chlorinator solution hose or by bubbling chlorine gas through distilled or tap water. The stability of the chlorine solution may be improved by storing in the dark or amber glass stoppered bottles. Even so, it will lose strength and must be standardized each day that it is used. Alternatively, household hypochlorite solution, which contains about 30,000–50,000 mg/L chlorine equivalent, may be diluted to suitable strength. This is more stable than a chlorine solution, but should not be used more than a week without restandardizing. The solution used for determining chlorine demand should preferably be the same kind of chlorine solution as is actually applied in plant treatment. For making temporary standards for calibrating a photometer, it is preferable that hypochlorite be used. Depending on the intended use, a suitable strength of chlorine solution will usually be between 100 and 1000 mg/L. If used for chlorine demand determination, it should be sufficiently strong so that the volume of treated portions will not be increased more than 5% by addition of the chlorine solution.

2 ml of acetic acid is added to 10–25 mL distilled water in a flask to which about 1 g potassium iodide is also added. It is sufficient for the analyst to estimate this on a spatula or small spoon, after first having familiarized himself with the quantity by one or more weighing. Measure into the flask a suitable volume of the chlorine solution.

Titration is carried out with standardized 0.025 N sodium thiosulfate titrant until the yellow iodine color is almost gone. 1–2 mL starch indicator solution is added and the titration is continued till the disappearance of the blue color. Blank solution is also determined by adding identical quantities of acid, KI, and starch indicator to a volume of distilled water corresponding to the sample used for titration.

$$\text{mg/mL Cl} = \frac{(A+B) \times N \times 35.45}{\text{mL sample}}$$

where A, mL titration for sample; B, mL titration for blank which may be positive or negative; and N, normality of $Na_2S_2O_3$.

18.4.7.1 Procedure

Ten equal portions of the sample is measured and amply mixed. Chlorine water is added to this so that first portion leaves no chlorine residual at the end of the contact period, especially if low demands are being studied. Increasing amounts of chlorine to the successive portions in the series is added. The increase in dosage between portions may be as low as 0.1 mg/L for determining low demands, and up to 1.0 mg/L or more for higher demands. Mixing while the chlorine solution is being added to the sample is imperative. It may be advantageous to dose the portions of the sample according to a staggered schedule that will permit the determination of chlorine residuals at the predetermined contact time.

The usual purpose of a chlorine demand test is to determine the amount of chlorine required to produce a specific free, combined, or total available chlorine residual after a definite, record time interval. It is recommended, therefore, that the testing be carried out at the end of the contact period corresponding to the contact time at the point of control. This may vary from a few minutes to several hours. To determine in advance the effect of chlorination, the plant contact time and temperature should be duplicated in the laboratory. Under some circumstances it is desirable to make several chlorine determinations after different periods of contact, such as 15, 30, and 60 min. Such a procedure will give an indication of the stability of the residual chlorine as related to time, which will be very useful information in plant control. The time of contact must be recorded. During the contact period the chlorinated samples must be protected from strong daylight.

18.4.8 Estimation of Fluoride

A fluoride concentration of approximately 1.0 mg/L is an effective preventive of dental caries without harmful effects on health. Fluoride may occur naturally in water or may be added in controlled amounts. Some fluorosis may occur when the fluoride level exceeds the recommended limits. In rare instances the fluoride concentration naturally occurring may approach 10 mg/L. Such waters should be defluoridated to reduce the fluoride content to the acceptable levels.

The accurate determination of fluoride in water supplies has increased in importance with the growth of the practice of fluoridation of supplies as a public health measure. The maintenance of a constant fluoride concentration is essential in maintaining the effectiveness and safety of the fluoridation procedure.

Among the many methods suggested for the determination of fluoride ion in water, the colorimetric methods are believed to be the most satisfactory at the present time. They are based on the reaction between fluoride and a zirconium-dye lake. The fluoride reacts with the dye lake, dissociating a portion of it into a colorless complex anion (ZrF_6^-) and the dye. As the amount of fluoride is increased, the color produced becomes progressively lighter or different in hue, depending on the reagent used.

Because all of the methods are subject to errors where interfering ions are present, it may be necessary to distill the sample as directed in preliminary step I or II prior to making the fluoride determination. When interfering ions are not present in excess of the tolerances of the method, the fluoride determination may be made directly without distillation.

18.4.8.1 SPADNS (Sodium 2-(ParastilfopheylAzo)–1,8-Dihydroxy-3,6-Naphthalene diSulfonate) Method

The reaction rate between fluoride and zirconium ions is influenced greatly by the acidity of the reaction mixture. By increasing the proportion of acid in the reagent, the reaction can be made practically instantaneous. Under such conditions, however, the effect on various ions differs from that in the conventional alizarin methods. The selection of dye for this rapid fluoride method is governed largely by the resulting tolerance to these ions.

Standard fluoride solution: SPADNS solution: 0.985 g SPADNS, sodium 2-(parastilfopheylazo)–1,8-dihydroxy-3,6-naphthalene disulfonate, also called 4,5-dihydroxy-3-(parasulfophenylazxo)-2,7-naphthalenedisulfonic acid trisodium salt is dissolved in distilled water and diluted to 500 mL. This solution is stable indefinitely if protected from direct sunlight.

Zirconyl-acid reagent: 0.133 g of zirconyl chloride octahydrate, $ZrOCl_2{\cdot}8H_2$, is dissolved in about 25 mL distilled water. 350 mL of concentrated HCl is added and dilute to 500 mL with distilled water.

Acid zirconyl-SPADNS reagent: Mix equal volumes of SPADNS solution and zirconyl-acid reagent to produce a single reagent which is stable for at least 2 years.

Reference solution: Add 10 mL SPADNS solution to 100 mL distilled water. Dilute 7 mL concentrated HCl to 10 mL and add to the diluted SPADNS solution. The resulting solution, used for setting the reference point (zero) of the spectrophotometer or photometer, is stable and may be reused indefinitely. This reference solution can be eliminated by using, if desired, one of the prepared standards as a reference.

18.4.8.1.1 Procedure

Preparation of standard curve: Prepare fluoride standards in the range of 0–1.40 mg/L by diluting appropriate quantities of the standard fluoride solution to 50 mL with distilled water. Pipet 5.00 mL each of SPADNS solution and zirconyl-acid reagent, or 10.00 mL of the mixed acid zirconyl-SPADNS reagent, to each standard and mix well, exercising care to avoid contamination during the process. Set the photometer to zero absorbance with the reference solution and obtain the absorbance readings of the standards immediately. Plot a curve of the fluoride-absorbance relationships. Prepare a new standard curve whenever a fresh batch of reagent is made up or a different standard temperature is desired. If no reference solution is used, set the photometer at some convenient point established with a prepared fluoride standard.

Sample pretreatment: If the sample contains residual chlorine, remove it by adding 1 drop (0.05 mL) of sodium arsenite solution for each 0.1 mg Cl and mix. (Sodium arsenite concentrations of 1300 mg/L produce an error of 0.1 mg/L at 1.0 mg/L F.)

Color development: Use a 50.0 mL sample or an aliquot diluted to 50 mL. Adjust the temperature of the sample to that used for the standard curve. Add 5.00 mL each of the SPADNS solution and zirconyl-acid reagent, or 10.00 mL of the acid zirconyl-SPADNS reagent; mix well, exercising care to prevent contamination during the process; and read the absorbance immediately or at any subsequent time, first setting the reference point of the photometer as above. If the absorbance falls beyond the range of the standard curve, repeat the procedure using a smaller sample aliquot.

18.4.8.1.2 Calculation

$$\text{mg/L F} = \frac{\text{A} \times 1000}{\text{mL sample}} \times \frac{\text{B}}{\text{C}}$$

where A, mg F determined photometrically. The ratio B/C applies only when a sample is distilled to a volume B, and an aliquot C taken from it for color development.

18.4.9 Estimation of Hardness

Hardness of water was understood to be a measure of the capacity of the water for precipitating soap. Soap is precipitated chiefly by the calcium and magnesium ions commonly present in water, but may also be precipitated by ions of other polyvalent metals, such as iron, aluminum, manganese, strontium, and zinc, and by hydrogen ions. Because all but the first two are usually present in insignificant concentrations in natural waters, hardness is defined as a characteristic of water which represents the total concentration of just the calcium and magnesium ions expressed as calcium carbonate. However, if present in significant amounts, other hardness-producing metallic ions should be included.

When the hardness is numerically greater than the sum of the "carbonate alkalinity" and the "bicarbonate alkalinity," that amount of hardness which is equivalent to the total alkalinity is called "carbonate hardness"; the amount of hardness in excess of this is called "noncarbonate hardness." When the hardness is numerically equal to, or less than, the sum of carbonate and bicarbonate alkalinity, hardness is attributed to carbonate hardness and there is no noncarbonate hardness. The hardness may range from zero to hundreds of milligrams per liter in terms of calcium carbonate, depending on the source and treatment to which the water has been subjected.

Selection of method: Two approaches are presented for the determination of hardness. Method A, hardness by calculation, is applicable to all waters and is considered to yield higher accuracy. If a complete mineral analysis is performed, the hardness can be reported by calculation. Method B, the EDTA titration method, which measures the calcium and magnesium ions, may be applied with appropriate modification to any kind of water. The procedure described affords a means of rapid analysis.

Reporting of results: When reporting hardness, the analyst should state either the ions determined or the method used, e.g., "hardness (Ca, Mg, Sr, Fe, Al, etc.)," "hardness (EDTA)."

18.4.9.1 Ethylenediaminetetraacetic Acid Titrimetric Method

Ethylenediamine tetraacetic acid and its sodium salts (abbreviated EDTA) from a chelated soluble complex when added to a solution of certain metal cations like calcium, magnesium, aluminium. If a small amount of a dye such as eriochrome black T is added to an aqueous solution containing calcium and magnesium ions at a pH of 10.0 ± 0.1, the solution will become wine red. If EDTA is then added as a titrant, the calcium and magnesium will be complexed. After sufficient EDTA has been added to complex all the magnesium and calcium, the solution will turn from wine red to blue. This is the endpoint of the titration. Magnesium ion must be present to yield a satisfactory endpoint in the titration. A small amount of complexometrically neutral magnesium salt of EDTA is therefore added to the buffer, a step which automatically introduces sufficient magnesium and at the same time obviates a blank correction.

The sharpness of the endpoint increases with increasing pH. The pH, however, cannot be increased indefinitely because of the danger of precipitating $CaCO_3$ or $Mg(OH)_2$, and because the dye changes color at high pH values. The pH value of 10.0 ± 0.1 recommended in this procedure is a satisfactory compromise. A limit of 5 min is set for the duration of the titration in order to minimize the tendency toward $CaCO_3$ precipitation (Table 18.3).

18.4.9.2 Reagents

18.4.9.2.1 Buffer Solution

1. Dissolve 16.9 g ammonium chloride, NH_4Cl, in 143 mL concentrated ammonium hydroxide, NH_4OH; add 1.25 g of magnesium salt of EDTA (this salt is available commercially) and dilute to 250 mL with distilled water.

Table 18.3 Maximum Concentrations of Interferences Permissible With Various Inhibitors[a]

Interfering Substance	Maximum Interference Concentration mg/L		
	Inhibitor I	Inhibitor II	Inhibitor III
Aluminum	20	20	20
Barium	÷	÷	÷
Cadmium	÷	20	÷
Cobalt	Over 20	0.3	**0+**
Copper	Over 30	20	0.3
Iron	Over 30	5	20
Lead	÷	20	÷
Manganese (Mn^{++})	÷	1	1
Nickel	Over 20	0.3	**0+**
Strontium	÷	÷	÷
Zinc	÷	200	÷
Polyphosphate	—	10	—

÷, Titrates as hardness.
+ inhibitor fails if substance present.
[a]Based on 25 mL aliquot diluted to 50 mL.

2. In the absence of the magnesium salt of EDTA, dissolve 1.179 g disodium salt of EDTA dihydrate (analytical reagent grade) and 0.780 g $MgSO_4 \cdot 7H_2O$ or 0.644 g $Mg\ Cl_2 \cdot 6H_2O$ in 50 mL distilled water. Add this solution to 16.9 g NH_4Cl and 143 mL concentrated NH_4OH with mixing, and dilute to 250 mL with distilled water. To attain the highest accuracy, adjust to exact equivalence through appropriate addition of a small amount of EDTA or magnesium sulfate or chloride.

Keep the solution in a plastic or resistant-glass container, tightly stoppered to prevent loss of NH_3 or pickup of CO_2. A frequently opened container should not hold more than a month's supply. The buffer solution is best dispensed by means of a bulb-operated pipet. Discard the buffer when 1 or 2 mL added to the sample fails to produce a pH of 10.0 ± 0.1 at the endpoint of the titration.

The dye eriochrome black T is the sodium salt of 1-(1-hydroxy-2-naphthylazo)-5-nitro-2-naphthol-4-sulfonic acid, number 203 in the color index. Many types of indicator solutions are advocated in the literature and for the most part are satisfactory. The prime difficulty with indicator solutions is their instability through aging, giving rise to indistinct endpoints in the EDTA titration. For example, alkaline solutions of the dye are sensitive to oxidants, and aqueous or alcoholic solutions are stable for only about a week. Dry mixtures of the dye and sodium chloride are stable. Prepared dry mixtures of the indicator and an inert salt are available commercially.

The following formulations have been widely used and are generally satisfactory:

1. 0.5 g of dye is mixed with 4.5 g hydroxylamine hydrochloride. This mixture is then dissolved in 100 mL of 95% ethyl or isopropyl alcohol.
2. 0.5 to 1.0 g dye is mixed in 100 g of an appropriate solvent such as 2″, 2.2′-nitrilotriethanol (also called triethanolamine) or 2-methoxyethanol (also called ethylene glycol monomethyl ether).
3. 0.5 g dye and 100 g NaCl are mixed together to prepare a dry powder mixture.

All indicator formulations tend to deteriorate, especially when exposed to moist air. If the endpoint color change is not clear and sharp, it usually means that an appropriate inhibitor is needed. If sodium cyanide inhibitor does not sharpen the endpoint, the indicator is probably at fault.

1. Analytical reagent grade disodium ethylenediamine tetraacetate dihydrate, also called (ethylenedinitrilo) tetraacetic aid disodium salt (EDTA), $Na_2H_2C_{10}H_{12}O_8N_2 \cdot 2H_2O$. Weigh 3.723 g of the dry reagent, dissolve in distilled water, and dilute to 1000 mL.

2. The technical grade of the disodium salt of EDTA dihydrate may also be used if the titrant is allowed to stand for several days and is then filtered. Dissolve 4.0 g of such material in 800 mL distilled water. Adjust the titrant so that 1.00 mL = 1.00 mg $CaCO_3$.

Standard calcium solution: Weigh 1000 g anhydrous calcium carbonate, $CaCO_3$, powder (primary standard or special reagent low in heavy metals, alkalis, and magnesium) into a 500 mL erlenmeyer flask. Place a funnel in the neck of the flask and add, a little at a time, 1 + 1 HCl until all the $CaCO_3$ has dissolved. Add 200 mL distilled water and boil for a few minutes to expel CO_2. Cool, add a few drops of methyl red indicator, and adjust to the intermediate orange color by adding 3 N NH_4OH or 1 + 1 HCl, as required. Transfer quantitatively to a 1 L volumetric flask and fill to the mark with distilled water. This standard solution is equivalent to 1.00 mg $CaCO_3$ per 1.00 mL.

18.4.9.2.2 Procedure

The aliquot of sample taken for the titration should require less than 15 mL of EDTA titrant. The duration of titration should not exceed 5 min measured from the time of the buffer addition.

Dilute 25.0 mL of sample to about 50 mL with distilled water in a porcelain casserole or other suitable vessel. Add 1–2 mL of buffer solution. Usually 1 mL will be sufficient to give a pH of 10.0–10.1. If in the titration a sharp endpoint color change is not obtained, it usually means that an inhibitor needs to be added at this point in the procedure or that the indicator has deteriorated.

Add 1 to 2 drops of indicator solution or an appropriate amount of dry-powder indicator formulation. Add the standard EDTA titrant slowly, with continuous stirring, until the last reddish tinge disappears from the solution, adding the last few drops at 3–5-s intervals. The color of the solution at the endpoint is blue under normal conditions.

18.4.9.2.3 Calculation

$$\text{Hardness (EDTA) as mg/L } CaCO_3 = \frac{A \times B \times 1000}{\text{mL Sample}}$$

where A, mL titration for sample; and B, mg $CaCO_3$ equivalent to 1.00 mL EDTA titrant.

18.4.10 Magnesium

Magnesium ranks eighth among the elements in order of abundance and is a common constituent of natural water supplies. Important contributors to the hardness of water, magnesium salts break down on heating to form deleterious scale in boilers. Concentrations in excess of 128 mg/L Ca also exert a cathartic and diuretic action. Chemical softening treatment or ion exchange is employed to reduce the magnesium and associated hardness to tolerable levels. The magnesium concentration may vary from zero to several hundred mg/L, depending on the source and the treatment of the water.

Magnesium can be determined by the gravimetric method only after prior removal of calcium salts and is generally determined by this method on the filtrate and washings from the gravimetric and permanganate calcium determination. By the photometric method, magnesium can be determined in the presence of calcium salts directly on the water sample. Both methods can be applied to all concentrations by the selection of suitable aliquots.

18.4.11 Estimation of Nitrogen (Ammonia)

Ammonia nitrogen is present in variable concentrations in many surface and ground waters. A product of microbiologic activity, ammonia nitrogen is sometimes accepted as chemical evidence of sanitary pollution when encountered in raw surface supplies. Its occurrence in ground supplies is quite general, as a result of natural reduction processes. Dechlorination in case of chlorination of drinking water samples, prior to analysis will convert all the forms of chloramines to ammonia.

18.4.12 Nitrogen (Nitrate)

Nitrate represents the most highly oxidized phase in the nitrogen cycle, and normally reaches important concentrations in the final stages of biologic oxidation. It generally occurs in trace quantities in surface water supplies, but may attain high levels in some groundwater. In excessive amounts, it contributes to the illness known as infant methemoglobinemia. A limit of 45 mg/L nitrate has accordingly been imposed on drinking water as a means of averting this condition. The nitrate concentration of most drinking water usually falls below 10 mg/L.

18.4.12.1 Ultraviolet Spectrophotometric Method

Measurement of the ultraviolet absorption at 220 mμ enables a rapid means of determining nitrate. The nitrate calibration curve follows Beer's law up to 11 mg/L N. Because dissolved organic matter may also absorb at 220 mμ, and nitrate does not absorb at 275 mμ, a second measurement is made at 275 mμ for the purpose of correcting the nitrate value. Filtration of the sample is done to remove possible interference from suspended particles. Acidification with 1 N hydrochloric acid is designed to prevent interference from hydroxide or carbonate concentrations up to 1000 mg/L as $CaCO_3$.

18.4.12.1.1 Procedure

Preparation of standard curve: Nitrate calibration standards in the range 0–0.35 mg N by diluting to 50 mL the following volumes of the standard nitrate solution: 0, 1.00, 2.00, 4.00, 7.00, and 35.0 mL are prepared. To each standard 1 mL, 1 N HCl is added, mixed, and measured at 220 mμ against redistilled water.

When nitrite, chromium, and anionic surfactants are known to be present in the sample, correction curves for each of the substance are prepared at 2 mg/L intervals up to 10 mg/L. Potassium nitrite, KNO_2, potassium dichromate, $K_2Cr_2O_7$, alkyl benzene sulfonate (or linear alkylate sulfonate), and redistilled water are commonly used. The absorbance given by varying concentrations of each substance at wavelength 220 mμ against redistilled water is measured, and plotted as separate curves for each of these interfering materials.

18.4.12.1.2 Calculation

Correction for dissolved organic matter: Convert the absorbance or transmittance measurement at 275 mμ into equivalent nitrate by reading the nitrate value form the standard calibration curve obtained at 220 mμ. Multiply the value by a suitable correction factor, which has been determined on a sufficient number of samples of the particular water. Deduct the organic correction from the gross nitrate result.

Correction for nitrite, hexavalent chromium, or surfactants: Deduct the equivalent nitrate values for each of these interfering substances from the gross nitrate result.

$$\text{mg/L nitrate N} = \frac{\text{net mg nitrate N} \times 1000}{\text{mL sample}}$$

$$\text{mg/L } NO_3 = \text{mg/L nitrate N} \times 4.4.3$$

18.4.13 Nitrogen (Nitrite)

Nitrate, an intermediate stage in the nitrogen cycle, may occur in water as a result of the biological decomposition of proteinaceous materials. When correlated with the concentration of other nitrogen forms, trace amounts of nitrite may indicate organic pollution. Nitrate may also be produced in water treatment plants or in the distribution system through the action of bacteria or other organisms on ammonia nitrogen fed at elevated temperatures in the combined residual chlorination of water. Nitrite can likewise enter a water supply through its use as a corrosion inhibitor in industrial process water. The nitrite concentration of a drinking water rarely exceeds 0.1 mg/L.

The nitrite concentration is determined through the formation of a reddish-purple azo dye produced at pH 2.0 to 2.5 by the coupling of diazotized sulfanilic acid with naphthylamine hydrochloride. The diazotization method is suitable for the visual determination of nitrite nitrogen in the range 1–25 μg/L N. Photometric measurements can be performed in the range 5–50 μg/L if a 5 cm light path and a green color filter are available. The color system obeys Beer's law up to 0.18 mg/L N or 0.6 mg/L NO_2, with a 1 cm light path at 520 mμ.

18.4.13.1 Procedure

If the sample contains suspended solids and color, add 2 mL aluminum hydroxide suspension to 100 mL sample, stir thoroughly, allow to stand for a few minutes, and filter, discarding the first portion of the filtrate. Coagulation with zinc sulfate and hydroxide may also be practiced as described under nitrogen (ammonia).

To 50.0 mL clear sample, which has been neutralized to pH 7, or to an aliquot diluted to 50.0 mL, add 1.0 mL EDTA solution and 1.0 mL sulfanilic acid reagent. Mix thoroughly. At this point, the pH of the solution should be about 1.4. After it has been standing 3–10 min, add 1.0 mL naphthylamine hydrochloride reagent and 1.0 mL sodium acetate buffer solution; mix well. At this point, the pH of the solution should be 2.0 to 2.5. Measure the reddish-purple color after 10–30 min.

Measure the absorbance at or near 520 mμ against a reagent blank, and run parallel checks frequently against known nitrite standards, preferably in the nitrogen range of the samples. Re-determine complete calibration curves following the preparation of new reagents.

18.4.13.2 Calculation

$$\text{mg/L nitrite N} = \frac{\mu\text{g nitrite N}}{\text{mL sample}}$$

$$\text{mg/L } NO_2 = \text{mg/L nitrate N} \times 3.29.$$

18.4.14 Estimation of Oxygen (Dissolved)

Adequate "dissolved oxygen" (DO) is necessary for the life of fish and other aquatic organisms. The DO concentration may also be associated with corrosivity of water, photosynthetic activity, and septicity.

The azide modification of the "iodometric method" is recommended for most conditions. The unmodified iodometric method can be used only with water containing less than 0.1 mg/L nitric nitrogen and less than 0.5 mg/L ferrous iron, and free of other possible interferences.

18.4.14.1 Oxygen Demand (Biochemical)

BOD in water is determined. Seeding with sewage organisms, and elimination of interferences from residual chlorine and other bactericidal substances should be carried out. The amount of pollution in the sample will govern the need for and the degree of dilution. Samples with low DO values can be aerated to increase the initial DO content above that required by the BOD. Air is bubbled through a diffusion tube into the sample for 5 min, or until the DO is at least 7 mg/L. On one portion of the aerated sample the DO is determined; another portion is seeded and incubated for the BOD determination.

18.4.14.2 Oxygen Demand (Chemical)

The "Chemical Oxygen Demand" (COD) test indicates the quantity of oxidizable compounds present in water, and will vary with water composition, concentration of reagent, temperature, period of contact, and other factors. In some instances, a rough correlation between BOD and COD has been established. Since chemical oxidation and biologic oxidation are different processes, the results may differ to a large degree.

18.4.14.3 Iodometric Method

Ozone liberates free iodine from a potassium iodide solution. For accurate results the solution should be alkaline during the absorption for ozone. In practice, solutions of potassium iodide quickly become alkaline during the process. After acidification, the liberated iodine is titrated with standard 0.005 N sodium thiosulfate using starch indicator.

Interference: Because ozonated water may contain manganese dioxide, ferric ion, nitrite, possibly peroxide, and other oxidation products, these interferences are avoided by passing the ozone through the gaseous phase into a potassium iodide solution for titration. The stability of ozone solutions decreases progressively at each increment in temperature above freezing and with each increment in pH above 3.0.

Minimum detectable concentration: ~0.03 mg/ozone.

Standard gas washing bottles and absorbers, 1 L and 500 mL capacities, with medium-permeability porous-plate diffusers at bottom.

Glass, stainless steel, or aluminum piping, for carrying ozonized air. Good quality Teflon tubing may also be used for short runs, but not rubber.

18.4.14.3.1 Reagents

Potassium iodide solution: Dissolve 20 g KI, free from iodine, iodate, and reducing agents, in 1 L of freshly boiled and cooled distilled water and stored in an amber bottle.

Sulfuric acid, 1 N: Add 28 mL concentrated H_2SO_4 slowly and cautiously to approximately 750 mL distilled water and dilute to 1 L.

Standard sodium thiosulfate, 0.1 N: Dissolve 25 g $Na_2S_2O_3 \cdot 5H_2O$ in 1 L of freshly boiled distilled water. Standardize against potassium biniodate (also called potassium hydrogen iodate) or potassium dichromate.

Standard sodium thiosulfate titrant, 0.005 N: Dilute the proper volume (approximately 50 mL) of 0.1 N sodium thiosulfate to 1000 mL. For accurate work, standardize this solution daily, using either 0.005 N potassium biniodate or potassium dichromate solution for the purpose. Standard sodium thiosulfate titrant, exactly 0.005 N, is equivalent to 0.120 mg ozone per 1.00 mL.

Starch indicator solution: To 5 g potato, arrowroot, or soluble starch, in a mortar, add a little cold distilled water and grind to a thin paste. Pour into 1 L boiling distilled water, stir, and allow to settle overnight. Use the clear supernatent. Preserve with 1.25 g salicylic acid, 4 g zinc chloride, or a combination of 4 g sodium propionate and 2 g sodium azide added to 1 L of starch solution.

Standard iodine 0.1 N: Dissolve 40 g KI in 25 mL distilled water; then add 13 g resublimed iodine and stir until dissolved. Dilute to 1 L and standardize against sodium arsenite, primary standard grade.

Standard iodine 0.005 N: Dissolve 16 g KI in a little distilled water in a 1 L volumetric flask, add the proper volume (approximately 50 mL) of 0.1 N iodine solution, and dilute to the mark. For accurate work standardize this solution daily. Store the solution in an amber bottle or in the dark. Protect from direct sunlight at all times and keep from all contact with rubber.

18.4.14.3.2 Procedure

Sample collection: Collect an 800 mL sample in a 1 L washing bottle with a porous diffuser at the bottom.

Ozone absorption: Pass a stream of pure air or nitrogen through the sample, and then through an absorber containing 400 mL potassium iodide solution. Continue for not less than 5 min at a rate of 0.2–1.0 L/min to insure that all ozone is swept from the sample and absorbed in the potassium iodide solution.

Titration: Transfer the potassium iodide solution to a 1 L beaker, rinse the absorber, and add 20 mL H_2SO_4 to produce a pH below 2.0. Titrate with 0.005 N sodium thiosulfate titrant until the yellow color of the liberated iodine is almost, discharged. Add 4 mL starch indicator solution to impart a blue color, and continue the titration carefully but rapidly to the endpoint at which the blue color just disappears. Long contact of iodine and starch develops a blue compound which is difficult to decolorize.

Blank test: Correct the result of the sample titration by determining the blank contributed by such reagent impurities as (1) the free iodine or iodate in the potassium iodide, which liberates extra iodine, or (2) the traces of reducing agents that might reduce some of the iodine which is liberated.

Take 400 mL potassium iodide solution, 20 mL H_2SO_4, and 4 mL starch indicator solution. Perform whichever one of the blank titrations below applies:

1. If a blue color occurs, titrate with 0.005 N sodium thiosulfate to the disappearance of the blue, and record the result.
2. If no blue color occurs, titrate with 0.005 N iodine solution until a blue color appears. Back-titrate with 0.005 N sodium thiosulfate and record the difference.

18.4.14.3.3 Calculation

$$\text{mg/L } O_3 = \frac{(A \pm B) \times N \times 24,000}{\text{mL sample}}$$

where A, mL titration for sample; B, mL titration for blank which may be positive or negative; and N, normality of $Na_2S_2O_3$.

18.4.15 Estimation of pH

The pH of most natural waters falls within the range 4–9. The majority of waters are slightly basic due to the presence of carbonate and bicarbonate. A departure from the norm for water could be caused by the entry of strongly acidic or basic industrial wastes. A relatively common practice is the pH adjustment of the treatment plant effluent for the purpose of controlling corrosion in the distribution system. pH is the logarithm of the reciprocal of the hydrogen ion concentration in moles per liter. pH enters into the calculation of carbonate, bicarbonate, and carbon dioxide, as well as of the corrosion or stability index; and into the control of water treatment processes. The practical pH scale extends from 0, very acidic, to 14, very alkaline, with the middle value (pH 7) corresponding to exact neutrality at 25°C. Whereas "alkalinity" and "acidity" express the total reserve or buffering capacity of a sample, the pH value represents the instantaneous hydrogen ion activity.

18.4.15.1 Effect of Temperature on pH Values of Buffer Solutions

	pH Value		
Temperature °C	**pH 4 Buffer**	**pH 7 Buffer**	**pH 9 Buffer**
0	4.00	6.98	9.46
5	4.00	6.95	9.40
10	4.00	6.92	9.33
15	4.00	6.90	9.28
20	4.00	6.88	9.23
25	4.01	6.86	9.18
30	4.02	6.85	9.14
35	4.02	6.84	9.10
38	4.03	6.84	9.08
40	4.04	6.84	9.07
45	4.05	6.83	9.04
50	4.06	6.83	9.01
55	4.08	6.83	8.99
60	4.09	6.84	8.96

18.4.16 Estimation of Phosphate

Phosphate occurs in traces in many natural waters, and often in appreciable amounts during periods of low biologic productivity. Traces of phosphate increase the tendency of troublesome algae to grow in reservoirs. Waters receiving raw or treated sewage, agricultural drainage, and certain industrial waters normally contain significant concentrations of phosphate. Also, phosphate is frequently added to domestic and industrial waters in various forms. Sometimes both orthophosphate and polyphosphates (molecularly dehydrated phosphates) will be found in the same sample. Trace amounts of phosphate may also be combined with organic matter. Such phosphate seldom exceeds a few tenths of a milligram per liter. Phosphate in its various forms may also appear in the suspended matter or sludge of the sample taken.

As a general rule, unless reported otherwise, only the soluble phosphate is considered. Phosphate analyses are made primarily to control chemical dosage, or as a means of tracing flow or contamination.

18.4.16.1 Stannous Chloride Method for Orthophosphate

18.4.16.1.1 Reagents

Ammonium molybdate reagent (I): Dissolve 25 g $(NH_4)_6\ Mo_7O_{24}\cdot 4H_2O$ in 175 mL distilled water. Cautiously add 280 mL concentrated H_2SO_4 to 400 mL distilled water. Cool, add the molybdate solution, and dilute to 1 L.

Stannous chloride reagent (I): Dissolve 2.5 g of a fresh supply of $SnCl_2{\cdot}2H_2O$ in 100 mL glycerol. Heat on a water bath and stir with a glass rod to hasten dissolution of reagent. This reagent is stable and requires neither preservatives nor special storage.

18.4.16.1.1.1 Standard Phosphate Solution

18.4.16.1.1.1.1 Reagents for Extraction

1. *Benzene-isobutanol solvent:* Mix equal volumes of benzene and isobutyl alcohol.
2. *Ammonium bolybdate reagent (II)*: Dissolve 40.1 g $(NH_4)_6Mo_7O_{24}{\cdot}4H_2O$ in approximately 500 mL distilled water. Slowly add 396 mL molybdate reagent (I). Cool and dilute to a liter.
3. *Alcoholic sulfuric acid solution*: Cautiously add 20 mL concentrated H_2SO_4 to 980 mL methyl alcohol with continuous mixing.
4. *Dilute stannous chloride reagent (II)*: Mix 8 mL stannous chloride reagent (I) with 50 mL glycerol. This reagent is stable for at least 6 months.

18.4.16.1.2 Procedure

To 100 mL sample containing not more than 0.6 mg PO_4, and free from color and turbidity, add 1 drop (0.05 mL) phenolphthalein indicator. If the sample turns pink, add strong-acid solution, dropwise, to discharge the color. If more than 5 drops are required, take a smaller sample and dilute to 100 mL with distilled water after first discharging the pink color with acid.

Add, with thorough mixing after each addition, 4.0 mL molybdate reagent (I) and 0.5 mL (10 drops) stannous chloride regent (I). The rate of color development and the intensity of color depend on the temperature of the final solution, each 1°C increase producing about 1% increase in color. Hence, samples, standards, and reagents should be within 2°C of one another and at a temperature between 20 and 30°C. After 10 min, but before 12 min, employing the same specific interval for all determinations, measure the color photometrically at 690 mμ and compare with a calibration curve, using a distil water blank.

A blank must always be run on the reagents and distilled water. In as much as the color that first develops progressively and later fades, it is essential that timing be the same for samples as for standards. At least one standard should be tested with each set of samples or once each day that tests are made. The calibration curve may deviate from a straight line at the upper concentrations of the 1–6 mg/L range.

Extraction: When increased sensitivity is desired or interferences need to be overcome, extract the phosphate as follows: pipet a suitable aliquot of sample into a 100 mL graduated extraction cylinder and dilute, if necessary, to 40 mL with distilled water. Add 50.0 mL benzene-isobutanol solvent and 15.0 mL molybdate reagent (II). Close at once and shake vigorously for exactly 15 s. Any delay increases the amount of polyphosphate, it present, which will be included in the orthophosphate value. Remove the stopper and withdraw 25.0 mL of separated organic layer, using a pipet and a safety aspirator. Transfer to a 50 mL volumetric flask, add 15–16 mL alcoholic sulfuric acid solution, swirl, add 10 drops (0.05 mL) dilute stannous chloride reagent (II), swirl, and dilute to the mark with alcoholic sulfuric acid. Mix thoroughly; after 10 min, but before 30 min, read against the blank at 625 mμ. Prepare the blank by carrying 40 mL distilled water through the same procedure as the sample. Read the PO_4 concentration from a calibration curve prepared by taking known phosphate standards through the same procedural steps as the samples.

18.4.16.1.3 Calculation

The results from the direct and the extraction procedures can be calculated by the following equation:

$$\text{mg/L } PO_4 = \frac{\text{mg } PO_4 \times 1000}{\text{mL sample}}$$

18.4.17 Total Phosphate

The total-phosphate content of the sample includes all the soluble orthophosphate and polyphosphates, and insoluble phosphates precipitated during storage. If any insoluble phosphates are present, for practical purposes they are assumed to be insoluble orthophosphate. It is understood that total phosphate is not to include insoluble phosphates that may have been present in the original water and removed in sampling, unless expressly requested; in that case, such insoluble phosphate will be reported separately. Condensed phosphates, such as pyro-, tripoly-, and higher-molecular-weight species, are not normally present in natural waters, but are frequently added in the course of water treatment. Polyphosphates do not respond appreciably to the orthophosphate tests but can be hydrolyzed to orthophosphate by boiling with acid. Also, the insoluble phosphates can be dissolved by boiling with acid. Then, with the proper combinations of filtration and boiling with acid, and the orthophosphate value by method A or B, both the polyphosphates can be determined as their equivalent PO_4.

If precipitate or turbidity is present in the bottled sample, two portions must be taken for analysis. One should consist of 100 mL of the filtered sample. The other portion should consist of 100 mL of thoroughly mixed unfiltered sample. To each of the two 100 mL portions, or aliquots diluted to 100 mL, add 1 drop of phenolphthalein indicator solution. If a red color develops, add strong-acid solution dropwise to just discharge the color. Then add 1 mL in excess to each. Boil gently for at least 90 min, adding distilled water to keep the volume between 25 and 50 mL. Cool, neutralize to a faint pink color with sodium hydroxide solution, and restore the portions to the original 100 mL volume with distilled water. Determine the orthophosphate content of each treated portion. This gives the total phosphate present as PO_4 in each portion. Determine the orthophosphate on the filtered original untreated sample by the same method.

18.4.17.1 Calculation

If no precipitate or turbidity is present in the bottled sample, then:

Total phosphate $= C_1$

Orthophosphate $= C_3$

Polyphosphate $= C_1 - C_3$

All results are expressed as mg/L PO_4.

If a precipitate or turbidity is presenting the bottled sample, then:

Total phosphate $= C_1$

Orthophosphate $= C_3 + C_1 - C_2$

Polyphosphate $= C_2 - C_3$

where C_1 is the value obtained on the unfiltered portion and C_2 is the value obtained on the unfiltered portion and C_2 is the value obtained on the filtered portion. All results are expressed as mg/L PO_4.

18.4.18 Estimation of Potassium

Potassium ranks seventh among the elements in order of abundance, yet its concentration in most drinking water is trivial, seldom reaching 20 mg/L. However, occasional brines may contain in excess of 100 mg/L potassium.

Storage of sample: Samples should not be stored in soft-glass bottles because of the possibility of contamination from leaching of the glass. Polyethylene or pyrex bottles are preferable.

18.4.18.1 Flame Photometric Method

Trace amounts of potassium can be determined in either a direct-reading or internal-standard type of flame photometer at a wavelength of 768 mμ.

18.4.19 Sodium

Sodium ranks sixth among the elements in order of abundance; therefore, it is present in most natural waters. The levels may vary from negligible to appreciable. Relatively high concentrations may be found in brines and hard waters softened by the sodium exchange process. The ratio of sodium to total cations is important in agriculture and human pathology. Soil permeability has been found to be detrimentally affected by a high sodium ratio, while certain diseases require water with a low sodium concentration. A limiting concentration of 2–3 mg/L is recommended in feed waters destined for high-pressure boilers. When necessary, sodium is removed by the hydrogen exchange process or distillation.

Trace amounts of sodium can be determined in either a direct-reading or internal-standard type flame photometer at a wavelength of 589 mμ. The sample is sprayed into a gas flame and excitation is carried out under carefully controlled and reproducible conditions. The desired spectral line is isolated by the use of interference filters or by a suitable slit arrangement in light-dispersing devices such as prisms or gratings. The intensity of light is then measured by a phototube potentiometer or other appropriate circuit. The intensity of light at 589 mμ is approximately proportional to the concentration of the element. The calibration curve may be linear but has a tendency to level off in the upper reaches. The optimum lithium concentration may vary among individual flame photometers operating on the internal standard principle and, therefore, must be ascertained for the instrument at hand. In those cases where the alignment of the wavelength dial with the prism is not precise in the available photometer, the exact wavelength setting, which may be slightly different than 589 mμ, can be determined from the maximum needle deflection and then used for the emission measurements.

18.4.19.1 Estimation of Specific Conductance

Specific conductance yields a measure of water's capacity to convey an electric current. This property is related to the total concentration of the ionized substances in water and the temperature at which the measurement is made. The nature of the various dissolved substances, their actual and relative concentrations, and the ionic strength of the water sample vitally affect the specific conductance. An aqueous system containing dissociated molecules will conduct an electric current. In a direct-current field, the positive ions migrate toward the negative electrode, while the negatively charge ions migrate toward the positive electrode. Most inorganic acids, bases, and salts (such as hydrochloric acid, sodium carbonate, and sodium chloride) are good conductors. Conversely, molecules of such organic compounds as sucrose and benzene do not dissociate in aqueous solution and, therefore, conduct a current very poorly, if at all.

18.4.19.2 Calculation

The cell constant, *C*, is equal to the product of the measured resistance, in ohms, of the standard potassium chloride solution and the specific conductance, in mhos per centimeter, of this standard solution; $C = R_{KCl} \times 0.001413$ if the measurement is made at 25°C.

The specific conductance (℧/cm) of the water sample at 25°C is equal to the cell constant, C, divided by the resistance, in ohms, of the sample, R_S, measured at 25°C:

$$\text{Specific Conductance} = \frac{C}{R_s}$$

The specific conductance of most water is so low that it is standard practice to express it in μ℧/cm (the numerical value expressed in μ℧/cm is 1,000,000 times as large as the numerical value expressed in ℧/cm).

If the temperature of measurement is not exactly 25°C, it may be more convenient to calculate the specific conductance at 25°C according to the equation:

$$\text{Specific Conductance} = \frac{1413 \times R_{KCl}}{R_s} \text{ micromhos/cm}$$

where R_{KCl} and R_S are measured at the same temperature, preferably near room temperature, and in the range from 20 to 30°C.

18.4.20 Estimation of Sulfates

Sulfate is widely distributed in nature and may be present in natural waters in concentrations ranging from a few to several thousand mg/L. Mine drainage wastes may contribute high sulfate by virtue of pyrite oxidation. Because sodium and magnesium sulfate exert a cathartic action, the recommended sulfate concentration in potable supplies is limited to 250 mg/L.

Sulfate ion is precipitated in a hydrochloric acid medium with barium chloride in such manner as to form barium sulfate crystals of uniform size. The absorbance of the barium sulfate suspension is measured by a nephelometer and the sulfate ion concentration is determined by comparison of the reading with a standard curve.

18.4.20.1 Reagents

Conditioning reagent: Mix 50 mL glycerol with a solution containing 30 mL concentrated HCl, 300 mL distilled water, 100 mL 95% ethyl or isopropyl alcohol, and 75 g sodium chloride.

Standard sulfate solution: Prepare a standard sulfate solution as described:

$$1.00\ \text{mL} = 0.01\ \text{mg}\ SO_4.$$

1. Prepare by diluting 10.41 mL of the standard 0.0200 N H_2SO_4 titrant specified in alkalinity.
2. Dissolve 0.1479 g anhydrous sodium sulfate, Na_2SO_4, in distilled water, and dilute to 1000 mL.

18.4.20.2 Procedure

Measure 100 mL of sample, or a suitable aliquot made up to 100 mL, into a 250 mL Erlenmeyer flask. Add exactly 5.00 mL of conditioning reagent and mix in the stirring apparatus. While the solution is being stirred, add a spoonful of barium chloride crystals and begin the timing immediately. Stir for exactly 1 min at a constant speed.

Immediately after the stirring period has ended, pour some of the solution into the cell of the photometer and measure the turbidity at 30-s intervals for 4 min. Maximum turbidity is usually obtained within 2 min and the readings remain constant thereafter for 3–10 min. Consider the turbidity to be the maximum reading obtained in the 4-min interval.

Estimate the sulfate concentration in the sample by comparing the turbidity reading with a standard curve secured by carrying out the procedure with standard sulfate solutions. A suitable range of standards extends from

0 to 40 mg/L of sulfate ion in 5 mg/L increments. Above 40 mg/L the accuracy of the method decreases and the suspensions of barium sulfate lose stability.

Correction should be made for the apparent turbidity of the samples by running blanks in which no barium chloride is added. Stability of the conditions should be checked by running a standard sample with every three or four unknown samples.

18.4.20.3 Calculation

$$\text{mg/L } SO_4 = \frac{\text{mg } SO_4 \times 1000}{\text{mL sample}}$$

18.4.21 Estimation of Turbidity

Clear water is important in those industries where the product is destined for human consumption. Beverage producers, food processors, and treatment plants drawing upon a surface water supply commonly rely on coagulation, settling, and filtration measures to insure an acceptable effluent.

Turbidity in water is caused by the presence of suspended matter, such as clay, silt, finely divided organic matter, plankton, and other microscopic organisms. Turbidity should be clearly understood to be an expression of the optical property of a sample, which causes light to be scattered and absorbed rather than transmitted in straight lines through the sample. Attempts to correlate turbidity with the weight concentration of suspended matter are impractical, as the size, shape, and refractive index of the particulate materials are of most importance optically but bear little direct relationship to the concentration and specific gravity of the suspended matter. The standard method for the determination of turbidity is the candle method. However, suspensions standardized by this method may be used, with or without dilution, in other instruments. Unfortunately, the results obtained using other instruments will not always agree closely with those obtained using the candle turbidimeter; and owing to fundamental differences in optical systems, the results obtained with different types of instruments will not always check closely with one another even if the instruments are pre-calibrated against the candle turbidimeter.

Turbidity measurements are based on the light path through a suspension which just causes the image of the flame of a standard candle to disappear, that is, to become indistinguishable against the general background illumination, when the flame is viewed through the suspension.

18.5 PHYSICAL AND CHEMICAL EXAMINATION OF WASTEWATER

18.5.1 Treatment Plant Effluents, and Polluted Waters

18.5.1.1 Collection of Samples

Method of sampling sewage, effluents and wastes: The standard method prescribes the collection of a composite sample over a 24-h period. In many cases it may be advisable to composite individual samples over a 1-, 2-, or 4-h period; to divide the sample to represent shift work; or to extend it to cover the complete cycle of operation including all special, variable, or periodic discharges at irregular intervals, as well as Saturday and Sunday cleanups.

Individual portions should be taken in a wide-mouth bottle having a diameter of at least 35 mm at the mouth and a capacity of at least 120 mL. These portions should be collected each hour, in some cases each half hour or even every 5 min and mixed at the end of the sampling period, or combined in a single bottle as collected. If preservatives are to be used, they should be added to the sample bottle initially, so that all portions of the composite are preserved as soon as collected. Analysis of individual samples may sometimes be necessary.

It is desirable, and often absolutely essential, to combine the individual samples in volumes proportionate to the volume of flow. A final volume of 2–3 L is sufficient for sewage, effluents, and wastes. Automatic sampling devices are available but should not be used unless the sample is preserved as described below. Sampling devices, including bottles, should be cleaned daily of all growths of sewage organisms.

Samples of sewage, effluents, and wastes should be analyzed as soon as possible after collection. Determinations of substances such as dissolved oxygen, soluble sulfides, and residual chlorine should be made at the source. No single method of preservation is entirely satisfactory, and the preservative should be chosen with due regard to the determinations that are to be made. All methods of preservation may be inadequate when applied to suspended matter. Formaldehyde affects so many of the determinations that its use is not recommended.

- *Dissolved oxygen*: Samples for this determination are preserved as directed under oxygen (dissolved).
- *Biochemical oxygen demand*: Samples must be free from all preservatives. When samples are composited, the individual or composite sample must be chilled immediately to 3 to 4°C and kept at this temperature during

the compositing period. In samples stored at room temperature, the BOD may drop 10–40% in 6 h, but in some instances it may rise.

- *Chemical oxygen demand*: Samples for determining COD by the dichromate method should be preserved by adding sufficient H_2SO_4 to obtain a final acidity of pH 2–3.

18.5.2 Azide Modification of Iodometric Method

The azide modification is used for most sewage, effluents, and streams, especially if they contain more than 0.1 mg/L nitrite nitrogen and not more then 1 mg/L ferrous ion. Other reducing or oxidizing materials should be absent. If 1 mL fluoride solution is added before acidifying the sample and there is no delay in titration, the method is also applicable in the presence of 100–200 mg/L ferric iron.

18.5.2.1 Reagents

Manganese sulfate solution: Dissolve 480 g $MnSO_4 \cdot 4H_2O$, 400 g $MnSO_4 \cdot 2H_2O$, or 364 g $MnSO_4 \cdot H_2O$ in distilled water, filter, and dilute to 1 L. When uncertainty exists regarding the water of crystallization, a solution of equivalent strength may be obtained by adjusting the specific gravity of the solution to a value of 1.270 at 20°C. The manganese sulfate solution should liberate not more than a trace of iodine when added to an acidified solution of potassium iodide.

Alkali-iodide-azide reagent: Dissolve 500 g sodium hydroxide, NaOH (or 700 g potassium hydroxide, KOH) and 135 g sodium iodide, NaI (or 150 g potassium iodide, KI) in distilled water and dilute to 1 L. To this solution add 10 g sodium azide, NaN_3, dissolved in 40 mL distilled water. Potassium and sodium salts may be used interchangeably. This reagent should not give a color with starch solution when diluted and acidified.

Sulfuric acid, concentrated: The strength of this acid is about 36 N. Hence, 1 mL is equivalent to about 3 mL of the alkali iodide-azide reagent.

Starch solution: Prepare an emulsion of 5–6 g potato, arrowroot, or soluble starch in a mortar or beaker with a small quantity of distilled water. Pour this emulsion into 1 L of boiling water, allow to boil a few minutes, and let settle overnight. Use the clear supernate. This solution may be preserved with 1.25 g salicylic acid per liter or by the addition of a few drops of toluene.

Sodium thiosulfate stock solution, 0.10 N: Dissolve 24.82 g $Na_2S_2O_3 \cdot 5H_2O$ in boiled and cooled distilled water and dilute to 1 L. Preserve by adding 5 mL chloroform or 1 g NaOH per liter.

Standard sodium thiosulfate titrant, 0.025 N: Prepare either by diluting 250.0 mL sodium thiosulfate stock solution to 1000 mL, or by dissolving 6.20 g $Na_2S_2O_3 \cdot 5H_2O$ in freshly boiled and cooled distilled water and diluting to 1000 mL. Standard sodium thiosulfate solution may be preserved by adding 5 mL chloroform or 0.4 g NaOH per liter. Standard sodium thiosulfate solution, exactly 0.0250 N, is equivalent to 0.200 mg DO per 1.00 mL.

Standardize with (1) biniodate or (2) dichromate:

1. *Standard potassium biniodate solution*, 0.025 N: A stock solution equivalent in strength to 0.1 N thiosulfate solution contains 3.249 g/L $KH(IO_3)_2$. The biniodate solution equivalent to the 0.025 N thiosulfate contains 0.8124 g/L $KH(IO_3)_2$ and may be prepared by diluting 250 mL stock solution to 1 L.

   Standardization: Dissolve apsproximately 2 g KI, free from iodate, in an Erlenmeyer flask with 100–150 mL distilled water; add 10 mL 1 + 9 H_2SO_4, followed by exactly 20.00 mL standard biniodate solution. Dilute to 200 mL and titrate the liberated iodine with the thiosulfate titrant, adding starch toward the end of the titration, when a pale straw color is reached. Exactly 20.00 mL 0.025 N thiosulfate should be required when the solutions under comparison are of equal strength. It is convenient to adjust the thiosulfate solution to exactly 0.0250 N.
2. *Standard potassium dichromate solution*, 0.025 N: Potassium dichromate may be substituted for biniodate. A solution equivalent to 0.025 N sodium thiosulfate contains 1.226 g/L $K_2Cr_2O_7$. The $K_2Cr_2O_7$ should be previously dried at 103°C for 2 h. The solution should be prepared in a volumetric flask.

Standardization: Same as with biniodate, except that 20.00 mL standard dichromate solution is used. Place in the dark for 5 min, dilute to approximately 400 mL, and titrate with 0.025 N thiosulfate solution.

Special reagent-potassium fluoride solution: Dissolve 40 g $KF \cdot 2H_2O$ in distilled water and dilute to 100 mL.

18.5.2.2 Procedure

To the sample as collected in a 250–300 mL bottle, add 2 mL manganese sulfate solution followed by 2 mL alkali-iodide-azide reagent, well bellow the surface of the liquid; stopper with care to exclude air bubbles and mix by inverting the bottle several times. When the precipitate settles, leaving a clear supernatant above the manganese hydroxide floc, shake again. With seawater, a 10-min period of contact with the precipitate will be required. When setting has produced at least 100 mL clear supernate, carefully remove the stopper and immediately add 2.0 mL concentrated H_2SO_4 by allowing the acid to run down the neck of the bottle, re-stopper, and mix by gentle inversion until dissolution is complete. The iodine should be uniformly distributed throughout the bottle before decanting the amount needed for titration. This should correspond to 200 mL of the original sample after correction for the loss of sample by displacement with the reagents has been made. Thus, when a total of 4 mL (2 mL each) of the manganese sulfate and alkaliiodide-azide reagents is added to a 300-mL bottle, the volume taken for titration should be $200 \times 300/(300 - 4) = 203$ mL.

Titrate with 0.025 N thiosulfate solution to a pale straw color. Add solution to a pale straw color. Add 1–2 mL freshly prepared starch solution and continue the titration to the first disappearance of the blue color. If the endpoint is overrun, the sample may be back-titrated with 0.025 N biniodate solution, which is added dropwise, or by an additional measured volume of sample. Correction for the amount of biniodate solution or sample should be made. Subsequent recolorations due to the catalytic effect of nitrite or to traces of ferric salts which have not been complexed with fluoride should be disregarded.

18.5.2.3 Calculation

Because 1 mL 0.025 N sodium thiosulfate titrant is equivalent to 0.2 mg DO, each milliliter of sodium thiosulfate titrant used is equivalent to 1 mg/L DO if a volume equal to 200 mL of original sample is titrated.

If the results are desired in milliliters of oxygen gas per liter at 0°C and 760 mm pressure, multiply mg/L DO by 0.698.

To express the results as percent saturation at 760 mm atmospheric pressure, the solubility data in Table 18.4 may be used. Equations for correcting the solubilities to barometric pressures other than mean sea level are given below the table.

The solubility of DO in distilled water at any barometric pressure, P (mmHg), temperature, t (°C), and saturated vapor pressure u (mmHg), for the given t, may be calculated between the temperature of 0 and 30°C by:

$$\text{mL/L DO} = \frac{(\text{P} - \text{u}) \times 0.678}{35 + \text{t}}$$

Although 2 mL quantities of the reagents insure better contact with less agitation, it is permissible to use 1 mL reagent quantities with 250 mL bottles.

Table 18.4 Solubility of Oxygen in Water Exposed to Water-Saturated Air[a]

Temp °C	Chloride Concentration in Water (mg/L) 0	5000	10,000	15,000	20,000	Difference per 100 mg Chloride
	Dissolved Oxygen (mg/L)					
0	14.6	13.8	13.0	12.1	11.3	0.017
1	14.2	13.4	12.6	11.8	11.0	0.016
2	13.8	1301	12.3	11.5	10.8	0.015
3	13.5	12.7	12.0	11.2	10.5	0.015
4	13.1	12.4	11.7	11.0	10.3	0.014
5	12.8	12.1	11.4	10.7	10.0	0.014
6	12.5	11.8	11.1	10.5	9.8	0.014
7	12.2	11.5	10.9	10.2	9.6	0.013
8	11.9	11.2	10.6	10.0	9.4	0.013
9	11.6	11.0	10.4	9.8	9.2	0.012
10	11.3	10.7	10.1	9.6	9.0	0.012
11	11.1	10.5	9.9	9.4	8.8	0.011
12	10.8	10.3	9.7	9.2	8.6	0.011
13	10.6	10.1	9.5	9.0	8.5	0.011
14	10.4	9.9	9.3	8.8	8.3	0.010
15	10.2	9.7	9.1	8.6	8.1	0.010
16	10.0	9.5	9.0	8.5	8.0	0.010
17	9.7	9.3	8.8	8.3	7.8	0.010
18	9.5	9.1	8.6	8.2	7.7	0.009
19	9.4	8.9	8.5	8.0	7.6	0.009
20	9.2	8.7	8.3	7.9	7.4	0.009
21	9.0	8.6	8.1	7.7	7.3	0.00*9
22	8.8	8.4	8.0	7.6	7.1	0.008
23	8.7	8.3	7.9	7.4	7.0	0.008
24	8.5	8.1	7.7	7.3	6.9	0.008
25	8.4	8.0	7.6	7.2	6.7	0.00/8
26	8.2	7.8	7.4	7.0	6.6	0.008
27	8.1	7.7	7.3	6.9	6.5	0.008
28	7.9	7.5	7.1	6.8	6.4	0.008
29	7.8	7.4	7.0	6.6	6.3	0.008
30	7.6	7.3	6.9	6.5	6.1	0.008
31	7.5					
32	7.4					
33	7.3					
34	7.2					
35	7.1					
36	7.0					
37	6.9					
38	6.8					
39	6.7					

Continued

Table 18.4 Solubility of Oxygen in Water Exposed to Water-Saturated Air[a]—cont'd

	Chloride Concentration in Water (mg/L)					Difference per 100 mg Chloride
	0	5000	10,000	15,000	20,000	
Temp °C	Dissolved Oxygen (mg/L)					
40	6.6					
41	6.5					
42	6.4					
43	6.3					
44	6.2					
45	6.1					
46	6.0					
47	5.9					
48	5.8					
49	5.7					
50	5.6					

[a]At a total pressure of 760 mmHg. Under any other barometric pressure, P (mm; or P′, in.), the solubility, S′ (mg/L), can be obtained from the corresponding value in the table by the equation.

$$S' = S\frac{P-p}{760-p}$$

in which S is the solubility at 760 mm (29.92 in.) and p is the pressure (mm) of saturated water vapor at the temperature of the water. For elevations less than 3000 ft and temperatures below 25°C, p can be ignored. The equation then becomes:

$$S' = S\frac{P}{760}S = \frac{P'}{29.92}$$

Dry air is assumed to contain 20.90% oxygen (calculations made by Whipple and Whipple (1911, J. Am. Chem. Soc., 33:362-365)), and between 30 and 50°C by:

$$\text{mL/L DO} = \frac{(P-u) \times 0.827}{49+t}$$

18.5.3 Biochemical Oxygen Demand

Incubation bottles, 250–300 mL capacity, with ground-glass stoppers are used. Bottles should be cleaned with a good detergent and thoroughly rinsed and drained before use. As a precaution against drawing air into the dilution bottle during incubation, a water seal is recommended. Satisfactory water seals are obtained by inverting the bottles in a water bath or adding water to the flared mouth of special BOD bottles. Air incubator or water bath, thermostatically controlled at 20 ± 1°C. All light should be excluded to prevent formation of DO by algae in the sample.

18.5.3.1 Reagents

Distilled water: Water used for solutions and for preparation of the dilution water must be of the highest quality, distilled from a block tin or all-glass still, contain less than 0.01 mg/L copper, and be free of chlorine, chloramines, caustic alklalinity, organic material, or acids.

Phosphate buffer solution: Dissolve 8.5 g potassium dihydrogen phosphate, KH_2PO_4, 21.75 g dipotassium hydrogen phosphate, $K_2H_2PO_4$, 33.4 g disodium hydrogen phosphate heptahydrate, $Na_2HPO_4 \cdot 7H_2O$, and 1.7 g ammonium chloride, NH_4Cl, in about 500 mL distilled water and dilute to 1 L. The pH of this buffer should be 7.2 without further adjustment. If dilution water is to be stored in the incubator, the phosphate buffer should be added just prior to using the dilution water.

Magnesium sulfate solution: Dissolve 22.5 g Mg $SO_4 \cdot 7H_2O$ in distilled water and dilute to 1 L.

Calcium chloride solution: Dissolved 27.5 g anhydrous $CaCl_2$ in distilled water and dilute to 1 L.

Ferric chloride solution: Dissolve 0.25 g $FeCl_3 \cdot 6H_2O$ in distilled water and dilute to 1 L.

Acid and alkali solutions, 1 N: For neutralization of waste samples which are either caustic or acidic.

Sodium sulfite solution, 0.025 N: Dissolve 1.575 g anhydrous Na_2SO_3 in 1000 mL distilled water. This solution is not stable and should be operated daily.

Seeding material: The selection of the proper seed is an important factor in the BOD determination. In many cases, particularly in food-processing wastes, a satisfactory seed may be obtained by using the supernatant liquor from domestic sewage which has been stored at 20°C for 24–36 h.

Many industrial wastes contain organic compounds, which are not amenable to oxidation by domestic-sewage seed. In these cases the analyst may use seed developed in the laboratory, or the receiving water collected below the point of discharge of the particular waste (preferably 2–5 miles below). The last two are the most likely possibilities. Receiving water used as a seed source will undoubtedly give the best estimate of the effect of a waste on such a water; but it must be collected at a point where there has been built up a biota capable of using for food the particular organic compounds present. In some cases this might entail the collection of a satisfactory seed many miles below the point of discharge of the waste, which might not be practical. With recurrent wastes not easily susceptible to biologic oxidation, it is usually more practical to build up an acclimated seed in the laboratory. This may be done by aerating and feeding sewage or receiving water with small daily increments of the particular waste, together with sewage, until a satisfactory seed is developed.

18.5.3.2 Procedure

Preparation of dilution water: The distilled water used should have been stored in cotton-plugged bottles for a sufficient length of time to become saturated with DO. The water may also be aerated by shaking a partially filled bottle or with a supply of clean compressed air. Situations may be encountered where it is desired to use stabilized river water to check stream performance with laboratory procedure. The distilled water used should be as near 20°C as possible and of the highest purity. Place the desired volume of distilled water in a suitable bottle and add 1 mL each of phosphate buffer, magnesium sulfate, calcium chloride, and ferric chloride solutions for each liter of water.

Seeding: If necessary, the dilution water is seeded by using the seed found to be the most satisfactory for the particular waste under study. Only past experience can determine the actual amount of seed to be added per liter. Seeded dilution water should be used the same day it is made.

18.5.3.3 Pretreatment

1. **Samples containing caustic alkalinity or acidity**: Neutralize to about pH 7.0 with 1 N H_2SO_4 or NaOH, using a pH meter or bromthymol blue as an outside indicator. The pH of the seeded dilution water should not be changed by the preparation of the lowest dilution of sample.
2. **Samples containing residual chlorine compounds**: If the samples are allowed to stand for 1–2 h, the residual chlorine will often be dissipated. BOD dilutions can then be prepared with properly seeded standard dilution water. Higher chlorine residuals in neutralized samples should be destroyed by adding sodium sulfite. The appropriate quantity of sodium sulfite solution is determined on a 100–1000 mL portion of the sample by adding 10 mL of 1 + 1 acetic acid or 1 + 50 H_2SO_4, followed by 10 mL potassium iodide solution (10 g in 100 mL) and titrating with 0.025 N sodium sulfite solution to the starch-iodide endpoint.

Add to a volume of sample the quantity of sodium sulfite solution determined by the above test, mix, and after 10–20 min test aliquot samples for residual chlorine to check the treatment. Prepare BOD dilutions with seeded standard dilution water.

3. **Samples containing other toxic substances**: Samples such as those from industrial wastes frequently require special study and treatment; for example, toxic metals derived from plating wastes.
4. **Samples supersaturated with DO**: Samples containing more than 9.17 mg/L DO at 20°C may be encountered during winter months or in localities where algae are actively growing. To prevent loss of oxygen during inclusion of these samples, the DO should be reduced to saturation by bringing the sample to about 20°C in a partly filled bottle and agitating it by vigorous shaking or by aerating with compressed air.

Dilution technique: Make several dilutions of the prepared sample so as to obtain the required depletions. The following dilutions are suggested: 0.1–1.0% for strong trade wastes, 1–5% for raw and settled sewage. 5–25% for oxidized effluents, and 25–100% for polluted river waters.

1. Carefully siphon standard dilution water, seeded if necessary, into a graduated cylinder of 1000–2000 mL capacity, filling the cylinder half full without entrainment of air. Add the quantity of carefully mixed sample to make the desired dilution and dilute to the appropriate level with dilution water. Mix well with a plunger-type mixing rod, avoiding entrainment of air. Siphon the mixed dilution into two BOD bottles, one for incubation and the other for determination of the initial DO in the mixture; stopper tightly and incubate for 5 days at 20°C. The BOD bottles should be water sealed by inversion incubator or by using a special water-seal bottle. Prepare succeeding dilutions of lower concentration in the same manner, or by adding dilution water to the unused portion for the preceding dilution.
2. The dilution technique may be greatly simplified when suitable amounts of sample are measured directly into bottles of known capacity with a large-tip volumetric pipet and the bottle is filled with just sufficient dilution water so that the stopper can be inserted without leaving air bubbles. Dilutions greater than 1:100 should be made by diluting the waste in a volumetric flask before it is added to the incubation bottles for final dilution.

Determination of DO: If the sample represents 1% or more of the lowest BOD dilution, determine DO on the undiluted sample. This determination is usually omitted on sewage and settled effluents known to have a DO content of practically zero. With samples having an immediate oxygen demand, a calculated initial DO should be used, in as much as such a demand represents a load on the receiving water.

Incubation: Incubate the blank dilution water and the diluted samples for 5 days at 20°C. Then determine the DO in the incubated samples and the blank, using the azide modification of the iodometric method (*A*). In special cases, other modifications may be necessary. Those dilutions showing a residual DO of at least 1 mg/L and a depletion of at least 2 mg/L should be considered the most reliable.

Seed correction: If the dilution water is seeded, determine the oxygen depletion of the seed by setting up a separate series of seed dilutions and selecting those resulting in 40–70% oxygen depletions in 5 days. One of these depletions is then used to calculate the correction due to the small amount of seed in the dilution water. Do not use the seeded blank for seed correction because the 5-day seeded dilution water blank is subject to erratic oxidation due to the very high dilution of seed, which is not characteristic of the seeded sample.

Dilution water control: Fill two BOD bottles with unseeded dilution water. Stopper and water-seal one of these for incubation. The other bottle is for determining the DO before incubation. The DO results on these two bottles are used as a rough check on the quality of the unseeded dilution water. The depletion obtained should not be used as a blank correction; it should not be more than 0.2 mL and preferably not more than 0.1 mL.

18.6 EXAMINATION OF INDUSTRIAL WASTEWATERS

Collection of samples: The sampling and the analysis of industrial wastes usually require greater care and attention to details than sampling and analyzing sewage and sewage effluents. Industrial wastes are subject to rapid changes even within a few minutes, owing to breakdowns, spillovers, floor washing, evaporator entrainment, and numerous other causes. The purpose of sampling and analysis may be to show the peak load concentration, the duration of peak loads or the occurrence of variation throughout the day. For these purposes, samples should be taken at appropriate intervals of 5, 10, 15, or 60 min and analyzed separately. On the other hand, if the purpose is to show the average loss during a shift to a 24-h period, the individual samples which are collected every few minutes or hourly should be composited in proportion to the flow; preserved, if necessary; and analyzed as soon as possible. Preservation of samples is difficult because almost all preservatives interfere with some of the tests. Immediate analysis is ideal. Storage at a low temperature (4°C) is perhaps the best way to preserve most samples until the next day.

18.6.1 Chemical Oxygen Demand

The COD determination provides a measure of the oxygen equivalent of that portion of the organic matter in a sample that is susceptible to oxidation by a strong chemical oxidant. The values are an important, rapid parameter for stream and industrial waste studies and control of waste treatment plants. In the absence of a catalyst, however, the method fails to include some organic compounds (such as acetic acid), which are biologically available to the stream organisms, while including some biologic compounds (such as cellulose), which are not a part of the immediate biochemical load on the oxygen assets of the receiving water. The carbonaceous portion of nitrogenous compounds can be determined, but there is no reduction of the dichromate by any ammonia in a waste or by any ammonia liberated for, the proteinaceous matter (except in the presence of chlorides). With certain wastes containing toxic substances, this test may be the only method for determining the organic load. Where wastes contain only readily available, organic bacterial food and not toxic matter, the results can be used to approximate the ultimate carbonaceous BOD values. The use of exactly the same technique each time is important because only a part of the organic matter is included, the proportion depending on the chemical oxidant used, the structure of the organic compounds, and the manipulative procedure. The dichromate reflux method has been selected for the COD determination because it has advantages over other oxidants in reproducibility applicability to a wide variety of samples, and ease of manipulation. The test will find its major usefulness in a plant for waste control purposes after many values have been obtained and correlated with some other important parameter or parameters.

Most types of organic matter are destroyed by a boiling mixture of chromic and sulfuric acids. A sample is refluxed with known amounts of potassium dichromate and sulfuric acid, and the excess dichromate is titrated with ferrous ammonium sulfate. The amount of oxidizable organic matter is proportional to the potassium dichromate consumed.

18.6.2 Reagents

Standard potassium dichromate solution, 0.250 N: Dissolve 12.259 g $K_2Cr_2O_7$, primary standard grade, previously dried at 103°C for 2 h, in distilled water and dilute to 1000 mL.

Sulfuric acid reagent, concentrated: H_2SO_4 containing 22 g silver sulfate, Ag_2SO_4, per 9-lb bottle (1–2 days required for dissolution).

- Corning *5000 or equal.*
- Corning 2360, 91548, or equal.

Standard ammonium sulfate titrate: Dissolve 98 g Fe $(NH_4)_2$ $(SO_4)_2 \cdot 6H_2O$ in distilled water. Add 20 mL concentrated H_2SO_4, cool, and dilute to 1000 mL. This solution must be standardized against the standard potassium dichromate solution daily.

Standardization: Dilute 10.0 mL standard potassium dichromate solution to about 100 mL. Add 30 mL concentrated H_2SO_4 and allow to cool. Titrate with the ferrous ammonium sulfate titrant, using 2 or 3 drops of ferroin indicator.

$$\text{Normality} = \frac{\text{mL } K_2Cr_2O_7 \times 0.25}{\text{mL } Fe(NH_4)_2(SO_4)_2}$$

Ferroin indicator solution: Dissolve 1.485 g 1,10-phenanthroline monohydrate, together with 0.695 g $FeSO_4 \cdot 7H_2O$ in water and dilute to 100 mL.

18.6.3 Procedure

Place 0.4 g $HgSO_4$, which may be measured conveniently in a refluxing flask. Add 20.0 mL sample, or an aliquot diluted to 20.0 mL with distilled water, and mix. Then add 10.0 mL potassium dichromate titrant. Carefully add 30 mL concentrated H_2SO_4 containing Ag_2 SO_4 with mixing. Pumice granules or glass beads should be added to the reflux mixture to prevent bumping, which can be severe and dangerous. The use of 0.4 g $HgSO_4$ is sufficient to complex 40 mg chloride ion, or 2 g/L when 20 mL of sample is used. If more chlorides are present, more $HgSO_4$ must be added to maintain a $HgSO_4$:Cl ratio of 10:1. If a precipitate develops, it does not adversely affect the determination.

Attach the flask to the condenser and reflux the mixture for 2 h; shorter reflux period may be used for a particular waste if it has been found to give the maximum COD. Cool and then wash down the condenser with distilled water. Dilute the mixture to about 100 mL, cool to room temperature, and titrate the excess dichromate with standard ferrous ammonium sulfate, using ferroin indicator. Generally 2 to 3 drops of the indicator are used. Although the quantity of ferroin is not critical, it should not vary on sequential samples. This, however, depends on the individual analyst. The color change is sharp, going, from blue-green is reddish-brown and should be taken as the endpoint although the blue-green color may reappear. A blank consisting of 20 mL distilled water instead of the sample, together with the reagents, is refluxed in the same manner.

Alternate procedure for other sample sizes: Sample size ranging from 10.0 to 50.0 mL may be used, providing the volumes, weights, and normalities for the other reagents are adjusted accordingly. Typical examples are given in the following table:

Sample Size (1.1 mL)	0.25 N Standard Dichromate (1.2 mL)	Concentrated $H_2(SO_4)$ With Ag_2SO_4 (1.3 mL)	$HgSO_4$ g	Normality of Fe $(NH_4)_2(SO_4)_2$	Final Volume Before Titration (mL)
10.0	5.0	15	0.2	0.05	70
20.0	10.0	30	0.4	0.10	140
30.0	15.0	45	0.6	0.15	210
40.0	20.0	60	0.8	0.20	280
50.0	25.0	75	1.0	0.25	350

Satisfactory results with be obtained providing these ratios are maintained and the procedures outlined are followed through the entire procedure. If larger samples are used, 500 mL Erlenmeyer refluxing flasks may be used to permit titration within the refluxing flask.

18.6.4 Calculation

$$\text{mg/L COD} = \frac{(a-b)\,c \times 8000}{\text{mL sample}}$$

where COD, chemical oxygen demand from dichromate; a, mL Fe $(NH_4)_2(SO_4)_2$ used for blank; b, mL Fe $(NH_4)_2(SO_4)_2$ used for sample; c, normality of Fe $(NH_4)_2(SO_4)_2$.

18.7 EXAMINATION OF SLUDGE AND BOTTOM SEDIMENTS IN WASTEWATER TREATMENT PROCESS AND IN POLLUTED RIVERS, LAKES, AND ESTUARIES

18.7.1 Sludge Volume Index

The "Sludge Volume Index" (SVI) is the volume in milliliters occupied by 1 g activated sludge after settling the aerated liquor for 30 min. A 1 L sample is collected at the outlet of the aeration tanks, and settled for 30 min in a 1000 mL graduated cylinder; the volume occupied by the sludge is reported as percent or mL. The sample is thoroughly mixed, or an original sample taken, and the suspended solids are determined and reported in percent by weight or mg/L.

$$SVI = \frac{\%\ \text{settling by volume}}{\%\ \text{suspended matter}}$$

$$SVI = \frac{\text{mL settled sludge} \times 1000}{\text{mg/L suspended matter}}$$

The standard deviation of the sludge volume index was determined as 1,69 on an average index of 72; a coefficient of variation of 2.35% ($n = 1$; 10×10).

18.7.2 Sludge Density Index

The "Sludge Density Index" (SDI) is the reciprocal of the sludge volume index (SVI) multiplied by 100, and is calculated from the data determined by the methods described above.

$$SDI = \frac{\text{percent suspended matter} \times 100}{\text{percent settling by volume}}$$
$$= \frac{\text{mg/L suspended matter}}{\text{mL settled sludge} \times 10}$$
$$= \frac{100}{SVI}$$

The standard deviation of the sludge density index was found to be 0.033 on an average index of 1.40; a coefficient of variation of 2.35% ($n = 1$; 10×10).

18.8 BACTERIOLOGIC EXAMINATIONS OF WATER TO DETERMINE ITS SANITARY QUALITY

The isolation of pathogenic bacteria or other microorganisms from water and sewage cannot be recommended as a routine practice, in as much as the techniques available at present are tedious and complicated. The results are not of major significance and may be confusing in a particular study of pollution. For many years the coliform group of bacteria, as defined below, has been used to indicate the pollution of water with wastes and thus the suitability of a particular water supply for domestic and dietetic uses.

The cultural reactions and characteristics of the coliform group have been studied extensively.

18.8.1 Media Specifications

The use of dehydrated media is recommended to provide uniformity of culture media. In preparing media from the basic ingredients, follow the directions given below.

18.8.1.1 Dilution Water

Buffered water: To prepare stock phosphate buffer solution, dissolve 34.0 g potassium dihydrogen phosphate, KH_2PO_4, in 500 mL distilled water, adjust to pH 7.2 with 1 N NaOH, and dilute to 1 L with distilled water.

Add 1.25 mL stock phosphate buffer solution to 1 L distilled water. Dispense in amounts that will provide 99 ± 2.0 mL or 9 ± 0.2 mL, after autoclaving for 15 min.

Peptone dilution water: Prepare a 10% solution of peptone in distilled water. Dilute a measured volume to provide a final 0.5% solution.

Dispense in amounts to provide 99 ± 2.0 mL or 9 ± 0.2 mL after autoclaving for 15 min.

18.8.1.2 Nutrient Broth

Add 3 g beef extract and 5 g peptone to each liter of distilled water. Heat slowly on a water bath, stirring until peptone has dissolved. Adjust the reaction so that the pH reading after sterilization will be between 6.8 and 7.0. Bring to a boil, cool to 25°C, make up lost weight with distilled water, and clarify as desired. Distribute in test tubes or other containers of size and volume desired.

18.8.1.3 Lactose Broth

To nutrient broth prepared as above, add 0.5% lactose. Adjust the reaction so that the pH reading after sterilization will be between 6.8 and 7.0, preferably 6.9. Place in fermentation tubes and sterilize in an autoclave. Cool rapidly after removal from the autoclave. If the above condition of exposure to heat cannot be fulfilled, prepare a 10–20% solution of lactose in distilled water and sterilize by heating at 121°C for 10 min. Add this solution to sterile nutrient broth in an amount sufficient to make a 0.5% lactose solution, and tube with proper precautions for preserving its sterility. It is permissible to add, by means of a sterile pipet, directly to a tube of sterile nutrient broth, enough of the sterile lactose solution to make the required

0.5% concentration. The tubes so prepared must be incubated at 35°C for 24 h as a test for sterility before they are used.

When fermentation tubes or other containers for the examination of 10- or 100-mL portions of sample are prepared, the lactose broth medium must be of such strength that the addition of that volume of sample to the medium in the fermentation tube will not reduce the concentration of ingredients in the mixture below that in the standard medium. Where dehydrated medium is used, the proper concentration of ingredients may be obtained by using the following table:

Inoculum (mL)	Amount of Medium in Tube (mL)	Volume of Medium + Inoculum (mL)	Dehydrated Lactose Broth Required (g/L)
1	10 or more	11 or more	13
10	10	20	26.0
10	20	30	19.5
100	50	150	39
100	35	135	50.1
100	20	120	78

18.8.1.4 Standard Plate Count

Preparation and dilution: The sample bottle shall be shaken vigorously 25 times, and the required portion shall be withdrawn at once with a standard sterile pipet to the petri dish, dilution bottle, or tube. If dilutions are made, the dilution bottle shall be likewise shaken 25 times before portions are removed.

Plating: A 1-, 0.1-mL, or other suitable volume of the sample or dilution to be used for plating should be placed in the petri dish first. It is recommended that dilutions be used in preparing volumes less than 1 mL; in the examination of sewage or turbid water, a 0.1-mL inoculum of the original sample shall not be measured but an appropriate dilution should be prepared. Not less than 10 mL of liquefied agar medium at a temperature of 43 to 45°C should be added to the water in the petri dish. The agar may be stored in a container providing maintenance of the proper temperature for no longer than 3 h and shall not be remelted. Tryptone glucose extract agar (or plate count agar) shall be used. The cover of the dish should be lifted just enough for the introduction of the pipet or the culture medium. The agar and the sample shall be thoroughly mixed and uniformly spread over the bottom of the dish by tilting and rotating the dish. The plates shall be solidified as rapidly as possible after pouring, and placed immediately in the appropriate incubator. Not more than 20 min should elapse between plating and pouring.

Incubation: Incubation for the standard plate count using an agar medium shall be at a temperature of 35°±0.5°C for 24±2h or at 20°±0.5°C for 48±3h. In the examination of chlorinated supplies where chlorination has not been effective and where the chlorine in the sample has been neutralized by the addition of sodium thiosulfate, coliform bacteria may not develop sufficiently to be detected in 24h, although in 48h, the count may be appreciable. Glass-covered dishes and plastic dishes shall be inverted in the incubator.

Counting: In preparing plates, such amounts of water should be planted as will give from 30 to 300 colonies on a plate. The aim should be always to have at least two plates giving the numbers between these limits, except as provided below. Ordinarily, it is not desirable to plant more than 1.0mL in a plate; therefore, when the total number of colonies developing from 1.0mL is less than 30, it is obviously necessary to disregard the rule above and to record the result as observed. With this exception, only plates showing 30 to 300 colonies should be considered in determining the standard plate count. The result as reported shall be the average of all plates falling within the limits. Counting shall be done with an approved counting aid, such as the Quebec colony counter. If such equipment is not available, counting may be done with one providing equivalent magnification and illumination. To avoid fictitious accuracy and yet to express the numerical results by a method consistent with the precision of the technique employed, the recorded number of bacteria per milliliter shall not include more than two significant figures. For example, account of 142 is recorded as 140, and a count of 155 as 160, whereas a count of 35 is recorded as 35. Counts shall be designated as "standard plate count at 35°C," or "standard pate count at 20°C."

18.8.2 Tests for Presence of Members of Coliform Group

The standard test for the coliform group may be carried out either by the multiple-tube fermentation technique (presumptive test, confirmed test, or completed test).

18.8.2.1 Presumptive Test

Lactose broth or lauryl tryptose broth maybe used in the "Presumptive Test."

18.8.2.1.1 Procedure

1. Inoculate a series of fermentation tubes ("primary" fermentation tubes) with appropriate graduated quantities (multiples and submultiples of

1 mL) of the water to be tested. The concentration of nutritive ingredients in the mixture of medium and added portion of sample must conform to the requirements given under media specifications. The portions of the water sample used for inoculating the lactose broth fermentation tubes will vary in size and number with the character of the water under examination, but in general should be decimal multiples and submultiples of 1 mL. These should be selected in accordance with the above discussion of the multiple-tube test.

2. Incubate the fermentation tubes at $35 \pm 0.5°C$. Examine each tube at the end of 24 ± 2 h and, if no gas has formed, again at the end of 48 ± 3 h. Record the presence or absence of gas formation at each examination of the tubes, regardless of the amount. More detailed records of the amount of gas formed, though desirable for the purpose of study, are not necessary for performing the standard tests prescribed.

Interpretation: Formation within 48 ± 3 h of gas in any amount in the inner fermentation tubes constitutes a positive presumptive test.

The appearance of an air bubble must not be confused with actual gas production. If the gas formed is a result of fermentation, the broth medium will become cloudy and active fermentation may be shown by continued appearance of small bubbles of gas throughout the medium outside of the inner fermentation tube when gently shaken. The absence of gas formation at the end of 48 ± 3 h of incubation constitutes a negative test. An arbitrary limit of 48 h for observation doubtless excludes from consideration occasional members of the coliform group which form gas very slowly; but for the purpose of a standard test based on the definition of the coliform group, exclusion of these occasional slow gas-forming organisms is satisfactory.

18.8.2.2 Confirmed Test

The use of confirmatory brilliant green lactose bile broth fermentation tubes or of endo or eosin methylene blue agar plates is permitted.

Procedure: Submit all primary fermentation tubes showing any amount of gas at the end of 24-h incubation to the confirmed test. If active fermentation appears in the primary fermentation tube before the expiration of the 24-h period of incubation, it is preferable to transfer to the confirmatory medium without waiting for the full 24-h period to elapse. If additional primary fermentation tubes show gas production at the end of 48-h incubation, these too shall be submitted to the "confirmed test."

18.8.2.3 Completed Test

The completed test is used as the next step following the confirmed test. It may be applied to the brilliant green lactose bile broth fermentation tubes showing gas in the confirmed test, or to typical or atypical colonies found on the plates of solid differential medium used for the confirmed test.

18.8.2.3.1 Procedure

1. If the brilliant green lactose bile broth tubes used for the confirmed test are to be employed for the completed tests, streak one or more endo or eosin methylene blue plates from each tube showing gas, as soon as possible after the appearance of gas. Incubate the plates at $35 \pm 0.5°C$ for 24 ± 2 h.
2. From each of these plates, or from each of the plates used for the confirmed test, fish one or more typical coliform colonies or, if no typical colonies are present, fish two or more colonies considered most likely to consist of organisms of the coliform group, transferring each fishing to a lactose broth fermentation tube or a lauryl tryptose broth fermentation tube and to a nutrient agar slant.

The use of a colony counter is recommended to provide optimum magnification to assist in fishing colonies from the pates of selective medium. When transferring colonies, take care to choose, if possible, well-isolated colonies separated by at least 0.5 cm from other colonies, and barely to touch the surface of the colony with the needle in order to minimize the danger of transferring a mixed culture.

The agar slants and secondary broth tubes incubated at $35 \pm 0.5°C$ for 24 ± 2 h or 48 ± 3 h, and Gram stained preparations from those corresponding to the secondary lactose broth tubes that show gas are examined microscopically.

Interpretation: The formation of gas in the secondary lactose broth tube and the demonstration of Gram-negative, non-spore-forming, rod-shaped bacteria in the agar culture may be considered a satisfactory completed test, demonstrating the presence of a member of the coliform group in the volume of sample examined. If, after 48 ± 3 h, gas is produced in the lactose and no spores on the slant, the test may be considered "completed" and the presence of coliform organisms demonstrated.

18.8.2.4 Gram-Stain Technique

The "completed test" for coliform group organisms includes the determination of Gram-stain characteristics of the organisms isolated, as discussed in the preceding.

18.8.2.4.1 Reagents

1. *Ammonium oxalate-crystal violet*: Dissolved 2 g crystal violet (85% dye content) in 20 mL 95% ethyl alcohol; dissolve 0.2 g ammonium oxalate monohydrate in 20 mL distilled water; mix the two solutions, ordinarily in equal parts. It is sometimes found, however, that this gives so concentrated a stain that Gram-negative organisms do not properly decolorize. To avoid this difficulty, the crystal violet solution may be diluted as much as 10 times, and the diluted solution mixed with an equal quantity of ammonium oxalate solution.
2. *Lugol's solution, Gram's modification*: Dissolve 1 g iodine crystals and 2 g potassium iodide in 300 mL distilled water.
3. *Counterstain*: Dissolve 2.5 g safranin dye in 100 mL 95% ethyl alcohol. Add 10 mL of the alcoholic solution of safranin to 100 mL distilled water.
4. Ethyl alcohol, 95%.

Procedure: Stain the smear for 1 min with the ammonium oxalate-crystal violet solution. Wash the slide in water; immerse in Lugol's solution for 1 min.

Wash the stained slide in water; blot dry. Decolorize with ethyl alcohol for 30 s, using gentle agitation. Blot and cover with counter stain for 10 s; then wash, dry, and examine as usual.

Cells which decolorize and accept the safranin stain are Gram-negative. Cells which do not decolorize but retain the crystal violet stain are Gram-positive.

18.9 SOIL QUALITY ASSESSMENT

Table 18.5 gives the various ions present in the soil and the components to be analyzed most often when soil analysis is carried out.

18.9.1 Pre-Treatment

After removal of living material (such as mosses, roots, etc.) and objects >2 cm, collected samples (preferably not less than 500 g fresh material) should be transported to the laboratory as soon as possible and should be air dried or dried at a temperature of 40°C. They can then be stored until analysis. The sample is subsequently crushed or milled to size <2 mm.

Drying: Spread the material in a layer not thicker than 15 mm. If necessary, the sample is crushed while still damp and friable and again after drying. Dry the complete sample in drying oven at a temperature of 40°C,

Table 18.5 Ionic Forms of Element in Soil

Element Name	Chemical Symbol	Ionic Soil Forms	Comments Concerning Forms of the Element and the Element's Importance in Soils
Aluminum	Al	Al^{3+} $Al(OH)_2^{+}$	Can be toxic to plants in strongly acid soils; occurs as various hydroxyl forms.
Boron	B	H_3BO_3	A water-soluble plant nutrient in small concentrations.
Cadmium	Cd	Cd^{2+}	High atomic weight ("heavy metal") retained in animals and people and is highly toxic.
Calcium	Ca	Ca^{2+}	Essential plant nutrient; the cation often most prevalent in non-acidic soils.
Carbon	C	HCO_3^{-} CO_3^{2-}	The basic element of organic substances (mostly made by living organisms); component of carbon dioxide (CO_2).
Chlorine	Cl	Cl	An essential plant nutrient occurring in small amounts except when it is a part of soluble salts.
Copper	Cu	Cu^{2+}	An essential plant nutrient; may be as Cu^{+}(cuprous) in poorly aerated soils.
Hydrogen	H	H^{+}	An essential plant nutrient; a small, active, strongly adsorbed, and chemically active ion.
Iron	Fe	Fe^{3+}	An essential plant nutrient of low solubility in most soils; may be as Fe^{2+} in minerals and poorly aerated soil; as iron oxide (Fe_2O_3) it causes the reddish and yellowish coloring in soils.
Lead	Pb	Pb^{2+}	Toxic heavy metal; also as PbO_2 in soil.
Magnesium	Mg	Mg^{2+}	An essential plant nutrient; similar in properties and reactions as in calcium.

Table 18.5 Ionic Forms of Element in Soil—cont'd

Element Name	Chemical Symbol	Ionic Soil Forms	Comments Concerning Forms of the Element and the Element's Importance in Soils
Manganese	Mn	Mn^{2+}	An essential plant nutrient; also as MnO_2 in soil.
Mercury	Hg	Hg^{2+}	Toxic heavy metal; also as HgO in soil.
Molybdenum	Mo	MoO_4^{2-}	An essential plant nutrient required in very small amounts.
Nickel	Ni	Ni^{2+}	Similar importance as is Cd.
Nitrogen	N	NO_3^- NH_4^+	An essential plant nutrient; necessary for proteins; in complex organic forms; both ionic forms are usable by plants.
Oxygen	O	O^{2-} OH^-	An essential plant nutrient; as free gaseous form, O_2, it is essential to all respiration
Phosphorus	P	HPO_4^{2-} $H_2PO_4^-$	An essential plant nutrient; forms many low solubility phosphates with Ca, Al, Fe, and other heavy metals.
Potassium	K	K^+	An essential plant nutrient; soluble in soils, except minerals forms are very insoluble.
Silicon	Si	Si^{4+}	Common in minerals holding oxygen's together; sands and quartz are mostly SiO_2
Sodium	Na	Na^{2+}	Not essential nutrient, although it may be for some plants; very soluble; part of "soluble salts;" causes sealing of soil.
Sulfur	S	SO_4^{2-}	An essential plant nutrient; forms S^{2-}(sulfide) form or toxic hydrogen sulfide gas (H_2S) in poorly aerated soil.
Zinc	Zn	Zn^{2+}	An essential plant nutrient; often deficient in calcareous and eroded or leveled soils.

until the loss in mass of the sample is not greater than 5% (m/m) per 24 h. Break down the size of larger clods (greater than 15 mm) to accelerate the drying process.

Removal of fraction <2 mm: Remove stones and large objects by hand picking and sieving (<2 mm). Minimize the amount of fine material adhered. Weigh separately the fraction not passing the 2 mm sieve for determination of coarse fragment content. Crush (not ground) the clods greater than 2 mm taking care that crushing of original particles is minimized. Homogenize the <2 mm fraction.

Sieving and milling: The organic sample is crushed or milled to size <2 mm.

18.9.2 Preparation of Soil Extract

(For pH, electrical conductivity, alkalinity, chlorides, iron, sulfates)
10 g of soil was weighed. To this, 100 mL of distilled water was added for making a suspension of 1:10 w/v dilution. Sample was filtered through Wattman number 40 filter paper and filtrate was collected.

18.9.2.1 Preparation of Soil Extract (for Hardness)

50 g of soil were taken in a flask. To this, 100 mL of 40% of ethyl alcohol was added and shaken well. The suspension was filtered through Wattman filter paper. The residue on the filter paper was washed with 40%. The residue was kept in a beaker and 100 mL of ammonium acetate solution was added and kept overnight. The solution was filtered and filtrate was collected.

18.9.2.2 Preparation of Soil Extract (for Nitrates)

50 g of soil was weighed into an Erlenmeyer flask, and to this 250 mL of extraction reagent was added and shaken well. 0.4 g of calcium hydroxide and 1 g of magnesium carbonate was added to it and filtered through filter paper. The filtrate was used for estimation of nitrates.

18.9.2.3 Preparation of Soil Extract (for Phosphates)

1 g of soil was taken in a 500 mL conical flask. To this, 200 mL of sulfuric acid (0.002 N) was added and kept for half an hour. The suspension was filtered through filter paper and filtrate was used for estimation of phosphorus.

18.9.2.4 Preparation of Soil Extract (for Organic Carbon)

The soil organic carbon by wet digestion method of Walkey and Black (1934) as described by Jackson (1967) which involves oxidation with dichromate and back titration of excess of dichromate with ferrous sulfate and is expressed as percentage. The electrical conductivity (EC) in 1:2 soil:water extract was determined using EC bridge and expressed as $cm/s \times 10^{-4}$.

The soil extract was prepared by taking 10 g of soil in 100 mL of water and after vigorous stirring the solution was filtered through Whatmann filter paper and the filtrate was used for the analysis.

18.9.2.5 Organic Matter

Organic matter in the soil was estimated by titrimetric method. Organic matter in sample gets oxidized with potassium chromate and sulfuric acid. The contents are titrated against ferrous ammonium sulfates. At the endpoint, the dull green color gets changed through turbid blue to the brilliant green. 0.5 g of soil was taken into a 500 mL conical flask and to this 10 mL potassium dichromate solution and 20 mL sulfuric acid were added. To this, 200 mL distilled water, 10 mL phosphoric acid and 0.2 g sodium fluoride were added. 1 ml diphenylamine were added to the solution and titrated against ferrous ammonium sulfate solution. At the endpoint, the dull green color gets changed through turbid blue to the brilliant green. Blank was done in a similar manner with distilled water.

18.9.2.5.1 Calculation

$$\text{OM (mg/g)} = (6.791/X)\,[1 - V1/V2] \times 10$$
$$\text{OM } (\%) = (6.791/X)\,[1 - V1/V2]$$
$$\text{Carbon } (\%) = (6.791/X \times 1.724)\,[1 - V1/V2]$$

where OM, organic matter; X, weight of soil in g; V1, volume of titrant used against sample (mL); V2, volume of titrant used against distilled water blank (mL).

18.10 AIR QUALITY ASSESSMENT

Primary ambient air quality standards define levels of air quality necessary, with an adequate margin of safety, to protect public health. Secondary ambient air quality standards define levels of air quality necessary to protect public welfare. The averaging time criterion implies that for some shorter time intervals, the concentration may rise above the standard without adverse effect. Although air pollutant concentrations are generally log-normally distributed, only the total suspended particulate standard is based on the geometric mean. Monitoring of the air quality in order to get a comprehensive idea of the quality of air (Table 18.6A).

18.10.1 Air Quality Network Requirements

Table 18.6A Number of Air Quality Monitoring Stations

Pollutant	Region Population	Minimum Number of Air Quality Monitoring Stations
Particulate matter	Less than 100,000	4
	100,000–1,000,000	4 + 0.6 per 100,000 population
	1000-001–5,000,000	7.5 + 0.25 per 100,000 population
	Above 5,000,000	12 + 0.16 per 100,000 population
SO_2	Less than 100,000	3
	100,001–500,000	2.5 + 0.5 per 100,000 population
	500,001–1,000,000	6 + 0.15 per 100,000 population
	Above 1,000,000	20
NO_2	Less than 100,000	4
	100,000–500,000	4 + 0.6 per 100,000 population
	Above 1,000,000	10
CO	Less than 100,000	1
	100,000–500,000	1 + 0.15 per 100,000 population
	Above 5,000,000	6 + 0.05 per 100,000 population

Region means the study area around the project boundary area decided in scoping.
Additional monitoring locations should be set up if sensitive sites such as places of archeological importance and biosphere reserves exist.
National Ambient Air Quality Series: NAAQMS/25/2003-04-CPCB.

18.10.2 Ambient Air Sampling for Chemical Pollutants

One of the most difficult tasks in the analysis of ambient air quality for various chemical pollutants like SPM, SO_x, NO_x, etc., is to obtain a genuine and representative sample of the ambient air. This is because the concentrations of pollutants in the ambient air are likely to be extremely small and vary from place to place. In general ambient air sampling devices consists of a sample collector, a flow meter, and a pump to draw the air sample into the system. The type of pollutant and the method of analysis to be adopted for pollutant analysis dictate the type of sampling system to be used. Mainly there are two types of collector; one that collects particulate pollutants and one that is used for gaseous pollutants (see Table 18.6B).

18.10.3 Particulate Sampling

The following methods are used for sampling and estimation of particulate matter. For collecting dust particles of 1 mμ or larger (40 μm) in the atmosphere, clean glass jars are kept in the area where dust fall is to be

Table 18.6B Methods of Measurements of Air Pollutants

Method	Variable Measured	Principle
Gravimetric	PM10, PM2.5	Particles are trapped or collected on filters, and the filters are weighed to determine the volume of the pollutant.
Atomic absorption spectrometry (AAS)	Greater than 60 metals or metalloid elements (e.g., Pb, Hg, Zn)	This technique operates by measuring energy changes in the atomic state of the analyte. Emitted radiation is a function of atoms present in the sample.
Spectrophotometry	SO_2, O_3	Measure the amount of light that a sample absorbs. The amount of light absorbed indicates the amount of analyte present in the sample.
Chemiluminescence	SO_2, O_3	Based upon the emission spectrum of an excited species that is formed in the course of a chemical reaction.
Gas chromatography (GC), flame ionization detector (FID)	VOC	Responds in proportion to number of carbon atoms in gas sample.
Gas chromatography, mass spectrometry (GC–MS)	VOC	Mass spectrometers use the difference in mass-to-charge ratio (m/z) of ionized atoms or molecules to separate them from each other.
Fourier transform infrared spectroscopy (FTIR)	CO, VOC, CH_4	Sample absorbs infrared radiation and difference in absorption is measured.

determined and after a few hours or days, the dust is collected from each jar and then weighed. The average weight of dust in each jar is estimated and the dust fall is expressed as weight of dust per unit area per unit time. Containers, generally conical plastic jars, 10–15 cm in diameter (Fig. 18.1), open at the top are used. The jars are kept in strategic locations throughout a community or in the vicinity of particulate sources under study. Grit and dust fall into the jars, which sometimes have water to hold the dust. After a one month exposure, the jars are collected and brought into the laboratory where their contents are analyzed. In most cases only the total particulate matter is determined, and the results are expressed in terms of tons per

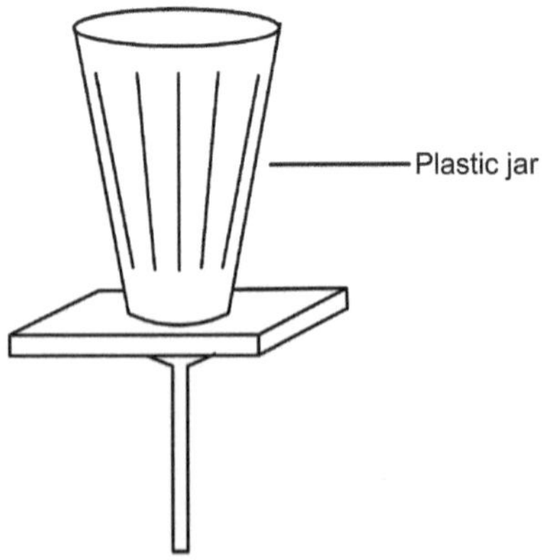

Figure 18.1 Dust fall collector.

square kilometer per month. Monthly isopleth maps can be constructed showing the variation of dust fall throughout the area.

18.10.4 High Volume Air Sampler

The most commonly used particulate matter sampler for ambient air sampling is the high volume sampler (Fig. 18.2). Various designs are available commercially, which work on the principle of collecting a large volume of air through suitable filter media. One specific unit makes use of specially fired glass-fiber filter, web-mounted on a rectangular aluminum filter holder. The capacity of the sampler is 60–70 cu ft/min; 84,000 cu ft of air may get filtered in 24 h. The collected particulate matter is dried and weighed, and then removed by washing for subsequent size separation and composition analysis.

High volume air samplers can be generally run as long as 24 h without an appreciable increase in airflow resistance. The multiple advantages are: relatively high airflow rates at low pressure drop, high particles storage capacity, and low moisture regain. A glass fiber medium is the most popular sampling device for ambient air quality monitoring because of the high collection efficiency and low cost.

A variety of analytical techniques are available for determining the concentrations of components of interest in collected particles.

Chemical methods are most satisfactory for the analysis of some elements present in particulates.

Sulfates and nitrates: These ions are often analyzed in the context of acid rain related studies. Until recently, analysis was done by

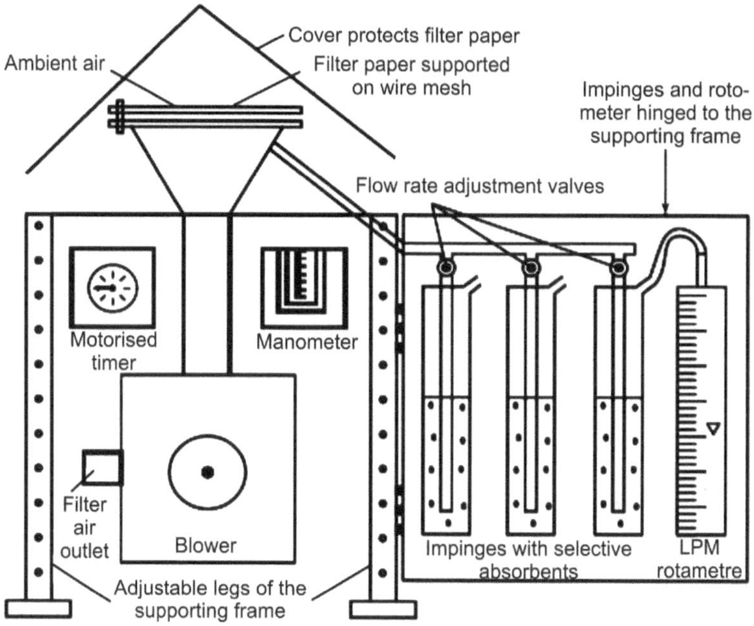

Figure 18.2 High volume air sampler

rather laborious procedures involving reduction of nitrate to nitrite, which was determined spectrophotometerically after diazotization and azo coupling reactions. A number of colorimetric reagents are available for sulfate and a barium chloride turbidimetric procedure was also in use. In advanced laboratories these methods were supplemented by ion chromatography, in which the anions are separated by ion exchange and detected most commonly by conductivity after suppression of the conductivity of the eluent.

Arsenic: Arsenic can be analyzed by the colorimetric silver diethyldithiocarbamate method. Zinc is used to reduce the arsenic in the acidified sample to gaseous arsine, AsH_3, which gets swept into an absorbing solution of silver diethyldithiocarbamate dissolved in pyridine. A soluble red complex is formed and its concentration can be monitored spectrophotometrically by measuring absorbance at 535 nm.

Lead: The lead in particulate matter can be determined by dithizone method. The sample is collected on a 0.45 mm membrane filter. If it is desired to trap gaseous lead compounds, a tube filled with I2 crystals is used after filtration. The filtered sample is digested with HNO_3, H_2SO_4, and

$HClO_4$ to solubilize the lead. Iodine crystals are dissolved in acid KI and reduced with sulfite:

$$I_3^- + H_2SO_3 + H_2O \rightarrow 3I^- + SO_4^{2-} + H^+$$

Again the lead is solubilized. Acidic Pb^{2+} from either source is made basic with NH_3 and extracted with a chloroform solution of dithizone.

Following careful removal of excess dithizone from the chloroform phase, the lead dithizone complex is determined spectrophotometrically by measuring the absorbance at 510 nm.

The following are the main criteria for adopting a sampling procedure for monitoring gaseous pollutants.

1. The sample collected should be representative with reference to time, location, and physicochemical conditions of the pollutant.
2. The volume of the sample collected should be appropriate in accordance with the analytical method adopted and the concentration of the pollutant present.
3. The frequency of sampling and the duration of sampling time (4 or 8 h) should be fixed keeping in mind the expected flow or 24 h average or variations of the pollutant levels.
4. The rate of sampling should correspond to maximum collection efficiency.
5. There should not be any physicochemical change for the pollutant in the process of sampling.

The common methods adopted for sampling of gaseous pollutants are.

Grab sampling: In this method the sample is collected by filling an evacuated flask or an inflatable bag. Plastic bags can be used for sampling and storing for analysis. Moisture condensation or diffusion through the walls of the bag will create some losses in the sample in this method. Using containers made of glass or stainless steel these losses can be minimized. These sample collectors can be initially evacuated and opened for sucking the air sample or filled with water and then used as a collector by draining away the water and sucking the ambient air.

Absorption in liquids: Using a liquid absorption medium for collecting gaseous samples is a versatile technique under use. By properly selecting the specific absorption liquid, the desired pollutant from air can be separated by its solubility in the media or a chemical reaction. Different types of collectors are available for this purpose. The most widely used collector is the impinger. In the impinger the gas stream is impinged at high velocity and to a flat surface, which may be the bottom of a collector or a specially designed plate. The midget impinger can handle flow rates of about 3 L/min while the Greenburgh–Smith type impingers can be used with flow rates of 30 L/min. The pollutant is estimated according to the method of selection of the liquid.

Adsorption on solid materials: In this method a packed column containing a finely divided solid adsorbent is used as a medium for the gaseous pollutant, which is sucked from the air passed through the thin column. The pollutants are concentrated and retained on the surface of the adsorbent. The common adsorbents used are activated charcoal or silica gel. The gases are desorbed by heating to the temperature of volatility of the trapped material or by sucking with inert gas or applying vacuum. The adsorbed gas can also be dissolved in a liquid solvent. Organic vapors trapped by this technique are analyzed by gas chromatography.

Freeze out sampling: In this method the air sample containing gaseous pollutants is passed through a series of cold traps, maintained at progressively lower temperatures. The pollutants are condensed in these traps which are brought to the laboratory, and after de-freezing the gaseous pollutants are estimated by gas chromatography, or IR or UV spectrophotometry, mass spectrophotometry or by wet chemical methods.

18.10.5 Stack Sampling of Gaseous Pollutants

In any stack sampling technique, obtaining a representative sample is the most important aspect to obtain reliable results. It will be difficult for the designer for air pollutant management if the data collected indicates either too high or too low a value compared to the true value. So the objectives of any stack sampling operation is to examine total quality and quantity, the physicochemical characteristics of the pollutant emitted from any stack and evaluate how any control device is working. The usual practices in determining stack sampling points are:

1. The sample site should be located *at least 8 times* the stack or the duct diameters downstream.
2. The sample site should be located at least two diameters upstream from any source of flow disturbance such as bends, fittings, or constrictions since any flow disturbance would cause a non-uniform particle concentration pattern.

18.10.6 Sulfur Dioxide

The methods for the estimation of sulfur dioxide are based on chemisorption, coulometric, conductometric, colorimetric, idometric, or turbidometric procedures.

18.10.7 West-Gaeke Method (Pararosaniline Method)

In this method sodium tetrachloromercurate, which absorbs SO_2 present in air quantitatively (collection efficiency 95%) at concentrations as low as 0.002 ppm and disallows its oxidation to SO_3. This is then allowed to react with formaldehyde and bleached pararosaniline causing the formation of a reddish purple "pararosaniline" and sulfuric acid which can be measured photometrically and SO_2 estimated. SO_2 in air in a wide range of concentration (from 0.002 to 5 ppm) can be estimated by this method. The sampling should be carried out at a flow rate of 200 mL–2.0 L/min for 30 min to 24 h . The duration and rate of sampling depends upon the concentration of SO_2 expected in the ambient air.

$$2\,NaCI + HgCI_2 \rightarrow 2Na^+ + [HgCI_4]^-$$
$$SO_2 + [HgCI_4]^- + H_2O \rightarrow [HgCI_2SO_3]^- + 2H^+ + 2CI^-$$

Pump 30–60 L of air through 10 mL of the scrubbing solution (13.6 g $HgCl_2$ and 7.5 g KCl in 1 L water) in a small impinger, at a rate of 1–2 L/min. Then add 1 mL of dilute pararosaniline reagent solution (4 mL of 1% aqueous solution with 6 mL of concentrated HCl and diluted to 100 mL with water) and 1 mL of 0.2% HCHO solution (5 mL of commercial 40% solution diluted to 1 L with water). After 20–30 min, measure the absorbance at 580 nm. Use a dilute solution of Na_2SO_3, counter-checked iodometrically as the standard. After the collection of a 1- or 24-h average sample in the field, the solution should be returned to the laboratory. Ammonic acid is to be added to reduce nitrogen dioxide, which otherwise would destroy some of the color reagent. The standard curve, for obtaining the unknown concentration, can be prepared with sodium metabisulphite in the range of 2–25 mg of SO_2. The concentration of SO_2 in mg of SO_2/m^3 can be calculated as follows:

$$SO_2 = g/m^3 = \frac{\mu g \text{ of } SO_2 \times 10^3}{\text{Total volume of air sampled (L)}}$$

When high concentration of $HgCl_2SO_3{}^{2-}$ is more than one site on the pararosaniline dye may react, and Beer's law will no longer be obeyed. This limits the method to solution concentrations of SO_2 between 0.05 and 5 mg/dm^3. The range of atmospheric concentrations may be extended by appropriate dilutions of concentrated samples from 0.002 to 5 ppm.

Nitrogen dioxide at levels exceeding 2 ppm constitutes a major interference with the original West–Gaeke method. It is possible to get rid of the interference by the addition of sulfamic acid (H_2NSO_3H), which acts as a reducing agent converting NO_2 to nitrogen gas.

18.10.8 Oxides of Nitrogen

The most common method of measuring the oxides of nitrogen is the colorimetric method of Griess–Saltzman. In this method NO_2 reacts with sulfanilic acid to from the diazonium salt which is coupled with N-(1 napthyl) ethylene diamine dihydrochloride to get a pink-colored dye complex. Within 15 min at room temperature itself the color development is complete. The developed color is measured at 550 nm. This method is also applicable to NO analysis or a combined analysis, by oxidizing NO to NO_2 by passing it through potassium permanganate solution. Analysis must be completed within an hour of sample collection. The method can be used to determine atmospheric concentrations of NO_2 from 40 to 1500 mg/m^3. High concentrations of SO_2 may bleach the reagent, giving low results, and the stoichiometry of the process is not completely clear, but it can be calibrated using permeation tubes and is easily modified for automated analysis. The standard curve can be made with sodium nitrite, ranging from 0.04 to 2.0 mg NO_2/mL. The concentration of NO_2 in the air can be calculated by the following formula:

$$\mu g\ NO_2/m^3 \text{ of air} = \frac{\mu g \text{ of } NO_2 \text{ in absorbent} \times 10^3}{\text{Volume of air sampled (L)} \times 0.82}$$

where the value 0.82 is a factor for collecting efficiency. The sampling rate is to be usually kept at 200 mL/min, and the sample is to be collected for 24 h for air containing NO_2 in the range of 20–750 mg/m^3.

18.10.9 Carbon Monoxide

For estimating carbon monoxide levels in the range of 1–50 ppm in ambient air, non-dispersive IR spectrophotometry is widely used. However, in this method carbon dioxide and water vapor which have absorption bands in the same region will interfere. The interference of water vapor can be eliminated by passing the sample through a drying agent. By interposing a carbon dioxide cell between the sample and reference cells and detecting the interference of carbon dioxide can be nullified.

18.10.10 Hydrocarbons

The commonly used methods for determining the hydrocarbon content of the atmosphere and of process streams include spectrographic, chromatographic, and flame ionization methods. All three methods find use for the comparatively high concentration in process streams.

18.11 BIOLOGICAL ASSESSMENT

18.11.1 Plankton Analysis

The physical and chemical characteristics of water affect the abundance, species composition, stability, and productivity of the indigenous populations of aquatic organisms. The biological methods used for assessing water quality includes collection, counting, and identification of aquatic organisms; biomass measurements; measurements of metabolic activity rates; toxicity tests; bioaccumulation; biomagnification of pollutants; and processing and interpretation of biological data. The work involving plankton analysis would help in:

1. Explaining the cause of color and turbidity and the presence of objectionable odor, tastes, and visible particles in waters.
2. The interpretation of chemical analyses.
3. Identifying the nature, extent, and biological effects of pollution.
4. Providing data on the status of an aquatic system on a regular basis.

Plankton: A microscopic community of plants (phytoplankton) and animals (zooplankton), found usually free-floating, swimming with little or no resistance to water currents, suspended in water, non-motile or insufficiently motile to overcome transport by currents, are called "plankton."

Phytoplankton (microscopic algae) usually occurs as unicellular, colonial, or filamentous forms and is mostly photosynthetic and is grazed upon by the zooplankton and other organisms occurring in the same environment. Zooplankton principally comprise of microscopic protozoans, rotifers, cladocerans, and copepods. The species assemblage of zooplankton also may be useful in assessing water quality.

The structure of photosynthetic populations in the aquatic ecosystems is dynamic and constantly changing in species composition and biomass distribution. An understanding of the community structure is dependent on the ability to understand the temporal distribution of the different species. Changes in species composition and biomass may affect photosynthetic rates, assimilation efficiencies, rates of nutrient utilization, grazing, etc.

Plankton, particularly phytoplankton, has long been used as indicators of water quality. Because of their short life spans, planktons respond quickly to environmental changes. They flourish both in highly eutrophic waters while a few others are very sensitive to organic and/or chemical wastes. Some species have also been associated with noxious blooms causing toxic conditions apart from the tastes and odor problems.

Plankton net: The plankton net is a field-equipment used to trap plankton. It has a polyethylene filter of a defined mesh size and a graduated measuring jar attached to the other end. A handle holds the net. The mesh size of the net determines the size range of the plankton trapped. The mesh number 30 of size 60 mm was used for collecting samples.

Sampling procedure: The manner in which sampling is done should conform to the objectives of the study. The "surface samples" (samples collected from the surface) are collected as close to the water surface as possible, mostly toward the center of the lake at regular monthly intervals. A known volume of the sample, 5 to 50 L is filtered and planktons are filtered and preserved for further analysis.

Labels: The sample label has the date, time of sampling, study area-lake name, and the volume measured and pasted on the containers of 50 mL capacity.

Preservation: The samples collected into the 100 mL polyethylene vials were preserved by adding suitable amounts of 1 mL chloroform to act as the narcotizing agent and 2 mL of 2% formalin for preservation and analyses.

Concentration technique: The plankton nets are used to collect samples for the qualitative and quantitative estimation of the plankton, by filtering a known volume of water (5–50 L) through the net depending on the plankton density of the tanks.

Qualitative and quantitative evaluation of plankton: Detailed analyses of phytoplanktonic populations are done by estimating the numbers in each species. The phytoplankton consisting of individual cells, filaments, and colonies are counted as individual cells. When colonies of species are counted, the average number of cells per colony is counted, and in filamentous algae, the average length of the filament has to be determined.

Sedimentation and enumeration by microscope: Preserved samples in bottles are mixed uniformly by gentle inversion and then exactly 1 mL of the sample is pipetted out into the S–C cell for analysis.

18.11.2 Microscope

18.11.2.1 Compound Microscope

A monocular compound microscope is used in the counting of plankton with different eyepieces such as 10×, 15× and 20×. The microscope is calibrated using plankton-counting squares.

18.11.3 Counting

18.11.3.1 Counting Cell: Sedgwick–Rafter (S–R) Cell

The Sedgwick–Rafter cell is a device used for plankton counting and is about 50 mm long by 20 mm wide and 1 mm deep. The cell is covered by a relatively thick cover slip and is calibrated to contain exactly 1.0 mL.

18.11.4 Method

18.11.4.1 Filling the Cell

The cover slip is placed diagonally across the S–R cell and filled with the sample carefully without air bubbles with a large-bore pipette. The sample is allowed to settle for about 5 min before the actual counting begins.

18.11.4.2 Strip Counting

A "strip" is the length of the cell that constitutes a volume approximately 50 mm long, 1 mm deep accounting to the volume of 25 mm^3, or 1/40 (2.5%) of the total cell volume. By moving the mechanical stage from left to right, the organisms can be examined in a systematic manner. By knowing the surface area of the portion counted in relation to that of the total, a factor is determined to expand the average counts of the plankton to the total area of the counting surface. This total area represents the number of organisms present per given volume of the sample. This volume expanded to an appropriate factor yields the organisms per liter of water for the lake.

The total number of planktons in the S–R cell is obtained by multiplying actual count in the strip by the number (enumeration factor) representing the portion of the S–R cell counted. The number of the strips counted is a function of the precision desired and the number of units (cells, colonies) for the strips measured. In this study, 500 cells were counted for estimation.

The plankton count in the S–R cell is got from the following:

$$\text{Number/ml} = \frac{\text{C X 1000 mm}}{\text{L X D X W X S}}$$

where C, number of organisms counted; L, length of each strip (S–R cell length) mm; D, depth of a strip (S–R cell depth) mm; W, width of a strip in mm; S, number of strips counted.

The plankton counts per strip are then determined by multiplying the actual count by the factor representing the counted portion of the whole S–R cell volume:

$$\text{Number/mL} = (\text{C})\left(1000\ \text{mm}^3\right) / (\text{L})(\text{D})(\text{W})(\text{S})$$

where C, number of organisms counted; L, length of each strip in mm (of S–R cell); D, depth of the strip in mm (S–R cell); W, width of the strip in mm (Whipple grid image width); S, number of strips counted.

Phytoplankton counting units: Some plankton are unicellular while others are multicellular (colonial), posing a problem for enumeration. For analysis, a colony of plankton is accounted as a single count. The large forms that cross two or more boundaries of the grid are counted separately at lower magnification and their number included in the total count.

18.12 QUALITY OF THE RESULTS AND PRESENTATION

Data reduction, validation, and reporting are the final steps in the data-generation process. The data obtained from an analytical instrument must first be subjected to the data reduction processes described in the applicable Standard Operating Procedures (SOP) before the final result can be obtained. Specify calculations and any correction factors, as well as the steps to be followed in generating the sample result, in the QA manual or SOP. Also specify all of the data validation steps to be followed before the final result is made available. Report results in standard units of mass, volume, or concentration as specified in the method or SOP. Report results below the method detection level in accordance with the procedure prescribed in the SOP. Ideally, include a statement of uncertainty with each result.

FURTHER READING

[1] H.R. Anderson, R.W. Atkinson, J.L. Peacock, L. Marston, K. Konstantinou, Meta-analysis of Time-series Studies and Panel Studies of Particulate Matter (PM) and Ozone (O_3), Report of a WHO task group, World Health Organization, Geneva, 2003.

[2] APHA, Standard Methods for Examination of Water and Wastewater, twentieth ed., American Public Health Association, Washington D.C, 1995.

[3] Cambridge, MA, Health Effects Institute: Revised Analyses of Time-series Studies of Air Pollution and Health: Revised Analyses of the National Morbidity, Mortality, and Air Pollution Study, Part II, 2003.

[4] D. Krewski, R.T. Burnett, M.S. Goldberg, K. Hoover, J. Siemiatycki, M. Abrahamowicz, et al., Particle Epidemiology Reanalysis Project, Health Effects Institute, Cambridge, MA, 2000.
[5] NEERI, Water and Wastewater Analysis – Manual, National Environmental Engineering Research Institute, Nagpur, 1988.
[6] A.K. Pandey, S.Z. Siddiqi, R. Rao, Physicochemical and biological characteristics of Husain sagar, an industrially polluted lake, Hyderabad, Proceedings of the Academy of Environmental Biology 2 (2) (1993) 161–167.
[7] R.K. Trivedy, P.K. Goel, Chemical and Biological Methods for Water Pollution Studies, Environmental Publication, Karad, Maharashtra, 1986.
[8] R.G. Wetzel, E. Likens, Limnological Analyses, second ed., W.B. Saunders Co., Philadelphia, PA, 1979.
[9] J.P. Higgins, S.G. Thompson, Quantifying heterogeneity in a meta-analysis, Statistics in Medicine 21 (2002) 1539–1558.

APPENDIX 1 INTERNATIONAL ENVIRONMENTAL LAW

These are the most commonly used abbreviations and acronyms for treaty research in international environmental law:

- Agenda 21: Program of Action for Sustainable Development (Agenda 21)
- Aarhus Convention: Convention on Access to Information, Public Participation in Decision-Making, and Access to Justice in Environmental Matters
- Basel Convention: Basel Convention on the Control of Transboundary Movements of Hazardous Wastes and Their Disposal
- Basel Ban Amendment: Protocol on Liability and Compensation for Damage Resulting from Transboundary Movements of Hazardous Wastes
- Biodiversity Convention: Convention on Biological Diversity
- Bunkers Convention: International Convention on Civil Liability for Bunker Oil Pollution Damage
- Cartagena Protocol: Cartagena Protocol on Biosafety to the Convention on Biological Diversity
- CBD: Convention on Biological Diversity
- CCAMLR: Convention on the Conservation of Antarctic Marine Living Resources
- CCSBT: Convention for the Conservation of Southern Bluefin Tuna
- CITES: Convention on the International Trade in Endangered Species of Wild Fauna and Flora
- CLC: Convention on Civil Liability for Oil Pollution Damage
- Climate Change Convention: United Nations Framework Convention on Climate Change
- Kyoto Protocol: Kyoto Protocol to the United Nations Framework Convention on Climate Change
- LRTAP: Geneva Convention on Long-Range Transboundary Air Pollution
- Montreal Protocol: Montreal Protocol on Substances that Deplete the Ozone Layer

- NAAEC: North American Agreement on Environmental Cooperation
- OLDEPESCA: Agreement Instituting the Latin American Organization for Fisheries Development
- OSPAR Convention: Convention for the Protection of the Marine Environment of the Northeast Atlantic
- Paris Convention: Paris Convention on Third-Party Liability in the Field of Nuclear Energy
- POPs: Persistent Organic Pollutants
- Rio Declaration: Rio Declaration on Environment and Development
- Stockholm Convention: Stockholm Convention on Persistent Organic Pollutants
- UNCCD: United Nations Convention to Combat Desertification
- UNCLOS: United Nations Convention on the Law of the Sea
- UNEP: United Nations Environment Program
- UNFCCC: United Nations Framework Convention on Climate Change
- Vienna Convention: Convention for the Protection of the Ozone Layer
- Wetlands Convention: Convention on Wetlands of International Importance, Especially as Waterfowl Habitat

Because of the many possible subtopics available for international environmental law, it is helpful to have a table of categories and correlating agreements. Following is a table of agreements and websites that are research based on the eight subtopics of hazardous waste, nuclear waste, ocean and marine sources, ozone and protection of the atmosphere, pollution, protection of species and wildlife, sustainable development, and trade and the environment.

Subtopics for International Environmental Law	**Conventions and Agreements**	**Websites for Research**
1. Hazardous Waste	*Basel Convention* ([a]1989) *Protocol on Liability and Compensation for Damage Resulting from Transboundary Movements of Hazardous Wastes* (1999)	Secretariat of the Basel Convention: UN Environmental Program (http://www.basel.int/text/text.html) "Basel Ban Amendment": Secretariat of the Basel Convention (http://www.basel.int/meetings/cop/cop5/docs/prot-e.pdf) United Nations Environmental Program: Protocol on Liability and Compensation (http://www.basel.int/pub/protocol.html)

Subtopics for International Environmental Law	Conventions and Agreements	Websites for Research
2. Nuclear waste	[b]*Convention on Nuclear Safety* (1994) *Convention on Third Party Liability in the Field of Nuclear Energy ["Paris Convention"]* (1960)	International Atomic Energy Agency Circular (http://www.iaea.org/Publications/Documents/Infcircs/Others/inf449.shtml) Reservations and Declarations to the Convention on Nuclear Safety (http://www.iaea.org/Publications/Documents/Conventions/nukesafety_reserv.pdf) Nuclear Energy Agency's Convention on Third Party Liability (http://www.nea.fr/html/law/nlparis_conv.html) Paris Convention Decisions, Recommendations, and Interpretations (http://www.nea.fr/html/law/paris-convention-dec-rec-int.pdf)
3. Ocean and Marine sources	*Agreement for the Establishment of the Indo-Pacific Fisheries Council* (1948) *Agreement Instituting the Latin American Organization for Fisheries Development* (1982) [b]*Convention for the Conservation of Salmon in the North Atlantic* (1982) *Convention for the Conservation of Southern Bluefin Tuna* (1993) *Convention for the Prevention of Marine Pollution from Land-Based Sources* (1974) *and "OSPAR" Convention* (1992)	Australian Treaty Series: AustLII (http://www.austlii.edu.au/au/other/dfat/treaties/1949/4.htm) Internet Guide to Int'l Fisheries Law (http://www.oceanlaw.net/texts/oldepesca.html) ENTRI Treaty Locator: CIESIN (Columbia) (http://sedac.ciesin.columbia.edu/entri/texts/salmon.north.atlantic.1982.html) Commission for the Conservation of Southern Bluefin Tuna (http://www.ccsbt.org/docs/about.html) OSPAR Commission Website (http://www.ospar.org/eng/html/welcome.html)

Continued

Subtopics for International Environmental Law	Conventions and Agreements	Websites for Research
4. Ozone and protection of the atmosphere	[b]*Convention for the Protection of the Ozone Layer ["Vienna Convention"]* (1985) [b]*Convention on Long-Range Transboundary Air Pollution* (1979) *Kyoto Protocol to the United Nations Framework Convention* (1997) [b]*Montreal Protocol* (1987) [b]*United Nations Framework Convention on Climate Change* (1992)	United Nations Environment Program (http://www.unep.ch/ozone/index.asp) United Nations Economic Commission for Europe (http://www.unece.org/env/lrtap/lrtap_h1.htm) UN FCCC Website (http://unfccc.int/essential_background/items/2877.php) UN Environment Program: Montreal Protocol (http://www.unep.ch/ozone/Treaties_and_Ratification/2B_montreal_protocol.asp) UN FCCC Website (http://unfccc.int/essential_background/convention/items/2627.php)
5. Pollution	*Stockholm Convention on Persistent Organic Pollutants* (2001) *International Convention for the Prevention of Pollution from Ships* (1973) *and 1978 Protocol* *International Convention on Civil Liability for Bunker Oil Pollution Damage* (2001) *International Convention on Civil Liability for Oil Pollution Damage and 1992 Protocol* (1969 and 1992)	Persistent Organic Pollutants Website (http://www.pops.int/default.htm) ENTRI: Environmental Treaties and Resource Indicators (http://sedac.ciesin.columbia.edu/entri/texts/pollution.from.ships.1973.html) AustLII: Australian Legal Information Institute (http://www.austlii.edu.au/au/other/dfat/treaties/notinforce/2002/25.html) CIESIN (Columbia University)/ENTRI (http://sedac.ciesin.columbia.edu/entri/texts/civil.liability.oil.pollution.damage.1969.html)

Subtopics for International Environmental Law	Conventions and Agreements	Websites for Research
6. Protection of species and wildlife	*Agenda 21: Programme of Action for Sustainable Development* (1992) *Cartagena Protocol* (2000) *Convention on Biological Diversity* (1992) [b]*North American Agreement on Environmental Cooperation* (1993) [b]*United Nations Convention to Combat Desertification in Countries Experiencing Serious Drought* (1994)	United Nations Sustainable Development (http://www.un.org/esa/sustdev/documents/agenda21/english/Agenda21.pdf) Biological Diversity Website (https://www.biodiv.org/doc/legal/cartagena-protocol-en.pdf) Convention on Biological Diversity Website (http://www.biodiv.org/convention/default.shtml) Commission for Environmental Cooperation (http://www.cec.org/pubs_info_resources/law_treat_agree/naaec/index.cfm?varlan=english) UNCCD Website (http://www.unccd.int/convention/menu.php)
7. Sustainable development	*Agenda 21: Programme of Action for Sustainable Development* (1992) [b]*Convention on the Conservation of Antarctic Marine Living Resources* (1980) [b]*Convention on Wetlands of International Importance especially as Waterfowl Habitat* (1971) *Rio Declaration on Environment and Development* (1992) *Antarctic Treaty (1959) and Protocol on Environmental Protection* (1991)	United Nations Sustainable Development (http://www.un.org/esa/sustdev/documents/agenda21/english/Agenda21.pdf) Convention on the Conservation of Antarctic Marine Living Resources Secretariat (http://www.ccamlr.org/pu/e/gen-intro.htm) Ramsar Convention on the Wetlands Website (http://www.ramsar.org/index_very_key_docs.htm) UN Environment Program (http://www.unep.org/Documents.multilingual/Default.asp?DocumentID=78&ArticleID=1163) Committee for Environmental Protection: Antarctic Treaty (http://www.cep.aq/)

Continued

Subtopics for International Environmental Law	Conventions and Agreements	Websites for Research
8. Trade and the environment	*Convention on Int'l Trade in Endangered Species of Wild Fauna and Flora ["CITES"]* (1973) *North American Agreement on Environmental Cooperation ["NAAEC"]* (1993) *Aarhus Convention* (1998)	CITES Official Documents (http://www.cites.org/eng/disc/text.shtml) Commission for Environmental Cooperation (http://www.cec.org/pubs_info_resources/law_treat_agree/naaec/index.cfm?varlan=english) United Nations Economic Commission for Europe (http://www.unece.org/env/pp/treatytext.htm)

[a]Dates referenced are the date of signature of the treaties.
[b]U.S. has signed and ratified the treaty.

APPENDIX 2

Criteria for raw water used for organized community water supplies (surface and ground water) primary parameters:

		Range/Limiting Value		
	Parameters	Use With Only Disinfection	Use After Conventional Treatment	Note
1.	pH	6.5 to 8.5	6.0 to 9.0	To ensure prevention of corrosion in treatment plant and distribution system, and interference in coagulation and chlorinating.
2.	Color Pt. scale Hz units	<10	<50	Color may not get totally removed during treatment.
3.	Suspended solids mg/L	<10	<50	High SS may increase the cost of treatment.
4.	Odor, dilution factor	<3	<10	May not be tackled during treatment.
5.	DO, (%saturation)	90–100	80–120	May imply higher chlorine demand.
6.	BOD, mg/L	<3	<5	Same as above.
7.	TKN, mg/L	<1	<3	Same as above.
8.	Ammonia, mg/L	<0.05	<1	Same as above.
9.	Fecal coliform MPN/100 mL	<200	<2000	Not more than 20% samples show greater than limit.
10.	EC, μm/hos/cm	<2000	<2000	High conductivity implies dissolved high solids making water unpalatable.
11.	Chloride, mg/L	<300	<300	May cause physiological impact and unpalatable taste.
12.	Sulfates, mg/L	<250	<250	May cause digestive problems.
13.	Phosphates, mg/L	<0.7	<1.0	May interfere with coagulation.

Continued

Parameters	Range/Limiting Value: Use With Only Disinfection	Range/Limiting Value: Use After Conventional Treatment	Note
14. Nitrate, mg/L	<50	<50	May cause methamoplobinemea.
15. Fluoride, mg/L	<1.0	<1.5	Higher value shall cause fluorosis and lower value shall carries.
16. Surfactants, mg/L	<0.2	<0.2	May impair treatability and cause foaming.

Additional parameters for periodic monitoring (seasonal; only to be done when there are known natural or anthropogenic sources in the upstream catchment region likely or apprehended to contribute, or other well-founded apprehensions):

Parameters	Desirable	Acceptable	Note
Dissolved iron mg/L	<0.3	<0.5	Affects taste and causes stains.
Copper, mg/L	—	<1.0	May cause liver damage.
Zinc, mg/L	—	<5.0	Cause bitter stringent taste.
Arsenic, mg/L	<0.01	<0.05	Cause hyperkeratosis and skin cancer.
Cadmium, mg/L	<0.001	<0.005	Toxic.
Total chromium, mg/L	<0.05	<0.05	Toxic.
Lead, mg/L	<0.05	<0.05	Physiological abnormality.
Selenium, mg/L	<0.01	<0.01	Toxic symptoms similar to arsenic.
Mercury, mg/L	<0.005	<0.0005	Carcinogenic and poisonous.
Phenols, mg/L	<0.001	<0.001	Toxic and causes taste and odor problem.
Cyanides, mg/L	<0.05	<0.05	Physiological abnormality.
PAH, mg/L	<0.0002	<0.0002	Carcinogenic.
Total pesticides, mg/L	<0.001	<0.0025	Tend to bioaccumulate and carcinogenic.

Source: Ecological Impact Assessment Series: EIAS/03/2002-03 Published by CPCB.

Use-based classification of surface waters in India:

Designated-Best-Use	Class of Water	Criteria
Drinking water source without conventional treatment but after disinfection	A	1. Total coliforms organism MPN/100 mL shall be 50 or less 2. pH between 6.5 and 8.5 3. Dissolved oxygen 6 mg/L or more 4. Biochemical oxygen demand 5 days 20°C 2 mg/L or less
Outdoor bathing (organized)	B	1. Total coliforms organism MPN/100 ml shall be 500 or less 2. pH between 6.5 and 8.5 3. Dissolved oxygen 5 mg/L or more 4. Biochemical oxygen demand 5 days 20°C 3 mg/L or less
Drinking water source after conventional treatment and disinfection	C	1. Total coliforms organism MPN/100 mL shall be 5000 or less 2. pH between 6 and 9 3. Dissolved oxygen 4 mg/L or more 4. Biochemical oxygen demand 5 days 20°C 3 mg/L or less
Propagation of wildlife and fisheries	D	1. pH between 6.5 and 8.5 2. Dissolved oxygen 4 mg/L or more 3. Free ammonia (as N) 1.2 mg/L or less
Irrigation, industrial cooling, controlled waste disposal	E	1. pH between 6.0 and 8.5 2. Electrical conductivity at 25°C micro mhos/cm Maximum 2250 3. Sodium absorption ratio maximum 26 4. Boron maximum 2 mg/L

Source: Guidelines For Water Quality Management–CPCB 2008.

APPENDIX 3

General Standards for Discharge of Effluents

S. No	Parameter	Standards: Inland Surface Water (a)	Public Sewers (b)	Land for Irrigation (c)	Marine Coastal Areas (d)
1	Color and odor				
2	Suspended solids mg/L, maximum	100	600	200	1. For process wastewater-100 2. For cooling water effluent 10% above total suspended matter of effluent
3	Particle size of suspended solids	Shall pass 850 μm IS sieve	–	–	1. Floatable solids maximum 3 mm 2. Settleable solids maximum 850 μm
4	pH Value	5.5–9.0	5.5–9.0	5.5–9.0	5.5–9.0
5	Temperature	Shall not exceed 5°C above the receiving water temperature	–	–	Shall not exceed 5°C above the receiving water temperature
6	Oil and grease mg/L maximum	10	20	10	20
7	Total Kjeldahl nitrogen (as NH_3), mg/L maximum	100	–	–	100

Continued

S. No	Parameter	Standards			
		Inland Surface Water	Public Sewers	Land for Irrigation	Marine Coastal Areas
		(a)	(b)	(c)	(d)
8	Free ammonia (as NH_3), mg/L maximum	5.0	–	–	5.0
9	Bio-chemical oxygen demand (5 days at 20°C), mg/L maximum	30	350	100	100
10	Chemical oxygen demand, mg/L maximum	250	–	–	250
11	Arsenic (as As), mg/L maximum	0.2	0.2	0.2	0.2
12	Mercury (as Hg), mg/L maximum	0.01	0.01	–	0.01
13	Lead (as Pb), mg/L maximum	0.1	1.0	–	2.0
14	Cadmium (as Cd), mg/L maximum	2.0	1.0	–	2.0
15	Hexavalent chromium (as Cr +6), mg/L maximum	2.0	1.0	–	2.0
16	Total chromium (as Cr), mg/L maximum	2.0	2.0	–	2.0
17	Copper (as Cu), mg/L maximum	3.0	3.0	–	3.0

General Standards for Discharge of Effluents

S. No	Parameter	Standards: Inland Surface Water (a)	Public Sewers (b)	Land for Irrigation (c)	Marine Coastal Areas (d)
18	Zinc (as Zn), mg/L maximum	5.0	15	–	15
19	Selenium (as Se), mg/L maximum	0.05	0.05	–	0.05
20	Fluoride (as F), mg/L maximum	2.0	15	–	15
21	Dissolved phosphates (as P), mg/L maximum	5.0	–	–	–
22	Sulfide (as S), mg/L maximum	2.0	–	–	5.0
23	Phenolic compounds (as C_6H_5OH), mg/L maximum	1.0	5.0	–	5.0
24	**Radioactive materials:**				
	(A) Alpha emitter micro curie/mL	10^{-7}	10^{-7}	10^{-7}	10^{-7}
	(B) Beta emitter micro curie/mL	10^{-6}	10^{-6}	10^{-6}	10^{-6}
25	Bio-assay test	90% survival of fish after 96 h in 100% effluent	90% survival of fish after 96 h in 100% effluent	90% survival of fish after 96 h in 100% effluent	90% survival of fish after 96 h in 100% effluent
26	Manganese (as Mn), mg/L	2	2	–	2

Continued

S. No	Parameter	Standards			
		Inland Surface Water	Public Sewers	Land for Irrigation	Marine Coastal Areas
		(a)	(b)	(c)	(d)
27	Iron (as Fe), mg/L	3	3	–	3
28	Vanadium (as V), mg/L	0.2	0.2	–	0.2
29	Nitrate nitrogen, mg/L	10	–	–	20

These standards shall be applicable for industries, operations, or processes other than those industries, operations, or process for which standards have been specified in the Environment Protection Rules, 1989.

G.S.R 422 (E) dated 19.05.1993 and G.S.R 801 (E) dated 31.12.1993 issued under the provisions of E (P) Act 1986.

APPENDIX 4

PRESUMPTIVE TEST

Inoculate lactose or lauryl tryptose fermentation tubes and incubate 24 ± 2 h at:

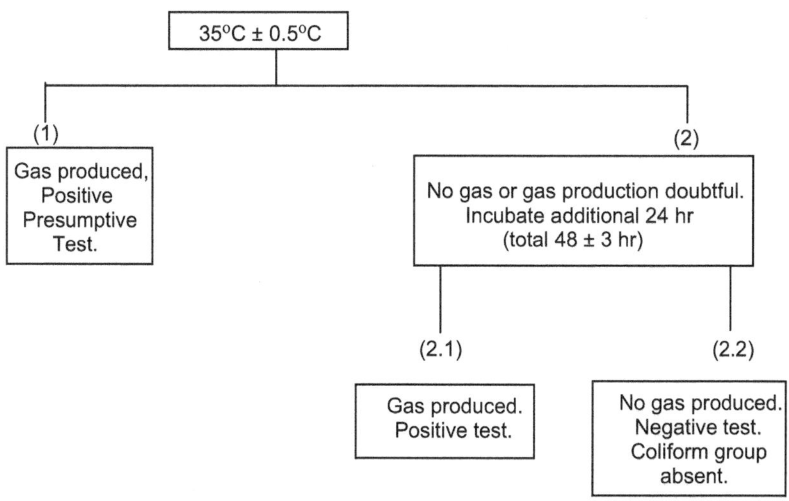

CONFIRMED TEST

Inoculate lactose or lauryl tryptose broth fermentation tubes Figs. A.4.1 and A.4.2 and Table A.4.1incubate 24 ± 2 h at:

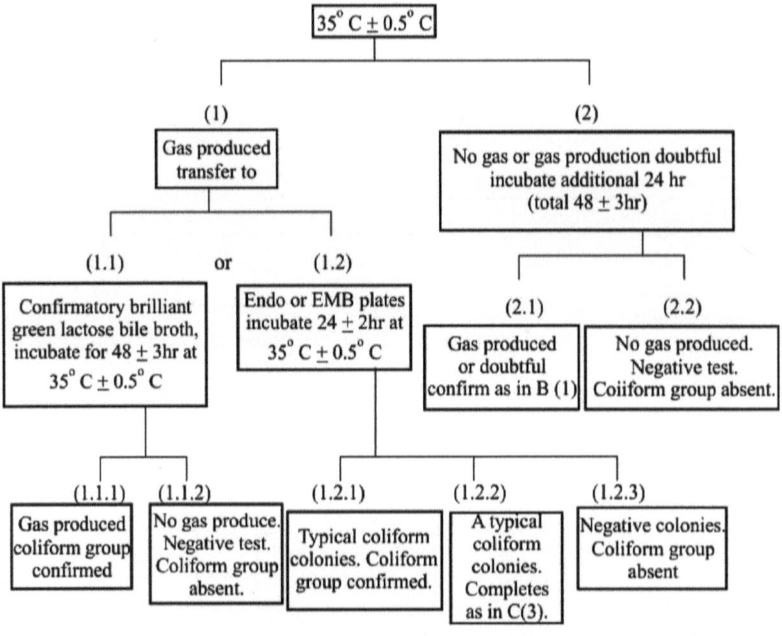

Figure A.4.1 Schematic outline of presumptive, confirmed, and completed tests.

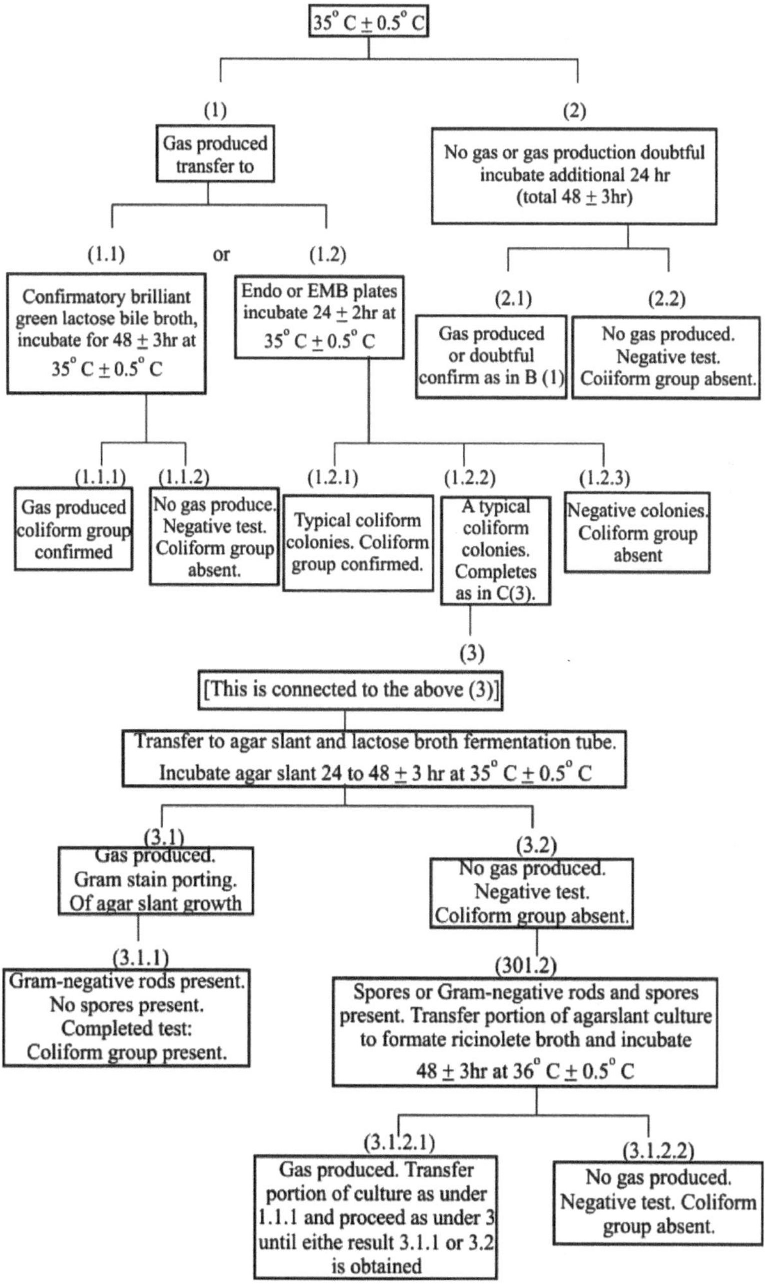

Figure A.4.2 Schematic outline of presumptive, confirmed, and completed tests.

Table A.4.1 MPN index and 95% confidence limits for various combinations of positive and negative results when five 10 mL portions, five 1 mL portions, and five 0.1 mL portions are used.

Number of Tubes Giving Positive Reaction Out of			MPN Index Per 100 mL	95% Confidence Limits		Number of Tubes Giving Positive Reaction Out of			MPN Index Per 100 mL	95% Confidence Limits	
5 of 10 mL Each	5 of 1 mL Each	5 of 0.1 mL Each		Lower	Upper	5 of 10 mL Each	5 of 1 mL Each	5 of 0.1 mL Each		Lower	Upper
0	0	1	2	<0.5	7	4	2	1	26	9	78
0	1	0	2	<0.5	7	4	3	0	27	9	80
0	2	0	4	<0.5	11	4	3	1	33	11	93
1	0	0	2	<0.5	7	4	4	0	34	12	93
1	0	1	4	<0.5	11	5	0	0	23	7	70
1	1	0	4	<0.5	11	5	0	1	31	11	89
1	1	1	6	<0.5	15	5	0	2	43	15	114
1	2	0	6	<0.5	15	5	1	0	33	11	93
2	0	0	5	<0.5	13	5	1	1	46	16	120
2	0	1	7	1	17	5	1	2	63	21	150
2	1	0	7	1	17	5	2	0	49	17	130
2	1	1	9	2	21	5	2	1	70	23	170
2	2	0	9	2	21	5	2	2	94	28	220
2	3	0	12	3	28	5	3	0	79	25	190
3	0	0	8	1	19	5	3	1	109	31	250
3	0	1	11	2	25	5	3	2	141	37	340
3	1	0	11	2	25	5	3	3	175	44	500
3	1	1	14	4	34	5	4	0	130	35	300
3	2	0	14	4	34	5	4	1	172	43	490
3	2	1	17	5	46	5	4	2	221	57	700
3	3	0	17	5	46	5	4	3	278	90	850
4	0	0	13	3	31	5	4	4	345	120	1000
4	0	1	17	5	46	5	5	0	240	68	750
4	1	0	17	5	46	5	5	1	348	120	1000
4	1	1	21	7	63	5	5	2	542	180	1400
4	1	2	26	9	78	5	5	3	918	300	3200
4	2	0	22	7	67	5	5	4	1609	640	5800

APPENDIX 5

National Ambient Air Quality Standards (NAAQS)

S. No	Pollutant	Time Weighted Average	Concentration in Ambient Air: Industrial, Residential, Rural, and Other Areas	Concentration in Ambient Air: Ecologically Sensitive Area (Notified by Central Government)	Methods of Measurement
(1)	(2)	(3)	(4)	(5)	(6)
1	Sulfur dioxide (SO_2), µg/m^3	Annual[1] 24 h[2]	50 80	20 80	• Improved West and Gaeke • Ultraviolet fluorescence
2	Nitrogen dioxide (NO_2), µg/m^3	Annual[1] 24 h[2]	40 80	30 80	• Modified Jacob and Hochheiser (Na-Arsenite) • Chemiluminescence
3	Particulate matter (size less than 10 µm) or PM_{10} µg/m^3	Annual[1] 24 h[2]	60 100	60 100	• Gravimetric • TOEM • Beta attenuation
4	Particulate matter (size less than 2.5 µm) or $PM_{2.5}$ µg/m^3	Annual[1] 24 h[2]	40 60	40 60	• Gravimetric • TOEM • Beta attenuation
5	Ozone (O_3) µg/m^3	8 h[2] 1 h[2]	100 180	100 180	• UV photometric • Chemiluminescence • Chemical method
6	Lead (Pb) µg/m^3	Annual[1] 24 h[2]	0.50 1.0	0.50 1.0	• AAS/ICP method after sampling on EPM 2000 or equivalent filter paper • ED-XRF using Teflon filter

7	Carbon monoxide (CO) mg/m^3	8 h[2] 1 h[2]	02 04	02 04	• Non-dispersive infrared (NDIR) spectroscopy
8	Ammonia (NH_3) $\mu g/m^3$	Annual[1] 24 h[2]	100 400	100 400	• Chemiluminescence • Indophenol blue method
9	Benzene (C_6H_6) $\mu g/m^3$	Annual[1]	05	05	• Gas chromatography-based continuous analyzer • Adsorption and desorption followed by GC analysis
10	Benzo(a)Pyrene (BaP); particulate phase only, ng/m^3	Annual[1]	01	01	• Solvent extraction followed by HPLC/GC analysis
11	Arsenic (As) ng/m^3	Annual[1]	06	06	• AAS/ICP method after sampling on EPM 2000 or equivalent filter paper
12	Nickel (Ni) ng/m^3	Annual[1]	20	20	• AAS/ICP method after sampling on EPM 2000 or equivalent filter paper

[1]Annual arithmetic mean of minimum 104 measurements in a year at a particular site taken twice a week 24 hourly at uniform intervals.

[2]24 hourly, 08 hourly, or 01 hourly monitored values, as applicable, shall be complied with 98% of the time in a year. 2% of the time, they may exceed the limits but not on two consecutive days of monitoring.

Note: Whenever and wherever monitoring results on two consecutive days of monitoring exceed the limits specified above for the respective category, it shall be considered adequate reason to institute regular or continuous monitoring and further investigation.

National Ambient Air Quality Standards, CPCB Notification dated 18 November 2009.

APPENDIX 6

Schedule
(See Paragraph 2 and 7)
List of Projects or Activities Requiring Prior Environmental Clearance

Project or Activity	Category With Threshold Limit			Conditions If any
	A	B		
(1)	(2)	(3)	(4)	(5)
1	**Mining Extraction of Natural Resources and Power Generation (for a Specified Production Capacity)**			
1(a)	Mining of minerals	≥50 ha of mining lease area Asbestos mining irrespective of mining area	<50 ha ≥5 ha of mining lease area	General conditions shall apply **Note**: Mineral prospecting (not involving drilling) are exempted provided that the concession areas have received previous clearance for physical survey
1(b)	Offshore and onshore oil and gas exploration, development, and production	All projects		**Note**: Exploration surveys (not involving drilling) are exempted provided that the concession areas have received previous clearance for physical survey
1(c)	River valley projects	1. ≥50 MW hydroelectric power generation 2. ≥10,000 ha of culturable command area	1. <50 MW ≥ 25 MW hydroelectric power generation 2. <10,000 ha of culturable command area	General conditions shall apply

Continued

Project or Activity	Category With Threshold Limit			Conditions If any
	A	B		
(1)	(2)	(3)	(4)	(5)
1(d)	Thermal power plants	≥500 MW (coal/lignite/naphta and gas based) ≥50 MW (pet coke diesel and all other fuels)	<500 MW (coal/lignite/napthand gas based); <50 MW ≥5 MW (pet coke, diesel and all other fuels)	General conditions shall apply
1(e)	Nuclear power projects and processing of nuclear fuel	All projects	–	
2	**Primary Processing**			
2(a)	Coal washeries	≥1 million ton/annum throughput of coal	<1 million ton/annum throughput of coal	General conditions shall apply (if located within mining area, the proposal shall be appraised together with the mining proposal)
2(b)	Mineral beneficiation	≥0.1 million ton/annum mineral throughput	<0.1 million ton/annum mineral throughput	General conditions shall apply (mining proposal with mineral beneficiation shall be appraised together for grant of clearance)
3	**Materials Production**			
3(a)	Metallurgical industries (ferrous and non-ferrous)	1. Primary metallurgical industry All projects 2. Sponge iron manufacturing ≥200 TPD 3. Secondary metallurgical processing industry All toxic and heavy metal producing units ≥20,000 tons/annum	Sponge iron manufacturing <200 TPD Secondary metallurgical processing industry 1. All toxic and heavy metal producing units <20,000 tons/annum 2. All other non-toxic secondary metallurgical processing industries >5000 tons/annum	General conditions shall apply for sponge iron manufacturing

Project or Activity	Category With Threshold Limit			
	A	B		Conditions If any
(1)	(2)	(3)	(4)	(5)
3(b)	Cement plants	≥1.0 million tons/ annum production capacity	<1.0 million tons/ annum production capacity. All stand-alone grinding units	General conditions shall apply
4	**Materials Processing**			
4(a)	Petroleum refining industry	All projects	–	–
4(b)	Coke oven plants	≥2,50,000	<2,50,000 and ≥25,000 tons/ annum	–
4(c)	Asbestos milling and asbestos based products	All projects	–	–
4(d)	Chlor-alkali industry	≥300 TPD production capacity or a unit located out side the notified industrial area/ estate	<300 TPD production capacity and located within a notified industrial area/estate	Specific conditions shall apply No new mercury cell-based plants will be permitted and existing units converting to membrane cell technology are exempted from this notification
4(e)	Soda ash industry	All projects	–	–
4(f)	Leather/skin/ hide processing industry	New projects outside the industrial area or expansion of existing units out side the industrial area	All new or expansion of projects located within a notified industrial area/estate	Specific conditions shall apply
5	**Manufacturing/Fabrication**			
5(a)	Chemical fertilizers	All projects	–	–
5(b)	Pesticides industry and pesticide-specific intermediates (excluding formulations)	All units producing technical grade pesticides	–	–

Continued

Project or Activity	Category With Threshold Limit			Conditions If any
	A	B		
(1)	(2)	(3)	(4)	(5)
5(c)	Petro-chemical complexes (industries based on processing of petroleum fractions and natural gas and/or reforming to aromatics)	All projects	–	–
5(d)	Man-made fibers manufacturing	Rayon	Others	General conditions shall apply
5(e)	Petrochemical based processing (processes other than cracking and reformation and not covered under the complexes)	Located outside the notified industrial area/estate	Located in a notified industrial area/estate	Specific conditions shall apply
5(f)	Synthetic organic chemicals industry (dyes and dye intermediates; bulk drugs and intermediates excluding drug formulations; synthetic rubbers; basic organic chemicals, other synthetic organic chemicals and intermediates)	Located outside the notified industrial area/estate	Located in a notified industrial area/estate	Specific conditions shall apply
5(g)	Distilleries	1. All molasses based distilleries 2. All cane juice/non-molasses based distilleries ≥30 KLD	All cane juice/non-molasses based distilleries ≤30 KLD	General conditions shall apply
5(h)	Integrated point industry	–	All projects	General conditions shall apply

Project or Activity	Category With Threshold Limit			Conditions If any
	A	B		
(1)	(2)	(3)	(4)	(5)
5(i)	Pulp and paper industry excluding manufacturing of paper from waste paper and manufacture of paper from ready pulp without bleaching	Pulp manufacturing and pulp and paper manufacturing industry	Paper manufacturing industry without pulp manufacturing	General conditions shall apply
5(j)	Sugar industry	–	≥5000 cane crushing capacity	General conditions shall apply
5(k)	Induction/arc furnaces/cupola furnaces 5 TPH or more	–	All projects	General conditions shall apply
6	**Service Sectors**			
6(a)	Oil and gas transportation pipeline (crude and refinery/ petrochemical products), passing through national parks/sanctuaries/ coral reefs/ecologically sensitive areas including LNG terminal	All projects		–
6(b)	Isolated storage and handling of hazardous chemicals (as per threshold planning quantity indicated in column 3 of schedule 2 and 3 of MSIHC rules 1989 amended 2000)	–	All projects	General conditions shall apply

Continued

Project or Activity	Category With Threshold Limit			Conditions If any
	A	B		
(1)	(2)	(3)	(4)	(5)
7	**Physical Infrastructure including Environmental Services**			
7(a)	Airports	All projects	–	–
7(b)	All ship breaking yards including ship breaking units	All projects	–	–
7(c)	Industrial estates/parks/ complexes/ area, export processing zones (EPZs), special economic zones (SEZs), biotech parks, leather complexes	If at least one industry in the proposed industrial estate falls under the Category A, entire industrial area shall be treated as Category A, irrespective of the area Industrial estates with area greater than 500 ha and housing at least one Category B industry	Industrial estates housing at least one Category B industry and area <500 ha. Industrial estates of area >500 ha and not housing any industry belonging to Category A or B	Special conditions shall apply **Note**: Industrial estate of area below 500 ha and not hosing any industry of category A or B does not require clearance
7(d)	Common hazardous waste treatment, storage, and disposal facilities (TSDFs)	All integrated facilities having incineration and landfill or incineration alone	All facilities having land fill only	General conditions shall apply
7(e)	Ports, harbors	≥5 million TPA of cargo handling capacity (excluding fishing harbors)	<5 million TPA of cargo handling capacity and/or ports/harbous ≥10,000 TPA of fish handling capacity	General conditions shall apply

Project or Activity	Category With Threshold Limit			Conditions If any
	A	B		
(1)	(2)	(3)	(4)	(5)
7(f)	Highways	1. New national highways; and 2. Expansion of national highways greater than 30km, involving additional right of way greater than 20m involving land acquisition and passing through more than one state	1. New state highways; and 2. Expansion of national/state highways greater than 30km involving additional right of way greater than 20m involving land acquisition.	General conditions shall apply
7(g)	Aerial rope ways		All projects	General conditions shall apply
7(h)	Common effluent treatment plants (CETPs)		All projects	General conditions shall apply
7(i)	Common municipal solid waste management facility (CMSWMF)		All projects	General conditions shall apply
8	**Building/Construction Projects/Area Development Projects and Townships**			
8(a)	Building and construction projects		≥20,000 sq m and <1,50,000 sq m of built-up area#	#(built up area for covered construction; in the case of facilities open to the sky, it will be the activity area)

Continued

Project or Activity	Category With Threshold Limit			Conditions If any
	A	B		
(1)	(2)	(3)	(4)	(5)
8(b)	Townships and area development projects		Covering an area ≥50 ha and or built up area ≥1,50,000 sq m ++	++All projects under item 8(b) shall be appraised as Category B 1

Note: **Specific Condition (SC)**: If any industrial estate/complex/export processing zones/special economic zones/biotech parks/leather complex with homogeneous type of industries such as items 4(d), 4(f), 5(e), 5(f), or those industrial estates with pre-defined set of activities (not necessarily homogeneous, obtains prior environmental clearance, individual industries including proposed industrial housing within such estates/complexes will not be required to take prior environmental clearance, so long as the Terms and Conditions for the industrial estate/complex are complied with (such estates/complexes must have a clearly identified management with the legal responsibility of ensuring adherence to the Terms and Conditions of prior environmental clearance, who may be held responsible for violation of the same throughout the life of the complex/estate).
Notification published in the Gazette of India, Extraordinary, Part-II, and Section 3, Sub-section (ii) Ministry of Environment and Forests, New Delhi 14 September, 2006.

APPENDIX 7 STANDARDS FOLLOWED IN ENVIRONMENTAL IMPACT ASSESSMENT STUDIES

Land Use/Land Cover Classification System Followed in EIA Process

Level I	Level II	Level III
1. Built-up land	1.1. Built-up land	1.1.1. Urban (towns and cities)
2. Agricultural land	2.1. Crop land (i) kharif (ii) rabi (iii) double cropped	2.1.1. Irrigated crop land 2.1.2. Unirrigated crop land
	2.2. Fallow	2.2.1. Fallow
	2.3. Plantation	2.3.1. Types of plantation, casuarina, coconut, tea, etc.
3. Forest	3.1 Evergreen/semi-evergreen	3.1.1. Dense/closed 3.1.2. Open
	3.2. Deciduous	
	3.3. Degraded scrub land	
	3.4. Forest blank	3.4.1. Degraded forest 3.4.2. Forest blank
	3.5. Forest plantation	3.5.1. Types of plantation e.g., teak, sal, etc.
	3.6. Mangrove	
4. Wastelands	4.1. Salt-affected land	
	4.2. Waterlogged land	
	4.3. Marshy/swampy land	
	4.4. Gullied/ravinous land	
	4.5. Land with or without scrub	
	4.6 Sandy area (coastal and desertic)	Minimum mappable unit IS 2.25 ha on 1:50,000 scale
	4.7. Barren rocky/stony waste/sheet rock areas	
5. Water bodies	5.1. River/stream	
	5.2 Lake/reservoir/tank/canal	

Continued

Level I	Level II	Level III
6. Others	6.1. Shifting cultivation	6.1.1. Current 6.1.2. Old/abandoned
	6.2. Grassland/grazing land	6.2.1. Grassland/grazing land
	6.3. Snow covered/glacial area	6.3.1. Snow covered/ glacial area
	6.4. Mining area	6.4.1. Mining dumps

Land use/Land cover categories at different levels and corresponding scales for mapping are as follows.
Level I—categories—1:1000,000 scale.
Level II—categories—1:250,000 scale.
Level III—categories—1:50,000 scale and 1:25,000 scale.
Description and classification of land use/land cover: NRSA—TR—LU & CD—01-90.

Criteria for Raw Water Used for Organized Community Water Supplies (Surface and Groundwater) Primary Parameters

		Range/Limiting Value		
	Parameters	Use With Only Disinfection	Use After Conventional Treatment	Note
1.	pH	6.5–8.5	6.0–9.0	To ensure prevention of corrosion in treatment plant and distribution system and interference in coagulation and chlorinating.
2.	Color pt. scale Hz units	<10	<50	Color may not be totally removed during treatment.
3.	Suspended solids mg/L	<10	<50	High SS may increase the cost of treatment.
4.	Odor, dilution factor	<3	<10	May not be tackled during treatment.
5.	DO, (%saturation)	90–100	80–120	May imply higher chlorine demand.
6.	BOD, mg/L	<3	<5	Same as above.
7.	TKN, mg/L	<1	<3	Same as above.
8.	Ammonia, mg/L	<0.05	<1	Same as above.
9.	Fecal coliform MPN/100 mL	<200	<2000	Not more than 20% samples show greater than limit.

	Parameters	Range/Limiting Value Use With Only Disinfection	Use After Conventional Treatment	Note
10.	EC, µm/hos/cm	<2000	<2000	High conductivity implies dissolved high solids making water unpalatable.
11.	Chloride, mg/L	<300	<300	May cause physiological impact and unpalatable taste.
12.	Sulfates, mg/L	<250	<250	May cause digestive problems.
13.	Phosphates, mg/L	<0.7	<1.0	May interfere with coagulation.
14.	Nitrate, mg/L	<50	<50	May cause methamoplobinemea.
15.	Fluoride, mg/L	<1.0	<1.5	Higher value shall cause fluorosis and lower value shall carries.
16.	Surfactants, mg/L	<0.2	<0.2	May impair treatability and cause foaming.

Additional parameters for periodic monitoring (seasonal; only to be done when there are known natural or anthropogenic sources in the upstream catchment region likely or apprehended to contribute or other well-founded apprehensions).

Parameters	Desirable	Acceptable	Note
Dissolved iron mg/L	<0.3	<0.5	Affects taste and cause stains.
Copper, mg/L	–	<1.0	May cause liver damage.
Zinc, mg/L	–	<5.0	Causes bitter stringent taste.
Arsenic, mg/L	<0.01	<0.05	Causes hyperkeratosis and skin cancer.
Cadmium, mg/L	<0.001	<0.005	Toxic.

Continued

Parameters	Desirable	Acceptable	Note
Total chromium, mg/L	<0.05	<0.05	Toxic.
Lead, mg/L	<0.05	<0.05	Physiological abnormality.
Selenium, mg/L	<0.01	<0.01	Toxic symptoms similar to arsenic.
Mercury, mg/L	<0.005	<0.0005	Carcinogenic and poisonous.
Phenols, mg/L	<0.001	<0.001	Toxic and cause taste and odor problem.
Cyanides, mg/L	<0.05	<0.05	Physiological abnormality.
PAH, mg/L	<0.0002	<0.0002	Carcinogenic.
Total pesticides, mg/L	<0.001	<0.0025	Trend to bioaccumulate and carcinogenic.

Ecological Impact Assessment Series: EIAS/03/2002-03 Published by CPCB.

Use Based Classification of Surface Waters in India

Designated Best Use	Class of Water	Criteria
Drinking water source without conventional treatment but after disinfection	A	1. Total coliforms organism MPN/100 mL shall be 50 or less 2. pH between 6.5 and 8.5 3. Dissolved oxygen 6 mg/L or more 4. Biochemical oxygen demand 5 days 20°C 2 mg/L or less
Outdoor bathing (organized)	B	1. Total coliforms organism MPN/100 mL shall be 500 or less 2. pH between 6.5 and 8.5 3. Dissolved oxygen 5 mg/L or more 4. Biochemical oxygen demand 5 days 20°C 3 mg/L or less

Designated Best Use	Class of Water	Criteria
Drinking water source after conventional treatment and disinfection	C	1. Total coliforms organism MPN/100 mL shall be 5000 or less 2. pH between 6 and 9 3. Dissolved oxygen 4 mg/L or more 4. Biochemical oxygen demand 5 days 20°C 3 mg/L or less
Propagation of wildlife and fisheries	D	1. pH between 6.5 and 8.5 2. Dissolved oxygen 4 mg/L or more 3. Free ammonia (as N) 1.2 mg/L or less
Irrigation, industrial cooling, controlled waste disposal	E	1. pH between 6.0 and 8.5 2. Electrical conductivity at 25°C micro mhos/cm maximum 2250 3. Sodium absorption ratio maximum 26 4. Boron maximum 2 mg/L

Guidelines for Water Quality Management–CPCB 2008.

National Ambient Air Quality Standards

S. No	Pollutants	Time Weighted Average	Concentration in Ambient Air: Industrial, Residential, Rural and Other Areas	Concentration in Ambient Air: Ecologically Sensitive Area (Notified by Central Government)	Methods of Measurement
1.	2.	3.	4.	5.	6.
1.	Sulfur dioxide (SO_2), μg/m^3	Annual[1] 24 h[2]	50 80	20 80	Improved West and Gaeke Ultraviolet fluorescence
2.	Nitrogen dioxide (NO_2), μg/m^3	Annual[1] 24 h[2]	40 80	30 80	• Modified Jacob and Hochhieser (Na-Arsenite) • Chemiluminescence
3.	Particulate matter (size less than 10 μm) or PM_{10}, μg/m^3	Annual[1] 24 h[2]	60 100	60 100	• Gravimetric • TOEM • Beta attenuation
4.	Particulate matter (size less than 2.5 μm) or $PM_{2.5}$, μg/m^3	Annual[1] 24 h[2]	40 60	40 60	• Gravimetric • TOEM • Beta attenuation
5.	Ozone (O_3), μg/m^3	8 h[2] 1 h[2]	100 180	100 180	• UV-photometric • Chemiluminescence • Chemical method
6.	Lead (Pb), μg/m^3	Annual[1] 24 h[2]	0.50 1.0	0.50 1.0	• AAS/ICP method after sampling on EPM 2000 or equivalent filter paper • ED-XRF using Teflon filter

7.	Carbon monoxide (CO), mg/m^3	8 h[2]	02	02	• Non-dispersive infrared (NDIR) spectroscopy
		1 h[2]	04	04	
8.	Ammonia (NH_3), µg/m^3	Annual[1]	100	100	• Chemiluminescence
		24 h[2]	400	400	• Indophenol blue method
9.	Benzene (C_6H_6), µg/m^3	Annual[1]	05	05	• Gas chromatography-based continuous analyzer • Adsorption and desorption followed by GC analysis
10.	Benzo(O)Pyrene (BaP) – Particulate phae only, ng/m^3	Annual[1]	01	01	• Solvent extraction followed by HPLC/GC analysis
11.	Arsenic (As), ng/m^3	Annual[1]	06	06	• AAS/ICP method after sampling on EPM 2000 or equivalent filter paper
12.	Nickel (Ni), ng/m^3	Annual[1]	20	20	• AAS/ICP method after sampling on EPM 2000 or equivalent filter paper

[1]Annual arithmetic mean of minimum 104 measurements in a year at a particular site taken twice a week 24 hourly at uniform intervals.

[2]24 hourly, 8 hourly, or 1 hourly monitored values, as applicable, shall be complied with 98% of the time in a year. 2% of the time, they may exceed the limits but not on two consecutive days of monitoring.

Whenever and wherever monitoring results on two consecutive days of monitoring exceed the limit specified above for the respective category, it shall be considered adequate reason to institute regular/continuous monitoring and further investigations.

CPCB notification Dated 18 November 2009.

Noise Ambient Air Quality Standards

		Limits in db (A) L_{eq}	
Area Code	**Category of Area**	**Day Time**	**Night Time**
A	Industrial area	75	70
B	Commercial area	65	55
C	Residential area	55	45
D	Silence zone	50	40

Day time shall mean from 6.00 a.m. to 10.00 p.m.
Night time shall mean from 10.00 p.m. to 6.00 a.m.
Silence zone is an area comprising not less than 100 m around hospitals, educational institutions, courts, religious places, or any other areas, which is declared as such by the competent authority.
Mixed categories of areas may be declared as one of the four above-mentioned categories by the competent authority.
dB(A) Leq denotes the time weighted average of the level of sound in decibels on scale A which is relatable to human hearing.
A "decibel" is a unit in which noise is measured.
"A", in dB(A) L_{eq}, denotes the frequency weighting in the measurement of noise and corresponds to frequency response characteristics of the human ear.
L_{eq}: It is an energy mean of the noise level over a specified period.
Noise pollution (Regulation and control) Rules, 2000.

INDEX

'*Note:* Page numbers followed by "f" indicate figures, "t" indicate tables and "b" indicates boxes.'

B

D

E

G

H

I

J

K

L

O

Z

Kolibri 360 GmbH
Ferdinand-Jühlke-Straße 7
99095 Erfurt, Deutschland
produktsicherheit@kolibri360.de